CONCEPTS OF MODERN ENGINEERING PHYSICS

[Strictly based on the B.E. B.Tech. Syllabus of Punjab Technical University]

A. S. VASUDEVA
Prof. & Head, Deptt. of Physics
Rayat Institute of Engineering &
Information Technology
Rail Majra, Distt. Nawawshahar
(Punjab)
Former Head, Department of Physics
and Applied Sciences
Guru Nanak Dev Engineering College
Ludhiana

KAMAL KISHORE
Department of Physics
DAV Institute of Engineering
and Technology
Jalandhar (Punjab)

S. CHAND & COMPANY LTD.
(An ISO 9001 : 2000 Company)
RAM NAGAR, NEW DELHI - 110 055

S. CHAND & COMPANY LTD.
(An ISO 9001 : 2000 Company)
Head Office : 7361, RAM NAGAR, NEW DELHI - 110 055
Phones : 23672080-81-82, 9899107446, 9911310888;
Fax : 91-11-23677446
Shop at: **schandgroup.com;** E-mail: **schand@vsnl.com**

Branches :

- 1st Floor, Heritage, Near Gujarat Vidhyapeeth, Ashram Road, **Ahmedabad**-380014. Ph. 27541965, 27542369, ahmedabad@schandgroup.com
- No. 6, Ahuja Chambers, 1st Cross, Kumara Krupa Road, **Bangalore**-560001. Ph : 22268048, 22354008, bangalore@schandgroup.com
- 238-A M.P. Nagar, Zone 1, **Bhopal** - 462 011. Ph : 4274723. bhopal@schandgroup.com
- 152, Anna Salai, **Chennai**-600002. Ph : 28460026, chennai@schandgroup.com
- S.C.O. 6, 7 & 8, Sector 9D, **Chandigarh**-160017, Ph-2749376, 2749377, chandigarh@schandgroup.com
- 1st Floor, Bhartia Tower, Badambadi, **Cuttack**-753 009, Ph-2332580; 2332581, cuttack@schandgroup.com
- 1st Floor, 52-A, Rajpur Road, **Dehradun**-248 001. Ph : 2740889, 2740861, dehradun@schandgroup.com
- Pan Bazar, **Guwahati**-781001. Ph : 2514155, guwahati@schandgroup.com
- Sultan Bazar, **Hyderabad**-500195. Ph : 24651135, 24744815, hyderabad@schandgroup.com
- Mai Hiran Gate, **Jalandhar** - 144008 . Ph. 2401630, jalandhar@schandgroup.com
- A-14 Janta Store Shopping Complex, University Marg, Bapu Nagar, **Jaipur** - 302 015, Phone : 2709153, jaipur@schandgroup.com
- 613-7, M.G. Road, Ernakulam, **Kochi**-682035. Ph : 2381740, cochin@schandgroup.com
- 285/J, Bipin Bihari Ganguli Street, **Kolkata**-700012. Ph : 22367459, 22373914, kolkata@schandgroup.com
- Mahabeer Market, 25 Gwynne Road, Aminabad, **Lucknow**-226018. Ph : 2626801, 2284815, lucknow@schandgroup.com
- Blackie House, 103/5, Walchand Hirachand Marg , Opp. G.P.O., **Mumbai**-400001. Ph : 22690881, 22610885, mumbai@schandgroup.com
- Karnal Bag, Model Mill Chowk, Umrer Road, **Nagpur**-440032 Ph : 2723901, 2777666 nagpur@schandgroup.com
- 104, Citicentre Ashok, Govind Mitra Road, **Patna**-800004. Ph : 2300489, 2302100, patna@schandgroup.com

First Edition 1999
First Revised Edition 2003
Second Revised Edition 2007

ISBN : 81-219-1878-2
Code : 10A 206

PRINTED IN INDIA
By Rajendra Ravindra Printers (Pvt.) Ltd., 7361, Ram Nagar, New Delhi-110 055 and published by S. Chand & Company Ltd. 7361, Ram Nagar, New Delhi-110 055

PREFACE TO SECOND REVISED EDITION

Although **Concepts of Modern Engineering Physics** was the first book covering the syllabi of Punjab Technical University, Jalandhar and it was accepted whole-heartedly by students and teachers alike. However, due to the repeated changes of syllabi of P.T.U. as it being a new University, the book had to be revised and some of the chapters became redundant as these were replaced by new topics. Though the book was revised with the additional chapters, the discarded chapters also formed the part of the book. In order to cater to the needs of the students as per the revised syllabi, the book needed thorough revision and also the chapters not included in the PTU syllabi had to be taken out of the book so that the book may appear to be exactly in the format of PTU syllabi. In order to make the book more useful, whole matter of the book was thoroughly revised and many solved problems which have been taken from the University papers of the last 7–8 years have been included. The book as such has been updated to cater to the needs of students of PTU and it is the authors earnest hope that the students will accept the book and find it more suitable for their final examinations.

The authors feel really obliged to Mr. Navin Joshi, General Manager (S&M), Mr. P.J. Singh, Manager, Jalandhar office; Mr. Surinder Singh, Manager, Chandigarh office, for their constant interest in the publication of the present book. We will fail in our duties if we do not extend our heart felt thanks to Mr. Ravindra Kumar Gupta CMD, for his encouragement in bringing out the new edition. We are also thankful to the editorial staff and other members who helped us to bring out the book with their whole hearted co-operation.

Though every care has been taken to avoid any misprints, omissions, mistakes etc. yet some of them might have been left due to oversight. The authors will personally feel obliged if any short coming is pointed out by students or teachers to the publishers or the authors. It will be our utmost effort to make the book as much useful as possible by incorporating the remedial measures to the suggestions given by the worthy readers.

A. S. Vasudeva
Kamal Kishore

REVISED SYLLABUS

CONCEPTS OF MODERN ENGINEERING PHYSICS

(w. e. f. 2004 for B. Tech. Ist Year of P.T.U. Jalandhar)

Ph-101, Engineering Physics

SYLLABUS (PART-A)

1. Electrostatics & Electrodynamics : Charge distribution, Integral/ Differential approach to Gauss's law, Faraday's law, Ampere's circuital law. Dielectric polarization, Electric displacement, 'Susceptibility and permittivity, Maxwells equation and their importance.

2. Magnetic Materials : Terminology and classification of magnetic materials. Types of magnetism (dia. para, ferro and ferri), magnetic anisotropy and magnetostriction, magnetic domains, hard ans soft magnetic materials, ferrites and their applications.

3. Lasers : Spontaneous and stimulated emission; Einstein's coefficients, population inversion and optical pumping; three-level and four-level lasers. Ruby, He-Ne, CO_2, Semiconductor lasers; holography.

4. Fibre Optics : Introduction, basic theory. Acceptance Angle, Numerical Aperture, Normalised Frequency, etc.; modes for propagation; material dispersion and pulse dispersion in optical fibre, fibre connectors, splicers and couplers, application of optical fibre.

PART-B

5. Theory of Relativity: Concepts of ether; Michelson Morley experiment; Einstein's postulates and Lorentz transformation equations; length, time and simultaneity in relativity; addition of velocity, variation of mass with velocity, Mass-energy relation, energy momentum relation.

6. X-Rays : Continuous X-ray spectra; production of characteristic X-ray spectra; Moseley's law; Bragg's law and its applications in crystallography; absorption of X-rays and Absorption coefficient; Non-destructive techniques (qualitative idea).

7. Quantum Theory : Need of Quantum mechanics. The Compton effect; matter waves, group and phase velocities, Uncertainty Principles and its application; time independent and time dependent Schrodinger wave equation, Eigen values and Eigen functions. Born's interpretation and normalization of wave function, orthogonal wave functions, applications of Schrodinger wave equation (particle in a box and harmonic oscillator).

8. Superconductivity : Introduction and magnetic properties of Superconductors (Meissner Effect); Type I and Type II Superconductors, thermodynamics of Superconductors, field penetration and London equations, BCS theory of superconductivity.

FUNDAMENTAL CONSTANTS

	Physical Quantity	Symbol	Value
(i)	Elementary charge	e	1.602×10^{-19} C
(ii)	Unit of mass	MeV/c	5.322 × 10–22 Kg.m/s
(iii)	Atomic mass unit	amu (u)	1.6605×10^{-27} Kg = 931.49 Me V/c^2
(iv)	Mass of electron	m_e	9.1094×10^{-31} Kg = 0.511 MeV/c^2
(v)	Mass of proton	m_p	1.672×10^{-27} Kg = 938.27 MeV/c^2
(vi)	Mass of neutron	m_n	1.6749×10^{-27} Kg = 939.57 McV/c^2
(vii)	Planck's constant	h	6.627×10^{-34} J.s
		$\hbar$	1.0546×10^{-34} J.s
		$\hbar$c	3.1615×10^{-26} J.m
(viii)	Bohr radius	ro	5.2918×10^{-11} m
(ix)	Permittivity of free space	ε_0	8.8542×10^{-12} C^2/Nm^2
		$1/4\ \pi\varepsilon_0$	8.987 ×109 Nm^2
(x)	Speed of light in vaccum	c	2.998×10^{8} m/s
(xi)	Permeability of vaccum	μ_0	$4\pi \times 10^{-7}$ N/m^2
(xii)	Avogadro'a number	N_0	6.023×10^{-23} $mole_{-1}$
(xiii)	Boltzman constant	k	1.38×10^{-23} J/K
(xviii)	Gravitational constant	G	6.673×10^{-11} Nm^2/Kg^2
(xiv)	Stefan's constant	σ	5.67×10^{-8} W/m^2K^4
(xviii)	Fine structure constant	α	0.00729 ~ 1/137
(xv)	Bohr magneton	μ_B	9.274×10^{-24} J/T
(xvi)	Nuclear magneton	μ_N	5.051×10^{-27} J/T
(xvii)	Rydberg constant	R	1.09737×10^{7} m^{-1}

CONTENTS

1

ELECTROSTATICS AND ELECTRODYNAMICS

1.1 Scalar and vector quantities

Physical quantities which can be measured have been classified into two categories.

1. **Scalar:** Which require only the magnitude to define them *e.g., mass, length, time ,density, speed, volume, temperature, work, energy, current, charge potential etc.*

2. **Vectors:** To define some quantities completely both magnitude and direction have to be specified *e.g.,* ; *Displacement, velocity, acceleration, current density, magnetic moment, electric and magnetic fields, momentum, force, torque, weight, etc.*

An important difference between scalar and vector quantities is that while *scalars* can the added, subtracted, multiplied or divided by *simple arithmatic,* however, for vector quantities, *vector algebra* is to the used for addition, subtraction, multiplication or division.

1.2. Representation of a Vector

A vector $\vec{A}$ with its components A_x, A_y and A_g along *x, y* and *z* directions respectively is represented as

$$\vec{A} = A_x\,\hat{\mathbf{i}} + A_y\,\hat{\mathbf{j}} + A_z\,\hat{\mathbf{k}}$$

The magnitude of vectors $\vec{A}$ is give by

$$|A| = \sqrt{A_x^2 + A_y^2 + A_z^2}$$

It the angle made by $\vec{A}$ with *x, y* and *z* directions are α, β and γ respectively then *direction cosines l, m and n for x, y and z axes* are

$$l = \cos\alpha = \frac{A_x}{|A|}$$

$$m = \cos\beta = \frac{A_y}{|A|}$$

$$n = \cos\gamma = \frac{A_z}{|A|}$$

$$\therefore \quad l^2 + m^2 + n^2 = \frac{\cos^2\alpha + \cos^2\beta + \cos^2\gamma}{|A|^2}$$

$$= \frac{A_x^2 + A_y^2 + A_z^2}{A^2} = 1$$

1.3. Product of Two Vectors

Product of two vectors can be a *salar or a vector*, so the vectors can be multiplied in two ways.

1. Dot or scalar product
2. Cross or vector product.

1. *Dot or Scalar product* : Dot product of two vectors $\vec{A}$ and $\vec{B}$ is represented by $\vec{A}.\vec{B}$. It is a scalar quantity whose magnitude is the product of magnitude of $\vec{A}$ and $\vec{B}$ and cosine of the angle between them.

From figure 1.1 vector $\vec{A}$ and $\vec{B}$ can the represented by $\vec{Oa}$ and $\vec{Ob}$, angle θ is the angle between two vectors.

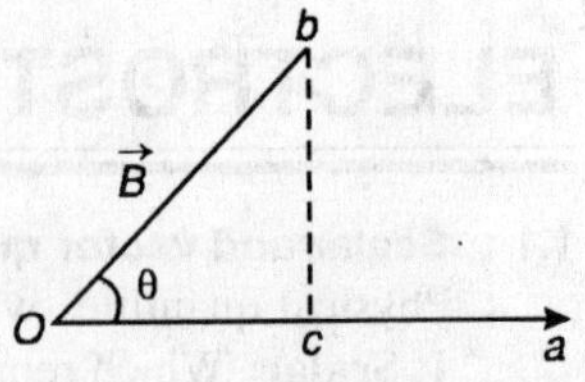

Fig. 1.1

$$\therefore \quad \vec{A}.\vec{B} = |A|\,|B| \cos\theta$$

(2) *Cross or Vector product* : Cross product of two vectors is represented by $\vec{A}\times\vec{B}$. It is a vector quantity whose magnitude is equal is the product of magnitude of $\vec{A}$ and $\vec{B}$ and the sine of angle by which $\vec{B}$ is inclined on $\vec{A}$. The direction is ⊥ to the plane containing $\vec{A}$ and $\vec{B}$. The direction is given by right hand sczew from *viz.*, positive direction is that in *which a right handed screw if rotated from first to second vector through a smaller angle between then advances.*

$$\vec{A}\times\vec{B} = \vec{C}$$

Where $$\vec{C} = |A|\,|B| \sin|\,\vec{A},\vec{B}\,|\,\hat{n}$$

Where $\hat{n}$ is a unit vector normal to the plane containing $\vec{A}$ and $\vec{B}$ as shown in figure 1.2.

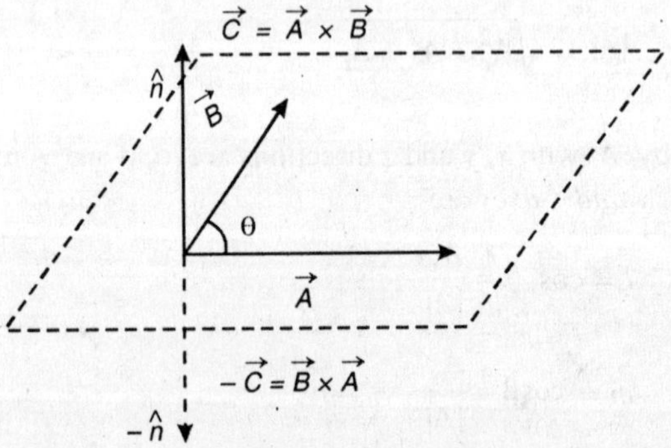

Fig. 1.2

It may be noted that unlike scalar product, vector product does not obey the commutative law because

$$\vec{A}\times\vec{B} = |A|\,|B| \sin\,(\vec{A}\times\vec{B})\hat{n}$$

and $$\vec{B}\times\vec{A} = |B|\,|A| \sin\,(\vec{B},\vec{A})\hat{n}$$

$$= -|A|\,|B| \sin \,(\vec{A}, \vec{B})\hat{n}$$

$$\therefore \qquad \vec{A} \times \vec{B} = -\vec{B} \times \vec{A}$$

Also if the two vectors $\vec{A}$ and $\vec{B}$ are represented by the two sides of a parallelogram, then $\vec{A} \times \vec{B}$ represents the *area of the parallelogram.*

If $\qquad \vec{A} = A_x\,\hat{i} + A_y\,\hat{j} + A_z\,\hat{k}$

and $\qquad \vec{B} = B_x\,\hat{i} + A_y\,\hat{j} + A_z\,\hat{k}$

Then $\vec{A} \times \vec{B}$ may the represented by

$$\vec{A} \times \vec{B} = \begin{vmatrix} \hat{i} & \hat{j} & \hat{k} \\ A_x & A_y & A_z \\ B_x & B_y & B_z \end{vmatrix}$$

Some physical examples of vector or cross product are

(*i*) Torque τ, which is represented by the moment of force $\vec{F}$ about a point whose position vector is $\vec{r}$ it is represented by

$$\vec{\tau} = \vec{r} \times \vec{F}$$

(*ii*) Force on a charge of moving with a velocity $\vec{v}$ in a magnetic field $\vec{B}$. It may be expressed as

$$\vec{F} = \vec{v} \times \vec{B}$$

(*iii*) Angular momentum of a rotating particle, which may be expressed as

$$\vec{L} = \vec{r} \times \vec{p}\,.$$

1.4 Scalar and Vector Fields : If a physical quantity has varying values at different points in space *i.e.,* the quantity has to be represented by *different coordinates* at different points in space so that the physical quantity may the expressed as a continuous function of position of the point in that region, such a physical quantity is said to be a *point function or function of the position.*

The region in which the function represents the physical quantity is called the field of the physical quantity.

The physical quantity may either the a scalar or a vector and so there may be two types of fields :

(*i*) Scalar field

(*ii*) Vector field

(*i*) **Scalar field:** If a scalar field like *temperature, energy, potential,* etc., is represented by a continuous function of position in that region, the field created is said to be *a scalar field.* Such a scalar function will have unique values at each point in the region. If the physical quantity depends on position (*x, y, z*) of a point in space, scalar field may then be expressed as :

$$\phi = \phi\ (x, y, z)$$

Thus the function ϕ (x, y, z) can be used to find the value of the scalar quantity ϕ at all the points of the regions.

An example of the scaler is the *distribution of temperature* in a metallic rod whose one end is hot while other is cold. Similarly distribution of gravitational potential in a gravitational field, distribution of electric or magnetic potentials in their respective fields are examples of scalars.

A graphical representation of scalar field shows that the function ϕ (x, y, z) has fixed and *same value* on each point of a surface but different values at different surfaces. Such a surface is called *equi- scalar surface.* In case the potential is same on a surface it is called *equi-potential* surface which for same fixed tempature on a surface, it is called *equi-temperature surface*.

For example in case of distribution of temperature in a metallic rod, with one end hot and other cold, we can consider isothermal surfaces $\perp$ to the direction of heat flow.

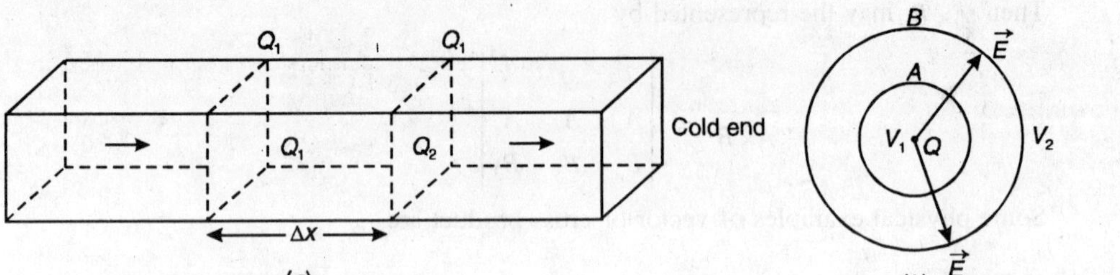

(*a*)
(Equi-thermal surfaces Q_1 and Q_2)

(*b*)
(Equi-potential surfaces V_1 and V_2)

Fig. 1.3

In a steady state, each isothermal will correspond to *definite temprature*. Similarly in case of distribution of charge, each sphere surrounding the charge will correspond to *equi-potential surface*, having the same fixed potential for a particular sphere.

(*ii*) **Vector field:** If a vector quantity $\vec{F}$ (like *velocity, force, electric field, magnetic field etc.*) is expressed by a continuous function of position (x, y, z) in a region $F(x, y, z)$, a *vector field* is said is exist in that region. It is important to note that at each point in this region, the magnitude and direction of the vector function is well defined and it *changes from point to point*. An important example of vector field is the *velocity distribution* in a fluid. Another example is variation of *electric and magnetic fields* etc.

Representation of vector field : A vector field may graphically be represented by a curve joining the direction of various points in that vector field. *The tangent drawn at any point on this curve gives the direction* of the vector at that point.

This curve is called vector line or flux line or line of flow of the vector field.

The magnitude of the vector at any point of the flux line is represented by taking a surface of unit area normal to the flux line considering the various points on it *such that the number of points is numerically equal to the magnitude of the vector at that point*. Now if we draw parallel flux lines to *represent the direction of vector function F (x, y, z)* and the *number of flux lines* through unit area normal to the flux line will represent the *magnitude of the vector function* at that point.

Figure 1.4 (*a*) represent the non uniform vector distribution in different layers of the fluid through a small tube in *stream line* flow condition. While Fig. 1.4 (*b*) represents the flux lines of a uniform electric field between two oppositely charged conductors.

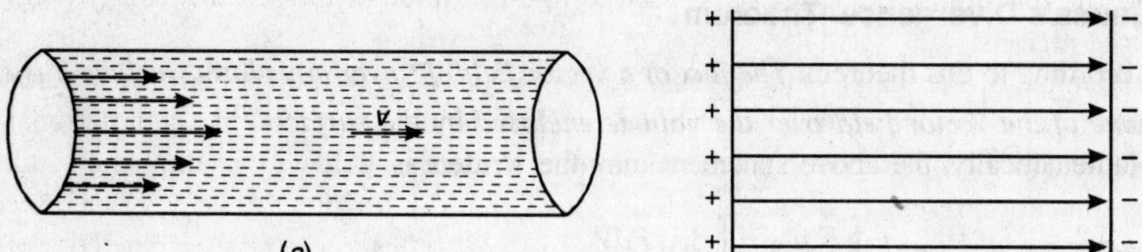

(a) (b)

Fig. 1.4

1.5. Gradient of a Scalar Field

If a scalar function at a point (x, y, z) in space is represented by $\phi_{x,y,z}$ (which means that $\phi_{x,y,z}$ is a scalar field), its cartesian components along three mutually $\perp$ axes x, y and z are respectively $\frac{\partial \phi}{\partial x}, \frac{\partial \phi}{\partial y}$ and $\frac{\partial \phi}{\partial z}$. The gradient of scalar function f (x, y, z) is then defined in the following way

$$\text{grad } \phi = \hat{\mathbf{i}} \frac{\partial \phi}{\partial x} + \hat{\mathbf{j}} \frac{\partial \phi}{\partial y} + \hat{\mathbf{k}} \frac{\partial \phi}{\partial z}$$

$$= \left(\hat{\mathbf{i}} \frac{\partial}{\partial x} + \hat{\mathbf{j}} \frac{\partial}{\partial y} + \hat{\mathbf{k}} \frac{\partial}{\partial z} \right) \phi = \vec{\nabla} \phi$$

Which means that the *gradient of a scalar function is a vector.*

Where $\vec{\nabla}$ is a vector operator and is called *del operator*

1.6. Physical Significance of Gradient of Scalar Field

In physics, gradient of a scalar field finds lot of significance. For example, the negative gradient of electric potential at a point defines the electric field at that point. Which means $\vec{E} = -\text{ grad } V = -\vec{\nabla} V.$

So the electric field at a point is a vector whose magnitude is equal to the maximum rate of variation of electric potential and direction is that maximum decrease in potential *i.e.,* right angles to the equipotential at that point (Fig 1.3 (*b*) It may be stated in other form as the electric potential is a scalar quantity whose *negative gradient gives us the electric field.*

In case of a potentiometer, a battery may be used to maintain a constant potential difference between the ends of potentiometer wire connecting positive terminal of the battery at one end and negative terminal at the other end. *A potential gradient is said to exist along the length of the wire.* Flow of current in the wire is *opposite to the direction of potential gradient.*

In the same way, when heat flows through a rod by heating one end and keeping the other end of the rod at room temprature. Heat flows through the rod and if θ_1 and θ_2 are the temperatures of the isothermals A and B Fig 1.3 (*a*) at distances x and $x + dx$ from hot end. Then temperature gradient is $\frac{\theta_2 - \theta_1}{dx} = -\frac{d\theta}{dx}$. So the temperature gradient will represent the direction of flow of heat. *Heat flows in a direction in which the variation of temperature is maximum i.e.,* perpendicular to the direction of isothermal surface from high temperature to low temperature. In the same way we can define the gravitational field $\vec{F} = -\vec{\nabla}\phi$ where ϕ corresponds to the gravitational potential.

1.7. Gauss's Divergence Theorem

According to this theorem. *The flux of a vector field $\vec{F}$ over any closed surface S is equal to the volume of the Vector field over the volume enclosed by the surface.*

Mathematically, the above statement may the written as

$$\oint \vec{F}.d\vec{s} = \int_v \text{div } \vec{F} dV \quad \text{... (1)}$$

We can prove the above theorem by considering a finite volume V of any shape enclosed by a surface S (as shown in Fig. 1.5 (a) in the region of any vector $\vec{F}$.

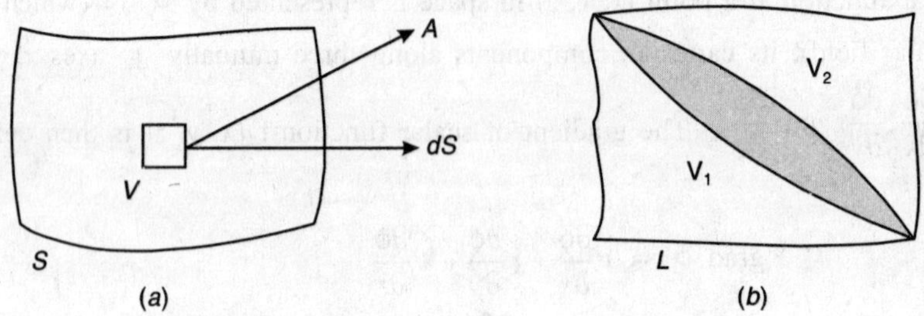

Fig. 1.5

The flux diverging from the surface S of volume V is given by

$$\phi = \oint_S \vec{F}.d\vec{S} \quad \text{... (1)}$$

Let us divide the volume V into two parts comprising volume V_1 and volume V_2 enclosed by surfaces S_1 and S_2 respectively

Flux diverging from surface S_1 is given by

$$\phi_1 = \oint_S \vec{F}.d\vec{S_1} \quad \text{... (2)}$$

and the flux diverging from surface S_2 is

$$\phi_2 = \oint_{S_2} \vec{F}.d\vec{S_2} \quad \text{... (3)}$$

Since the shaded part of Fig. 1.5 (b) is common to both the volumes field and also the flux coming out of both volumes being equal and opposite will cancel each other when considered together, The rest of be flux will be same in both the cases so we have

$$\phi = \oint_S \vec{F}.d\vec{S} = \int_{S_1} \vec{F}.d\vec{S_1} + \int_{S_2} \vec{F}.d\vec{S_2} \quad \text{... (4)}$$

It we divide the volume V into large number of volumes V_1, V_2, V_3,........ V_n enclosed by surfaces S_1, S_2, S_3, ... S_n respectively, then we have

$$\phi = \oint_S \vec{F}.d\vec{S} = \int_{S_1} \vec{F}.d\vec{S_1} + \int_{S_2} \vec{F}.d\vec{S_2} + ... + \int_{S_3} \vec{F}.d\vec{S_2} + ... \int_{S_n} \vec{F}.\vec{S_n}$$

$$= \sum_{i=1}^{n} \int_S \vec{F}.d\vec{S_i}$$

Which may be written as

$$\oint_S \vec{F}.d\vec{S} = \sum_{i=1}^{n} V_i \frac{\int_{S_i} \vec{F}.d\vec{S_2}}{V_i} \quad \text{.... (5)}$$

If n is sufficiently large, then volume V_i becomes infinitely small *i.e.*, n tends to ∞ and V_i tends to zero.

In that limit we may write

$$\underset{V_i \to 0}{\text{Lim}} \frac{\int_{S_i} \vec{F}.d\vec{S_2}}{V_i} = \text{div}\,\vec{F} \quad \text{... (6)}$$

Now converting the summation into integration and writing dv for small volume, equation. (5) may the written in that limit as

$$\int_{S_i} \vec{F}.d\vec{S} = \int_v \text{div}\,\vec{F}.dV \quad \text{... (7)}$$

This is the *Gauss's divergence Theorem* and is useful for converting the **volume integral of the divergence of vector field into the surface integral of vector field and vice-versa.**

1.8. Derivation of Poisson's and Laplace's Equation from Gauss's Law.

(i) *Poisson's equation*

Differential form of gauss's law may the written as

$$\vec{\nabla}.\vec{E} = \frac{\rho}{\varepsilon_0} \quad \text{(in S.I. Units)} \quad \text{... (1)}$$

Where E is the electric field intensity and ρ the volume density of the charge.

By definition, the electric field at a point may the written as the negative gradient of electric potential ϕ, *i.e.*

$$\vec{E} = -\,\text{grad}\,\phi = -\vec{\nabla}\phi$$

So the eq. (1) may the written as

$$\text{div}\,\vec{E} = \vec{\nabla}.\vec{E} = \vec{\nabla}.(-\nabla\phi)$$

$$= -\nabla^2\phi = \frac{\rho}{\varepsilon_0}$$

$$\therefore \quad \nabla^2\phi = \frac{-\rho}{\varepsilon_0}. \quad \text{... (2)}$$

or

$$\left(\frac{\partial^2\phi}{\partial x^2} + \frac{\partial^2\phi}{\partial y^2} + \frac{\partial^2\phi}{\partial z^2}\right) = -\rho/\varepsilon_0$$

Which is *Poison's equation* and it holds good for each point in space.

Laplace's equation : If the enclosed charge density is zero *i.e.*, the point of observation is in empty space, then $\rho = 0$ and we have

$$\nabla^2\phi = 0$$

or $$\frac{\partial^2 \phi}{\partial x^2}+\frac{\partial^2 \phi}{\partial y^2}+\frac{\partial^2 y}{\partial z^2}=0 \qquad \text{... (3)}$$

Which is Laplace's equation and the operator ∇^2 known as *Laplace's Operator is*

1.9. Significance of Laplace's Equation

Laplace's equation is an important differential equation and a function which satisfies the above equation is called *spherical harmonic* which has a very important property *viz.*

The average value of a function $\phi(x, y, z)$ satisfying Laplace's equation over any spherical surface is equal to its value at the centre of spherical surface.

We have already concluded that *electric potential and intensity* at any point due to spherical shell of charge is the same as if the whole charge were concentrated at its centre, in accordance with the Laplace's equation.

1.10. Stoke's Theorem

Stoke's theorem states that the *line integral of the tangential component of a vector point function $\vec{F}$ taken around a closed curve C is equal to the surface integral of the normal component of $\vec{F}$ taken over any surface S having C as it boundary.*

The above theorem may also be stated as the *flux of the curl of a vector field $\vec{F}$ over any surface S of any shape is equal to the line integral of the vector field $\vec{F}$ over the boundary of the surface.*

Mathematically, the above statement may the written as

$$\int_S \text{curl } \vec{F} . d\vec{S} = \oint_C \vec{F} . d\vec{r} \qquad \text{... (1)}$$

Where $d\vec{S}$ represents infinite element of area S and $d\vec{r}$ is an infinite small element of boundary C of the surface.

As is clear from above; this theorem helps to *convent the surface integral of the curl of the vector field $\vec{F}$ to the line integral of vector field and vice versa.*

In order to derive Stoke's theorem, let us consider a *surface S with C as is boundary.* Let us now calculate the line integral of the vector function $\vec{F}$ round the boundary C of Surface S.

The line integral of $\vec{F}$ around the boundary C of surface S

$$= \oint_C \vec{F} . d\vec{r}$$

Let us divide the surface S into two parts of surfaces S_1 and S_2 having boundaries C_1 and C_2 respectively as shown in Fig. 1.6 (*b*)

The line integral of $\vec{F}$ around the boundary of S_1

$$= \oint_{C_1} \vec{F} . d\vec{r_1}$$

and the line integral of $\vec{F}$ around the boundary of S_2

$$= \oint_{C_2} \vec{F} . d\vec{r_2}$$

As the thick line LM is common boundary to both the surfaces and is traversed in opposite directions in the two integrals (Fig.1.6 *b*), the integral of A along this common boundary being equal and opposite, cancel each other when considered together.

The rest of the boundaries C_1 and C_2 is identical to the original boundary C. Thus obviously

$$\oint_C \vec{F}\, d\vec{r} = \oint_{C_1} \vec{F}\, d\vec{r}_1 + \oint_{C_2} \vec{F}.d\vec{r}_2 \quad \text{... (2)}$$

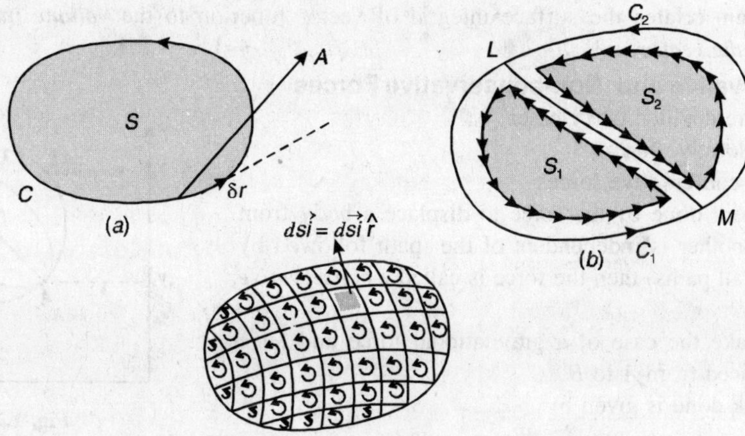

Fig. 1.6

Now divide the surface S into a large number of parts $S_1, S_2, ..., S_i, ... S_n$ having boundaries $C_1, C_2, ... C_i, ... C_n$ respectively as shown in fig 1.6 (*c*). In that case

$$\oint \vec{F}\, d\vec{r} = \oint_{C_1} \vec{F}\, d\vec{r}_1 + \oint_{C_2} \vec{F}.d\vec{r}_2 + ... \oint_{C_i} \vec{F}.d\vec{r}_i + ... \oint_{C_n} \vec{F}.d\vec{r}_n$$

$$= \sum_{i=1}^{n} \oint \vec{F}.d\vec{r}_i$$

Which may the written in the form

$$\oint_C \vec{F}.d\vec{r} = \sum_{i=1}^{n} S_i \oint_{C_i} \frac{\vec{F}.d\vec{r}_i}{S_i} \quad \text{... (4)}$$

If we consider n to be sufficiently large, in that case S_i becomes infinitely small *i.e.*, if $n \to \infty$, $S_i \to 0$ and in that limit we may write from the definition of curl of vector field $\vec{F}$ as

$$\lim_{S_i \to 0} \frac{\oint_{C_i} \vec{F}.d\vec{r}_i}{S_i} = \text{curl } \vec{F}.\hat{n}.$$

Now converting summation into integration and writing $d\vec{S}$ for infinitely small area S, So the eq. (4) in the limit form may the written as

$$\oint_C \vec{F}.d\vec{r} = \oint_C \text{curl } \vec{F}.\hat{n}d\vec{S}$$

or $$\oint_C \vec{F}.d\vec{r} = \oint_C \text{curl } \vec{F}.d\vec{S}$$

This equation is known as **Stoke's theorem**.

The basic difference between Stoke's theorem and Gauss's theorem is that whereas Stokes theorem relates the *line integral* of a vector function to *surface integral of the curl of a vector*, Gauss's theorem relates the surface integral of vector function to the *volume integral of the devergence of the vector*.

1.11. Conservative and Non-conservative Forces

Forces are devided two classes

(*i*) Consevative forces

(*ii*) Non-conservative forces

If the work done by the force to displace a body from one point to another is independent of the path followed by the body (for all paths) then the force is called a *conservative force*.

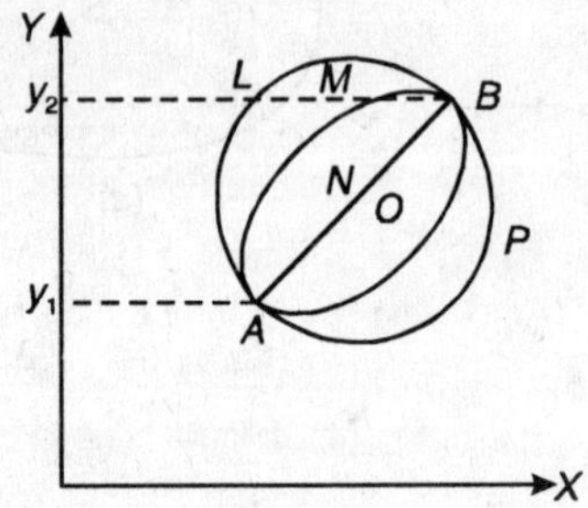

Fig. 1.7

Let us take the case of a gravitational force and let a body be displaced from *A* to *B*.

The work done is given by

$$W_{AB} = -mg\,(y_2 - y_1)$$

As is evident the work done by gravitational force depends on the coordinates of *A* and *B* and is independent of path followed *vig*., via *L, M , N, O, P* etc. So the gravitational force is an example of *conservative force*.

Other examples of conservative forces are *electrostatic force, restoring force of a spring* etc.

In the case of restoring force of spring, work done is given by

$$W = \frac{1}{2}k\,(x_1^2 - x_2^2)$$

Where x_1 and x_2 are measured with respect to the equilibrium position $x = 0$.

A conservative force and a non-conservative force can the described by using the concept of *closed path*. When the starting point and the end point is same, it is called *closed path* and if the net work done by a force around any closed path is zero, then it is called *conservative force*, otherwise a *non-conservative force* An example of non-conservative force is *fractional forces*. Non consevative force is also called a *dissipative force*. In case of conservative force, the representation is given by

$$\oint \vec{F}.d\vec{S} = 0$$

Where $\oint$ represents a *closed path*

In case of non-conservative force $\oint \vec{F}.d\vec{S} \neq 0$. In case of frictional force, he motion is opposed due to friction. When a block is moved in clockwise or anticlockwise direction against friction in that case $\oint \vec{F}.d\vec{S} \neq 0$.

NUMERICAL PROBLEMS

Example 1. *The potential function in an electric field is represented as* $V_{(x,y,z)} = C(x^2+y^2+z^2)$ *where C is an arbitrary constant. Show that the electric field is radial.*

Solution. Electric field $\vec{E} = -\nabla V = -\left[\hat{\mathbf{i}}\frac{\partial V}{\partial x}+\hat{\mathbf{j}}\frac{\partial V}{\partial y}+\hat{\mathbf{k}}\frac{\partial V}{\partial z}\right]$

$$= -\left[\hat{\mathbf{i}}\frac{\partial}{\partial x}\{C(x^2+y^2+z^2)\}+\hat{\mathbf{j}}\frac{\partial}{\partial y}\{C(x^2+y^2+z^2)\}+\hat{\mathbf{k}}\frac{\partial}{\partial z}\{C(x^2+y^2+z^2)\}\right]$$

$$= -[\hat{\mathbf{i}}2Cx+\hat{\mathbf{j}}2Cy+\hat{\mathbf{k}}2Cz] = -2C(\hat{\mathbf{i}}x+\hat{\mathbf{j}}y+\hat{\mathbf{k}}z)$$

$= -2C\vec{r}$ where $\vec{r}$ is the position vector.

So, the direction of electric field $\vec{E}$ is opposite to $\vec{r}$, *i.e.,* towards the origin. **So the field is radial.** **Ans.**

Example 2. *If* $V = 2x^2 - 3y^2 + z^2$ *represents the electrostatic potential at a point, find the electric field intensity at a point (3, 2, – 3)*

Solution. Electric field intensity is given by

$$\vec{E} = -\vec{\nabla}V = -\left(\hat{\mathbf{i}}\frac{\partial V}{\partial x}+\hat{\mathbf{j}}\frac{\partial V}{\partial y}+\hat{\mathbf{k}}\frac{\partial V}{\partial z}\right)$$

$$= -\left[\hat{\mathbf{i}}\frac{\partial}{\partial x}(2x^2-3y^2+z^2)+\hat{\mathbf{j}}\frac{\partial}{\partial y}(2x^2-3y^2+z^2)+\hat{\mathbf{k}}\frac{\partial}{\partial z}(2x^2-3y^2+z^2)\right]$$

$$= -[\hat{\mathbf{i}}(4x)+\hat{\mathbf{j}}(-6y)+\hat{\mathbf{k}}(2z)] = -4x\hat{\mathbf{i}}+6y\hat{\mathbf{j}}-2z\hat{\mathbf{k}}$$

At a point (3, 2, – 3).

$$\vec{E} = -4(3)\hat{\mathbf{i}}+6(2)\hat{\mathbf{j}}-2(-3)\hat{\mathbf{k}}$$

or $\vec{E} = -12\hat{\mathbf{i}}+12\hat{\mathbf{j}}+6\hat{\mathbf{k}}$. **Ans.**

Example 3. *A closed surface S encloses a volume V. Calculate the value of div* $\vec{F}$ *and* $\iint_S \vec{F}.d\vec{S}$ *for a vector field* $\vec{F} = 3x\hat{\mathbf{i}}+4y\hat{\mathbf{j}}+2z\hat{\mathbf{k}}$.

Solution. By Gauss's divergence theorem, we have

$$\iint_S \vec{F}.d\vec{S} = \iiint_V (\text{div}\,\vec{F})\,dV$$

Now, div $\vec{F} = \frac{\partial F_x}{\partial x}+\frac{\partial F_y}{\partial y}+\frac{\partial F_z}{\partial z}$

$$= \frac{\partial}{\partial_x}(3x)+\frac{\partial}{\partial y}(4y)+\frac{\partial}{\partial z}(2z) = 3+4+2 = 9$$

$\therefore \iint_S \vec{F}.d\vec{a} = \iiint_V 9\,dV = 9V.$ **Ans.**

Example 4. *In a conducttion phenomen of flow of heat through a rod, under the steady state, isothermal surface is defined by the equation* $\theta = 9x - 5y + 6z$. *Calculate the temperature field and show it to be uniform .*

Solution. By definition, temp. field is given by

$$\text{Temp. field} = \nabla\phi = \left(\hat{\mathbf{i}}\frac{\partial\theta}{\partial x} + \hat{\mathbf{j}}\frac{\partial\theta}{\partial y} + \hat{\mathbf{k}}\frac{\partial\theta}{\partial z}\right)$$

$$= \hat{\mathbf{i}}\frac{\partial}{\partial x}(9x-5y+6z) + \hat{\mathbf{j}}\frac{\partial}{\partial y}(9x-5y+6z) + \hat{\mathbf{k}}\frac{\partial}{\partial z}(9x-5y+6z)$$

$$= \hat{\mathbf{i}}(9) + \hat{\mathbf{j}}(-5) + \hat{\mathbf{k}}(6)$$

$$= 9\hat{\mathbf{i}} + 5\hat{\mathbf{j}} + 6\hat{\mathbf{k}}$$

This is uniform field as it is independent of x, y, z. Ans.

Example 5. *The electric field intensity at a point of a closed surface is expressed as* $\vec{E} = (2\hat{\mathbf{i}} + 3\hat{\mathbf{j}} - 2\hat{\mathbf{k}})$ *volt/metre. Calculate the electric flux through a small element of area* $\vec{E} = (\hat{\mathbf{i}} + \hat{\mathbf{j}} - \hat{\mathbf{k}})\, m^2$

Solution. Electric flux is given by

$$d\phi = \vec{E}.d\vec{S} = (2\hat{\mathbf{i}} + 3\hat{\mathbf{j}} - 2\hat{\mathbf{k}}).(\hat{\mathbf{i}} + \hat{\mathbf{j}} - \hat{\mathbf{k}})$$

$$= 2 + 3 + (-2)(-1) = 2 + 3 + 2 = \mathbf{7\ volt \times m.} \quad \textbf{Ans.}$$

Example 6. *Prove that the divergence of vector field which obeys the inverse square law is zero.*

Solution. An inverse square vector field is given by.

$$\vec{F} \propto \frac{1}{r^2}\hat{r} \quad \text{or} \quad \vec{F} = \frac{K}{r^2}\hat{r} = \frac{K}{r^3}.\vec{r}$$

where $\vec{r} = \hat{\mathbf{i}}x + \hat{\mathbf{j}}y + \hat{\mathbf{k}}z$ is the position vector and K is a constant.

So we have $\vec{F} = \dfrac{K}{r^3}(\hat{\mathbf{i}}x + \hat{\mathbf{j}}y + \hat{\mathbf{k}}z)$

$$\therefore \quad \text{Div } \vec{F} = \vec{\nabla}.\vec{F} = \frac{\partial F_x}{\partial x} + \frac{\partial F_y}{\partial y} + \frac{\partial F_z}{\partial x} = \frac{\partial}{\partial x}\left(\frac{Kx}{r^3}\right) + \frac{\partial}{\partial y}\left(\frac{Ky}{r^3}\right) + \frac{\partial}{\partial z}\left(\frac{Kz}{r^3}\right)$$

$$= \left\{\frac{K}{r^3} + Kx\left(\frac{-3}{r^4}\right)\frac{\partial r}{\partial x}\right\} + \left\{\frac{K}{r^3} + Ky\left(\frac{-3}{r^4}\right)\frac{\partial r}{\partial y}\right\} + \left\{\frac{K}{r^3} + Kz\left(\frac{-3}{r^4}\right)\frac{\partial r}{\partial z}\right\}$$

$$= \frac{3K}{r^3} - \frac{3K}{r^4}\left\{x\frac{\partial r}{\partial x} + y\frac{\partial r}{\partial y} + z\frac{\partial r}{\partial z}\right\}$$

Now $r^2 = x^2 + y^2 + z^2$

Differentiating w.r.t. x.

$$\therefore \quad 2r\frac{\partial r}{\partial x} = 2x \quad \text{or} \quad \frac{\partial r}{\partial x} = \frac{x}{r},$$

Similarly, $\dfrac{\partial r}{\partial y} = \dfrac{y}{r}$ and $\dfrac{\partial r}{\partial z} = \dfrac{z}{r}$

So
$$\nabla.\vec{F} = \frac{3K}{r^3} - \frac{3K}{r^4}\left[x\frac{x}{r} + y\frac{y}{r} + z\frac{z}{r}\right]$$

$$= \frac{3K}{r^3} - \frac{3K}{r^5}(x^2 + y^2 + z^2)$$

$$= \frac{3K}{r^3} - \frac{3K}{r^5}(r^2) = \mathbf{0} \ \textbf{Ans.}$$

Example 7. *Show that a vector field $\vec{F} = KR^n \hat{R}$ is an irrotational field for all values of n, but is a solenoidal field for n = – 2. Here K is a constant.*

Solution.

$$\vec{F} = KR^n \hat{\text{R}} = KR^{n-1} \vec{R}$$

By definition, for an irrotational field, the value of *curl $\vec{F}$ must be zero*

Now

$$\text{curl } \vec{F} = \text{curl } (KR^{n-1} \vec{R}) = \vec{\nabla} \times [KR^{n-1}(\hat{i}x + \hat{j}y + \hat{k}z)]$$

$$= \begin{vmatrix} \hat{\mathbf{i}} & \hat{\mathbf{j}} & \hat{\mathbf{k}} \\ \dfrac{\partial}{\partial x} & \dfrac{\partial}{\partial y} & \dfrac{\partial}{\partial z} \\ KR^{n-1}x & KR^{n-1}y & KR^{n-1}z \end{vmatrix}$$

Which means

$$\text{curl } \vec{F} = K\left[\hat{\mathbf{i}}\left\{\frac{\partial}{\partial y}(R^{n-1}z) - \frac{\partial}{\partial z}(R^{n-1}y)\right\}\right.$$

$$\left. + \hat{\mathbf{j}}\left\{\frac{\partial}{\partial z}(R^{n-1}x) - \frac{\partial}{\partial x}(R^{n-1}z)\right\} + \hat{\mathbf{k}}\left\{\frac{\partial}{\partial x}(R^{n-1}y) - \frac{\partial}{\partial x}(R^{n-1}x)\right\}\right]$$

$$= K\left[\hat{\mathbf{i}}\left\{z(n-1)R^{n-2}\frac{\partial R}{\partial y} - y(n-1)R^{n-2}\frac{\partial R}{\partial z}\right\} + \hat{\mathbf{j}}(\ldots.) + \hat{\mathbf{k}}(\ldots.)\right]$$

$$= K[\hat{\mathbf{i}}\{(n-1)R^{n-3}(zy - yz)\} + \hat{\mathbf{j}}\{(n-1)R^{n-3}(xz - zx)\} + \hat{\mathbf{k}}\{(n-1)R^{n-3}(yx - xy)\}]$$

$$= 0.$$

But for a solenoidal field, the value of div $\vec{F}$ *should be zero.*

$$\text{div } \vec{F} = \text{div } (KR^{n-1} \vec{R}) = \vec{\nabla}.[KR^{n-1}(\hat{\mathbf{i}}x + \hat{\mathbf{j}}y + \hat{\mathbf{k}}z)]$$

$$= \frac{\partial}{\partial_x}(KR^{n-1}x) + \frac{\partial}{\partial y}(KR^{n-1}y) + \frac{\partial}{\partial z}(KR^{n-1}z)$$

$$= K\left[\left\{R^{n-1} + x(n-1)R^{n-2}\frac{\partial R}{\partial x}\right\} + \left\{R^{n-1} + y(n-1)R^{n-2}\frac{\partial R}{\partial y}\right\} + \left\{R^{n-1} + z(n-1)R^{n-2}\frac{\partial R}{\partial z}\right\}\right]$$

$$= K\left[3R^{n-1} + (n-1)R^{n-2}\left\{x\frac{\partial R}{\partial x} + y\frac{\partial R}{\partial y} + z\frac{\partial R}{\partial z}\right\}\right]$$

$$= K\left[3R^{n-1} + (n-1)R^{n-2}\left\{x\left(\frac{x}{R}\right) + y\left(\frac{y}{R}\right) + z\left(\frac{z}{R}\right)\right\}\right]$$

$$= KR^{n-1}(3+n-1) = K(n+2)R^{n-1}$$

Which means div $\vec{F} = 0$ if $n + 2 = 0$

or $n = -2$ **Ans.**

Example 8. *If* $\vec{r}$, *position vector is* $\vec{r} = \hat{\mathbf{i}}x + \hat{\mathbf{j}}y + \hat{\mathbf{k}}z$, *find :*

(*i*) grad r^n, (*ii*) div $r^n\hat{r}$, and (*iii*) curl $r^n\vec{r}$.

Solution. We have $\vec{r} = \hat{\mathbf{i}}x + \hat{\mathbf{j}}y + \hat{\mathbf{k}}z$ $\quad\therefore r^2 = x^2 + y^2 + z^2$

also $\dfrac{\partial r}{\partial x} = \dfrac{x}{r}, \dfrac{\partial r}{\partial y} = \dfrac{y}{r}, \dfrac{\partial r}{\partial z} = \dfrac{z}{r}$

(*i*) grad $r^n = \vec{\nabla} r^n = \hat{\mathbf{i}}\dfrac{\partial}{\partial x}r^n + \hat{\mathbf{j}}\dfrac{\partial}{\partial y}r^n + \hat{\mathbf{k}}\dfrac{\partial}{\partial z}r^n$

$$= \hat{\mathbf{i}}nr^{n-1}\frac{\partial r}{\partial x} + \hat{\mathbf{j}}nr^{n-1}\frac{\partial r}{\partial y} + \hat{\mathbf{k}}nr^{n-1}\frac{\partial r}{\partial z}$$

$$= nr^{n-1}\left[\hat{\mathbf{i}}\frac{\partial r}{\partial x} + \hat{\mathbf{j}}\frac{\partial r}{\partial y} + \hat{\mathbf{k}}\frac{\partial r}{\partial z}\right]$$

$$= nr^{n-1}\left[\hat{\mathbf{i}}\frac{x}{r} + \hat{\mathbf{j}}\frac{y}{r} + \hat{\mathbf{k}}\frac{z}{r}\right]$$

$$= nr^{n-2}(\hat{\mathbf{i}}x + \hat{\mathbf{j}}y + \hat{\mathbf{k}}z) = nr^{n-2}\vec{r} \quad \textbf{Ans.}$$

(*ii*) div $r^n\hat{r} = \text{div}\, r^n\left(\dfrac{\vec{r}}{r}\right) = \text{div}\, r^{n-1}\vec{r}$

$$= \vec{\nabla}.[r^{n-1}(\hat{\mathbf{i}}x + \hat{\mathbf{j}}y + \hat{\mathbf{k}}z)]$$

$$= \left(\hat{\mathbf{i}}\frac{\partial}{\partial x} + \hat{\mathbf{j}}\frac{\partial}{\partial y} + \hat{\mathbf{k}}\frac{\partial}{\partial z}\right).(\hat{\mathbf{i}}r^{n-1}x + \hat{\mathbf{j}}r^{n-1}y + \hat{\mathbf{k}}r^{n-1}z)$$

$$= \frac{\partial}{\partial x}(r^{n-1}x) + \frac{\partial}{\partial y}(r^{n-1}y) + \frac{\partial}{\partial z}(r^{n-1}z)$$

$$= \left[r^{n-1} + x(n-1)r^{n-2}\frac{\partial r}{\partial x}\right] + \left[r^{n-1} + y(n-1)r^{n-2}\frac{\partial r}{\partial y}\right] + \left[r^{n-1} + z(n-1)r^{n-2}\frac{\partial r}{\partial z}\right]$$

$$= 3r^{n-1} + (n-1)r^{n-2}\left[x\frac{\partial r}{\partial x} + y\frac{\partial r}{\partial y} + z\frac{\partial r}{\partial z}\right]$$

$$= 3r^{n-1} + (n-1)r^{n-2}\left[x\left(\frac{x}{r}\right) + y\left(\frac{y}{r}\right) + z\left(\frac{z}{r}\right)\right]$$

$$= 3r^{n-1} + (n-1)r^{n-2}\frac{(x^2+y^2+z^2)}{r}$$

$$= 3r^{n-1} + (n-1)r^{n-1}$$

$$= (3+n-1)r^{n-1} = \mathbf{(n+2)r^{n-1}}. \textbf{ Ans.}$$

(*iii*) curl $r^n \vec{r} = \vec{\nabla}\times(r^n \vec{r})$

$$= \left(\hat{\mathbf{i}}\frac{\partial}{\partial x}+\hat{\mathbf{j}}\frac{\partial}{\partial y}+\hat{\mathbf{k}}\frac{\partial}{\partial z}\right)\times[r^n(\hat{\mathbf{i}}x+\hat{\mathbf{j}}y+\hat{\mathbf{k}}z)]$$

$$= \begin{vmatrix} \hat{\mathbf{i}} & \hat{\mathbf{j}} & \hat{\mathbf{k}} \\ \frac{\partial}{\partial x} & \frac{\partial}{\partial y} & \frac{\partial}{\partial z} \\ r^n x & r^n y & r^n z \end{vmatrix}$$

$$= \hat{\mathbf{i}}\left[\frac{\partial}{\partial y}(r^n z)-\frac{\partial}{\partial z}(r^n y)\right]+\hat{\mathbf{j}}\left[\frac{\partial}{\partial z}(r^n x)-\frac{\partial}{\partial x}(r^n z)\right]+\hat{\mathbf{k}}\left[\frac{\partial}{\partial x}(r^n y)-\frac{\partial}{\partial y}(r^n x)\right]$$

$$= \hat{\mathbf{i}}\left[r^n 0 + znr^{n-1}\frac{\partial r}{\partial y} - r^n\times 0 - ynr^{n-1}\frac{\partial r}{\partial z}\right]+\ldots+\ldots$$

$$= \hat{\mathbf{i}}\left[nzr^{n-1}\left(\frac{y}{r}\right)-nyr^{n-1}\left(\frac{z}{r}\right)\right]+\ldots+\ldots$$

$$= \hat{\mathbf{i}}[nr^{n-2}(zy-yz)]+\hat{\mathbf{j}}[nr^{n-2}(xz-zx)]+\hat{\mathbf{k}}[nr^{n-2}(yx-xy)]$$

$$= \mathbf{0} \quad \textbf{Ans.}$$

Example 9. *A vector field is given as* $\vec{F} = y\hat{\mathbf{i}}+(x^2+y^2)\hat{\mathbf{j}}+(yz+zx)\hat{\mathbf{k}}$. Find: (*i*) div $\vec{F}$, (*ii*) curl $\vec{F}$.

Solution. (*i*) div $\vec{F} = \frac{\partial F_x}{\partial x}+\frac{\partial F_y}{\partial y}+\frac{\partial F_z}{\partial z}$

$$= \frac{\partial}{\partial x}(y)+\frac{\partial}{\partial y}(x^2+y^2)+\frac{\partial}{\partial_z}(yz+zx)$$

$$= 0+2y+y+x$$

$$= \boldsymbol{x+3y}. \textbf{ Ans.}$$

(*ii*) curl $\vec{F} = \begin{vmatrix} \hat{\mathbf{i}} & \hat{\mathbf{j}} & \hat{\mathbf{k}} \\ \frac{\partial}{\partial x} & \frac{\partial}{\partial y} & \frac{\partial}{\partial z} \\ F_x & F_y & F_z \end{vmatrix} = \begin{vmatrix} \hat{\mathbf{i}} & \hat{\mathbf{j}} & \hat{\mathbf{k}} \\ \frac{\partial}{\partial x} & \frac{\partial}{\partial y} & \frac{\partial}{\partial z} \\ y & x^2+y^2 & yz+zx \end{vmatrix}$

$$= \hat{\mathbf{i}}\left[\frac{\partial}{\partial y}(yz+zx)-\frac{\partial}{\partial z}(x^2+y^2)\right]+\hat{\mathbf{j}}\left[\frac{\partial}{\partial z}(y)-\frac{\partial}{\partial x}(yz+zx)\right]+\hat{\mathbf{k}}\left[\frac{\partial}{\partial x}(x^2+y^2)-\frac{\partial}{\partial y}(y)\right]$$

$$= \hat{\mathbf{i}}[z-0]+\hat{\mathbf{j}}[0-z]+\hat{\mathbf{k}}[2x-1] = z\hat{\mathbf{i}}-z\hat{\mathbf{j}}+(2x-1)\hat{\mathbf{k}} \quad \textbf{Ans.}$$

Example 10. *If* ϕ *and* Ψ *are the functions of position (x, y, z), show that: grad* $(\phi\Psi) = \phi$ *grad* $\Psi + \Psi$ grad ϕ.

Solution. grad $(\phi\Psi) = \vec{\nabla}(\phi\Psi) = \left(\hat{\mathbf{i}}\frac{\partial}{\partial x} + \hat{\mathbf{j}}\frac{\partial}{\partial y} + \hat{\mathbf{k}}\frac{\partial}{\partial z}\right)(\phi\Psi)$

So $\vec{\nabla}(\phi\Psi) = \hat{\mathbf{i}}\frac{\partial}{\partial x}(\phi\Psi) + \hat{\mathbf{j}}\frac{\partial}{\partial y}(\phi\Psi) + \hat{\mathbf{k}}\frac{\partial}{\partial z}(\phi\Psi)$

$$= \hat{\mathbf{i}}\left[\phi\frac{\partial\Psi}{\partial x} + \Psi\frac{\partial\phi}{\partial x}\right] + \hat{\mathbf{j}}\left[\phi\frac{\partial\Psi}{\partial y} + \Psi\frac{\partial\phi}{\partial y}\right] + \hat{\mathbf{k}}\left[\phi\frac{\partial\Psi}{\partial z} + \Psi\frac{\partial\phi}{\partial z}\right]$$

$$= \phi\left[\hat{\mathbf{i}}\frac{\partial\Psi}{\partial x} + \hat{\mathbf{j}}\frac{\partial\Psi}{\partial y} + \hat{\mathbf{k}}\frac{\partial\Psi}{\partial z}\right] + \Psi\left[\hat{\mathbf{i}}\frac{\partial\phi}{\partial x} + \hat{\mathbf{j}}\frac{\partial\phi}{\partial y} = \hat{\mathbf{k}}\frac{\partial\phi}{\partial z}\right]$$

$$= \phi\vec{\nabla}\Psi + \Psi\vec{\nabla}\phi = \phi\,\text{grad}\,\Psi + \Psi\,\text{grad}\,\phi.$$ **Ans.**

Example 11. *If* ϕ *is a scalar field and* $\vec{A}$ *is a vector field, prove that*

$$\text{curl}\,(\phi\vec{A}) = \phi\ \text{curl}\ \vec{A} + (\text{grad}\ \phi) \times \vec{A}$$

or
$$\vec{\nabla}\times(\phi\vec{A}) = \phi\vec{\nabla}\times\vec{A} + \vec{\nabla}\phi\times\vec{A}$$

Solution. curl $(\phi\vec{A}) = \vec{\nabla}\times(\phi\vec{A})$

$$\therefore \quad \vec{\nabla}\times(\phi\vec{A} = \begin{vmatrix} \hat{\mathbf{i}} & \hat{\mathbf{j}} & \hat{\mathbf{k}} \\ \frac{\partial}{\partial x} & \frac{\partial}{\partial y} & \frac{\partial}{\partial z} \\ \phi A_x & \phi A_y & \phi A_z \end{vmatrix}$$

$$= \hat{\mathbf{i}}\left[\frac{\partial}{\partial y}(\phi A_z) - \frac{\partial}{\partial z}(\phi A_y)\right] + \hat{\mathbf{j}}\left[\frac{\partial}{\partial z}(\phi A_x) - \frac{\partial}{\partial x}(\phi A_z)\right] + \hat{\mathbf{k}}\left[\frac{\partial}{\partial x}(\phi A_y) - \frac{\partial}{\partial y}(\phi A_x)\right]$$

$$= \hat{\mathbf{i}}\left[A_z\frac{\partial\phi}{\partial y} + \phi\frac{\partial A_z}{\partial y} - A_y\frac{\partial\phi}{\partial z} - \phi\frac{\partial A_y}{\partial z}\right] + \ldots + \ldots$$

or
$$= \phi\left[\hat{\mathbf{i}}\left(\frac{\partial A_z}{\partial y} - \frac{\partial A_y}{\partial z}\right) + \hat{\mathbf{j}}\left(\frac{\partial A_x}{\partial z} - \frac{\partial A_z}{\partial x}\right) + \hat{\mathbf{k}}\left(\frac{\partial A_y}{\partial x} - \frac{\partial A_x}{\partial y}\right)\right]$$

$$+ \hat{\mathbf{i}}\left(A_z\frac{\partial\phi}{\partial y} - A_y\frac{\partial\phi}{\partial z}\right) + \hat{\mathbf{j}}\left(A_x\frac{\partial\phi}{\partial z} - A_z\frac{\partial\phi}{\partial x}\right) + \hat{\mathbf{k}}\left(A_y\frac{\partial\phi}{\partial x} - A_x\frac{\partial\phi}{\partial y}\right)$$

$$\therefore \quad \vec{\nabla}\times\phi\vec{A} = \phi\vec{\nabla} + (\vec{\nabla}\phi)\times\vec{A} = \phi\ \text{curl}\ \vec{A} + (\text{grad}\ \phi) \times \vec{A}$$ **Ans.**

Example 12. *Prove that div* $(\vec{A}\times\vec{B}) = \vec{B}\,.$ *curl* $\vec{A} - \vec{A}\,.$ *curl* $\vec{B}$.

Hence show that if $\vec{A}$ *and* $\vec{B}$ *are the ir-rotational fields, then* $\vec{B}\times\vec{A}$ *is a solenoidal field.*

Solution. Let $\vec{A}\times\vec{B} = \vec{F}$, so we have

$$\therefore \quad \vec{F} = \vec{A}\times\vec{B} = \begin{vmatrix} \hat{\mathbf{i}} & \hat{\mathbf{j}} & \hat{\mathbf{k}} \\ A_x & A_y & A_z \\ B_x & B_y & B_z \end{vmatrix}$$

$$= \hat{\mathbf{i}}(A_yB_z - A_zB_y) + \hat{\mathbf{j}}(A_zB_z - A_xB_z) + \hat{\mathbf{k}}(A_xB_y - A_yB_x)$$

$$= \hat{\mathbf{i}}V_x + \hat{\mathbf{j}}V_y + \hat{\mathbf{k}}V_z$$

Hence, div $(\vec{A}\times\vec{B})$ = div $\vec{F} = \vec{\nabla}.\vec{F} = \dfrac{\partial F_x}{\partial x} + \dfrac{\partial F_y}{\partial y} + \dfrac{\partial F_z}{\partial z}$

$$= \frac{\partial}{\partial x}(A_yB_z - A_zB_y) + \frac{\partial}{\partial y}(A_zB_x - A_xB_z) + \frac{\partial}{\partial z}(A_xB_y - A_yB_x)$$

$$= \left(A_y\frac{\partial B_z}{\partial x} + B_z\frac{\partial A_y}{\partial y} - B_y\frac{\partial A_z}{\partial x} - A_z\frac{\partial B_y}{\partial x}\right) + ... + ...$$

$$= B_x\left(\frac{\partial A_x}{\partial_y} - \frac{\partial A_y}{\partial z}\right) + B_y\left(\frac{\partial A_x}{\partial z} - \frac{\partial A_z}{\partial x}\right) + B_z\left(\frac{\partial A_y}{\partial x} - \frac{\partial A_x}{\partial y}\right) - A_x\left(\frac{\partial B_z}{\partial y} - \frac{\partial B_y}{\partial z}\right)$$

$$- A_y\left(\frac{\partial B_x}{\partial z} - \frac{\partial B_z}{\partial x}\right) - B_x\left(\frac{\partial A_z}{\partial y} - \frac{\partial A_y}{\partial z}\right) + B_yA_z\left(\frac{\partial B_y}{\partial x} - \frac{\partial B_x}{\partial y}\right)$$

$$= (\hat{\mathbf{i}}B_x + \hat{\mathbf{j}}B_y + \hat{\mathbf{k}}B_z).\left[\hat{\mathbf{i}}\left(\frac{\partial A_z}{\partial y} - \frac{\partial A_y}{\partial z}\right) + \hat{\mathbf{j}}\left(\frac{\partial A_x}{\partial z} - \frac{\partial A_z}{\partial x}\right) + \hat{\mathbf{k}}\left(\frac{\partial A_y}{\partial x} - \frac{\partial A_x}{\partial y}\right)\right]$$

$$- (\hat{\mathbf{i}}A_x + \hat{\mathbf{j}}A_y + \hat{\mathbf{k}}A_z).\left[\hat{\mathbf{i}}\left(\frac{\partial B_z}{\partial y} - \frac{\partial B_y}{\partial z}\right) + \hat{\mathbf{j}}\left(\frac{\partial B_x}{\partial z} - \frac{\partial B_z}{\partial x}\right) + \hat{\mathbf{k}}\left(\frac{\partial B_y}{\partial x} - \frac{\partial B_x}{\partial y}\right)\right]$$

$$= \vec{B}.\text{curl}\,\vec{A} - \vec{A}.\text{curl}\,\vec{B}\,.$$

Now if $\vec{A}$ and $\vec{B}$ are the irrotational fields (which means curl $\vec{A} = 0$ and curl $\vec{B} = 0$) then div $\vec{A}\times\vec{B} = 0$. *i.e.,* $\vec{A}\times\vec{B}$ **will be a solenoidal field. Ans.**

Example 13. *If* $\vec{A} = x^2z\hat{\mathbf{i}} - 2y^3z^2\hat{\mathbf{j}} + xy^2z\hat{\mathbf{k}}$, *find the values of (i) div* $\vec{A}$, *(ii) curl* $\vec{A}$, *at a point (2, 2, 2)*

Solution. (*i*) div $\vec{A} = \vec{\nabla}.\vec{A} = \dfrac{\partial A_x}{\partial_x} + \dfrac{\partial A_y}{\partial_y} + \dfrac{\partial A_z}{\partial_z}$

$$= \frac{\partial}{\partial x}(x^2z) + \frac{\partial}{\partial y}(-2y^3z^2) + \frac{\partial}{\partial z}(xy^2z)$$

$$= 2xz - 6y^2z^2 + xy^2$$

At the point (2, 2, 2)

div $\vec{A} = 2\,(2)\,(2) - 6\,(2)^2\,(2)^2 + 2\,(2)^2 = 8 - 96 + 8 = -80$ **Ans.**

$$(ii)\ \text{curl}\ \vec{A} = \nabla\times\vec{A} = \begin{vmatrix} \hat{\mathbf{i}} & \hat{\mathbf{j}} & \hat{\mathbf{k}} \\ \frac{\partial}{\partial x} & \frac{\partial}{\partial y} & \frac{\partial}{\partial_z} \\ A_x & A_y & A_z \end{vmatrix} = \begin{vmatrix} \hat{\mathbf{i}} & \hat{\mathbf{j}} & \hat{\mathbf{k}} \\ \frac{\partial}{\partial x} & \frac{\partial}{\partial y} & \frac{\partial}{\partial_z} \\ x^2z & -2y^3z^2 & xy^2z \end{vmatrix}$$

$$= \hat{\mathbf{i}}\left[\frac{\partial}{\partial y}(xy^2z) - \frac{\partial}{\partial z}(-2y^3z^2)\right] + \hat{\mathbf{j}}\left[\frac{\partial}{\partial z}(x^2z) - \frac{\partial}{\partial x}(xy^2z)\right] + \hat{\mathbf{k}}\left[\frac{\partial}{\partial x}(-2y^3z^2) - \frac{\partial}{\partial y}(x^2z)\right]$$

$$= \hat{\mathbf{i}}[2xyz + 4y^3z] + \hat{\mathbf{j}}[x^2 - y^2z] + \hat{\mathbf{k}}[0-0]$$

$$= \hat{\mathbf{i}}(2xyz + 4y^3z) + \hat{\mathbf{j}}(x^2 - y^2z)$$

At the point (2, 2, 2)

$$\text{Curl}\ \vec{A} = \hat{\mathbf{i}}(2\times2\times2\times2 + 4\times2^3\times2) + \hat{\mathbf{j}}(2^2 - 2^2\times2)$$

$$= \hat{\mathbf{i}}(16+64) + \hat{\mathbf{j}}(4-8)$$

$$= 80\hat{\mathbf{i}} - 4\hat{\mathbf{j}}$$ **Ans.**

Example 14. *Show that* $\vec{F} = (x+y)\hat{\mathbf{i}} + (x+z)\hat{\mathbf{j}} + (y-z)\hat{\mathbf{k}}$ *represents a conservative field.*

Solution. For a conservative field, curl $\vec{F}$ must be zero

Now, curl $\vec{F} = \nabla\times\vec{F}$

$$= \begin{vmatrix} \hat{\mathbf{i}} & \hat{\mathbf{j}} & \hat{\mathbf{k}} \\ \frac{\partial}{\partial x} & \frac{\partial}{\partial y} & \frac{\partial}{\partial z} \\ x+y & x+y & y-z \end{vmatrix}$$

$$= \hat{\mathbf{i}}\left[\frac{\partial}{\partial y}(y-z) - \frac{\partial}{\partial z}(x+z)\right] + \hat{\mathbf{j}}\left[\frac{\partial}{\partial z}(x+y) - \frac{\partial}{\partial x}(y-z)\right] + \hat{\mathbf{k}}\left[\frac{\partial}{\partial x}(x+z) - \frac{\partial}{\partial y}(x+y)\right]$$

$$= \hat{\mathbf{i}}(1-1) + \hat{\mathbf{j}}(0-0) + \hat{\mathbf{k}}(1-1) = 0.$$

Which means that $\vec{F}$ is conservative field. Ans.

Example 15. If $\vec{A} = -y\hat{\mathbf{i}} + x\hat{\mathbf{j}}$.

Find: (i) curl $\vec{A}$ and (ii) $\oint \vec{A}\, d\vec{l}$ for a closed curve $x^2 + y^2 = r^2, z = 0$. Hence prove the Stocke's theorem.

Solution. (*i*) curl $\vec{A} = \begin{vmatrix} \hat{\mathbf{i}} & \hat{\mathbf{j}} & \hat{\mathbf{k}} \\ \frac{\partial}{\partial x} & \frac{\partial}{\partial y} & \frac{\partial}{\partial z} \\ -y & x & 0 \end{vmatrix} = \hat{\mathbf{k}}[1-(-1)] = 2\hat{\mathbf{k}}.$ **Ans.**

(ii) $\oint \vec{A}.d\vec{1} = \oint (A_x dx + A_y dy + A_z dz) = \int (-y dx + x dy)$

$$= -\int y dx + \int x dy.$$

For the curve $x^2 + y^2 = r^2$, $x = r \cos \theta$ and $y = r \sin \theta$

$\therefore$ $dx = -r \sin\theta \, d\theta$ and $dy = r \cos\theta \, d\theta$

Hence, $\oint \vec{A}.d\vec{1} = \int_0^{2\pi} (r^2 \sin^2\theta + r^2 \cos^2\theta) d\theta = \int_0^{2\pi} r^2 d\theta$

$$= 2\pi r^2$$

Now from Stoke's theorem,

$$\oint \vec{A}.d\vec{1} = \iint_S \text{curl}\, \vec{A}.d\vec{a}$$

$$= \iint_S^S 2\hat{k}.da\hat{k} = \iint_S 2da$$

$= 2a = 2(\pi r^2) = 2\pi r^2$ [since area of circle is $a = \pi r^2$]

Which proves the Stoke's theorem.

ADDITIONAL TYPICAL PROBLEMS

Example 1. *Prove that (i) grad r =* $\hat{n}$, *where* $\hat{n}$ *is a unit vector.*

(ii) grad (1/r) $= -\dfrac{\vec{r}}{r^3}$, *where* $\vec{r}$ *is a position vector.*

Solution. In this ease, $\vec{r}$ is a position vector, which may the written as

$\vec{r} = xi + yi + zk$, therefore, $|\vec{r}| = r = \sqrt{(x^2 + y^2 + z^2)}$

[$\because$ If $\vec{A} = A_x i + A_y j + A_z k$, so $A = \sqrt{(A_x^2 + A_y^2 + A_z^2)}$

(*i*) grad $r = \hat{n}$

$\therefore$ Taking L.H.S grad $r = \vec{\nabla} (x^2 + y^2 + z^2)^{1/2}$

$$= \left(i\frac{\partial}{\partial x} + j\frac{\partial}{\partial y} + k\frac{\partial}{\partial z}\right)(x^2 + y^2 + z^2)^{1/2}$$

or grad $r = i\dfrac{\partial}{\partial x}(x^2 + y^2 + z^2)^{1/2} + j\dfrac{\partial}{\partial y}(x^2 + y^2 + z^2)^{1/2} + k\dfrac{\partial}{\partial z}(x^2 + y^2 + z^2)^{1/2}$ ... (*i*)

Let us consider the fresh term, $\dfrac{\partial}{\partial x}(x^2 + y^2 + z^2)^{1/2}$

Differentiating it partially with respect to x, we get

$$\therefore \quad \frac{\partial}{\partial x}(x^2 + y^2 + z^2)^{1/2} = \left(\frac{1}{2}\right)(x^2 + y^2 + z^2)^{(1/2)-1}.(2x)$$

$$= \frac{x}{(x^2 + y^2 + z^2)^{1/2}}$$

Again, differentiating partially other terms w.r.t. y and z, we have

$$\frac{\partial}{\partial y}(x^2+y^2+z^2)^{1/2} = \frac{y}{(x^2+y^2+z^2)^{1/2}} \text{ and } \frac{\partial}{\partial z}(x^2+y^2+z^2)^{1/2} = \frac{z}{(x^2+y^2+z^2)^{1/2}}$$

Putting these values in equation (*i*), we have

$$\nabla r = \frac{xi}{(x^2+y^2+z^2)^{1/2}} + \frac{yj}{(x^2+y^2+z^2)^{1/2}} + \frac{zk}{(x^2+y^2+z^2)^{1/2}}$$

$$= \frac{xi+yj+zk}{(x^2+y^2+z^2)^{1/2}} = \frac{\vec{r}}{r} = \hat{n}, \text{ where } \hat{n} \text{ is a unit vector.} \quad (\because \vec{r} = \hat{n}r)$$

= R.H.S. Hence proved.

(*ii*) $\quad \text{grad}\left(\frac{1}{r}\right) = -\frac{\vec{r}}{r^3}$

Taking $\quad \text{L.H.S.} = \vec{\nabla}\left(\frac{1}{r}\right) = \left(i\frac{\partial}{\partial x} + j\frac{\partial}{\partial y} + k\frac{\partial}{\partial z}\right)\left(\frac{1}{(x^2+y^2+z^2)^{1/2}}\right)$

or $\quad \nabla\left(\frac{1}{r}\right) = i\frac{\partial}{\partial x}(x^2+y^2+z^2)^{-1/2} + j\frac{\partial}{\partial y}(x^2+y^2+z^2)^{-1/2} + k\frac{\partial}{\partial z}(x^2+y^2+z^2)^{-1/2} \quad \text{... (ii)}$

Taking the first term $\frac{\partial}{\partial x}(x^2+y^2+z^2)^{-1/2}$ and differentiating partially w.r.t. x, we get

$$\therefore \quad \frac{\partial}{\partial x}(x^2+y^2+z^2)^{-1/2} = \left(-\frac{1}{2}\right)(x^2+y^2+z^2)^{-1/2-1}.(2x) = -\frac{x}{(x^2+y^2+z^2)^{3/2}}$$

Again, differentiating partially other terms of equation (*i*) w. r. t. y and a, we have

$$\frac{\partial}{\partial y}(x^2+y^2+z^2)^{-1/2} = -\frac{y}{(x^2+y^2+z^2)^{3/2}} \textit{ and } \frac{\partial}{\partial z}(x^2+y^2+z^2)^{-1/2} = -\frac{z}{(x^2+y^2+z^2)^{3/2}}$$

Putting these values in equation (*i*), we get

$$\text{grad}\left(\frac{1}{r}\right) = -\frac{xi}{(x^2+y^2+z^2)^{3/2}} - \frac{yj}{(x^2+y^2+z^2)^{3/2}} - \frac{zk}{(x^2+y^2+z^2)^{3/2}}$$

$$= -\left[\frac{xi+yj+zk}{(x^2+y^2+z^2)^{3/2}}\right] = -\frac{\vec{r}}{(r^2)^{3/2}} = -\frac{\vec{r}}{r^3}$$

R.H.S. Hence proved. **Ans.**

Example 2. *If A is a constant vector, prove grad* $(\vec{r}.\vec{A}) = \vec{A}$ *where* $\vec{r}$ *is a position vector.*

Solution. As is given $\vec{A}$ is a constant vector, which means that, its components A_x, A_y and A_z are constant and independent of x, y and z. $\vec{r}$ is a position vector which depends on x, y and z, which may be written as

$$\vec{r} = xi + yj + zk \text{ and } \vec{A} = A_x i + A_y j + A_z k$$

$$\therefore \quad \vec{r}.\vec{A} = (xi + yj + zk).(A_x i + A_y j + A_z k)$$

$$= xA_x + yA_y + zA_z \qquad (\because \vec{A}.\vec{B} = A_xB_x + A_yB_y + A_zB_z)$$

Now, $$\text{grad}(\vec{r}.\vec{A}) = \left(i\frac{\partial}{\partial x} + j\frac{\partial}{\partial y} + k\frac{\partial}{\partial z}\right)(x A_x + y A_y + z A_z)$$

$$= i A_x.\frac{\partial x}{\partial x} + j A_y\frac{\partial y}{\partial y} + k A_z\frac{\partial z}{\partial z} \qquad (\because A_x, A_y \text{ and } A_z \text{ are constants})$$

$$= A_x i + A_y j + A_z k = \vec{A} \quad \textbf{Ans.}$$

Example 3. *If $\phi(x, y, z) = 3(x^2y - y^2x)$; calculate grad at the point (1, – 2, –1).*

Solution. $$\text{grad } \phi = \left(i\frac{\partial}{\partial x} + j\frac{\partial}{\partial y} + k\frac{\partial}{\partial z}\right)\phi$$

In this case $\phi(x, y, z) = 3x^2y - 3y^2x$

$\therefore$ $$\text{grad } \phi(x, y, z) = \text{grad }(3x^2y - 3y^2x)$$

$$= i\frac{\partial}{\partial x}(3x^2y - 3y^2x) + j\frac{\partial}{\partial y}(3x^2y - 3y^2x) + k\frac{\partial}{\partial z}(3x^2y - 3y^2x)$$

$$= (6xy - 3y^2)\, i + (3x^2 - 6yx)\, j + (0)\, k$$

At the point (1, – 2, – 1), we have

$$\text{grad } \phi = [6 \times (1) \times (-2) - 3(-2)^2]\, i + [3(1)^2 - 6(-2)(1)]\, j$$

$$= [-12 - 12]\, i + [3 + 12]\, j$$

$$= \mathbf{-24i + 15j} \quad \textbf{Ans.}$$

Example 4. *Find grad r^n, wher r is the distance of any point (x, y, z) from the origin (0, 0, 0).*

Solution. We have

$$\text{grad } \phi = \vec{\nabla}\phi = \left(i\frac{\partial}{\partial x} + j\frac{\partial}{\partial y} + k\frac{\partial}{\partial z}\right)\phi$$

$$\text{grad } r^n = \left(i\frac{\partial}{\partial x} + j\frac{\partial}{\partial y} + k\frac{\partial}{\partial z}\right)r^n$$

As given, $\vec{r}$ is a position vector, which means

$$\vec{r} = xi + yj + zk \quad \therefore\ r = \sqrt{(x^2 + y^2 + z^2)} \qquad (\because A = \sqrt{(A_x^2 + A_y^2 + A_z^2)}\,]$$

or $$r^n = (x^2 + y^2 + z^2)^{n/2}$$

or $$\nabla r^n = \left(i\frac{\partial}{\partial x} + j\frac{\partial}{\partial y} + k\frac{\partial}{\partial z}\right)(x^2 + y^2 + z^2)^{n/2}$$

or $$\nabla r^n = i\frac{\partial}{\partial x}(x^2 + y^2 + z^2)^{n/2} + j\frac{\partial}{\partial y}(x^2 + y^2 + z^2)^{n/2} + k\frac{\partial}{\partial z}(x^2 + y^2 + z^2)^{n/2}$$

Let us consider the first term, $\frac{\partial}{\partial x}(x^2 + y^2 + z^2)^{n/2}$ and differentiating it w.r.t. x, we get

$$\frac{\partial}{\partial x}(x^2+y^2+z^2)^{n/2} = \left(\frac{n}{2}\right)(x^2+y^2+z^2)^{(n/2)-1}.(2x)$$

$$= nx\,(x^2+y^2+z^2)^{(n-2)/2}$$

Again, differentiating partially other terms of equation (*i*) w.r.t. *y* and *z*, we have

$$\frac{\partial}{\partial y}(x^2+y^2+z^2)^{n/2} = ny\,(x^2+y^2+z^2)^{(n-2)/2}$$

and $$\frac{\partial}{\partial z}(x^2+y^2+z^2)^{n/2} = nz\,(x^2+y^2+z^2)^{(n-2)/2}$$

Putting these values is equation (*i*) we have

$$\text{grad } r^n = (x^2+y^2+z^2)^{(n-2)/2}\, n\,x\,i + (x^2+y^2+z^2)^{(n-2)/2}\, n\,y\,j + (x^2+y^2+z^2)^{(n-2)/2}\, n\,z\,k$$

$$= n\,(x^2+y^2+z^2)^{(n-2)/2}\,(x\,i + y\,j + z\,k)$$

$$= n\,(r^2)^{(n-2)/2}\,\vec{r} = n\,r^{n-2}\,\vec{r} \quad \textbf{Ans.}$$

[**Imp:** Students may ncte that this is a general result and may he used for particular cases, *e.g.*

If $n = 1$, grad $r = \frac{\vec{r}}{r}$; If $n = -1$, grad $\left(\frac{1}{r}\right) = -\frac{\vec{r}}{r^3}$ and If $n = -3$, grad $\left(\frac{1}{r^3}\right) = \left(\frac{-3\,\vec{r}}{r^5}\right)$ etc.]

Example 5. *If the electric potential at any point due to a charge q is given by V = q/r, then show by vector method that the electric intensity* $\vec{E}$ *is given by* $\vec{E} = \frac{q}{r^3}\vec{r}$.

Solution. In the form Electric intensity $\vec{E}$ can be represented as the negative gradient of potential *V*, in the form

$$\vec{E} = -\text{ grad } V$$

In this case $V = \frac{q}{r}$, where $\vec{r}$ is the position vector and *r* its magnitude.

$$\therefore \quad \vec{E} = -\text{ grad }(q/r) = -q\text{ grad }(1/r) = -q\nabla\left(\frac{1}{r}\right).$$

For evaluating grad (1/*r*) see Example (4) so we have

the solution, grad $\left(\frac{1}{r}\right) = -\frac{\vec{r}}{r^3}$; $\quad \therefore\ \vec{E} = -q\left(-\frac{\vec{r}}{r^3}\right) = \frac{q}{r^3}\vec{r}$ **Ans.**

Example 6. *The electric potential at any point in the x-y plane is given by*

$$\phi = 5x\,(x^2+y^2)^{1/2} + y(x^2+y^2)^{-1/2}$$

Find the cartesian components of the electric field intensity at that point.

Solution. We have $\vec{E} = (E_x i + E_y j + E_z k) = -\text{grad }\phi = -i\frac{\partial\phi}{\partial y} - j\frac{\partial\phi}{\partial y} - k\frac{\partial\phi}{\partial z}$

$$\therefore \quad E_x = -\frac{\partial\phi}{\partial x},\ E_y = -\frac{\partial\phi}{\partial y},\ E_z = -\frac{\partial\phi}{\partial z}$$

In this case $\phi = 5x(x^2 + y^2)^{1/2} + y(x^2 + y^2)^{-1/2}$

$$\therefore \quad E_x = -\frac{\partial}{\partial x}[5x(x^2 + y^2)^{1/2} + y(x^2 + y^2)^{-1/2}]$$

$$= -[5(x^2 + y^2)^{1/2} + 5x\left(\frac{1}{2}\right)(x^2 + y^2)^{-1/2}(2x) + y\left(-\frac{1}{2}\right)(x^2 + y^2)^{-3/2}(2x)]$$

$$= -[5(x^2 + y^2)^{1/2} + 5x^2(x^2 + y^2)^{-1/2} - xy(x^2 + y^2)^{-3/2}]$$

So, the x-component of electric field will the

$$\mathbf{E_x = -[5(x^2 + y^2)^{1/2} + 5x^2(x^2 + y^2)^{-1/2} - xy(x^2 + y^2)^{-3/2}]} \quad \textbf{Ans.}$$

Similarly, $$E_y = -\frac{\partial}{\partial y}[5x(x^2 + y^2)^{1/2} + y(x^2 + y^2)^{-1/2}]$$

$$= -[5x\left(\frac{1}{2}\right)(x^2 + y^2)^{-1/2}(2y) + (x^2 + y^2)^{-1/2} + y\left(-\frac{1}{2}\right)(x^2 + y^2)^{-3/2}(2y)]$$

$$= -[5xy(x^2 + y^2)^{-1/2} + (x^2 + y^2)^{-1/2} - y^2(x^2 + y^2)^{-3/2}]$$

So, y component of electric field is given by

$$\mathbf{E_y = -[5xy(x^2 + y^2)^{-1/2} + (x^2 + y^2)^{-1/2} - y^2(x^2 + y^2)^{-3/2}]} \quad \textbf{Ans.}$$

Example 7. *A region is specified by the potential function given by, $\phi = 4x^2 + 3y^2 - 9z^2$ calculate electric field strength at any point (3, 4, 5) in this region.*

Solution. Electric field is given by, $\vec{E} = -\text{grad}\,\phi = -\left(i\frac{\partial\phi}{\partial x} + j\frac{\partial\phi}{\partial y} + k\frac{\partial\phi}{\partial z}\right)$

In this case $\phi = 4x^2 + 3y^2 - 9z^2$

$$\therefore \quad \frac{\partial\phi}{\partial x} = 8x, \frac{\partial\phi}{\partial y} = 6y \text{ and } \frac{\partial\phi}{\partial z} = -18z$$

$$\therefore \quad \vec{E} = -(8xi + 6yj - 18zk)$$

At any point (3, 4, 5) We have the value of $\vec{E}$

$$\vec{E} = -(8 \times 3\,i + 6(4)\,j - 18(5)\,k] = \mathbf{-24j - 24j + 90k} \quad \textbf{Ans.}$$

Example 8. *If $\vec{r}$ is the position vector of a point, then find grad (log r).*

Solution. Now, grad $\phi = \vec{\nabla}\phi = \left(i\frac{\partial}{\partial x} + j\frac{\partial}{\partial y} + k\frac{\partial}{\partial z}\right)\phi$

Since $\vec{r}$ is a position vector, which means, $\vec{r} = xi + yj + zk \quad \therefore r = \sqrt{(x^2 + y^2 + z^2)}$

$$\therefore \quad \log r = \log(x^2 + y^2 + z^2)^{1/2} = \frac{1}{2}\log(x^2 + y^2 + z^2)$$

and $$\text{grad}(\log r) = \vec{\nabla}(\log r) = \frac{1}{2}\vec{\nabla}\log(x^2 + y^2 + z^2)$$

or grad (log r)

$$= \frac{1}{2}\left\{i\frac{\partial}{\partial x}[\log(x^2 + y^2 + z^2)] + j\frac{\partial}{\partial y}[\log(x^2 + y^2 + z^2)] + k\frac{\partial}{\partial z}[\log(x^2 + y^2 + z^2)]\right\} \quad ...(i)$$

Let us consider the first term, $\frac{\partial}{\partial x}$ [log $(x^2 + y^2 + z^2)$] and differentiating partially

$$\frac{\partial}{\partial x} [\log (x^2 + y^2 + z^2)] = \frac{1}{(x^2+y^2+z^2)}\frac{\partial}{\partial x}(x^2 + y^2 + z^2) \quad \left(\because \frac{d}{dx}\log(x) = \frac{1}{x}\right)$$

$$= \frac{2x}{(x^2+y^2+z^2)}$$

Again, differentiating partially other terms of equation (*i*) w.r.t y and z, we have

$$\frac{\partial}{\partial y} [\log (x^2 + y^2 + z^2)] = \frac{2y}{(x^2+y^2+z^2)} \text{ and } \frac{\partial}{\partial z} [\log (x^2 + y^2 + z^2)] = \frac{2z}{(x^2+y^2+z^2)}$$

Putting these values in equation (*i*), we get

$$\nabla (\log r) = \frac{1}{2}\left[\frac{2xi}{(x^2+y^2+z^2)} + \frac{2yj}{(x^2+y^2+z^2)} + \frac{2zk}{(x^2+y^2+z^2)}\right]$$

$$\text{or grad } (\log r) = \frac{xi+yj+zk}{(x^2+y^2+z^2)} = \frac{\vec{\mathbf{r}}}{\mathbf{r}^2} \quad \textbf{Ans.}$$

Example 9. *A vector function has the following components* $E_x = 6xy$; $E_y = 3x^2 - 3y^2$ *and* $E_z = 0$, *show that div* $\vec{E} = 0$.

Solution. We have the relation for div. A as

$$\text{div } \vec{A} = \vec{\nabla}.\vec{A} = \frac{\partial A_x}{\partial x} + \frac{\partial A_y}{\partial y} + \frac{\partial A_z}{\partial z}$$

In this case $\vec{A} = \vec{E}$ and $E_x = 6xy$; $E_y = 3x^2 - 3y^2$ and $E_z = 0$

$$\therefore \quad \text{div } \vec{E} = \frac{\partial E_x}{\partial x} = \frac{\partial E_y}{\partial y} + \frac{\partial E_z}{\partial z} = \frac{\partial}{\partial x}(6xy) + \frac{\partial}{\partial y}(3x^2 - 3y^2) + \frac{\partial}{\partial y}(0)$$

or $\quad$ div $\vec{E} = 6y - 6y = \mathbf{0}$. **Ans.**

Example 10. If $\vec{r}$ is the position vector of a point then show that

(*i*) div $\vec{r} = 3$; $\quad$ (*ii*) div $(\vec{r}/r^3) = 0$;

(*iii*) div $(r^n \vec{r}) = (3 + n)r^n$ and (*iv*) div $r^4 \vec{r}) = 7r^4$

Solution. Sin $\vec{r}$ is a positron vector, that is, $r = xi + yj + zk$, so,

$$r = \sqrt{(x^2+y^2+z^2)}$$

Also div $\vec{A} = \frac{\partial A_x}{\partial x} + \frac{\partial A_y}{\partial y} + \frac{\partial A_z}{\partial z}$

(*i*) *div* $\vec{r} = 3$

In this case $A_x = x$, $A_y = y$ and $A_z = z$

$$\therefore \quad div\ \vec{r} = \frac{\partial x}{\partial x} + \frac{\partial y}{\partial y} + \frac{\partial z}{\partial z} = \mathbf{1 + 1 + 1 = 3} \ \textbf{Ans.}$$

(*ii*) $div\ (\vec{r}/r^3) = 0$

In this case $\vec{A} = \dfrac{\vec{r}}{r^3} = \dfrac{xi + yj + zk}{(x^2 + y^2 + z^2)^{3/2}}$

$$\therefore A_x = \frac{x}{(x^2 + y^2 + z^2)}, A_y = \frac{y}{(x^2 + y^2 + z^2)^{3/2}} \text{ and } A_z = \frac{z}{(x^2 + y^2 + z^2)^{3/2}}$$

$$\therefore \quad \text{div } \vec{A} = \text{div}\left(\frac{\vec{r}}{r^3}\right)$$

$$= \frac{\partial}{\partial x}\left[\frac{x}{(x^2 + y^2 + z^2)^{3/2}}\right] + \frac{\partial}{\partial y}\left[\frac{y}{(x^2 + y^2 + z^2)^{3/2}}\right] + \frac{\partial}{\partial z}\left[\frac{z}{(x^2 + y^2 + z^2)^{3/2}}\right] \quad \dots (i)$$

Let us consider the first term, $\dfrac{\partial}{\partial x}\left[\dfrac{x}{(x^2 + y^2 + z^2)^{3/2}}\right]$ and differentiate it partially w.r.t. x, we have

$$\frac{\partial}{\partial x}\left[\frac{x}{(x^2 + y^2 + z^2)^{3/2}}\right] = \frac{\partial}{\partial x}[x(x^2 + y^2 + z^2)^{-3/2}]$$

$$= x.\frac{\partial}{\partial x}(x^2 + y^2 + z^2)^{-3/2} + (x^2 + y^2 + z^2)^{-3/2}\frac{\partial}{\partial x}(x)$$

$$= x.\left(\frac{-3}{2}\right)(x^2 + y^2 + z^2)^{-(3/2)-1}(2x) + (x^2 + y^2 + z^2)^{-3/2}$$

$$= -3x^2(x^2 + y^2 + z^2)^{-5/2} + (x^2 + y^2 + z^2)^{-3/2}$$

Again, differentiating partially other terms of equation (*i*) w.r.t. y and z, we get

$$\frac{\partial}{\partial y}[y(x^2 + y^2 + z^2)^{-3/2}] = -3y^2(x^2 + y^2 + z^2)^{-5/2} + (x^2 + y^2 + z^2)^{-3/2}$$

and $$\frac{\partial}{\partial z}[z(x^2 + y^2 + z^2)^{-3/2}] = -3z^2(x^2 + y^2 + z^2)^{-5/2} + (x^2 + y^2 + z^2)^{-3/2}$$

Putting these values in equation (1), we have

$$\text{div } (\vec{r}/r^3) = -3x^2(x^2 + y^2 + z^2)^{-5/2} + (x^2 + y^2 + z^2)^{-3/2} - 3y^2(x^2 + y^2 + z^2)^{-5/2}$$
$$+ (x^2 + y^2 + z^2)^{-3/2} - 3z^2(x^2 + y^2 + z^2)^{-5/2} + (x^2 + y^2 + z^2)^{-3/2}$$

or $$\text{div } (\vec{r}/r^3) = (-3x^2 - 3y^2 - 3z^2)(x^2 + y^2 + z^2)^{-5/2} + 3(x^2 + y^2 + z^2)^{-3/2}$$
$$= -3(x^2 + y^2 + z^2)(x^2 + y^2 + z^2)^{-5/2} + 3(x^2 + y^2 + z^2)^{-3/2}$$

$$= -3(x^2+y^2+z^2)^{-3/2} + 3(x^2+y^2+z^2)^{-3/2} = 0$$

Which is R.H.S. hence proved

(iii) div $(r^n \vec{r}) = (3+n)r^n$

Now $\quad r^n = (x^2+y^2+z^2)^{n/2}$ and $\vec{r} = xi + yj + zk$

So $\quad \vec{A} = r^n \vec{r} = (xi+yj+zk)(x^2+y^2+z^2)^{n/2}$

$\therefore A_x = x(x^2+y^2+z^2)^{n/2}$; $A_y = y(x^2+y^2+z^2)^{n/2}$; $A_z = z(x^2+y^2+z^2)^{n/2}$

$$= \text{div}(r^n \vec{r}) = \frac{\partial}{\partial x}[x(x^2+y^2+z^2)^{n/2}] + \frac{\partial}{\partial y}[y(x^2+y^2+z^2)^{n/2}] + \frac{\partial}{\partial z}[z(x^2+y^2+z^2)^{n/2}] \quad \text{... } (i)$$

Considering the first term, $\frac{\partial}{\partial x}[x(x^2+y^2+z^2)^{n/2}]$ and differentiates it partially w.r.t. x, we have

$$\frac{\partial}{\partial x}[x(x^2+y^2+z^2)^{n/2}] = x\frac{\partial}{\partial x}(x^2+y^2+z^2)^{n/2} + (x^2+y^2+z^2)^{n/2}\frac{\partial}{\partial x}(x)$$

$$= (x).\left(\frac{n}{2}\right)(x^2+y^2+z^2)^{n/2-1}.(2x) + (x^2+y^2+z^2)^{n/2}$$

$$= nx^2(x^2+y^2+z^2)^{(n-2)/2} + (x^2+y^2+z^2)^{n/2}$$

Again, differentiating other terms partially w.r.t. y and z, we obtain

$$\frac{\partial}{\partial y}[y(x^2+y^2+z^2)^{n/2}] = ny^2(x^2+y^2+z^2)^{(n-2)/2} + (x^2+y^2+z^2)^{n/2}$$

and $\frac{\partial}{\partial z}[z(x^2+y^2+z^2)^{n/2}] = nz^2(x^2+y^2+z^2)^{(n-2)/2} + (x^2+y^2+z^2)^{n/2}$

Putting these values in equation (*i*), obtain

$$\text{div } (r^n \vec{r}) = nx^2(x^2+y^2+z^2)^{(n-2)/2} + (x^2+y^2+z^2)^{n/2} + ny^2(x^2+y^2+z^2)^{(n-2)/2}$$
$$+ (x^2+y^2+z^2)^{n/2} + nz^2(x^2+y^2+z^2)^{(n-2)/2} + (x^2+y^2+z^2)^{n/2}$$
$$= (nx^2+ny^2+nz^2)(x^2+y^2+z^2)^{(n-2)/2} + 3(x^2+y^2+z^2)^{n/2}$$
$$= n(x^2+y^2+z^2)^{n/2} + 3(x^2+y^2+z^2)^{n/2}$$
$$= (n+3)(x^2+y^2+z^2)^{n/2}$$
$$= (n+3)(r^2)^{n/2}$$

$= (3+n)r^n = R.H.S.$, ***Hence proved.***

(iv) div $(r^4 \vec{r}) = 7r^4$

Now $(r^4 \vec{r}) = (x^2+y^2+z^2)^{4/2}(xi+yj+zk) = (x^2+y^2+z^2)^2(xi+yj+zk)$

$$\therefore \text{ L.H.S.} = \text{div } r^4\vec{r} = \frac{\partial}{\partial x}[(x^2+y^2+z^2)^2x]\frac{\partial}{\partial y}[(x^2+y^2+z^2)^2y] + \frac{\partial}{\partial z}[(x^2+y^2+z^2)z]$$

$$= 2x(x^2+y^2+z^2)^{2-1}(2x)+(x^2+y^2+z^2)^2+2y(x^2+y^2+z^2)^{2-1}.(2y)$$
$$+ (x^2+y^2+z^2)^2+2z(x^2+y^2+z^2)^{2-1}(2z)+(x^2+y^2+z^2)^2$$
$$= (4x^2+4y^2+4z^2)(x^2+y^2+z^2)+3(x^2+y^2+z^2)^2$$
$$= 4(x^2+y^2+z^2)^2+3(x^2+y^2+z^2)^2$$
$$= 7\ (x^2+y^2+z^2)^2$$
$= 7\ (r^2)^2 = \mathbf{7r^4}$ = R.H.S. **Hence proved.**

Example 11. *Find the constant 'a' for which the vector* $\vec{A} = (x + 3y)i + (y - 2z)j + (x + az)k$ *is solenoidal.*

Solution. A vector $\vec{A}$ is said to be solenoidal if its divergence is zero, it means div $A = 0$

In this case $A = (x + 3y)i + (y - 2z)j + (x + az)k = A_x\ i + A_y j + A_z k$

Also div $\vec{A} = \vec{\nabla}.\vec{A} = \dfrac{\partial A_x}{\partial x}+\dfrac{\partial A_y}{\partial y}+\dfrac{\partial A_z}{\partial z}$

$\therefore$ div $\vec{A} = \dfrac{\partial}{\partial x}(x+3y)+\dfrac{\partial}{\partial y}(y-2z)+\dfrac{\partial}{\partial z}(x+az)$

or div $A = 1 + 1 + a = 2 + a$

For a solenoidal vector div $\vec{A} = 0$, therefore $2 + a = 0$

Hence $\boldsymbol{a = -2}$ **Ans.**

Example 12. *Prove that*

(a) div $(S\vec{A}) = S$ *div* $\vec{A} + \vec{A}$ *. grad S (b) div* $(\vec{A}+\vec{B})$ = *div* $\vec{A}$ + *div* $\vec{B}$

(c) div grad S = $\nabla^2\ S$

Solution. We have div (*a*) $(S\vec{A})$ = S div $\vec{A}+\vec{A}$. **grad** *S*

Also div $\vec{A} = \vec{\nabla}.\vec{A} = \dfrac{\partial}{\partial x}A_x+\dfrac{\partial}{\partial y}A_y+\dfrac{\partial}{\partial z}A_z$

In this case is a scalar and $\vec{A} = A_x i + A_y j + A_z k$

$\therefore$ L.H.S. = div $(S\vec{A}) = \vec{\nabla}.(S\vec{A}) = \dfrac{\partial}{\partial x}(SA_x)+\dfrac{\partial}{\partial y}(SA_y)+\dfrac{\partial}{\partial z}(SA_z)$

$$= S\frac{\partial A_x}{\partial x}+A_x\frac{\partial S}{\partial x}+S\frac{\partial A_y}{\partial y}+A_y\frac{\partial S}{\partial y}+S\frac{\partial A_z}{\partial z}+A_z\frac{\partial S}{\partial z}$$

$$= S\left(\frac{\partial A_x}{\partial x}+\frac{\partial A_y}{\partial y}+\frac{\partial A_z}{\partial z}\right)+\left(A_x\frac{\partial S}{\partial x}-A_y\frac{\partial S}{\partial y}+A_z\frac{\partial S}{\partial z}\right)$$

$$= S(\vec{\nabla}.\vec{A})+(A_x i+A_y j+A_z k).\left(i\frac{\partial S}{\partial x}+j\frac{\partial S}{\partial y}+\frac{\partial S}{\partial z}\right)$$

$= S$ div $\vec{A}+\vec{A}.\nabla S$

$= S \text{ div } \vec{A} + \vec{A} \cdot \text{grad } S =$ R.H.S., Hence proved

(b) $(\vec{A}+\vec{B}) = \text{div } \vec{A} + \text{div } \vec{B}$, where $\vec{A} = Axi + Ayj + Azk$ and $\vec{B} = Bxi + Byj + Bzk$

$\therefore\ (\vec{A}+\vec{B}) = (A_x + B_x)i + (A_y + B_y)j + (A_z + B_z)k$

$\therefore$ Let us take L.H.S. as

$$= \text{div } (\vec{A}+\vec{B}) + \vec{\nabla}.(\vec{A}+\vec{B}) = \frac{\partial}{\partial x}(A_x + B_x) + \frac{\partial}{\partial y}(A_y + B_y) + \frac{\partial}{\partial z}(A_z + B_z)$$

$$\text{or div } (\vec{A}+\vec{B}) = \frac{\partial A_x}{\partial x} + \frac{\partial B_x}{\partial x} + \frac{\partial A_y}{\partial y} + \frac{\partial B_y}{\partial y} + \frac{\partial A_z}{\partial z} + \frac{\partial B_z}{\partial z}$$

$$= \left(\frac{\partial A_x}{\partial x} + \frac{\partial A_y}{\partial y} + \frac{\partial A_z}{\partial z}\right) + \left(\frac{\partial B_x}{\partial x} + \frac{\partial B_y}{\partial y} + \frac{\partial B_z}{\partial z}\right)$$

$= \text{div } \vec{A} + \text{div } \vec{B}$ which is R.H.S.

$\therefore$ ***div*** $(\vec{A}+\vec{B})$ **= div** $\vec{A}$ **+ div** $\vec{B}$.

$\therefore$ L.H.S. = div (gad S) = $\vec{\nabla} \cdot \text{grad } S = \vec{\nabla}.(\vec{\nabla} S)$

Also, grad $S = i\frac{\partial S}{\partial x} + j\frac{\partial S}{\partial y} + k\frac{\partial S}{\partial z}$ and div $\vec{A} = \frac{\partial A_x}{\partial x} + \frac{\partial A_y}{\partial y} + \frac{\partial A_z}{\partial z}$

In this case, $\vec{A}$ = grad S, $\therefore\ A_x = \frac{\partial S}{\partial x}, A_y = \frac{\partial S}{\partial y}, A_z = \frac{\partial S}{\partial z}$

$$\therefore\ \text{div (grad S)} = \left(i\frac{\partial}{\partial x} + j\frac{\partial}{\partial y} + k\frac{\partial}{\partial z}\right) \cdot \left(i\frac{\partial S}{\partial_x} + j\frac{\partial S}{\partial y} + k\frac{\partial S}{\partial z}\right)$$

$$= \frac{\partial}{\partial x}\left(\frac{\partial S}{\partial x}\right) + \frac{\partial}{\partial y}\left(\frac{\partial S}{\partial y}\right) + \frac{\partial}{\partial z}\left(\frac{\partial S}{\partial z}\right) = \frac{\partial^2 S}{\partial x^2} + \frac{\partial^2 S}{\partial y^2} + \frac{\partial^2 S}{\partial z^2}$$

$$= \left(\frac{\partial^2}{\partial x^2} + \frac{\partial^2}{\partial y^2} + \frac{\partial^2}{\partial z^2}\right)S$$

= $\nabla^2 S$ **Which is R.H.S. Hence proved Ans.**

Example 13. *Find the value of* $\iint \vec{r}.\hat{n}dS$, *where S is a closed surface and* $\vec{r}$ *is the position vector.*

Solution. From the Gauss-divergence theorem, we have

$$\iint_S \vec{A}.d\vec{S} = \iiint_V div\, A\, dV$$

In this case, $d\vec{S} = \hat{n}dS$ and $\vec{A} = \vec{r} = xi + yj + zk$

$$\therefore\ \iint_S \vec{r}.\hat{n}dS = \iiint_V \text{div } \vec{r}dV = \iiint_V \left(\frac{\partial x}{\partial x} + \frac{\partial y}{\partial y} + \frac{\partial z}{\partial z}\right) dx\,dy\,dz$$

$$= \iiint_V (1+1+1)dx\,dy\,dz = 3\iiint_V dx\,dy\,dz = 3V \textbf{ Ans.}$$

where V is the volume enclosed by the surface S.

Example 14. *Using Gauss-divergence theorem, evaluate* $\iint_S \vec{F}.d\vec{S}$, *where* $\vec{F} = 4xzi - y^2j + yzk$ *and S is the surface of the cube bounded by x = 0, x = 1, y = 0, y = 1, z = 0, z = 1.*

Solution. From Gauss-divergence thorem, we have

$$\iint_S \vec{F}.d\vec{S} = \iiint_V \text{div } \vec{F}.dV = \iiint_V \left(\frac{\partial F_x}{\partial x} + \frac{\partial F_y}{\partial y} + \frac{\partial F_z}{\partial z}\right) dx\,dy\,dz$$

In this case $\vec{F} = 4xzi - y^2j + yzk$

$$\therefore \quad \iint_S \vec{F}.d\vec{S} = \iiint_V \left[\frac{\partial}{\partial x}(4xz) + \frac{\partial}{\partial y}(-y^2) + \frac{\partial}{\partial z}(yz)\right] dx\,dy\,dz$$

$$= \iiint_V (4z - 2y + y)\,dx\,dy\,dz$$

The surface S of the cube is bounded by $x = 0$, $x = 1$, $y = 0$, $y = 1$, $z = 1$

$$\therefore \quad \iint_S \vec{F}.d\vec{S} = \int_0^1\int_0^1\int_0^1 (4z - y)\,dx\,dy\,dz = \int_0^1\int_0^1 \left(\frac{4z^2}{2} - yz\right)_0^1 dx\,dy = \int_0^1\int_0^1 (2-y)\,dx\,dy$$

$$= \int_0^2 \left(2y - \frac{y^2}{2}\right)_0^2 dx = \int_0^1 \frac{3}{2} dx$$

$$\iint_S \vec{F}.d\vec{S} = \frac{3}{2}\int_0^1 dx = \frac{3}{2}(x)_0^1 = \frac{3}{2} = \mathbf{1.5} \textbf{ Ans.}$$

Example 15. *Prove* $\iint_S (\text{curl } \vec{F}).d\vec{S} = 0$ *for any closed surface S.*

Solution. From Gauss-divergence theorem

$$\iint_S \vec{A}.d\vec{S} = \iiint_V \text{div } \vec{A}\,dV$$

In this case, $\vec{A} = \text{curl } \vec{F}$

$$\therefore \quad \iint_S (\text{curl } \vec{F})d\vec{S} = \iiint_V \text{div}(\text{curl } \vec{F})dV$$

but div (curl $\vec{F}$) = $\vec{\nabla}$.curl $\vec{F}$ = $\vec{\nabla}.(\vec{\nabla} \times \vec{F})$

$$\vec{\nabla}.(\vec{\nabla} \times \vec{F}) = \begin{vmatrix} \frac{\partial}{\partial x} & \frac{\partial}{\partial y} & \frac{\partial}{\partial z} \\ \frac{\partial}{\partial x} & \frac{\partial}{\partial y} & \frac{\partial}{\partial z} \\ F_x & F_y & F_z \end{vmatrix} = 0$$

So, $\quad \iint_S (\text{curl } \vec{F}).d\vec{S} = \mathbf{0}$ **Ans.**

Example 16. *Verify Stoke's theorem for* $\vec{F} = (2x-y)\,\vec{i} - yz^2\,\vec{j} - y^2z\,\vec{k}$, *where S is the upper half surface of sphere* $x^2+y^2+z^2 = 1$ *and C is its boundary.*

Solution. Boundary of the sphere is a circle of radius one and centre at (0, 0, 0) while lies in the x-y plane. Let $x = \cos t$, $y = \sin t$, $z = 0$ and $0 \le t < 2\pi$ be the parametric equations of the boundary, so we have

$$\vec{\nabla}\times\vec{F} = \vec{i}\left(\frac{\partial}{\partial y}F_z - \frac{\partial}{\partial z}F_y\right) + \vec{J}\left(\frac{\partial}{\partial z}F_x - \frac{\partial}{\partial x}F_z\right) + \vec{k}\left(\frac{\partial}{\partial x}F_y - \frac{\partial}{\partial y}F_x\right)$$

In this case $\vec{F} = (2x-y)\,\vec{i} - yz^2\,\vec{j} - y^2z\,\vec{k}$

$\therefore \qquad F_x = 2x-y, F_y = -yz^2$ and $F_z = -y^2z$

$$\therefore \qquad \vec{\nabla}\times\vec{F} = \vec{i}\left[\frac{\partial}{\partial y}(-y^2z) - \frac{\partial}{\partial z}(-yz^2)\right] + \vec{j}\left[\frac{\partial}{\partial z}(2x-y) - \frac{\partial}{\partial x}(-y^2z)\right]$$

$$+ \vec{k}\left[\frac{\partial}{\partial x}(-y^2z) - \frac{\partial}{\partial y}(2x-y)\right]$$

$$= \vec{i}[-2yz+2yz] + \vec{j}[0-0] + \vec{k}[0+(1)] = \vec{k} \qquad \text{... (i)}$$

and $\oint_C \vec{F}.d\vec{r} = \oint_C [(2x-y)dx - yz^2dy - y^2zdz] \qquad [\because \vec{F}.d\vec{r} = F_x dx + F_y dy + F_z dz]$

Putting $\qquad x = \cos t$, $y = \sin t$ and $z = 0$

$\therefore \qquad dx = -\sin t\, dt, \quad dy = \cos t\, dt$

$$\therefore \qquad \oint \vec{F}.d\vec{r} = \int_0^{2\pi} (2\cos t - \sin t)(-\sin t\, dt) - 0 - 0$$

$$\text{or} \qquad \oint \vec{F}.d\vec{r} = \pi$$

$$\oint\oint_S \left(\vec{\nabla}\times\vec{F}\right).\hat{n}\,dS = \oint\oint_S \vec{k}.\hat{n}\,dS = \oint\oint_R dx\,dy \qquad (\because \text{From eqn. } (i))$$

where $R = dx\,dy$ and it is the projection of surface S on x-y plane

$$\text{So } \oint\oint_R dx\,dy = \int_{x=-1}^{0}\int_{\sqrt{1-x^2}}^{0} dy\,dx = 4\int_0^1\int_0^{\sqrt{1-x^2}} dy\,dx$$

$$= 4\int_0^1 \sqrt{(1-x^2)}\,dx = \pi \qquad \text{... (ii)}$$

Hence, the Stoke's theorem $\oint \vec{\mathbf{F}}.d\vec{\mathbf{r}} = \int\int_S \left(\vec{\nabla}\times\vec{\mathbf{F}}\right).d\vec{\mathbf{S}}$, is verified.

12. Electric Fields and Gauss's Law

We can show from the following experiment that there are *two kinds* of chargers: if a glass rod is rubbed with silk and hanged from a long silk thread as shown in Fig. 1.8, and if a second glass rod is rubbed with silk and held near the rubbed end of the first rod, the rods will repel each other. On the other hand, a hard-rubber rod rubbed with fur will attract the glass rod. Two hard-rubber rods rubbed with fur will repel each other. We explain these facts by saying that rubbing a rod gives it an *electric charge* and that the charges on the two rods exert forces on each other. Clearly the charges on the glass and on the hard rubber must be *different in nature*.

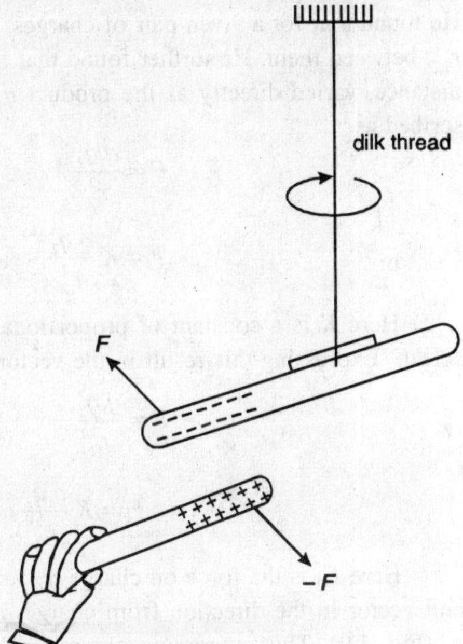

Fig. 1.8

Benjamin Franklin (1706-1790), the first American physicist, who named the kind of electricity that appears on the glass *positive* and the kind that appears on the hard rubber *negative*; these names have been retained to this day. We can sum up these periments by saying that *like charges repel and unlike charges attract.*

Electric effects are not limited to glass rubbed with silk or to hard rubbed rubbed with fur. Any substance rubbed with any other substance under suitable conditions will become *charged* to some extent; by comparing the known change with achengel glass rod or a charged hard-rubber rod, it can be labelled as either *positive or negative*.

The modern view of bulk matter is that, in its normal or neutral state, it contains equal amounts of positive and negative electricity. If two bodies, like glass and silk are rubbed together, a small amount of charge is transferred from one to the other, upsetting the electric neutrality of each. In this case the glass would becomes positive, and the silk negative.

1.13. Conductors and Insulators

A metal rod held in the hand and rubbed with fur will not seem to develop a charge. It is possible to charge such a rod, however, if it is furnished with a glass or hard-rubber handle and if the metal is not touched with the hands while rubbing it. The explanation is that metals, the human body, and the earth are *conductors of electricity* and that glass, hard rubber, plastics, etc., are insulators (also called *dielectrics*).

In conductors electric charges are free to move through the material, whereas in insulators they are not. Although there are no perfect insulators, the insulating ability of fused quartz is about 10^{25} times as great as that of copper, so that for many practical purposes some materials behave as if they were perfect *insulators*.

A class of materials called *semiconductors* is intermediate between *conductors* and *insulators* in its ability to conduct electricity. Among the elements, *silicon and germanium* are well-known examples. In semiconductors the electrical conductivity can often be greatly increased by adding very small amounts of other elements; traces of arsenic or boron are often added to silicon for this purpose. Semiconductors have many practical applications, among which is their use in the construction of transistors.

1.14. Coulomb's Law

It can be easily demonstrated that like charges repel each other, and unlike charges attract each other. Quantitative investigations regarding the dependence of forces of attraction and repulsion on the magnitude of the charges and the distance between them were carried out by Coulomb. He found that for a given pair of charges, this force varied inversely as the square of the distance r_{12}, between them. He further found that the force between two charges q_1 and q_2 held at a given distance, varied directly as the product $q_1 q_2$. Thus mathematically, Coulomb's law may be described as

$$F \propto \frac{q_1 q_2}{r_{12}^2}$$

or

$$F = K \frac{q_1 q_2}{r_{12}^2}$$

Here K is a constant of proportionality and its value depends upon the system of unit we select. Expressing this result in the vector form, we get,

$$F \propto \frac{q_1 q_2}{r_{12}^2}$$

$$\therefore \quad F_2 = K \frac{q_1 q_2}{r_{12}^2} \hat{r}_{12}$$

Here F_2 is the force on charge q_2, and r_{21} is the unit vector in the direction from charge q_1 to charge q_2 (Fig. 1.9). This expression automatically includes is itself the fact that like charges repel and unlike charges attract each other, and the force is along the line joining the two charges. It also reveals that these forces obey Newton's law

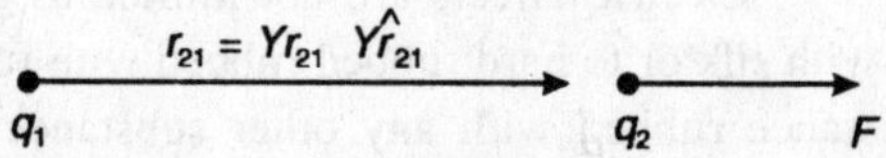

Fig. 1.9

$$F_2 = -F_1 \qquad \text{... (3)}$$

Coulomb's law holds strictly for point charges. But, for experimental investigations, charged spheres coated with good conductors of electricity can be employed. From the argument of symmetry one can conclude that charge is equally distributed at the surface of such a sphere and thus its overall effect can be calculated as if the charge were located at its centre. But this assumption is true only if the distance r is much larger than the size of the spheres. If the spheres are quite close, then due to mutual forces, the charge distribution will not be uniform and, hence, we cannot take it to be located at the centre.

In S.I. units (rationalised M.K.S. units)

$$F = \frac{1}{4\pi\varepsilon_0} \frac{q_1 q_2}{r_{12}^2} \qquad \text{... (4)}$$

The unit of charge is coulomb with symbols C. A. *coulomb is defined as that amount of charge which flows through a given cross-section of a wire in one second when the wire is carrying a current of one ampere.* Thus is Eqn. (4) everything has been defined except the constant to ε_0 known as the *permittivity* of free space, which is adjusted in such a way that Eqn. 4 fits in the experiment. The experimental value of ε_0 is :

$$\varepsilon_0 = 8.854 \times 10^{-12}\ \text{C}^2/\text{N m}^2$$
$$= 8.854 \times 10^{-12}\ \text{farad/m} \qquad \text{... (5)}$$

The unit farad/m for ε_0 will be explained in the chapter of condensers. The form of the Coulomb's law equation, and the method of evaluation of the constant of proportionality, are similar to that being done in *gravitation*. But the *electric force* is much stronger than the *gravitational forces. A charged glass rod can lift a piece of paper against the gravitational pull of such a huge mass i.e., the earth.*

Example 1. *Calculate the force between two charges of one coulomb each, when they are at distance of l m.*

Solution.
$$F = \frac{1}{4\pi\varepsilon_0} \times \frac{1 \times 1}{(1)^2} N$$

$$= \frac{1}{4\pi \times 8.854 \times 10^{-12}} \text{N} = \mathbf{8.986 \times 10^9\ N} \quad \textbf{Ans.}$$

This force is equal to the force with which **9.2 × 10^5** (about one million) metric tons of mass are attracted by the earth.

Example 2. *Find the radius of the water drops which will remain just suspended with a charge of l electron in the Earth's field of 3 volt/cm. (e = 4.8 × 10^{-10} e.s.u)*

Solution. We have the relation

$$eE = mg$$

$$4.8 \times 10^{-10} \times \frac{3}{300} = \frac{4}{3}\pi r^3 \times 981$$

$$r = \mathbf{1.04 \times 10^{-5}\ cm.} \quad \textbf{Ans.}$$

Example 3. *The distance between an electron and a proton in hydrogen atom is 5.3 × 10^{-10} meters, calculate the magnitude of (a) electrical force (b) gravitational force them (elementary charge = 1.602 × 10^{-19} coulomb, mass of electron = 9.11 × 10 31 kgm, mass of Proton = 1.67 × 10^{-27} kgm, G = 6.67 × 10^{-11} M.K.S. units.*

Solution. Electrical force between the two charges

$$= \frac{1}{4\pi\varepsilon_0} \times \frac{q_1 q_2}{r^2}$$

$$= 9 \times 10^9 \times \frac{1.602 \times 10^{-19} \times 1.602 \times 10^{-19}}{(5.3 \times 10^{-10})^2}$$

[$\because$ $q_1 = q_2$ = charge on electron or proton *i.e.*, the elementary charge]

= 8.212 × 10^{-10}N Newton

$$\text{Gravitational Force} = G\frac{M.m}{d^2}$$

$$= 6.67 \times 10^{-11} \times \frac{1.67 \times 10^{-27} \times 9.11 \times 10^{-31}}{(5.3 \times 10^{-11})^2}$$

$$= \mathbf{3.6 \times 10^{-47}\ Newtons} \quad \textbf{Ans.}$$

which shows that electrostatic force is 10^{37} times the gravitational force.

Example 4. *Two parallel plates are kept in air 1 cm apart and a p.d. of 100 volts is applied. If there be an electron resting on the plate at lower potential, find (i) the force on the electron in newtons (ii) K.E. and velocity of this electron when it reaches the other plate (e = 1.6 × 10^{-19} coulomb, m = 9.1 × 10^{-31} kgm)*

Solution. Electric intensity $E = \frac{V}{d} = \frac{100}{0.01} = 10^4$ V/m

Force $$F = \frac{1}{4\pi\varepsilon_0} \times eE = 9 \times 10^9 \times 1.6 \times 10^{-19} \times 10^4$$

$= 1.44 \times 10^{-5}$ N

Acceleration $$a = \frac{F}{m} = \frac{1.44 \times 10^{-5}}{9.1 \times 10^{-31}} = 1.582 \times 10^{25} \text{ m/sec}^2$$

Velocity $$v = \sqrt{2ad} = \sqrt{2 \times 1.582 \times 10^{25} \times 0.01}$$

$= 5.536 \times 10^{11}$ m/sec

$$\text{K.E.} = \frac{1}{2}mv^2 = \frac{1}{2} \times 9.1 \times 10^{-31} \times (5.536 \times 10^{11})^2$$

$= \mathbf{1.314 \times 10^{-7}}$ **joules Ans.**

Example 5. If the total positive and the total negative charges in a copper penny each of value 1.3×10^5 coulomb are separated to a distance such that their force of attraction is 4.5 N. How far apart must they be?

Solution. We have

$$F = \frac{1}{4\pi\varepsilon_0} \frac{q_1 q_2}{r^2}$$

Putting $q_1 q_2 = q$ and solving for r we get

$$r = q\sqrt{\frac{1/4\pi\varepsilon_0}{F}}$$

$$= 1.3 \times 105 \text{ coul} \sqrt{\frac{9.0 \times 10^{-9} \text{N} - \text{m}^2/\text{coul}^2}{4.5 \text{ N}}}$$

$= \mathbf{1.8 \times 108}$ **m Ans.**

If more than two charges are present, Eq. (4) holds for every pair of charges. Let the charges be q_1, q_2 and q_3, etc., we calculate the force exerted on any one (say q_1) by all the others from the vector equation

$$\vec{F_1} = \vec{F_{12}} + \vec{F_{13}} + \vec{F_{14}} + ...$$

where $\vec{F_{12}}$, for example, is the force exerted on q_1 by q_2

1.15. Charge is Quantized

Earlier electric charge was thought of as a continuous fluid, an idea that was useful for many purposes. The atomic theory of matter, however, has shown that fluids themselves, such as water and air, are not continuous but are made up of atoms. Experiment shows that the "electric fluid" is not continuous either, but that it is made up of integral multiples of a certain minimum electric charge. This *fundamental charge*, to which we give the symbol *e*, has the magnitude 1.60210 $\times 10^{-19}$ coul. Any physically existing charge *q*, no matter what its origin, can be written as *ne* where *n* is a positive or a negative integer.

When a physical property such as charge exists in discrete "packets" rather than in continuous amounts, the property is said to be *quantized*. Quantization is basic to modern physics. The existence of atoms and of particles like the electron and the proton indicates that mass is quantized also.

The quantum of charge e is so small that the "graininess" of electricity does not show up in large-scale experiments, just as we do not realize that the air we breathe is made up of atoms. In an ordinary 220 volt, 100 watt light bulb, for example, 3×10^{-18} *elementary charges* enter and leave the filament every second

1.16. Matter and Charge

Matter as we ordinarily experience it can be regarded as composed of three ends of elementary particles, *proton, neutorn*, and *electron*, then masses and charges are given below. None that the masses of the *neutron* and he *proton* are approximately equal but that the electron is $\frac{1}{1840}$ times smaller.

Particle	*Symbol*	*Charge*	*Mass*
Proton	p	$+e$	1.67252×10^{-27} kg
Neutron	n	0	1.67482×10^{-27} kg
Electron		e^-	9.1091×10^{-31} kg

Atoms are made up of a dense, positively charged *nucleus*, surrounded by a cloud of electrons; The nucleus varies in radius from about 1×10^{-15} meter for hydrogen to about 7×10^{-15} meter for the heaviest atoms. The outer diameter of the electron cloud, that is, the diameter of the atom, lies in the range $1–3 \times 10^{-10}$ meter, about 10^5 times larger than the nuclear diameter.

Example 6. *A point charge of + 3.0 × 10^{-6} coul is 12 cm distant from a second charge of – 1.5 × 10^{-6} coul. Calculate the magnitude and direction of the force of each charge.*

Solution. Magnitude of the force on each charge is

$$F = \frac{q_1 q_2}{4\pi\varepsilon_0 r^2}$$

$$= \frac{(3\times10^{-6}\text{C})(1.5\times10^{-6}\text{C})}{(0.12\text{m})^2} \times 9\times10^9 \text{N}-\text{m}^2/\text{C}^2$$

$$= \mathbf{2.8\ N\ \ Ans.}$$

The force will be attractive.

Example 7. *How for apart must two protons be if the electrical repulsive force on either is equal to its weight?*

Solution. Let the charge on each proton be q. Coul and let d be the distance between them. Then electrical repulsive force on either is given by

$$F = \frac{q^2}{4\pi\varepsilon_0 d^2} = mg$$

or

$$d = \frac{q}{\sqrt{4\pi\varepsilon_0\, mg}}$$

$$= \frac{0.6\times10^{-19}\text{C}}{\sqrt{\dfrac{9\times10^9\text{N}-\text{m}^2/\text{C}^2}{1.67\times10^{-27}\text{kg.}\times9.8\text{m}^{-2}}}}$$

$$= 0.119\text{ m} = \mathbf{11.9\ cm.\ \ Ans.}$$

Example 8. *In the radioactive decay of U^{238}, the centre of emerging α-particles is at a certain instant 9×10^{-15} m from the centre of the residual nucleus Th^{234}. At this instant*

(*a*) what is the force on the α-particle
(*b*) what is the acceleration ?

Solution. (*a*) We have the coulomb force given by the relation

$$F = \frac{Z_1 Z_2 e^2}{4\pi\varepsilon_0 r^2}$$

where $Z_1 \rightarrow$ At. No. of α-particle
and $Z_2 \rightarrow$ At. No. of Thorium nucleus
$r \rightarrow$ distance of separation

$$\therefore \quad F = \frac{2 \times 90 \times (1.6 \times 10^{-19} \text{C})^2}{(9 \times 10^{-15} \text{m})^2} \times 9 \times 10^9 \text{N} - \text{m}^2/\text{C}^2$$

$$= \mathbf{512\ N.} \quad \textbf{Ans.}$$

(*b*) Mass of α-particle $= 4 \times 1.67 \times 10^{-27}$ kg
$M = 6.69 \times 10^{-27}$ kg

$$\therefore \quad \text{Acceleration} = \frac{\text{Force}}{\text{Mass}} = \frac{512\text{N}}{6.69 \times 10^{-27}\text{ kg}} = \mathbf{7.7 \times 10^{-28}\ ms^{-2}} \textbf{ Ans.}$$

Example 9. *Find the ratio between the electrical force F_e and gravitational force F_g between two electrons.*

Solution.

$$\frac{F_e}{F_g} = \frac{Q^2/4\pi\varepsilon_0 r^2}{G\dfrac{m^2}{r^2}}$$

$$= \frac{Q^2}{4\pi\varepsilon_0 G m^2}$$

$$= \frac{(9 \times 10^9 \text{Nm}^2 \text{C}^2) \times (1.60 \times 10^{-19} \text{C}^2)}{(6.67 \times 10^{-11} \text{Nm}^2 \text{kg}^{-2}) \times (9.1 \times 10^{-31} \text{kg})^2}$$

$$= 4.16 \times 10^{42}$$

$$\therefore \quad F_e = \mathbf{4.16 \times 10^{42} Fg} \quad \textbf{Ans.}$$

Which shows that electrical force is about 10^{42} times stronger than the gravitational force for two electrons.

The significance of Coulomb's law goes far beyond the description of the forces acting between charged balls or rods. This law, when incorporated into the structure of quantum physics, correctly describes

(*i*) the electric forces that bind the electrons of an atom to its nucleus,
(*ii*) the forces that bind atoms or molecules together to form molecules, and
(*iii*) the forces that bind atoms or molecules together to form includes or liquids. Thus most of the forces of our daily experience that are not gravitational in nature are electrical.

In the atomic *nucleus* we encounter a new force which is neither gravitational nor electrical in nature. This strong attractive force, which binds together the *protons and neutrons* that make up the nucleus is called simply the nuclear force. If this force were not present, the nucleus would fly apart at once because of the strong Coulomb repulsion force that acts between its *protons*.

Example 10. *What repulsive Coulomb force exists between two protons in a nucleus of copper? Assume a separation of 4.0×10^{-15} meter.*

Solution. From Coulomb's law,

$$F = \frac{1}{4\pi\varepsilon_0}\frac{q_1 q_2}{r^2}$$

$$= \frac{(9.0\times10^9 \text{N} - \text{m}^2/\text{coul}^2)(1.6\times10^{-19}\text{ coul})^2}{(4.0\times10^{-15})^2}$$

$$= \mathbf{14\ N\ \ Ans.}$$

This enormous repulsive force must be more than compensated for by the strong attractive nuclear forces. This example, combined with example 4, shows that nuclear binding, forces are much stronger than *atomic binding forces*. *Atomic binding forces* are, in turn, much stronger than *gravitational forces* for the same particles separated by the same distance.

1.17. Conservation of Charges

When a glass rod is rubbed with silk, a positive charge appears on the rod. Measurement shows that a negative charge of equal magnitude appears on the silk. This suggests that rubbing does not create charge but merely transfers it from one object to another, disturbing slightly the electrical neutrality of each. This hypothesis of he *conservation of charge* has been verified experimentally both for large scale events and at the atomic and nuclear level; no exceptions have ever been found.

An example of charge conservation is found in radioactive decay, of which the following process is typical:

$$U^{238} = Th^{234} + He^4$$

The radioactive "parent" nucleus, U^{238}, contains 92 protons (that is, its atomic number $Z = 92$). It disintegrates spontaneously by emitting an α-particle (He^4; Z = 2) transmitting itself into the nucleus Th^{234} with $Z = 90$. Thus the amount of charge present before disintegration (+ 92e) is the same as that present after the disintegration.

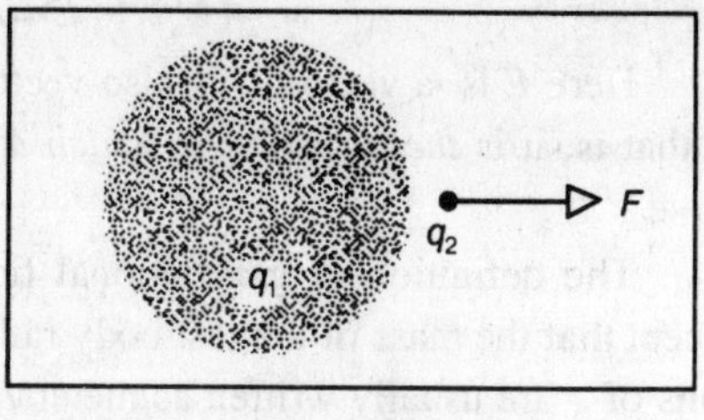

Fig. 1.10

Another example of charge conservation is found in nuclear reactions, of which the bombard-ment of Ca^{44} with cyclotron accelerated protons is typical. In a particular collision a neutron may emerge from the nucleus, leaving Sc^{44} as a "residual" nucleus:

$$_{20}Ca^{44} + p = {}_{21}Sc^{44} + n$$

The sum of the atomic numbers before the reaction (20 + 1) is exactly equal to the sum of the atomic numbers after the reaction (21 + 0), So in this case also *charge is conserved.*

1.18. The Electric Fields

With every point in space near the earth is associted with a *gravitational field strength* vector g. This is a the gravitational acceleration that a test body, placed at that point and released, would experience. If m is the mass of the body and F the gravitational force acting on it, g is given by

$$g = F/m$$

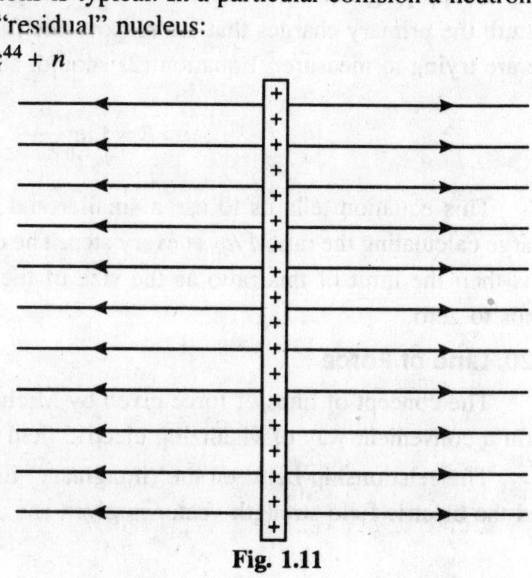
Fig. 1.11

This is an example of a vector field. For points near the surface of the earth the field is often taken as uniform; that is, g is constant at all points.

The flow of water in a river provides another example of a vector field called a *flow field.* Every point in the water has associated with it a vector quantity, the velocity v with which the water flows past the point. If g and v do not change with time, the corresponding field is described as *stationary.* In the case of the river, note that even though the water is moving, the vector v at any point does not change with time for steady flow conditions.

The space surrounding a charged rod is affected by the presence of the rod, and we speak of an *electric field* in this space. In the same way we speak of magnetic field in the space around a *bar magnet.*

We can explain the electric fields as under:

1. Charge q_1 in Fig 1.10 sets up an electric field in the space around itself. This field is represented by the shading in the figure; later we shall show how to represent electric fields more effectively.
2. The field acts on charge q_2; this shows up in the force f that q_2 experiences.

1.19. The Electric Field Strength *E*

To define the electric field we place a small test body carrying a test charge q_0 (assumed positive for convenience) at the point in space that is to be examined, and we measure the electrical force F (if any) that acts on this body. The *electric field strength E* at the point is defined as

$$E = F/q_0 \quad \text{... (2)}$$

Here E is a vector, F is also vector, q_0 being a scalar. The direction of E is the direction of F, that is, it is *the direction in which a resting positive charge placed at the point would tend to move.*

The definition of gravitational field strength g is much like that of electric field strength except that the mass of the test body rather than its charge is the property of interest. Although the units of g are usually written as meters/sec^2, they could also be written as N/kg (Eq. 1) ; those for E are N/coul. Thus both g and E are expressed as a force divided by a property (mass or charge) of the test body.

In applying Eq. (2), we must use a test charge as small as possible. A large test charge might disturb the primary charges that are responsible for the field, thus changing the very quantity that we are trying to measure. Equation (2) should, strictly be replaced by

$$E = \lim_{q_0 \to 0} \frac{F}{q_0} \quad \text{... (3)}$$

This equation tells us to use a smaller and smaller test charge calculating the ratio F/q_0 at every step. The electric field E is then the limit of this ratio as the size of the test charge tends to zero.

1.20. Line of Force

The concept of lines of force given by Michael Faraday, form a convenient way of visulizing electric field patterns.

The relationship between the (imaginary) lines of force and the electric field strength vector is given as:

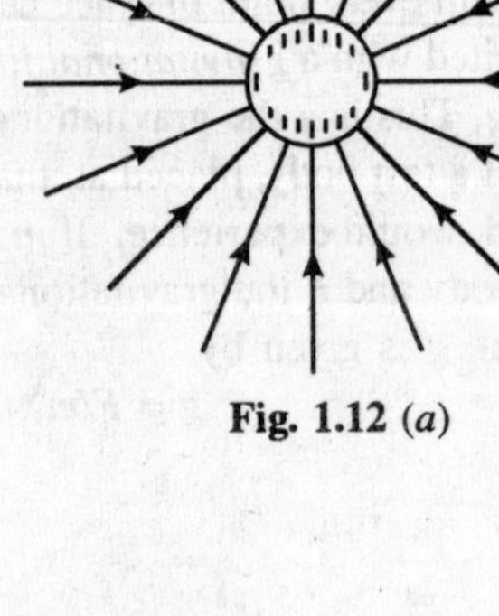

Fig. 1.12 (*a*)

1. The tangent to a line of force at any point gives the *direction* of E at that point.
2. The lines of force are drawn so that the number of lines per unit cross-sectional area is proportional to the *magnitude* of E. Where the lines are close together E is large and where they are far apart E is small.

Figure 1.11 shows the lines of force for a uniform sheet of positive charge. We assume that the sheet is infinitely large, which, for a sheet of finite dimensions is equivalent to considering only those points whose distance from the sheet is small as compared to the distance to the nearest edge of the sheet. A positive test charge, released in front of such a sheet, would move away from the sheet along a perpendicular line. So the electric field strength vector at any point near the sheet must be at right angles to the sheet. The lines of force are uniformly spaced, which means that E has the same magnitude for all points near the sheet.

Figure 1.12 shows the lines of force for a negatively charged sphere. From symmetry the lines must be along radii. They point inward because a free positive charge would be accelerated in this direction. The electric field E is not constant but decreases with increasing distance from the charge. This is evident in the lines of force, which are farther apart at greater distances. From symmetry, E is the same for all points that lie at the same distance from the centre of the charge.

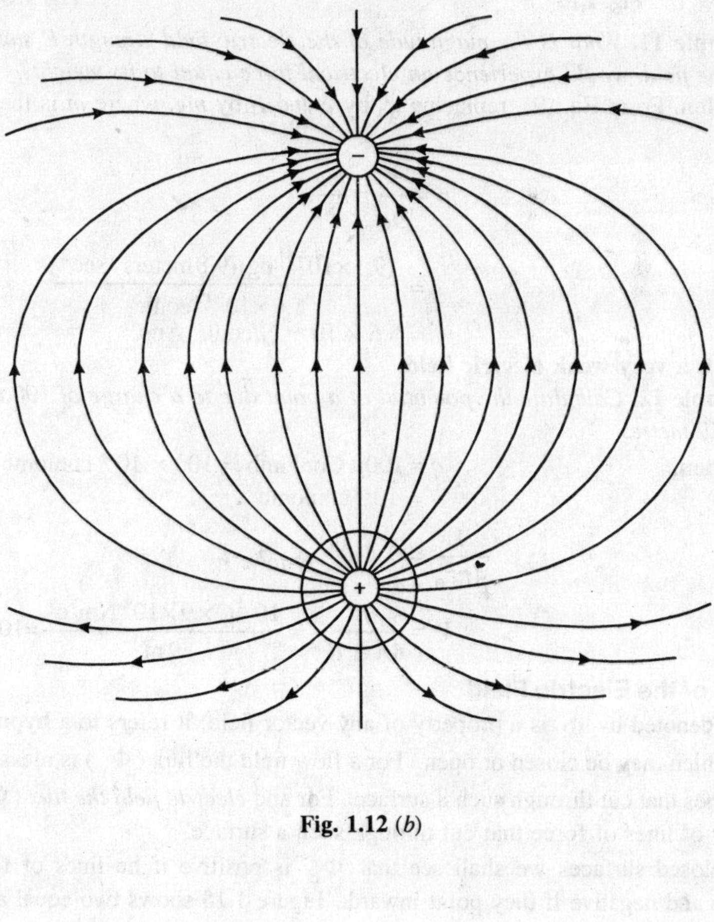

Fig. 1.12 (*b*)

Figures 1.13 and 1.14 show the lines of force for two equal like charges and for two equal unlike charges, respectively.

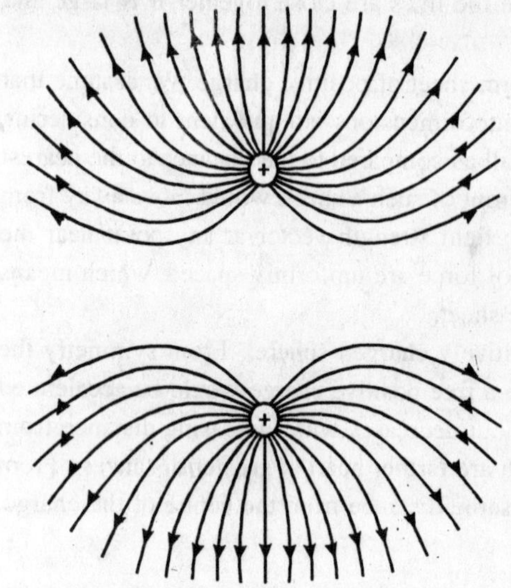

Fig. 1.13

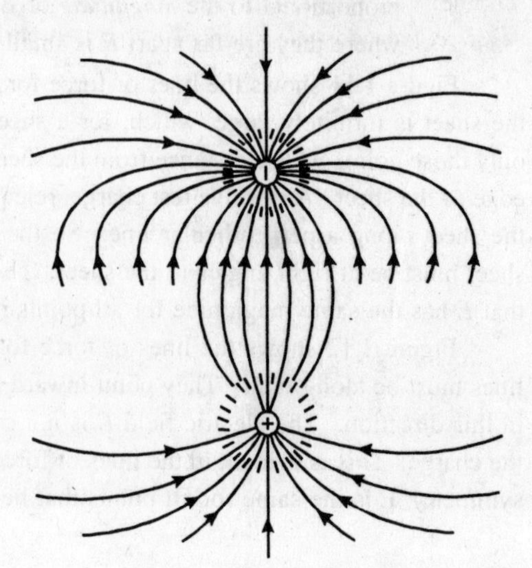

Fig. 1.14

Example 11. *What is the magnitude of the electric field strength E such that an electron, placed in the field, would experience an electrical force equal to its weight?*

Solution. From Eq. (2), replacing q_0 by e and F by mg, where m is the electron mass, we have

$$E = \frac{F}{q_0} = \frac{mg}{e}$$

$$= \frac{(9.1 \times 10^{-31}\,\text{kg})(9.8\,\text{meters / sec}^2)}{1.6 \times 10^{-19}\,\text{coul.}}$$

$$= 5.6 \times 10^{-11}\ \text{N/coul.}\ \textbf{Ans.}$$

This is a very weak electric field.

Example 12. *Calculate the potential at a point due to a charge of 100 micro coulomb at a distance of 9 metres.*

Solution. $q = 100\mu$ Coulomb $= 100 \times 10^{-6}$ coulomb

$= 10^{-4}$ coulomb

and $$\frac{1}{4\pi\varepsilon_0} = 9 \times 10^9\ \text{N-m}^2\,\text{c}^{-2}$$

$$V = \frac{E}{4\pi \in_0 R} = \frac{10^{-4}\,C \times 9 \times 10^9\ \text{Nm}^2\text{c}^2}{9\,\text{m}} = 10^5\ \text{Volt}\ \textbf{Ans.}$$

1.21. Flux of the Electric Field

Flux denoted by Φ is a property of any vector field; it refers to a **hypothetical surface** in the field, which may be closed or open. For a flow field the flux (Φ_v) is measured by the number of streamlines that cut through such a surface. For and *electric field the flux* (Φ_E) is measured by the number of lines of force that cut through such a surface.

For closed surfaces we shall see that Φ_E is positive if he lines of force point outward everywhere and negative if they point inward. Figure 1.15 shows two equal and opposite charge and their lines of force. Curves S_1, S_2, S_3 and S_4 are the intersections with the plane of the figure of four hypothetical closed surfaces. From the statements just given, Φ_E is positive for surface S_1 and negative for S_2. The flux of the electric field is important because Gauss's law, *one of the four basic equation of electromagnetism*, is expressed in terms of it.

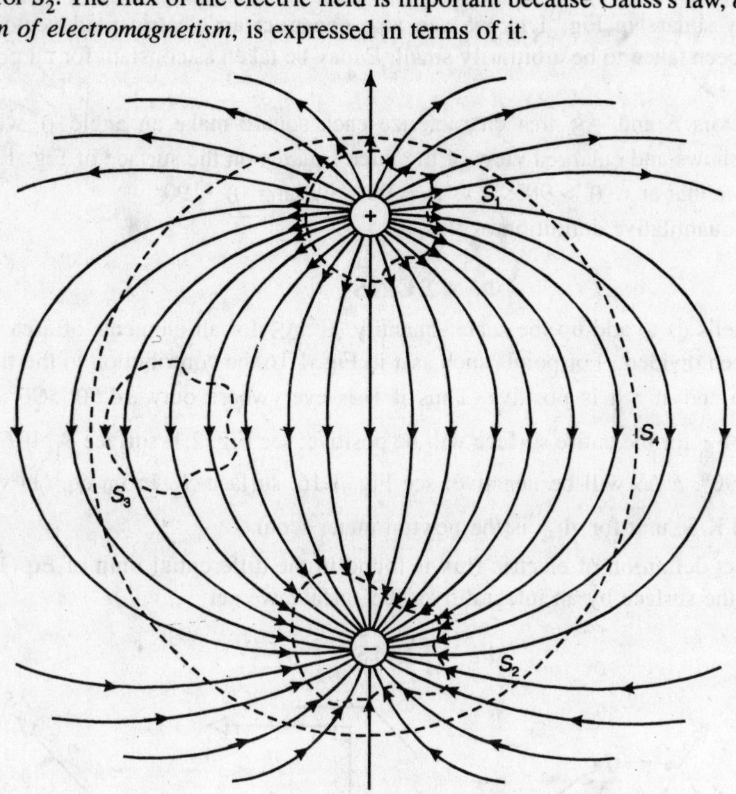

Fig. 1.15

To define Φ_E precisely, consider Fig. 1.16 which shows an arbitrary closed surface immersed in electric field. Let the surface be divided into elementary squares ΔS, each of which is small

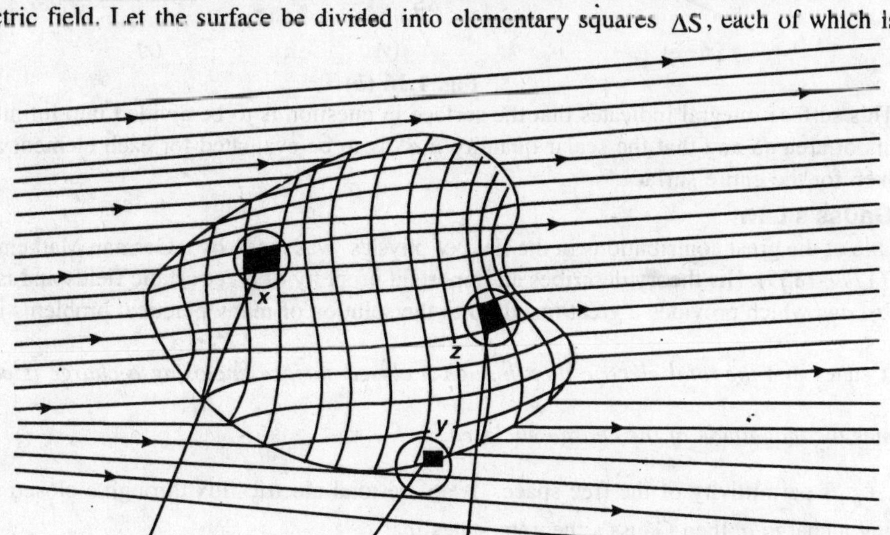

Fig. 1.16 (*a*)

enough so that it may be considered to be plane. Such an element of area can be represented as a vector ΔS, whose magnitude is the area ΔS; the direction of ΔS is taken as the outward-drawn normal to the surface.

At every square in Fig. 1.15 we can also construct an electric field vector *E*. Since the squares have been taken to be arbitrarily small, *E* may be taken as constant for all points in a given square.

The vectors *E* and ΔS that characterize each square make an angle θ with each other. Figure 1.16*b* shows and enlarged view of the three squares on the surface of Fig. 1.16 (*a*) marked *x*, *y*, and *z* Note that at *x*, $\theta > 90°$; at *y*, $\theta = 90°$; and at *z*, $\theta < 90°$.

A semi-quantitative definition of flux is

$$\Phi_E \cong \Sigma \vec{E} . \Delta \vec{S} \qquad \text{... (1)}$$

Which tells us to add up the scalar quantity $E . \Delta S$ for all elements of area into which the surface has been divided. For points such as *x* in Fig. 1.16 the contribution to the flux is negative; at *y* it is zero and at *z* it is positive. Thus if E is everywhere outward, θ >90°, ***E*.Δ*S*** will be positive, and Φ_E for the entire surface will be positive; see Fig. 2.1, surface S_1. If *E* is everywhere inward, $\theta > 90°$, *E*.Δ*S* will be negative; see Fig. 1.16, surface S_2. From eq. (1) we see that the appropriate M.K.S unit for Φ_E is the newton meter²/coul.

The exact definition of electric flux is found in the differential limit of Eq (1) by replacing the sum over the surface by an integral over the surface we get:

$$\Phi_E = \oint \vec{E} . d\vec{S} \qquad \text{... (2)}$$

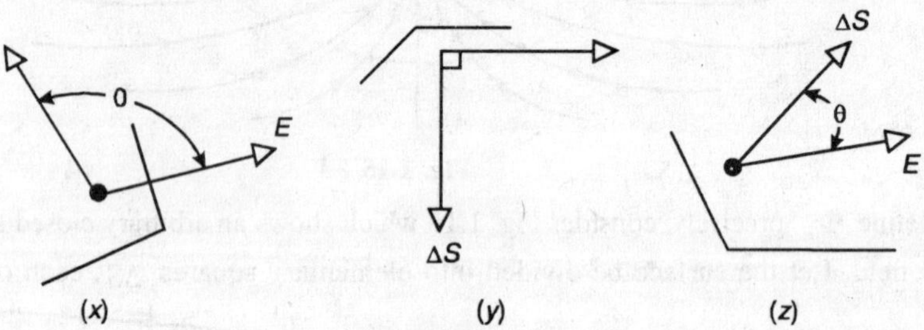

Fig. 1.16 (*b*)

This surface integral indicates that the surface in question is to be divided into infinitesimal elements of area *dS* and that the scalar quantity *E.dS* is to be evaluated for each element and the sum taken for the entire surface.

1.22. Gauss's Law

One of the great contributions in the field of physics was made by a German Mathematician Gauss (1777-1855). His theory describes an important property of electrostatic fields and is called *Gauss's Law*; which provides a great insight into the solution of many practical problems in electrostatics.

It states that *the total electric flux through a closed surface enclosing a charge is equal to* $\frac{1}{\varepsilon_0}$ *times the magnitude of the charge enclosed.*

where, ε_0 is permittivity of the free space. If Φ be total electric flux through a closed surface enclosing a charge *q*, then Gauss's theorem states that

$$\Phi = \frac{q}{\varepsilon_0} \qquad \text{... (1)}$$

Let us consider a closed surface S enclosing a charge q. Let area element $d\vec{S}$ be at a distance r from the charge q enclosed in the closed surface S. If $\vec{E}$ is electric field at the area element, then electric flux passing through the area element is given by the relation

$$d\Phi = \oint_S \vec{E}.d\vec{S}$$

The total electric flux through the closed surface S will be

$$\Phi = \vec{E}.d\vec{S}$$

Then, Gauss's theorem may also have the form

$$\oint_S \vec{E}.d\vec{S} = \frac{q}{\varepsilon_0} \quad \text{... (2)}$$

where $\oint_S$ represents integral over a *closed surface*.

Proof of Gauss's Law

Let θ be the angle between the directions of $\vec{E}$ and dS, then

$$\vec{E}.d\vec{S} = EdS \cos\theta$$

Also
$$E = \frac{1}{4\pi\varepsilon_0}.\frac{q}{r^2}$$

where E is the magnitude of electric field due to charge q at an area element (say dS)

$$\therefore \quad \oint \vec{E}.d\vec{S} = \oint \frac{1}{4\pi\varepsilon_0}.\frac{q}{r^2}.dS\cos\theta$$

$$= \frac{q}{4\pi\varepsilon_0}\oint\frac{dS\cos\theta}{r^2} = \frac{q}{4\pi\varepsilon_0}.d\omega$$

where $d\omega$ is the solid angle subtend at O due to area element dS.

$$\therefore \quad \oint \vec{E}.d\vec{S} = \frac{q}{4\pi\varepsilon_0}.4\pi = \frac{q}{\varepsilon_0}$$

which proves Gauss's theorem.

Mathematically Gauss's Law may be stated as

If a closed surface enclosed a charge, then the surface integral of the electric field due to the enclosed charge over a closed surface is equal to $1/\varepsilon_0$ *times the charge enclosed.*

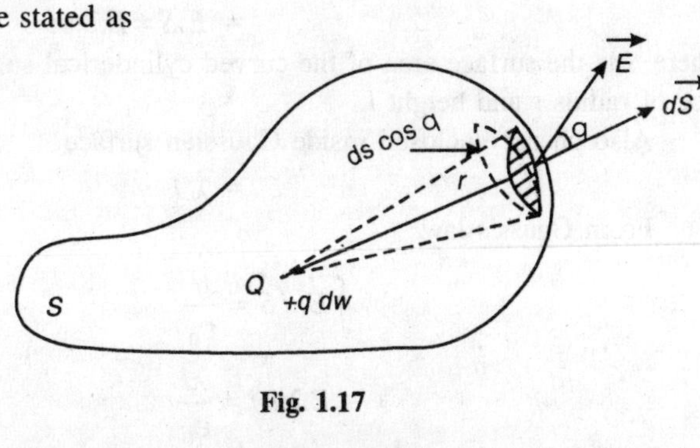

Fig. 1.17

Gauss's law is very important especially in situations where calculation of electric field with the application of Coulomb's law of principle of superposition, becomes difficult. In these cases a closed surface is chosen which, encloses the source charge. Such a surface is called **Gaussian surface. An imaginary closed surface enclosing a charge is called Gaussian surface.**

1.23. Force between Two Point Charges in Vacuum

Consider two point charges q_1 and q_2 present in vacuum and separated by a distance q with point charge q_1 as centre and radius to r, draw a sphere. It will be Gaussian surface enclosing charge q_1. Let E be the magnitude of electric field at point charge q_2, then we have the force experienced by point charge q_2 due to the electric field of charge q_1 as

$$F = q_2 E \quad \text{....(1)}$$

Now, according to Gauss's theorem,

$$\int \vec{E} d\vec{S} = \frac{q_1}{\varepsilon_0}$$

$$\therefore \quad \frac{q_1}{\varepsilon_0} = \int \vec{E}.d\vec{S} = E\int dS = E.4\pi r^2$$

or $$E = \frac{1}{4\pi\varepsilon_0}.\frac{q_1}{r^2} \quad \text{... (2)}$$

From equations (1) and (2), we have

$$F = \frac{1}{4\pi\varepsilon_0}.\frac{q_1 q_2}{r^2}$$

1.24. Applications of Gauss's Law

(*i*) **Electric field due to a line charge (field due to a thin wire).** Let us consider a thin infinitely long wire or a rod (cylinderical conductor). Let it have a uniform linear charge density λ and let P be the point at a distance r from the line charge. To find $\vec{E}$ *at P, let us select a cylinderical surface of radius r and height l as a Gaussian surface. As* $\vec{E}$ *is every where parallel to the two ends of the cylinder, they will not contribute any electric flux.*

Symmetry of the problem suggests that $\vec{E}$ is equal in magnitude and directed outward (As line charge is positive) at every point of the curved cylinderical surface.

It is this curved cylinderical surface which contributes to the electric flux

$$\oint \vec{E}.d\vec{S}$$

$$= \vec{E}.\vec{S} = E.2\pi rl$$

where S is the surface area of the curved cylinderical surface of radius r and height l.

Also charge enclosed inside Gaussian surface

$$= \lambda.l$$

From Gauss's law,

$$\oint \vec{E}.d\vec{S} = \frac{q}{\varepsilon_0}$$

$$\therefore \quad E.2\pi rl = \frac{\lambda l}{\varepsilon_0}$$

or $$E = \frac{\lambda}{2\pi\varepsilon_0 r}$$

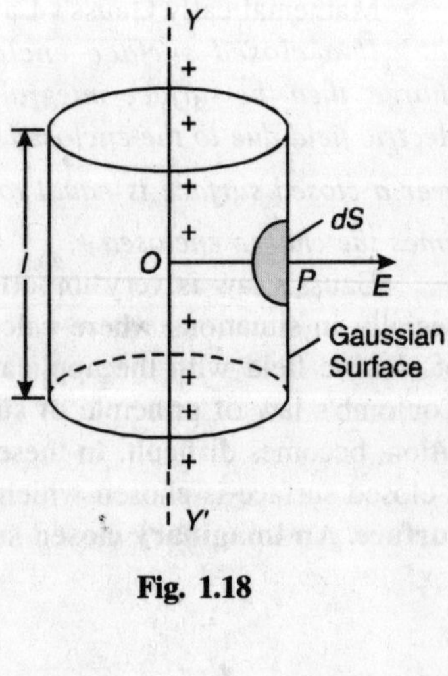

Fig. 1.18

So the intensity of electric field of a line charge is inversely proportional to the distance r from the line.

It may be noted that $\vec{E}$ will be radially outward if the charge is positive and radially inward if the charge is negative.

Instead of a wire, if a cylinderical conductor of radius R is considered. Then if the charge is distributed throughout the volume of the cylinder with charge density ρ, in that case $q = (\pi r^2.l).\rho$

$$\therefore \qquad E.2\pi rl = \frac{(\pi R^2 l)\rho}{\varepsilon_0}$$

$$\text{or} \qquad E = \frac{\rho R^2}{2\varepsilon_0 r}$$

(*ii*) **Electric field due to an infinite plane sheet of charge.** Consider an infinite thin plane sheet of of positive charge having a uniform surface charge density σ on both sides of the sheet. By symmetry, it follows that the field is perpendicular to the plane sheet of charge and in outward direction. To find electric field due to the plane sheet of charge at any point P distant r from it, choose a cylinder of area of cross-section S through the point P as the Gaussian surface [Fig. 1.19]. Since electric lines of force are parallel to the curved surface of the cylinder, the flux due to electric field of the plane sheet of charge passes only through the two circular caps of the cylinder. If E is magnitude of electric field at point P, then electric flux crossing through the Gaussian surface is given by

$$\oint \vec{E}.d\vec{S} = \oint_{S_1} \vec{E}.d\vec{S} + \oint_{S_2} \vec{E}.d\vec{S}$$

$$= \int_{S_1} E\,dS + \int_{S_2} \vec{E}.d\vec{S}$$

$$= ES + ES = 2\,ES \qquad \text{... (1)}$$

Charge enclosed inside the Gaussian surface

$$q = \sigma.S \qquad \text{... (2)}$$

From Gauss's law

$$\oint \vec{E}.d\vec{S} = \frac{q}{\varepsilon_0} \qquad \text{... (3)}$$

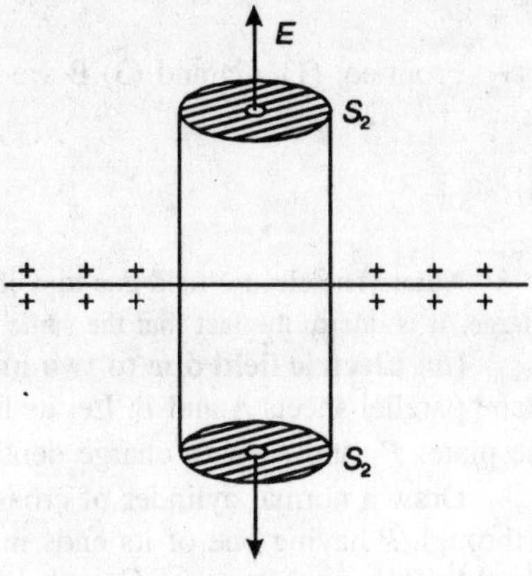

Fig. 1.19

From eq. (1), (2) and (3) we get

$$2\,ES = \frac{\sigma S}{\varepsilon_0}$$

$$\text{or} \qquad E = \frac{\sigma}{2\varepsilon_0}$$

which means that the magnitude of the field is independent of the distance from the sheet.

In case sheet is negatively charged, $\vec{E}$ points normally towards the sheet.

(*iii*) **Electric field at any point due to infinite charged conducting plate.** When a charge is given to a conducting plate, the same is distributed over the entire surface of the plate. If the plate is of uniform size and thickness, the surface density (σ) of the plate is uniform and is same on the two sides of the surfaces. To find the electric field at P, draw a cylinder (Gaussian surface) through P which is normal to the surface of the plate. Let S be the cross-sectional area of the cylinder. One end of the cylinder is inside the conductor and so does not contribute to the electric flux as no charge exists inside the conductor. Also no contribution of flux is given by the curved surface of this cylinder as $\vec{E}$ is parallel to this surface. So, it is only the end face around P which contributes towards electric flux. So, we have

$$\oint \vec{E}.d\vec{S} = ES \qquad \text{... (1)}$$

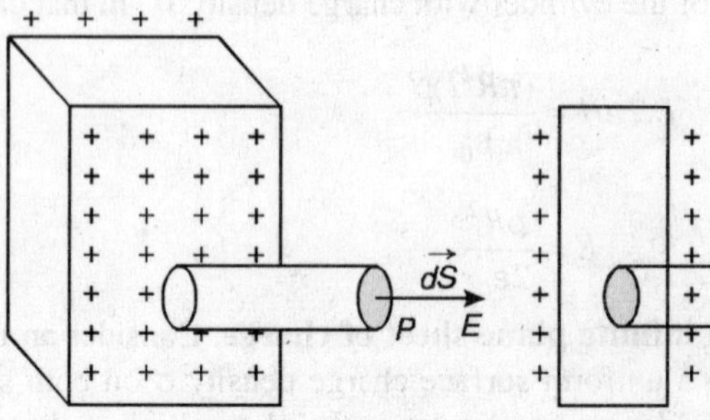

Fig. 1.20

Charge enclosed inside Gaussian surface is

$$q = \sigma.S \qquad \text{... (2)}$$

From Gauss's theorem

$$\oint \vec{E}.d\vec{S} = \frac{q}{\varepsilon_0} \qquad \text{... (3)}$$

From eq. (1), (2) and (3) B we get $E.S = \dfrac{\sigma.S}{\varepsilon_0}$

or
$$E = \frac{\sigma}{\varepsilon_0}$$

Note: The electric field due to a charged conducting plate is two times the field due to plane sheet of charge. It is due to the fact that the same charge in present on its both sides.

(*iv*) **Electric field due to two infinite parallel sheets of charge.** Let us consider two infinite plane parallel sheets A and B. Let us first consider a point P lying anywhere in the region between the plates P_1 and P_2 with charge densities $+\sigma$ and $-\sigma$.

Draw a normal cylinder of cross- sectional area S through P having one of its ends in the conducting plate A. Then we have from Gauss's law (as in case of electric field due to an infinite plane sheet)

$$\oint_S \vec{E}.d\vec{S} = \frac{q}{\varepsilon_0}$$

or
$$ES = \frac{\sigma S}{\varepsilon_0}$$

or
$$E = \frac{\sigma}{\varepsilon_0}$$

So, the field between the two planes is uniform *and independent of their separation*. This gives us a way to produce uniform electric field in a limited region of space.

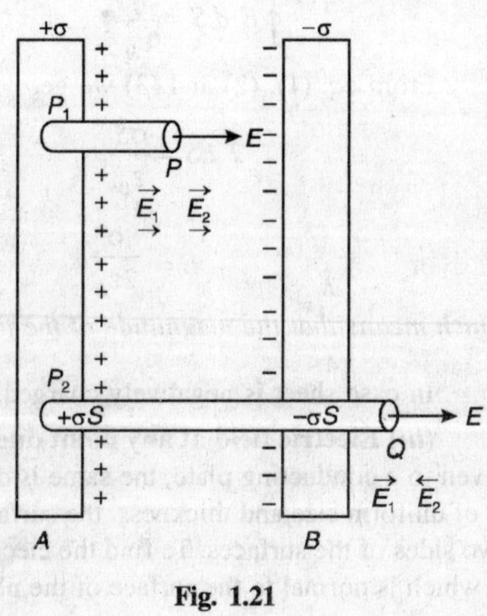

Fig. 1.21

Again let Q be a point *lying outside* the planes.

Draw a cylinder of area of cross-section S through q with one its face lying in plane A. The surface of cylinder which contributes towards the flux is the *end face passing through the point Q.*

$$\therefore \quad \oint_S \vec{E}.d\vec{S} = \oint_S \vec{E}.d\vec{S} = ES \quad \text{... (1)}$$

Charge inside Gaussian surface is

$$q = \sigma S - \sigma S = 0 \quad \text{... (2)}$$

From Gauss's law

$$\oint_S \vec{E}.d\vec{S} = \frac{q}{\varepsilon_0} \quad \text{... (3)}$$

From eqns. (1), (2) and (3), we get

$$ES = 0 \quad \text{or} \quad E = 0$$

Which shows that the electric field intensity due to two oppositely charged infinite plates is σ/ε_0 at any point between the plates and is *zero* for all external points.

(*v*) **Electric field due to a uniformly charged spherical shell.** Consider a thin spherical shell of radius R and centre O. Let q be charge on the spherical shell. The electric field due to charged spherical shell is radial and is spherically symmetric. Let us find electric field at point P at a distance r from the centre of the spherical shell.

(*a*) *When point P is inside* the spherical shell. Draw a Gaussian surface through the point P and it will be a sphere of radius r and centre of (Fig. 1.22)

Let E be the electric field at P due to the charge q on spherical shell. As is evident the field due to spherical shell will be *spherically symmetric*. Which means that at every point on the surface of spherical shell, the field has same magnitude and is normal to the surface. So, total flux through the Gaussian surface will be

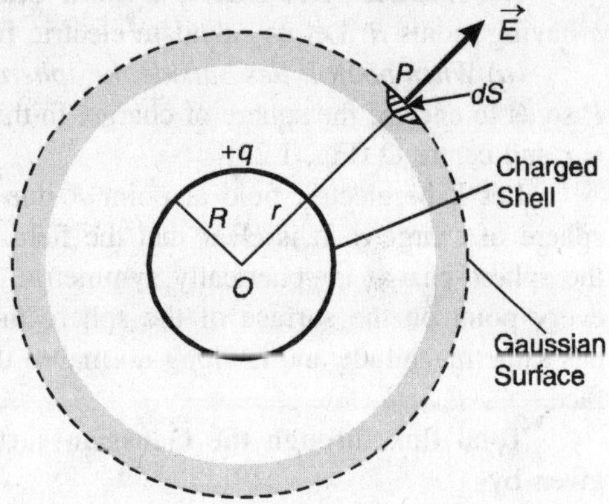

Fig. 1.22

$$\oint_S \vec{E}.d\vec{S} = \oint_S \vec{E}.d\vec{S} = \oint_S \vec{E}.d\vec{S} = E.4\pi r^2$$

Since the charge enclosed by the Gaussian surface is q, according to the Gauss's theorem

$$E.4\pi r^2 = \frac{q}{\varepsilon_0} \qquad (r > R)$$

or

$$E = \frac{1}{4\pi\varepsilon_0}\frac{q}{r^2} \quad \text{... (1)}$$

From equation (1), it follows that for the points outside the charged spherical shell, the shell behaves as though the charge on the shell, were concentrated at its centre. If σ is uniform surface charge, density of the spherical shell, then

$$q = 4\pi R^2 \sigma$$

put q in equation (1), we get,

$$E = \frac{\sigma}{\varepsilon_0} \frac{R^2}{r^2} \quad \text{... (2)}$$

(*b*) *When P is on the surface of spherical shell.* In this case the Gaussian surface through P will just enclose the charged spherical shell. So from Gauss's law we have

$$E.4\pi R^2 = \frac{q}{\varepsilon_0} \qquad [\because r = R]$$

or
$$E = \frac{q}{4\pi\varepsilon_0 R^2} = \frac{1}{4\pi\varepsilon_0} \cdot \frac{q}{R^2} \quad \text{... (3)}$$

Since $q = 4\pi R^2 \sigma$, the eq. (3) will become

$$E = \frac{1}{4\pi\varepsilon_0} \cdot \frac{4\pi R^2 \sigma}{R^2} = \frac{\sigma}{\varepsilon_0} \quad \text{... (4)}$$

(*c*) When *P Lies Inside the Spherical Shell.* In this case, the Gaussian surface through the point P will not enclose any charge and so from Gauss's theorem,

$$E.4\pi r^2 = \frac{0}{\varepsilon_0} = 0$$

$$\therefore \quad E = 0$$

(*vi*) **Electric Field Due to a Solid Charged Sphere.** Consider an isolated sphere of charge q having radius R. Let us calculate electric field at point P at a distance r from its centre O.

(*a*) *When point P lies outside the sphere of charge* : Draw the Gaussian surface through point P so as to enclose the sphere of charge. In this case, the Gaussian surface is a sphere of radius $OP = r$ and centre O (Fig. 1.23)

Let E be electric field at point P due to the sphere of charge q. It is clear that the field due to the sphere charge is spherically symmetric, *i.e.*, at every point on the surface of the sphere the field has same magnitude and is along normal to the surface.

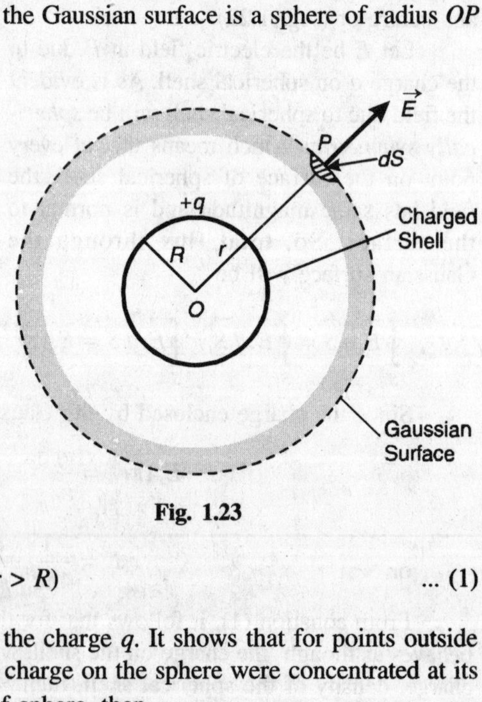

Fig. 1.23

Total flux through the Gaussian surface is given by

$$\oint_S \vec{E}.d\vec{S} = \oint_S E.dS = \oint_S \vec{E}.d\vec{S} = E.4\pi r^2$$

From Gauss's theorem

$$E.4\pi r^2 = \frac{q}{\varepsilon_0}$$

or
$$E = \frac{1}{4\pi\varepsilon_0} \frac{q}{r^2} \quad (r > R) \quad \text{... (1)}$$

which is the same as that a point at distance r from the charge q. It shows that for points outside the sphere of charge, the sphere behaves as if all the charge on the sphere were concentrated at its centre. If ρ be the uniform charge volume density of sphere, then

$$q = \frac{4}{3} \pi R^3 . \rho \quad \text{... (2)}$$

Putting the value of q in eq. (1), we get;

$$E = \frac{1}{4\pi\varepsilon_0} \cdot \frac{4\pi R^3 . \rho}{3r^2}$$

$$= \frac{\rho R^3}{3\varepsilon_0 r^2} \qquad \text{... (3)}$$

(b) When point P lies on the surface of charged sphere. In such a case, the area of the Gaussian surface is equal to surface area of the sphere of charge *i.e.,* equal to $4\pi R^2$. If E electric field at point P on the surface of the sphere of charge then from Gauss's law

$$E.4\pi R^2 = \frac{q}{\varepsilon_0} \qquad (r = R)$$

or

$$E = \frac{1}{4\pi\varepsilon_0} \frac{q}{R^2} \qquad \text{... (4)}$$

Since, $q = \frac{4}{3}\pi R^3 \rho$, the equation (4) will get the form

$$E = \frac{\rho R}{3\varepsilon_0} \qquad \text{... (5)}$$

(c) When point P lies inside the charged sphere. In such a case, Gaussian surface is spherical, whose centre is O and radius is $OP = r$ as shown in Fig. 1.24 If q' is charge enclosed by the Gaussian surface, then we have,

$$E.4.\pi r^2 = \frac{q'}{\varepsilon_0} \qquad (r < R)$$

or

$$E = \frac{1}{4\pi\varepsilon_0} \cdot \frac{q'}{r^2} \qquad \text{... (6)}$$

Now,

$$q' = \frac{q}{\frac{4}{3}\pi R^3} \times \frac{4}{3}\pi r^3 = q\frac{r^3}{R^3}$$

Putting for q' in equation (6), we get

$$E = \frac{1}{4\pi\varepsilon_0} \frac{qr}{R^3} \qquad \text{... (7)}$$

Since, $q = \frac{4}{3}\pi R^3 \rho$, the equation gets the form

$$E = \frac{\rho r}{3\varepsilon_0} \qquad \text{... (8)}$$

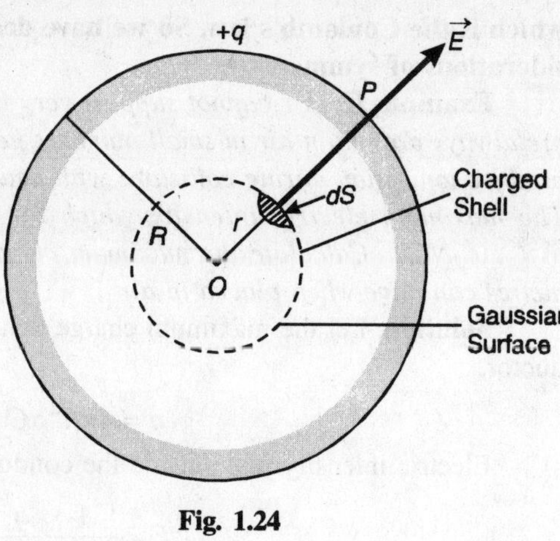

Fig. 1.24

1.25. Gauss's Law and Coulomb's Law

Coulomb's law can be deduced from Gauss's law and symmetry considerations. To do so, let us apply Gauss's law to an isolated point charge q as in Fig. 1.25. Although Gauss's law holds for any surface whatever, information can most readily be extracted for a spherical surface r centred on the charge. The advantage of this surface is that, from symmetry, E must be normal to it and must have the same magnitude for all points on the surface.

In fig. 1.25 both E and dS at any point on the Gaussian surface are directed radially outward. The angle between them is zero and the quantity $E.dS$ become simply EdS. Gauss's law (eq. 4) thus reduced to

$$\varepsilon_0 \oint E.dS = \varepsilon_0 \oint E.dS = q$$

Because E is constant for all points on the sphere it can be taken out of the integral sign, giving us

$$\varepsilon_0 E \oint .dS = q$$

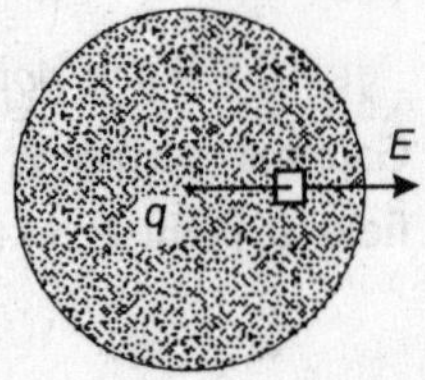

Fig. 1.25

where the integral is simply the area of the sphere* So we get

$$\varepsilon_0 E(4\pi r^2) = q$$

or
$$E = \frac{1}{4\pi\varepsilon_0}\frac{q}{r^2} \quad \text{... (5)}$$

Equation (5) gives the magnitude of the electrical field strength E at any point a distance r from an isolated point charge q. The direction of E is already known from symmetry.

Let us put a second point charge q_0 at the point at which E calculated. The magnitude of the force that acts on it is given by

$$F = Eq_0$$

Combining with Eq. (5) we get

$$F = \frac{1}{4\pi\varepsilon_0}\frac{qq_0}{r^2} \quad \text{... (6)}$$

which is the Coulomb's law. So we have deduced Coulomb's law from Gauss's law and considerations of symmetry.

Example 13. *Air cannot support very intense electric fields. The charged particles, which are always present in air in small numbers get accelerated to very high velocities. These, in turn, produce ionization during collisions with neutral molecules. Due to this air becomes conducting. The maximum electric intensity which air can withstand at atmospheric pressure is about* 3×10^6 NC^{-1}. *Calculate the maximum charge density which a metallic conductor of radius R metres can have when placed in air.*

Solution. Let the maximum charge density be σ C m^{-2}. Therefore, total charge on the conductor,

$$q = 4\pi R^2 \sigma C$$

Electric intensity just outside the conductor's surface;

$$E = \frac{1}{4\pi\varepsilon_0}\frac{q}{R^2}$$

$$= \frac{1}{4\pi\varepsilon_0}\frac{4\pi R^2\sigma}{R^2}$$

$$= \frac{\sigma}{\varepsilon_0}$$

As the maximum value of E can be 3×10^6 NC^{-1},

$$3 \times 10^6 \text{ NC}^{-1} = \frac{1}{8.85 \times 10^{-12} \text{C}^2\text{N}^{-1}\text{m}^{-2}} \times \sigma \text{Cm}^{-2}$$

or, $$\sigma = 3\times10^6 \times 8.85\times10^{-12}$$
$$= 26.6 \times 10^{-6}$$

$\therefore$ maximum charge density = $\mathbf{26.6 \times 10^{-6}\ C\ m^{-2}}$ **Ans.**

Note that it is independent of the radius of the sphere. In fact this holds for conductors of other shapes also.

Let us assume that the sphere is of copper, and is positively charged. The total number of atoms in the ball is

$$n = \frac{\text{Mass of the ball} \times \text{Avogardro's number}}{\text{molecular weight of copper}}$$

$$= \frac{4}{3}\pi R^3 \frac{\rho}{M} \times N$$

$$= \frac{4}{3}\pi R^3 \times 10^6\,\text{cm}^3 \times 8.95\ \text{g cm}^{-3} \times \frac{6.022\times10^{23}\,\text{mole}^{-1}}{63.54\text{gmole}^1}$$

$$= 3.55 \times 10^{29} R^3$$

Maximum charge the sphere can hold is given by

$$q_m = 4\pi R^2 \sigma$$
$$= 4\pi R^2 \times 26.6\times10^{-6} C$$
$$= 3.34\times10^{-4} R^2 C$$

Now, charge on one electron is 1.602×10^{-19} C. Therefore, the number of electrons to be removed from the sphere to charge it to maximum possible value is,

$$n_c = \frac{3.34 \times 10^{-4} R^2}{1.602 \times 10^{-19}}$$
$$= 2.09 \times 10^{15} R^2$$

If each atom of copper contributes one free electron. The fraction of the total free electrons removed is,

$$\frac{n_e}{n} = \frac{2.09\times10^{15} R^2}{3.55\times10^{29} R^3}$$
$$= \frac{5.89}{R}\times10^{-15}$$

If the radius of the sphere is 0.01 m, this fraction becomes 5.89×10^{-18}. Thus, though the sphere is charged to the maximum extent, the number of electrons transferred makes a negligibly small fraction of he total number of free electrons present. This is even less than 1 in 10^{-12}.

Example 14. *Two parallel plates of area 20 cm² each are separated by a distance of 10 mm. If one plate charies a charge + 2 × 10⁻⁹ C, and the other carries a charge – 2 × 10⁻⁹ C, calculate the electric field in the space between them.*

Solution. Surface charge density $= \frac{2\times10^{-9}}{20\times10^{4}} \text{C m}^{-2}$

$$= 10^{-6}\ \text{C m}^{-2}$$

$\therefore$ The electric field $= \frac{10^{-6}\text{Cm}^{-2}}{8.85\times10^{-12}\text{C}^2\text{N}^{-1}\text{m}^{-2}}$

$$= \mathbf{1.13 \times 10^5\ NC^{-1}}$$ **Ans.**

1.26. Mechanical Force at the Surface of a Charged Conductor — Electrostatic Pressure

Let us take a conductor of any shape and concentrate on any one point P (fig. 1.26). Let the charge density around P be σ (σ will vary from point to point at the surface of the conductor). The electric field just outside P will be normal to the surface of the conductor. If it is not so, *i.e.*, E is inclined just near the surface, then it will have a component along the surface of the conductor, which will make the charges move and redistribute themselves. As along as the charges are stationary, E will not have any component along the surface and, hence, will be normal to it. Let us draw a Gaussian surface as shown in Fig. 1.26 (*a*). The only contribution to the electric flux would be through the end cap outside the conductor. So we have.

$$ES = \sigma S / \varepsilon_0$$

or,
$$E = \sigma / \varepsilon_0 \quad \text{... (1)}$$

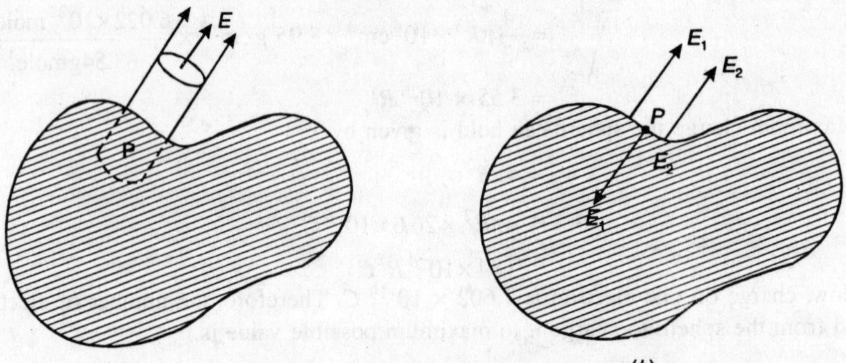

Fig. 1.26

From eqn. (1) we can conclude that this is a general result which holds good for conductors of any shape and under any circumstances.

Now, let us separate this small area S around P from the rest of the conductor and calculate the force on it. It should be clearly understood that when we have to calculate the force on a charge placed along with other charges, the electric field at the site of the charge under consideration should be calculated by excluding that charge from the assembly. Now we have two systems of the charges—charges around P, and charges at the rest of the conductor.

Just outside P, let the field due to charges around P be E_1 and that due to rest of the charges be E_2 [Fig. 1.26 (*b*)]. Thus :

$$E = E_1 + E_2 \quad \text{... (2)}$$

Inside the conductor, total electric field is zero. Therefore,

$$0 = E_1 - E_2 \quad \text{... (3)}$$

From Eqs. (2) and (3), we get,

$$E_2 = E_1 = E/2 = \frac{\sigma}{2\varepsilon_0} \quad \text{... (4)}$$

$\therefore$ Force on charge around $P = \sigma S \times \sigma / 2\varepsilon_0$

or, Force per unit area $= \sigma^2 . S / 2\varepsilon_0$ ... (5)

Thus every part of the surface is acted upon by a force directed outwards. The surface is, therefore, under outward electric pressure.

So the mechanical pressure on the surface will be given by

$$P = \frac{\sigma^2}{2\varepsilon_0}$$

Also $$E = \frac{\sigma}{\varepsilon_0} \quad \therefore \quad \sigma = \varepsilon_0 . E$$

$$\therefore \quad P = \frac{(\varepsilon_0 E)^2}{2\varepsilon_0} = \frac{\varepsilon_0 E^2}{2}$$

Mechanical Energy/unit volume :

The force on unit surface area of a conductor is $P = \frac{\varepsilon_0 E^2}{2}$. If due to this force, the conductor of surface area S moves out through a distance dx, then work done is given by

$$W = (PS).dx = \left(\frac{\varepsilon_0 E^2}{2}\right) S\,dx$$

$\therefore$ The energy stored/unit volume is given by

$$U_v = \frac{W}{S.dx} = \frac{\varepsilon_0 E^2}{2} \quad \text{... (7)}$$

where U_v is called **energy density**.

Example 15. *A soap bubble of radius R has been given a charge q. Find out the excess pressure inside the bubble.*

Solution. The pressure inside a soap bubble of radius R is greater than outside by an amount $p = 4T/R$, where T is the surface tension of the solution. But if the dubble carries a charge q, $\sigma = q / 4\pi R^2$, and there is an additional outward pressure of $\sigma^2 / 2\varepsilon_0$ or $q^2 / 32\pi^2\varepsilon_0 R^4$. This helps the pressure excess to stretch out the bubble. Therefore,

$$P + \frac{q^2}{32\pi^2\varepsilon_0 R^4} = \frac{4T}{R} \quad \text{or} \quad P = \frac{4}{3}\left[T - \frac{q^2}{128\pi^2\varepsilon_0 R^3}\right]$$

Thus the charge, in effect, reduces surface tension. A similar treatment can be applied to a liquid drop. Because due to surface tension effect, the tendency of drops to form on uncharged particles is very small. So if the particle are charged, the *supersaturated vapour would condense on them more readily*. A very important application of this is in the detection of charged particles, *e.g.*, with the help of a *cloud chamber*.

1.27. Electric Potential

The electric field around a charged rod can be described not only by a (vector) electric field strength E but also by a scalar quantity, the *electric potential V*. These quantities are intimately related, and often it is only a matter of convenience which is used in a given problem.

To find the *electric potential difference* between two points, A and B in an electric field, we move a test charge q_0 from A to B, always keeping it in equilibrium and we measure the work W_{AB} that must be done by the agent moving the charge. The electric potential difference is defined from

$$V_B - V_A = \frac{W_{AB}}{q_0} \quad \text{... (1)}$$

The work W_{AB} may be (*a*) positive (*b*) negative, or (*c*) zero. In these cases the electric potential at B will be (*a*) higher, (*b*) lower, or (*c*) the same as the electric potential at A.

The MKS unit of potential difference that follows from Eq. (1) is the joule/coul. The combination occurs so often that a special unit, the *volt*, is used to represent it; that is,

1 volt = 1 joule/coul.

Usually point A is chosen to be a large (taken as ∞) distance from all charges and the electric potential V_A at this infinite distance is arbitrarily taken to be zero. This allows to define the *electric potential* at a point. Putting $V_A = 0$ in Eq. (1) and dropping the subscripts leads to

$$V = \frac{W}{q_0} \quad \text{... (2)}$$

Bearing in mind the assumptions made about the reference position, we see from Eq (2) that V near an isolated positive charge is positive because positive work must be done by an outside agent to push a (positive) test charge in from unfinity. Similarly, the potential near an isolated negative charge is negative, because an outside agent must exert a retaining force (that is, must do negative work on), a (positive) test charge as it comes in from infinity. Electric potential as defined in Eq. (2) is a scalar because W and q_0 in that equation are scalars.

Both W_{AB} and $V_B - V_A$ in Eq. (1) are independent of the path followed in moving the test charge from point A to point B. If this were not so, point B would not have unique electric potential (with respect to point A as a deemed reference position) and the concept of potential would have limited usefulness.

There is a strong analogy between electrostatic forces and gravitational forces, based on the fact that their fundamental laws are inverse square *laws i.e.,*

$$F_E = \frac{1}{4\pi\varepsilon_0}\frac{q_1 q_2}{r^2} \quad \text{and} \quad F_g = G\frac{m_1 m_2}{r^2}$$

Thus we can define the gravitational potential V_g the relation from

$$V_g = \frac{W}{m}$$

where W is the work required to move a test body of mass m from infinity to the point in question. Gravitational equipotential surfaces can also be constructed; which will be everywhere at right angles to the gravitational field strength vector g. For a uniform gravitational field, such as that near the surface of the earth, these surfaces are horizontal planes. This correlates with the facts that no net work is required to move a body of mass m, between two points with the same elevation and the same net work is required to move a body along any path starting on a given horizontal surface and ending on another.

1.28. Potential and Field Strength

Let A and B in Fig. 1.27 be two points in a uniform electric field E, set up by an arrangement of charges not shown, and let A be a distance d in the field direction from B. Assume that a positive test charge q_0 is moved, by an external agent and without acceleration, from A to B along the straight line connecting them.

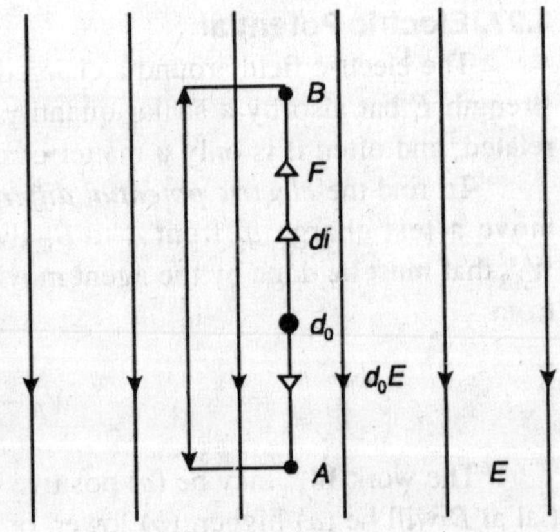

Fig. 1.27

The electric force on the charge is q_0E and points downwards. In order to move the charge in the way we have described we must counteract this force by applying an external force F of the same magnitude but directed upward. The work W done by the agent that supplies this force is

$$W_{AB} = Fd = q_0\, Ed \quad \text{... (3)}$$

Substituting this in Eq. (1) we get

$$V_B - V_A = \frac{W_{AB}}{q_0} = Ed \qquad \text{... (4)}$$

This equation shows the connection between potential difference and field strength for a simple special case. Note from this equation that another M.K.S. Unit for E is the volt/meter.

In fig. 1.28, B has a higher potential than A. This is reasonable because an external agent would have to do positive work to push a positive test charge from A to B. Figure 1.29 would be used as it stands to illustrate the act of lifting a body from A to B in the uniform gravitational field near the earth's surface.

We can find the connection between V and E in the more general case in which the field is *not* uniform and in which the test body is moved along a path that is *not* straight, as in Fig. 1.29. The electric field exerts a force q_0E on the test charge, as shown. To keep the test charge from accelerating, an external agent must apply a force F choosen to be exactly equal to $-q_0E$ for all position of the test body.

If the external agent causes the test body a displacement dl along the path from A to B, the element of work done by the external agent will be Fdl. To find the total work W_{AB} done by the external agent in moving the test charge from A to B, we sum up (integrate) the work contributions path is divided. We get

$$W_{AB} = \int_A^B F.dl = -q_0 \int_A^B E.dl \qquad \text{... (5)}$$

Such an integral is called a line integral. Note that we have substituted

$$F = q_0E$$

Putting the expression for W_{AB} into Eq (1) we get

$$V_B - V_A = \frac{W_{AB}}{q_0} = \int_A^B E.dl \qquad \text{... (6)}$$

If point A is taken to be infinitely distant and the potential V_A at infinity is taken to be zero, this equation gives the potential V at point B and given by

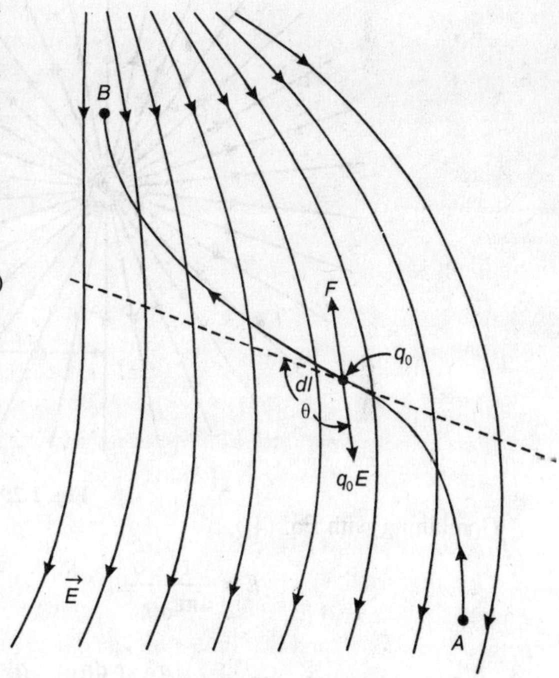

Fig. 1.28

$$V = \int_\infty^B E.dl \qquad \text{... (7)}$$

The equations (6) and (7) allow us to calculate the potential difference between any two points (or the potential at any point) if E is known at various points in the field.

1.29. Potential due to a Point Charge

Figure 1.29 shows two points A and B near an isolated point charge q. For simplicity we assume that A, B, and q lie on a straight line. Let us compute the potential difference between points A and B, assuming that a test charge q_0 is moved without acceleration along a radial line from A to B.

In Fig 1.29 E points to the right and dl, which is always in the direction of motion, points to the left. Therefore, in Eq. 6,

$$Edl = E \cos 180° \, dl = - Edl$$

However, as we move a distance dl to the left, we are moving in the direction of decreasing r because r is measured from q as an origin. So we have

$$dl = - dr$$

Combining these we get

$$E.dl = E \, dr$$

Substituting this in Eq. (6) we have

$$V_B - V_A = -\int_A^B E.dl = -\int_{r_A}^{r_B} E.dl$$

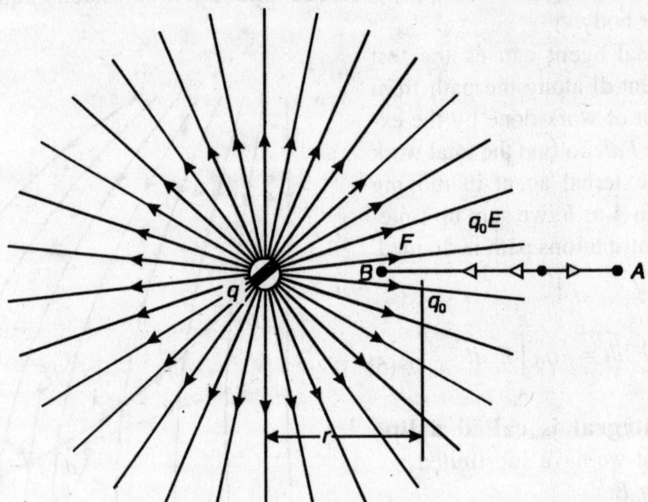

Fig. 1.29

Combining with Eq. (4)

$$E = \frac{1}{4\pi\varepsilon_0} \frac{q}{r^2}$$

we have $$V_B - V_A = -\frac{q}{4\pi\varepsilon_0} \int_{r_A}^{r_B} \frac{dr}{r^2} = \frac{q}{4\pi\varepsilon_0} \left(\frac{1}{r_B} - \frac{1}{r_A} \right) \quad \text{... (8)}$$

Choosing reference position A to be at infinity (that is, letting $r_A \to \infty$). Choosing $V_A = 0$ at this position, and dropping the subscript B we get

$$V = \frac{1}{4\pi\varepsilon_0} \frac{q}{r} \quad \text{... (9)}$$

This equation shows clearly that *equipotential surfaces for an isolated point charge are spheres concentric with the point change.* A study of the derivation will show that this relation also holds for points external to spherically symmetric charge distributions.

Example 16. *What is the electric potential at the surface of a gold nucleus? The radius is 6.6×10^{-15} meter and the atomic number Z = 79.*

Solution. The nucleus, assumed spherically symmetrical, behaves electrically for external points as if it were a point charge. Thus we can use Eq. 9 or, recalling that the proton charge is 1.6 $\times 10^{-19}$ coul.

$$V = \frac{1}{4\pi\varepsilon_0}\frac{q}{r} = \frac{(9.0\times10^9 \text{ nt} - \text{m}^2/\text{coul}^2)(79)(1.6\times10^{-19}\text{ coul})}{6.6\times10^{-15}\text{ meter}}$$

$$V = \mathbf{1.7 \times 10^7} \textbf{ volts. Ans.}$$

1.30 A Group of Point Charges

The potential at any point due to a group of point charges is found by (*i*) calculating the potential V_n due to each charge, as if the other charges were not present, and (*ii*) adding the quantities so obtained, we may write eq. (9) as,

$$V = \sum_n V_n = \frac{1}{4\pi\varepsilon_0}\sum_n \frac{q_n}{r_n} \qquad \text{... (10)}$$

where q_n is the value of the nth charge and r_n is the distance of this charge from the point in question. The sum used to calculate V is an *algebraic sum* and not a vector sum like the one used to calculate E for a group of point charges (eq. 6), which gives an *advantage of using potential over electric field strength field strength for making calculations.*

If the charge distribution is continuous, rather than being a collection of points, the sum in Eq. 10 is replaced by an integral, and given by

$$V = \int dV = \frac{1}{4\pi\varepsilon_0}\int \frac{dq}{r} \qquad \text{... (11)}$$

where dq is a differential element of the charge distribution, r is its distance from the point at which V is to be calculated, and dV is the potential it establishes at that point.

Example 17. *What is the potential at the centre of the square of Fig. 1.30? Assume that $q_1 = +1.0 \times 10^{-8}$ coul, $q_2 = -2.0 \times 10^{-8}$ coul, $q_3 = +3.0 \times 10^{-8}$ coul, $q_4 = +2.0 \times 10^{-8}$ coul, and $a = 1.0$ metre.*

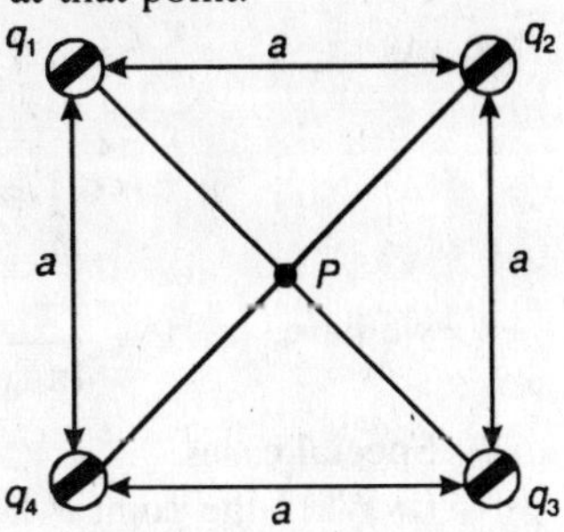

Fig. 1.30

Solution. The distance r each charge from P is $a/\sqrt{2}$ or 0.71 metre. from Eq. 10.

$$V = \sum_n V_n = \frac{1}{4\pi\varepsilon_0}\frac{q_1+q_2+q_3+q_4}{r}$$

$$= \frac{(9.0\times10^9 \text{N}-\text{m}^2/\text{C}^2)(1.0-2.0+3.0+2.0)\times10^{-8}\text{C}}{0.71\text{m}}$$

= **500 volts Ans.**

1.31. Potential Due to an Electric Dipole

Let AB be an electric dipole of length $2a$ and let P be any point where $OP = r$ and let θ be the angle between r and the dipole axis.

In ΔOAC, $\qquad \cos\theta = \frac{OC}{OA} = \frac{OC}{a}$

$\therefore \qquad OC = a\cos\theta$

similarly, $\qquad OD = a\cos\theta$

if $\qquad r >> a,$

then $PA = PC = PO + OC = r + a\cos\theta$

and $PB = PD = PO - OD = r - a\cos\theta$

If V be the potential at P due to electric dipole then

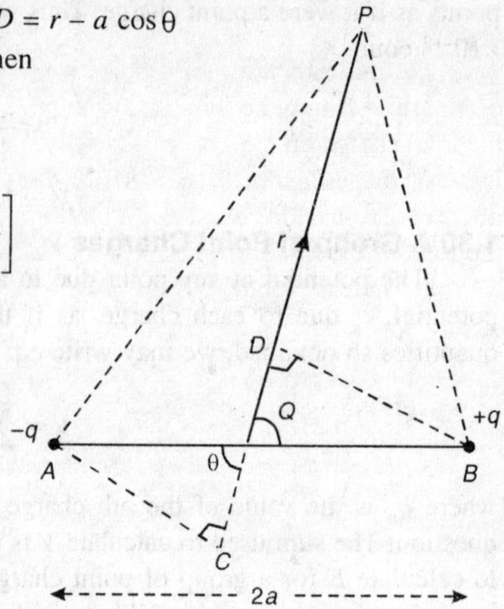

Fig. 1.31

$$V = \frac{1}{4\pi\varepsilon_0}\left(\frac{q}{PB} - \frac{q}{PA}\right)$$

$$= \frac{1}{4\pi\varepsilon_0}\left[\frac{1}{(r - a\cos\theta)} - \frac{1}{(r + a\cos\theta)}\right]$$

$$= \frac{q}{4\pi\varepsilon_0}\left[\frac{r + a\cos\theta - r + a\cos\theta}{r^2 - a^2\cos^2\theta}\right]$$

$$= \frac{q}{4\pi\varepsilon_0}\left[\frac{2a\cos\theta}{r^2 - a^2\cos^2\theta}\right]$$

$$= \frac{1}{4\pi\varepsilon_0}\frac{2qa\cos\theta}{r^2 - a^2\cos^2\theta}$$

$$= \frac{1}{4\pi\varepsilon_0}\frac{p\cos\theta}{r^2 - a^2\cos^2\theta}$$

where $p = 2qa$ = dipole moment of electric dipole

$$\therefore \quad V = \frac{1}{4\pi\varepsilon_0}.\frac{p\cos\theta}{r^2\left(1 - \frac{a^2}{r^2}\cos\theta\right)}$$

As $r >> a$

$\therefore$ $\frac{a}{r} << 1$, so neglect $\frac{a^2}{r^2}\cos^2\theta$ as compared to 1,

we have, $$V = \frac{1}{4\pi\varepsilon_0}.\frac{p\cos\theta}{r^2} \quad \text{... (1)}$$

Special cases

(*i*) When the point P lies on the axial line of the dipole on the side of positive charge,

$\theta = 0, \quad \therefore \cos\theta = 1$

So from eq. (1)

$$V = \frac{1}{4\pi\varepsilon_0}.\frac{p}{r^2} \quad \text{... (2)}$$

(*ii*) When the point P lies on the axial line of the dipole but on the side of negative charge, $\theta = 180°$ or $\cos\theta = -1$

$$\therefore \quad V = \frac{1}{4\pi\varepsilon_0}.\frac{p}{r^2} \quad \text{... (3)}$$

(*iii*) When the point P lies on the equitorial line of the dipole, $\theta = 90°$ or $\cos\theta = 0$

$$\therefore \quad V = 0 \quad \text{...(4)}$$

which means that no work is done in bringing a charge from infinity to the dipole along the perpendicular bisector of the dipole.

1.32. Electric Quadrupole

An electric quadrupole consists of *two electric dipoles placed end to end along the same line.*

OA and *OB* are two dipoles each of length *a*. It may be noted that though the total charge on the system as a whole is zero, *the potential and intensity are not zero*. Let us consider a point *P* along the axis of the quadrupole

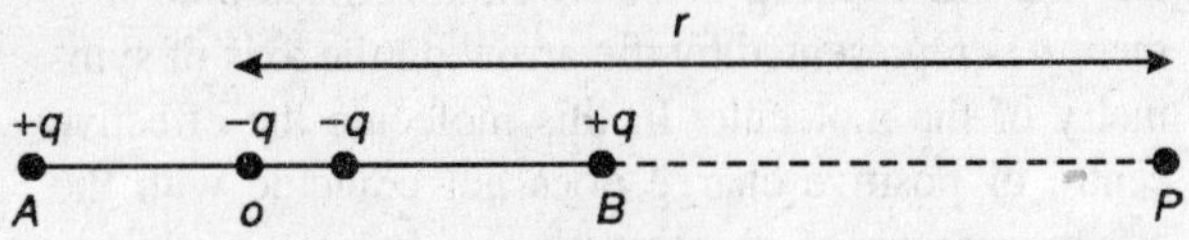

Fig. 1.32

It is clear from the Fig 1.32 (*b*) the potential at the point *P* is given by,

$$V = \frac{1}{4\pi\varepsilon_0}\left[\frac{q}{(r-a)}\frac{q}{r}\frac{q}{r}+\frac{q}{r+a}\right]$$

$$= \frac{q}{4\pi\varepsilon_0}\left[\frac{r(r+a)-2(r^2-a^2)+r(r-a)}{r(r^2-a^2)}\right]$$

$$= \frac{q}{4\pi\varepsilon_0}\left[\frac{r^2+ra-2r^2+2a^2+r^2-ra}{r(r^2-a^2)}\right]$$

$$= \frac{q}{4\pi\varepsilon_0}\cdot\frac{2a^2}{r(r^2-a^2)}$$

$$= \frac{1}{4\pi\varepsilon_0}\cdot\frac{2qa^2}{r(r^2-a^2)}$$

Now $Q = 2qa^2$, the quadrupole moment of charge distribution

$$\therefore \qquad V = \frac{1}{4\pi\varepsilon_0}\cdot\frac{Q}{r(r^2-a^2)} = \frac{1}{4\pi\varepsilon_0}\cdot\frac{Q}{r^3(1-a^2/r^2)}$$

If $r >> a$, $\frac{q}{r} << 1$

$\therefore$ $\frac{a^2}{r^2}$ can be neglected as compared to 1

$$\therefore \qquad V = \frac{1}{4\pi\varepsilon_0}\cdot\frac{Q}{r^3} \qquad \text{... (5)}$$

So in case of

(*i*) Monopole (single charge)

$$V \propto \frac{1}{r}$$

(*ii*) Dipole

$$V \propto \frac{1}{r^2}$$

(*iii*) Quadrupole

$$V \propto \frac{1}{r^3}$$

Many molecules have electric dipole moments. That for H_2O in its vapour state is 6.1×10^{-30} coulum. Figure 1.33 is a representation of this molecule, showing the three nuclei and the surrounding electron cloud. The dipole moment p is represented by the arrow on the axis of symmetry of the molecule. In this molecule the effective centre of positive charge does not coincide with the effective centre of negative charge. It is precisely because of this separation that the dipole moment exists.

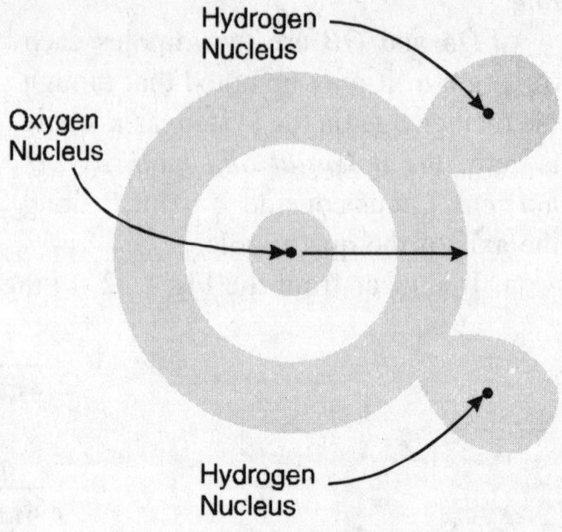

Fig. 1.33

Atoms, and many molecules, *do not* have permanent dipole moments. However, dipole moments may be induced by placing any atom or molecule in an external electric field. The action of the field (Fig. 1.33) is to separate the centres of positive and of negative charge. We say that the atom becomes polarized and acquires an *induced electric dipole moment.* Induced dipole moments disappear when the electric field is removed.

Electric dipoles are important in situations other than atomic and molecular ones. *Radio and radar antennas* are often in the form of a metal wire or rod in which electrons surge back and forth periodically. At a certain time one end of the wire or rod will be negative and the other end positive. Half a cycle later the polarity of the ends is exactly reversed. This is an *oscillating electric dipole*. It is so named because its dipole moment changes in a *periodic way* with time.

1.33 Electrical Potential Energy

If we raise a body from the earth's surface, the work that we do against the earth's gravitational attraction is stored as potential energy in the system (earth + body). If we release the body, the stored potential energy changes steadily into kinetic energy as the body drops. After the body comes to rest on the earth, this kinetic energy and just before the time of contact equal in magnitude to the originally stored potential energy, is transformed into heat energy in the system (earth + body).

A similar situation exists in electrostatics. Consider two charges q_1 and q_2 at a distance r apart, as in fig 1.34. If we increase the separation between them, and external agent must do work that will be positive if the charges are opposite in sign and negative otherwise. The energy represented by this work can be thought of as stored in the system $q_1 + q_2$ as *electric potential energy*. This energy, like all varieties of potential energy, can be transformed into other forms. If q_1 and q_2, in mechanics the concept of *potential energy* (of compressed springs falling masses, etc.) is more commonly used than than the concept *potential*. In electrostatics the reverse is true, electric potential being perhaps a more common concept than electric potential energy. So a student must be careful not to confuse these two different quantities, potential and potential energy for example, are charge of opposite sign and we release them, they will accelerate toward each other, transforming the stored potential energy into kinetic energy of the accelerating masses. The analogy to the earth + the body system is exact, except for the fact that *electric forces may be either attractive or repulsive, whereas gravitational forces are always attractive*.

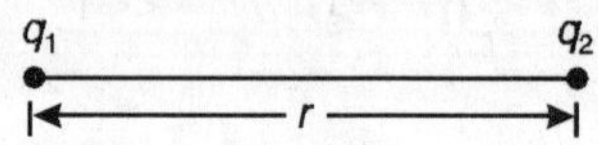

Fig. 1.34

We define the electric potential energy of a system of point charges as *the work required to assemble this system of charges by bringing them in from an infinite distance. We assume that the charges are all at rest, when they are infinitely separated, that is, they have no initial kinetic energy.*

In fig. 1.34 let us imagine q_2 removed to infinity and at rest. The *electric* potential at the original site of q_2, caused by q_1, is given by Eq. (9),

or $$V = \frac{1}{4\pi\varepsilon_0} \cdot \frac{q_1}{r}$$

If q_2 is moved in from infinity to the original distance r, the work required is, from the definition of electric potential, that is, from Eq. (2),

$$W = Vq_2$$

Combining these two equations and knowing that this work W is precisely the *electric* potential energy U of the system $q_1 + q_2$, we get

$$U = (W) = \frac{1}{4\pi\varepsilon_0} \cdot \frac{q_1 q_2}{r_{12}} \qquad \text{... (6)}$$

The subscript or r emphasizes that the distance involved is that between the point charge q_1 and q_2.

For systems containing more than two charges the procedure is to compute the potential energy for every pair of charges separately and to add the results algebraically. This procedure rests on a physical picture in which (*a*) charge q_1 is brought into position, (*b*) q_2 is brought from infinity to its position near q_1, (*c*) q_2 is brought from infinity to its position near q_1 and q_2 etc.

Example 18. *Three charges are arranged as in fig. 1.35 What is their mutual potential energy? Assume that q = 1.0 × 10⁻⁷ coul and a = 10 cm.*

Solution. The total energy of the configuration is the sum of the energies of each pair of particles and given by

$$U = U_{12} + U_{12} + U_{23}$$

$$= \frac{1}{4\pi\varepsilon_0}\left[\frac{(+q)(-4q)}{a} + \frac{(+q)(+2a)}{a} + \frac{(-4q)(+2q)}{a}\right]$$

$$= -\frac{10}{4\pi\varepsilon_0}\frac{q^2}{a}$$

$$= -\frac{(9.0 \times 10^9\,\text{N} - \text{m}^2/\text{coul}^2)(10)(1.0 \times 10^{-7}\,\text{coul})^2}{0.10\text{metre}}$$

$= \mathbf{-9.0 \times 10^{-3}}$ **joule. Ans.**

(The fact that the total energy is negative means that negative work would have to be done to assemble this structure, starting with the three charges separated and at rest at infinity. Expressed otherwise, 9.0×10^{-3} joule of work must be done to dismantle this structure, i.e. removing the charges to an infinite separation from one another.)

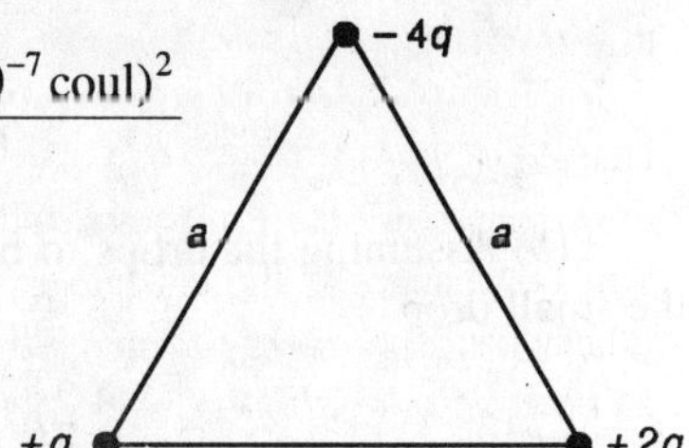

Fig. 1.35

Example 19. *The protons in a nucleus of U^{238} are 6.0×10^{-15} metre apart. What is their mutual electric potential energy ?*

Solution. we have the relation

$$U = \frac{1}{4\pi\varepsilon_0}\frac{q_1 q_2}{r} = \frac{(9.0 \times 10^9\,\text{N} - \text{m}^2/\text{coul}^2)(1.6 \times 10^{-19}\,\text{coul})^2}{6.0 \times 10^{-15}\,\text{metre}}$$

$= \mathbf{3.8 \times 10^{-14}}$ **joule = $\mathbf{2.4 \times 10^5}$ eV. Ans.**

Example 20. *Calculate (a) the electric potential established by the nucleus of a hydrogen atom at the mean distance of the circulating electron (r = 5.3 × 10⁻¹⁹ metre) (b) the electric potential energy of the atom when the electron is at its radius and (c) the kinetic energy of the* *electron assuming it to be moving in a circular orbit of this radius centred on the nucleus. (d) How much energy is required to ionize the hydrogen atom ? Express all energies in electron volts.*

Solution. (*a*) We have the relation for electric potential as

$$V = \frac{q}{4\pi\varepsilon_0} = \frac{9\times10^9 \text{N}-\text{m}^2/\text{c}^2}{5.3\times10^{-11}\text{m}}\times(1.6\times10^{-19}\text{C})$$

$= \mathbf{27.1\ volts.\ Ans.}$

(*b*) Electric potential energy of the atom is given by

$U = eV = -27.1$ eV. **Ans.**

(*c*) From the relation of electrostatic force balancing centriopetal force, we have

$$\frac{mv^2}{r} = \frac{e^2}{4\pi\varepsilon_0 r^2}$$

or $$mv^2 = \frac{e^2}{4\pi\varepsilon_0 r} = 27.1 \text{ eV}$$

$\therefore$ $$K.E. = \frac{1}{2}nv^2 = \frac{27.1}{2} = 13.55 \text{ eV. } \textbf{Ans.}$$

(*d*) Total Energy = Kinetic Energy + Potential Energy

= (13.55 – 27.1) eV

= **– 13.55 eV Ans.**

which is the energy required to ionize hydrogen atom.

Example 21. *A spherical drop of water carrying a charge of* 3×10^{-6} *coul has a potential of 500 volts at its surface. (a) What is the radius of the drop? (b) If two such drops of the same charge and radius combine to form a single spherical drop. What is the potential at the surface of the new drop formed ?*

Solution. (*a*) We have the relation

$$V = \frac{q}{4\pi\varepsilon_0 r}$$

$\therefore$ $$r = \frac{q}{4\pi\varepsilon_0 V} = \frac{(9\times10^9 \text{N}-\text{m}^2)}{500\text{volts}}\times3\times10^{-6}\text{C}$$

(*b*) Assuming the drops to be incompressible, the volume of the new drop will be twice of the small drop

i.e., $$\frac{4}{3}\pi R^3 = 2\times\frac{4}{3}\pi r^3$$

$\therefore$ Radius of new drop $R = 2^{1/3}\, r$

and charge of new drop $Q = 2q$

$\therefore$ potential at the surface of new drop will be

$$V_0 = \frac{Q}{4\pi\varepsilon_0 R} = \frac{2q}{4\pi\varepsilon_0 2^{1/3} r} = \frac{2^{2/3} q}{4\pi\varepsilon_0 r}$$

$= 2^{2/3}\, V$

$\therefore$ $V_0 = 2^{2/3} \times 500 =$ **790 volts. Ans.**

Example 22. *If the earth had a net charge equivalent to 1 electron/m² of surface area (a) What would the earth's potential be ? (b) What would the electric field due to the Earth be just outside its surface ? (Earth's radius =* 6.4×10^6 *metres).*

Solution. (*a*) Total charge on the Earth's surface

$$q = (-e)4\pi r^2$$

Where $-e$ is the electron charge and r is the radius of the Earth. The potential is given by

$$V = \frac{q}{4\pi\varepsilon_0 r} = -\frac{4\pi r^2 .e}{4\pi\varepsilon_0 r} = \frac{er}{\varepsilon_0}$$

$$= -\frac{1.6\times10^{-19}\text{C}\times6.4\times10^{-6}\text{m}}{8.9\times10^{-12}\text{C/N}-\text{m}^2}$$

$$= -\mathbf{0.115\ volt.\ Ans.}$$

(*b*) Electric field due to Earth just outside its surface is given by

$$E = \frac{q}{4\pi\varepsilon_0 r^2} = \frac{4\pi r^2 .e}{4\pi\varepsilon_0 r^2} = -\frac{e}{\varepsilon_0}$$

$$= -\frac{1.6\times10^{-19}\text{C}}{8.9\times10^{-12}\text{C}^2/\text{N}} = \mathbf{1.8 \times 10^{-6}\ N\ \ Ans.}$$

Negative sign shows that the direction of the field is radially inward.

Example 23. *Two metal spheres are 3.0 cm in radius and carry charges of + 1.0 × 10*$^{-8}$ *coul and –3.0 × 10*$^{-8}$ *could respectively, assumed to be uniformily distributed. If their centres are 2.0 metres apart calculate (a) the potential of the point hal way between their centres and (b) the potential of each sphere.*

Solution. The charges are assumed to be located at the centres of each sphere. The midway point will be 1.0 metre from either centres and the potential at mid point is given by

$$V = \frac{q_1}{4\pi\varepsilon_0 r_1} + \frac{q_2}{4\pi\varepsilon_0 r_2}$$

But in this case $r_1 = r_2 = r_0$ (say)

$$\therefore \qquad V = \frac{1}{4\pi\varepsilon_0 r_0}[q_1 + q_2]$$

$$= \frac{(9\times19^9\text{N}-\text{m}^2/\text{c}^2)(1\times10^{-8}-3\times10^5)}{(1.0\text{m})}$$

$$= -\mathbf{180\ volts.\ \ Ans.}$$

(*b*) The potential at the surface of sphere 1 is due to its own charge q_1 plus that of charge q_2 of sphere 2 acting at a distance r.

$$\therefore \qquad V = \frac{q_1}{4\pi\varepsilon_0 R} + \frac{q_2}{4\pi\varepsilon_0 r}$$

$$= (9\times10^9\text{N}-\text{m}^2/\text{c}^2)\left[\frac{1\times10^{-8}\text{C}}{0.03\text{m}} - \frac{3\times10^{-8}\text{C}}{2.0\text{m}}\right]$$

$$= \mathbf{2.865\ volts.\ \ Ans.}$$

The potential at the surface of sphere 2 is due to its own charge q_2 plus that of q_1 of sphere 1 acting at a distance r.

$$V = \frac{q_2}{4\pi\varepsilon_0 R} + \frac{q_1}{4\pi\varepsilon_0 r}$$

$$= (9\times10^9\text{N}-\text{m}^2/\text{c}^2)\left[\frac{-3\times10^{-8}\text{C}}{0.03\text{m}} + \frac{1\times10^{-8}\text{C}}{2}\text{m}\right]$$

$$= -\mathbf{8955\ volts.\ \ Ans.}$$

Example 24. *What is the potential gradient in volts/meter at a distance of 10^{-12} meter from the centre of the gold nucleus ? What is the gradient at the nuclear surface ?*

Solution. Potential gradient at a distance r is given by the relation

$$\frac{q}{4\pi\varepsilon_0 r^2} = \frac{79\times1.6\times10^{-19}\,\text{C}}{(10^{-12}\,\text{m})^2}\times9\times10^9\,\text{N}-\text{m}/\text{c}^2$$

$$= 1.14 \times 10^{17} \text{ volts/meter}$$

The potential gradient at the surface of gold nucleus of radius $R = 5 \times 10^{-15}$ is

$$= \frac{q}{4\pi\varepsilon_0 R^2} = \frac{79\times1.6\times10^{-19}\,\text{C}}{(5\times10^{-15}\,\text{m})^2}\times9\;10^9\,\text{N}-\text{m}/\text{c}^2$$

$$= \mathbf{4.55 \times 10^{21}\ V/m}$$ **Ans.**

Example 25. *A sphere of radius 10 cm is charged to a potential of 10 e.s.u. Calculate the outward pull per unit area.*

Solution. $V = 10$ e.s.u., $r = 10$ cm, $V = \frac{q}{r}$

$\therefore$ $q = V, r = 100$ e.s.u.

Also $E = \frac{q}{r^2} = \frac{100}{10^2} = 1$ unit

Force/unit area $= \frac{KE^2}{8\pi}$ in C.G.S. units

here, $K = 1, E = 1$

$$F = \frac{1\times1^2}{8\pi} = \frac{1}{8\pi} = \mathbf{0.0397\ dynes/cm^2}$$ **Ans.**

Example 26. *Surface density of charge of a conductor in air is 4 stat coulomb/cm². Calculate the outwards force/unit area on the conductor in (i) C.G.S. units (ii) M.K.S. units.*

Solution. (*i*) In C.G.S. units

$$F = \frac{2\pi\sigma^2}{K}$$

$\sigma = 4$ stat coulomb/cm², $K = 1$

$\therefore$ $$F = \frac{2\pi\times4^2}{1} = \mathbf{100.57\ dynes/cm^2}.$$ **Ans.**

(*ii*) In M.K.S. or S.I. Units

$$F = \frac{\sigma^2}{2\varepsilon_r\varepsilon_0}$$

Here, $$\sigma = \frac{4\times10^4}{3\times10^9} = 1.33\times10^{-5} \text{ coulomb/m}^2$$

$\varepsilon_r = 1$ for air and $\varepsilon_0 = 8.85 \times 10^{-12}$ coul²/newton-m²

$\therefore$ $$\sigma = \frac{(1.33\times10^{-5})^2}{2\times1\times8.85\times10^{-12}}$$

$$= \mathbf{10.057\ Newton/m^2}.$$ **Ans.**

Example 27. *Protons in the cosmic rays strike the earth's upper atmosphere at a rate, averaged over the earth's surface of 0.15 protons/cm-sec. What total current does the earth receive beyond its atmosphere in the form of incident cosmic ray protons ? The earth's radius is 9.4 × 10⁴ metres.*

Solution. No. of protons/m² falling over earth's surface

$$n = 0.15 \times 10^4 = 1500$$

No. of protons received by the total earth's surface having radius R

$$N = 4\pi R^2 n$$

$$= (4\pi)(6.4\times10^6)^2 \times 1500$$

$$= 7.72 \times 10^{17}/\text{sec}$$

Now each proton has a charge $q = 1.6 \times 10^{-19}$ Coulomb, so total current received by earth.

$$t = N\,q = (7.72 \times 10^{17} \times 1.6 \times 10^{-19}) \text{ amp.}$$

$$= \mathbf{0.1235\ amp} \quad \textbf{Ans.}$$

Example 28. *Each of two small sphares is charged positively, the combined charge being 5.0 × 10⁻⁵ coul. If each sphere is repelled from other by a force 1.0 newton when spheres are 2.0 metres apart, how is the total charge distributed between the spheres.*

Solution. Let the two charges be q and $(5 \times 10^{-5} - q)$ respectively.

$$\therefore \qquad F = \frac{q_1 \times q_2}{4\pi\varepsilon_0 r^2}$$

or

$$1\text{N} = \frac{q\times(5\times10^{-5} - q)\times(9\times10^9\,\text{N}-\text{m}^2/\text{coul}^2)}{(2\ \text{metre})^2}$$

or

$$= 9 \times 10^{-9} q^2 - 45 \times 10^4\, q + 4 = 0$$

Which is a quadratic equation and solving it we get

$$q_1 = \mathbf{1.2 \times 10^{-5}\ coul}$$

and

$$q_2 = \mathbf{3.8 \times 10^{-5}\ coul.} \quad \textbf{Ans.}$$

Example 29. *A thin walled metal sphere has a radius 25 cm and carries a charge of 2.0 × 10⁻⁷ coul. find E for a point (a) inside the sphere (b) just outside the sphere (c) 3.0 metres from the centre of the sphere.*

Solution. (*a*) Since no charge is enclosed within the charged metal sphere, so E for a point inside the sphere is zero.

(*b*) for calculating the field just outside the sphere we may consider the total charge to be concentrated at the centre.

$$E = \frac{q}{4\pi\varepsilon_0 r^2} = \frac{(9\times10^9\text{N}-\text{m}^2/\text{coul}^2)\times(2\times10^{-7}\text{coul})}{(0.25\,\text{m})^2}$$

$$E = \frac{q}{4\pi\varepsilon_0 r^2} = \frac{(9\times10^9\ \text{N}-\text{m}^2\text{koul}^2)\times(2\times10^{-7}\ \text{coul})}{3\ \text{m}^2}$$

$$= \mathbf{2.9 \times 10^4\ N/coul.} \quad \textbf{Ans.}$$

Example 30. *An infinite charge sheet has a surface density σ of 1.0 × 10⁻⁷ coulomb/m². How far apart are the equipotential surfaces whose potential differ by 5.0 volts?*

Solution. The field intensity is given by

$$E = \frac{\sigma}{2\varepsilon_0} = \frac{1.0\times10^{-7}\text{coulomb}/\text{m}^2}{2\times8.9\times10^{-12}\text{coulomb}^2/\text{N}-\text{m}^2}$$

$$= \mathbf{5.6 \times 10^3\ N/coulomb}$$

Let the equipotentials be dS metres apart,

∴ we have

$$E.dS = dV$$

or

$$dS = \frac{dV}{E} = \frac{5.9 \text{ volt}}{5.6 \times 10^3 \text{N / coulomb}}$$

$$= 0.89 \times 10^{-4} \text{ metres}$$

$$= \textbf{0.89 mm. Ans.}$$

Example 31. *A charge of 10^{-8} coulomb can be produced by simple rubbing. To what potential would such a charge raises an insulated conducting sphere of 10 cm radius ?*

Solution.

$$V = \frac{E}{4\pi\varepsilon_0 R} = \frac{10^{-8}\text{coulomb} \times 9 \times 10^9 \text{N} - \text{m}^2 / \text{coulomb}^2}{0.1 \text{ m}}$$

$$= \textbf{900 volts Ans.}$$

Example 32. *Consider a point charge with $q = 1.5 \times 10^{-8}$ coulomb (a) What is the radius of equipotential surface having a potential of 30 volts (b) Are surfaces whose potentials differ by a constant amount (say 1.0 volt) evenly spaced in radius?*

Solution. (*a*) Radius can be calculated from the relation

$$R = \frac{q}{4\pi\varepsilon_0 V} = \frac{(1.5 \times 10^{-8} \text{coul}) \times 9 \times 10^9 \text{N} - \text{m}^2 / \text{coulomb}^2}{(30 \text{volts})}$$

$$= \textbf{4.5 metres. Ans.}$$

(*b*) We have the potential given by

$$V = \frac{q}{4\pi\, \varepsilon_0 R}$$

∴

$$dV = -\frac{q}{4\pi\, \varepsilon_0} \frac{dR}{R^2}$$

or

$$dR = -\frac{4\pi\varepsilon_0}{q}.dV.R^2$$

From which it is clear that *dR* depends on *R*, so the surface whose potentials differ by a constant amount are not evenly spaced.

Example 33. *Calculate the potential energy of the charges shown in Fig 1.36*

Solution. Let us assume that charge q_1 is already present. When we bring q_2 to B then,

$$\text{Work done} = \frac{q_1}{4\pi\varepsilon_0 r_{12}} q_2$$

Now, the potential at C due to both q_1 and q_2

$$= \frac{1}{4\pi\varepsilon_0}\left[\frac{q_1}{r_{13}} + \frac{q_2}{r_{23}}\right]$$

Therefore, work done in bringing charge q_3 at C in the presence of both q_1 and q_2

$$= \frac{1}{4\pi\varepsilon_0}\left[\frac{q_1}{r_{13}} + \frac{q_2}{r_{23}}\right] q_3$$

We can also think of it as the sum of the work needed when q_1 is present alone, and that needed when q_2 is present alone. Therefore, total work done in assembling this arrangement of three charges is,

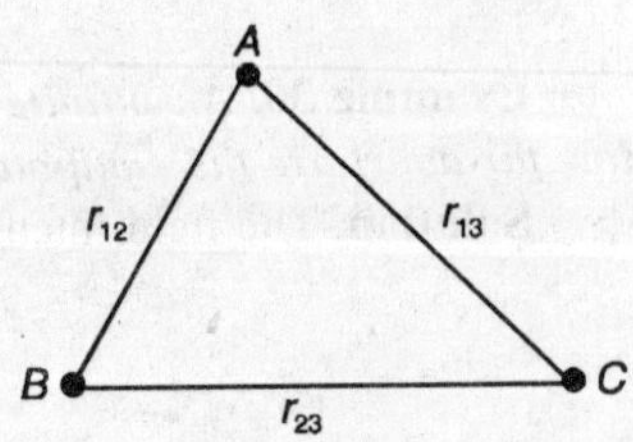

Fig. 1.36

$$W = \frac{1}{4\pi\varepsilon_0}\left[\frac{q_1q_2}{r_{12}} + \frac{q_1q_3}{r_{13}} + \frac{q_2q_3}{r_{23}}\right]$$

Example 34. *Two conducting spheres of radii R_1 and R_2 are at the same potential. Compare their charges, surface charge densities, and electric fields just outside them.*

Solution. Let q_1 and q_2, respectively, be the charges on the spheres 1 and 2. Therefore:

$$\frac{1}{4\pi\varepsilon_0}\frac{q_1}{R_1} = \frac{1}{4\pi\varepsilon_0}\frac{q_2}{R_2}$$

or $$\frac{q_1}{q_2} = \frac{R_1}{R_2}$$

Thus, charges are proportional to the radii. Now surface charge density on sphere 1,

$$\sigma_1 = q_1 / 4\pi R_1^{\,2}$$

Surface charge density on sphere 2,

$$\sigma_2 = q_2\ \pi R_2^{\,2}$$

$$\frac{\sigma_1}{\sigma_2} = \frac{q_1}{q_2} \times \frac{R_2^2}{R_1^2}$$

$$= \frac{R_1}{R_2} \times \frac{R_2^2}{R_1^2} = \frac{R_2}{R_1}$$

Therefore, the surface charge densities are inversely proportional to the radii.

Electric field just outside sphere 1,

$$E_1 = q_1 / 4\pi\varepsilon_0 R_1$$

Electric field just outside sphere 2,

$$E_2 = q_2 \pi\varepsilon_0 R_2$$

$$\therefore \quad \frac{E_1}{E_2} = \frac{R_2}{R_1}$$

Hence, the electric fields just out *side the two sphere as* are also inversely proportional to the radii.

Differential form of Gauss's law

Gauss's law in differential form for continuous large distribution is generally used, when we have to apply Gauss's law in the immediate vicinity of the field of charge. Differential form Gauss's law states that **the divergence of Electric field $\vec{R}$ at any point is space is equal $\frac{1}{\epsilon_0}$ times the volume charge density ρ at that point.**

The above statement is in consonance with the important equation of maxwell's equations.

We can write the above statement mathematically as

$$\text{div } \vec{E} = \frac{\rho}{\epsilon_0}$$

or $$\vec{\nabla}.\vec{E} = \frac{\rho}{\epsilon_0}$$

Where $\rho \longrightarrow$ Volume charge deusily *i.e.*, charge per unit volume

and $\epsilon_0 \longrightarrow$ permittively of free space.

In order to get an expression for Gauss's law in differential form, we can use Gauss's divergence theorem that relates the **volume integral with the surface integral**.

Let us Consider continuous charge distribution equation with varial volume charge density P, Consider a volume V enclosed by surfaces in this region. Let dv be an infinite small volume element enclosed by the surface ds, then from Gauss's law of continues charge distribution.

$$\oint \vec{E}.d\vec{S} = \frac{1}{\epsilon_0}\int_V \rho dV \quad \text{... (1)}$$

Using Gauss's divergence theorem *i.e.*,

$$\oint \vec{E}.d\vec{S} = \int_V (\vec{\nabla}.\vec{E})dV \quad \text{... (2)}$$

Compare eq. (1) and (2), we get

$$\int_V (\vec{\nabla}.\vec{E})dv = \frac{1}{\epsilon_0}\int_V \rho dv$$

or $$\int_V (\vec{\nabla}.\vec{E} - \frac{\rho}{\epsilon_0})dv = 0$$

Since we have considered the volume as arbitrary, so

$$\vec{\nabla}.\vec{E} = \frac{\rho}{\epsilon_0} \text{ or } \vec{\nabla}.\vec{E} = \rho/\epsilon_0 \quad \text{... (3)}$$

Eq. (3) is the differential form Gauss's theorem for continues charge distribution is M.K.S. units.

Physical meaning of the above statement of Gauss's law is that it **relates electric field at a point in space to the charge distribution ρ at that point in space**. In C.G.S. (Gaussian System) Gauss's law is expressed as

$$\vec{\nabla}.\vec{E} = 4\pi\rho \cdot$$

1.34 Poisson's and Laplace's Equations

In electrostatics, Poisson's and Laplace's equations are of fundamental significance. The intensity of electric field $\vec{E}$ and electric potential V (scalar function) at any point in an electric field are related all the equation,

$$\vec{E} = -\text{ grad V} = -\vec{\nabla}V \quad \text{... (1)}$$

where $$\vec{\nabla} = i\frac{\partial}{\partial x} + j\frac{\partial}{\partial y} + k\frac{\partial}{\partial z}$$

For a continuous distribution of charge of volume charge density ρ, we can express the differential form of Gauss's law

$$\text{div } \vec{E} = \frac{\rho}{\varepsilon_0} \quad \text{... (2)}$$

Putting the value of $\vec{E}$ from equation (1) in (2) we have

$$\text{div } (-\text{ grad V}) = \frac{\rho}{\varepsilon_0} \text{ or div grad V} = -\frac{\rho}{\varepsilon_0} \quad \text{... (3)}$$

Now div grad $V = \vec{\nabla}.\vec{\nabla} = \left(i\frac{\nabla}{\partial x}+j\frac{\nabla}{\partial y}+k\frac{\partial}{\partial z}\right).\left(i\frac{\partial}{\partial x}+j\frac{\partial}{\partial y}+k\frac{\partial}{\partial z}\right) = \frac{\partial^2}{\partial x^2}+\frac{\partial^2}{\partial y^2}+\frac{\partial^2}{\partial z^2}$

$$\left(\because \vec{A}.\vec{B} = A_xB_x + A_yB_y + A_zB_z\right)$$

$$= \vec{\nabla}^2$$

∇^2 is called Laplacian operator or just the Laplacian or del squared.

In terms of Laplacian operator equation (3) may be written as

$$\nabla^2 V = -\frac{\rho}{\epsilon_0} \qquad \text{... (4)}$$

The equation (4) is called the *Poisson's equation in M.K.S system. It relates the charge density ρ to the second derivatives of the potential V.*

In C.G.S. Gaussian system, the Poisson's equation may be expressed in the form

$$\nabla^2 V = -4\pi\rho \qquad \text{... (5)}$$

and in cartesian coordinates, Poisson's equation may be written as

$$\frac{\partial^2 V}{\partial x^2}+\frac{\partial^2 V}{\partial y^2}+\frac{\partial^2 V}{\partial z^2} = -\frac{\rho}{\varepsilon_0} \qquad \text{... (6)}$$

1.35 Laplace's Equation

In charge free regions, *i.e.,* regions where no charge exists *i.e.,* the Poisson's equation reduces to Laplace's equation as

$$\nabla^2 V = 0$$

Which may be written as

$$\text{div. grad } V = 0 \qquad \text{...(7)}$$

So we may write Laplace's equation in cartisian co-ordianties

$$\frac{\partial^2 V}{\partial x^2}+\frac{\partial^2 V}{\partial y^2}+\frac{\partial^? V}{\partial z^2} = 0 \qquad \text{... (8)}$$

In spherical polar coordinates (r, θ, ϕ) and cylindrical polar coordinates (r, θ, z) we may express the Poisson's equation as

$$\nabla^2 V = \frac{1}{r^2}\frac{\partial}{\partial r}\left(r^2\frac{\partial V}{\partial r}\right)+\frac{1}{r^2\sin\theta}\frac{\partial}{\partial\theta}\left(\sin\theta\frac{\partial V}{\partial\theta}\right)+\frac{1}{r^2\sin^2\theta}\frac{\partial^2 V}{\partial\phi^2} = -\frac{\rho}{\varepsilon} \qquad \text{... (9)}$$

$$\nabla^2 V = \frac{1}{r}\frac{\partial}{\partial r}\left(r^2\frac{\partial V}{\partial r}\right)+\frac{1}{r^2}\frac{\partial^2 V}{\partial\theta^2}+\frac{\partial^2 V}{\partial z^2} = -\frac{\rho}{\varepsilon} \qquad \text{... (10)}$$

Also in terms of spherical polar coordinates (r, θ, ϕ) and cylindrical polar coordinates (r, θ, z) Laplace's equation may be expressed in the form.

$$\nabla^2 V = \frac{1}{r^2}\frac{\partial}{\partial r}\left(r^2\frac{\partial V}{\partial r}\right)+\frac{1}{r^2\sin\theta}\frac{\partial}{\partial\theta}\left(\sin\frac{\partial V}{\partial\theta}\right)+\frac{1}{r^2\sin^2\theta}-\frac{\partial^2 V}{\partial\phi^2} = 0 \qquad \text{... (11)}$$

and

$$\nabla^2 V = \frac{1}{r}\frac{\partial}{\partial r}\left(r\frac{\partial V}{\partial r}\right)+\frac{1}{r^2}\frac{\partial^2 V}{\partial\theta^2}+\frac{\partial^2 V}{\partial z^2} = 0 \qquad \text{... (12)}$$

We can use Laplace's equation to find potential is free space. Laplace equation can be applied to bodies where volume charge density $\rho = 0$ or where charges are distributed on the surface of conducting bodies.

Laplace equation has universal applications and is used in various branches of physics like *gravitation, heat flow, magnetism and surface tension etc.*

ADDITIONAL PROBLEMS

Example 1. *How much electric flux will come out through a surface, $\vec{S} = 10j$ kept in an electric field,*

$$\vec{E} = 2i + 4j + 7k\ ?$$

(*ii*) *If 1 coulomb charge is placed at the centre of a cube of side 10 cm, calculate the flux coming out of any face of the cube.*

Solution : (*i*) The relation for electric flux coming out through a surface $\vec{S}$ may be written as, $\phi = \vec{E}.\vec{S}$

In this case $\vec{E} = 2i + 4j + 7k$ and $\vec{S} = 10j$

So $\phi = (2i + 4j + 7k).(10j) = 4 \times 10 =$ **40 units. Ans.**

(*ii*) As the charge is situated at the centre of a cube so, the electric flux through each of the face will be equal. Now from Gauss's law

The total flux through entire cube of six faces, $\phi = \dfrac{q}{\varepsilon_0}$

So, the electric flux through each of the faces, $\phi_1 = \dfrac{1}{6}\left(\dfrac{q}{\varepsilon_0}\right) = \dfrac{q}{6\varepsilon_0}$

In this case, $q = 1$ coulomb and $\varepsilon_0 = 8.85 \times 10^{-12}$ Coul.^{2}/(N-m^2)

$$\therefore \quad \phi_1 = \frac{1}{6\varepsilon_0} = \frac{1}{6 \times 8.85 \times 10^{-12}} = \mathbf{1.883 \times 10^{10}\ N\text{-}m^2/coul^2\ Ans.}$$

Example 2. *The potential at a certain distance from a point charge is 600 volts and the electric field is 200 N/C.*

(i) What is the distance of the point charge?

(ii) What is the magnitude of the charge?

Solution. Let d be the distance of the point from the point charge at which the potential V = 600 Volts.

So we have $E = V/d$ or $d = V/E$

$$\therefore \quad d = \frac{600\ \text{volts}}{200\ \text{N/C}} = 3\ \text{m}$$

(*ii*) Also we have the relation for electric potential.

(*ii*) $$V = \frac{1}{4\pi\varepsilon_0}\frac{q}{d}; \quad \therefore \quad q = Vd\frac{1}{(1/4\pi\varepsilon_0)}$$

In this case $V = 600$ volts, $d = 3$ m and $1/4\pi\varepsilon_0 = 9.0 \times 10^9$ N-m²/C².

$$q = \frac{600 \times 3 \times 1}{9.0 \times 10^9} = \mathbf{2 \times 10^{-11}} \textbf{ coulomb Ans.}$$

Example 3. *A point charge of q coulomb is located at the centre of a regular tetrahedron of side 'a' metre. Calculate the flux through one of the faces.*

Solution. In case of a regular tetrahedron it has four faces, each of which is an equilateral triangle of side 'a' In the present problem the point charge is situated at the centre so the electric flux through each of the face will be equal. From Gauss's law, the electric flux through the entire tetahedron or through all four faces is q/ε_0. so we have

$$\text{The electric flux through each face} = \frac{1}{4}\left(\frac{q}{\varepsilon_0}\right) = \frac{q}{4\varepsilon_0} \text{ Ans.}$$

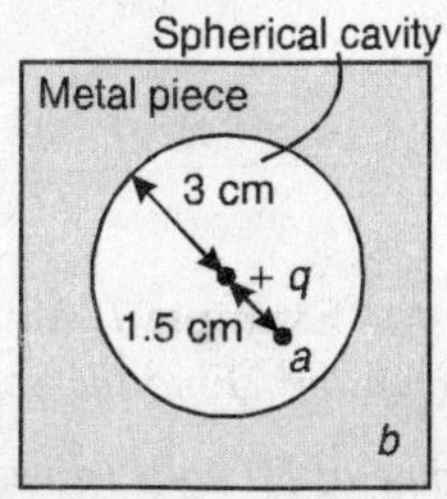

Fig. 1.37

Example 4. *A point charge $q = 2.0 \times 10^{-7}$ coulomb is placed at the centre of a spherical cavity of radius 3.0 cm is a metal piece. Find the electric intensities at 'a' and 'b' shown in fig. 1.37.*

Solution. (*i*) In the present case Gaussian surface is a sphere of radius '*a*'. As the charge is placed at *O*, the electric intensity *E* at a point '*a*', distant *r* (= 1.5 cms.) from the centre of a spherical cavity may be written as

$$E = \frac{1}{4\pi\varepsilon_0} \frac{q}{r^2}$$

In this case, $q = 2.0 \times 10^{-7}$ colomb, $r = 1.5$ cm = 0.015 m and

$$\frac{1}{4\pi\varepsilon_0} = 9 \times 10^9 \text{ N-m}^2/\text{Coul.}$$

$$\therefore \qquad E = 9 \times 10^9 \times \frac{2.0 \times 10^{-7}}{(0.015)^2} = \mathbf{8.0 \times 10^6} \textbf{ N/Coul Ans.}$$

(*ii*) As the spherical cavity is in a metal piece and the charge always resides on the outer surface so the electric intensit at *b* is zero, Ans.

Example 5. *Five thousand lines of force enter a volume of space and three thousand lines leave it. How much charge is contained in it ? (permittivity of free space, $\varepsilon_0 = 8.85 \times 10^{-12}$ coulomb b²/newton-metre).*

Solution. Lets *b* be the charge contained in a volume of space, then from Gauss's law in electrostatics, the net electric flux ϕ linked with the given volume of space may be written as

$$\phi = \frac{q}{\varepsilon_0} \qquad \text{or} \qquad q = \varepsilon_0 \phi$$

In this case, the number of lines of force passing normally through the surface

$$\phi = 5000 - 3000 = 2000$$

$$\therefore \qquad q = 8.85 \times 10^{-12} \times 2000 = \mathbf{1.77 \times 10^{-8}} \textbf{ Coulomb. Ans.}$$

Example 6. *A large plane charge sheet having surface charge density $\sigma = 2.0 \times 10^{-6}$ C/m² lies in the X–Y plane. Find the flux of the field through a circular area of radius 1 cm lying completely in the region where x, y, z all are positive and with its normal making an angle of 60° with Z-axis [$\varepsilon_0 = 8.83 \times 10^{-12}$] coulomb² / (Newton-metre²)].*

Solution. The magnitude of intensity of electric field on either side, near a plane sheet of charge having surface charge density σ, is given by the relation.

$$E = \frac{\sigma}{2\varepsilon_0}$$

and is in a direction normally away from positive charged sheet.

In this case, the electric field is along the Z-axis and the angle between the normal.

to the given area or area or area Vector $\Delta \vec{S}$ and field vector $\vec{E}$ is 60°.

So, the flux through a circular area = $\vec{E}.\Delta \vec{S} = E\Delta S \cos 60°$

Which may be written as

$$= \frac{\sigma}{2\varepsilon_0}(\pi r^2)\cos 60° = \frac{2.0\times10^{-6}\times3.14\times(1\times10^{-2})^2\times10}{2\times8.85\times10^{-12}}$$

ϕ = **17.74 N-m²/Coul Ans.**

Example 7. *An infinite charged sheet has a surface charge density of* σ = *1.0 × 10⁷ Coulomb /m². Find the separation between those equipotential surfaces which have potential difference of 50 volts (given* $\varepsilon_0 8.85\times10^{-12} C^2/N-m^2$ *).*

Solution. The intensity of electric field due to an infinite sheet of charge having surface charge density σ may be written as

$$E = \frac{\sigma}{2\varepsilon_0} \qquad \text{... (1)}$$

Let d be the separation between two equipotential surfaces is an electric field $\vec{E}$ that has a potential difference V, so we have

$$E = \frac{V}{d}; \qquad \text{... (2)}$$

$$\frac{V}{d} = \frac{\sigma}{2\varepsilon_0} \text{ or } d = \frac{2\varepsilon_0 V}{\sigma} \text{ From eq. (1) and (2), we get}$$

In this case $\varepsilon_0 = 8.85 \times 10^{-12}$ C²/N -m², V = 5.0 volt and $\sigma = 1.0 \times 10^{-7}$ C/m²

$$\therefore \qquad d = \frac{2\times8.85\times10^{-12}\times5.0}{1.0\times10^{-7}} = \mathbf{88.5\times10^{-5}} \textbf{ metre Ans.}$$

Example 8. *A thin spherical shell of metal has a radius of 0.25 metre and carries a charge of 0.2 micro-coulomb. Calculate the electric intensity at a point (i) inside the shell, (ii) just outside the shell and (iii) 3.0 metre from the centre of the shell.*

(ε_0 = 8.85 × 10⁻¹² Coul². N⁻¹ m⁻²).

Solution. (*i*) *The electric intensity at a point inside a thin charged spherical shell is zero* a at any point inside a closed shell Gaussian surface encloses no charge.

(*ii*) At a point just outside the shell, the shell is identical to the point charge placed at the centre. So, if R is the radius of the shell, then the electric intensity at a point just outside the shell is same as if the point is situated at the shell, which means

$$E = \frac{1}{4\pi\varepsilon_0}.\frac{q}{R^2}$$

In this case $\varepsilon_0 = 8.85 \times 10^{-12}$ Coul.^{2}/(N-m^2), $q = 0.2$ micro-coulomb $= 0.2 \times 10^{-6}$ coulomb and $R = 0.25$ m.

$$\therefore \quad E = \frac{1}{4 \times 3.14 \times 8.85 \times 10^{-12}} \times \frac{0.2 \times 10^{-6}}{(0.25)^2} = \mathbf{2.88 \times 10^4} \textbf{ N/Coulomb Ans.}$$

(*iii*) The electric intensity at a point at a distance r (= 3.0 m) ($r > R$) from the centre is may be written as

$$E = \frac{1}{4\pi\varepsilon_0} \frac{q}{r^2}$$

$$\therefore \quad E = \frac{1}{4 \times 3.14 \times 8.85 \times 10^{-12}} \cdot \frac{0.2 \times 10^{-6}}{(3.0)^2} = \textbf{200 N/Coul Ans.}$$

Example 9. *Four grams of gold are beaten into a thin leaf of area 1 sq. m and then placed upon a conductor. Calculate the charge density required by the conductor so that the piece of gold is just lifted up (* $\varepsilon_0 = 8.85 \times 10^{-12}$ *C²/N -m²)*

Solution. We have the upward force on the thin leaf of gold due to the charge on the conductor given by

$$\text{force/area} = \frac{\sigma^2}{2\varepsilon_0} \text{N/m}^2$$

the downward force acting on the foil due to its weight per square metre will be

$$mg = 4 \times 10^{-3} \times 9.8 \text{ N/m}^2$$

For juslitying of gold piecl, $\dfrac{\sigma^2}{2\varepsilon_0} = mg$ or $\dfrac{\sigma^2}{2\epsilon_2} = 4 \times 10^{-3} \times 9.8$

or $\qquad \sigma^2 = 2 \times 4 \times 10^{-3} \times 9.8 \times \varepsilon_0$

$$\therefore \quad \sigma = (2 \times 4 \times 10^{-3} \times 9.8 \times 8.85 \times 10^{-12})^{1/2} = \mathbf{8.33 \times 10^{-7}} \textbf{ coulomb/m}^2 \textbf{ Ans.}$$

Example.10. *From what distance should a 100 eV electron be fired toward a large metal plate having surface charge density of – 2.0 ×* 10^{-6} *Coulomb/m², so that it 'just' fails to strike the plate ? [e = 1.6 ×* 10^{-19} *Coulomb, 1 eV = 1.6 ×* 10^{-19} *joule and* $\varepsilon_0 = 8.85 \times 10^{-12}$ *C²/N–m²].*

Solution. Let an electron be fired from a distance x toward a large metal plate which has a surface charge density σ, that case we have the relation.

$$E = \frac{\sigma}{\varepsilon_0} \text{ (for a metallic conductor)}$$

As the surface charge density is negative, the so we have the repulsive force on electric as

$$F = eE = \frac{e\sigma}{\varepsilon_0} = \frac{1.6 \times 10^{-19} \times 2.0 \times 10^{-6}}{8.85 \times 10^{-12}} \textbf{ Newton} = \mathbf{3.62 \times 10^{-14}} \textbf{ N.}$$

The work done by the electron is traversing a distance x against this force may be written as

$$W = F.x = 3.62 \times 10^{-14}\, x$$

Now initial energy of electron is given by

$$K = 100 \text{ eV} = 100 \times 1.6 \times 10^{-19} \text{ joule}$$

The electron just fails to strike the plate *if the work done by it is equal to its initial kinetic energy K.*

So we have

$$W = F.x = K$$

$$\therefore \quad 3.62 \times 10^{-14}\, x = 100 \times 1.6 \times 10^{-19}$$

or $$x = \frac{100 \times 1.6 \times 10^{-19}}{3.62 \times 10^{-14}} = \mathbf{4.42 \times 10^{-4}} \textbf{ metre Ans.}$$

Example 11. *If the electric field near the earth's surface be 300 volt/m directed downward, what is the surface density of charge on earth's surface?*

Solution. We have the relation.

$$E = \frac{\sigma}{\varepsilon_0} \text{ or } \sigma = \varepsilon_0 E$$

In this case, ε_0 8.85×10^{-12} $C^2/N - m^2$ and E = 300 vlots/m

$$\therefore \quad = 8.85 \times 10^{-12} \times 300$$

$$\sigma = \mathbf{2.655 \times 10^{-9}\ C/m^2} \textbf{ Ans.}$$

Example 12. *The surface density of charge on a conductor in dry air is 2.67×10^{-5} C/m^2. Calculate the force acting on the surface per unit area due to this charge.*

Solution. We have the relation for force active on surface of a charged density σ as $F = \dfrac{\sigma^2}{2\varepsilon_0}$

In this case σ = 2.67×10^{-5} Coulomb/m² and $\varepsilon_0 = 8.85 \times 10^{-12}$ F/m

$$F = \frac{(2.67 \times 10^{-12})}{2 \times 8.85 \times 10^{-12}} = \mathbf{40.276\ N/m^2} \textbf{ Ans.}$$

Example 13. *Show that the potential functions*
(*i*) $V = x^2 + y^2 + z$ and (*ii*) $x^2 + y^2 + 2z^2$ satisfy Laplace's equation.

Solution. In cartesian coordinate laplace's equation is written as

In this case. $$\nabla^2 V = \frac{\partial^2}{\partial x^2} + \frac{\partial^2 V}{\partial y^2} + \frac{\partial^2 V}{\partial z^2} = 0$$

$$V = x^2 + y^2 + z$$

$$\nabla^2 (x^2 - y^2 + z) = \frac{\partial^2}{\partial x^2}(x^2 - y^2 + z) + \frac{\partial^2}{\partial y^2}(x^2 - y^2 + z) + \frac{\partial^2}{\partial z^2}(x^2 - y^2 + z)$$

$$= 2 - 2 + 0 = 0$$

Which means that function **$V = x^2 - y^2 + z$ satisfies Laplace's equation. Ans.**

(*ii*) In this case $V = x^2 + y^2 + 2z^2$

So we have $\nabla^2\ (x^2 + y^2 - 2z^2)$

$$= \frac{\partial^2}{\partial x^2}(x^2 + y^2 - 2z^2) + \frac{\partial^2}{\partial y^2}(x^2 + y^2 - 2z^2) + \frac{\partial^2}{\partial z^2}(x^2 + y^2 - 2z^2)$$

$$= 2 + 2 - 2 \times 2 = 0$$

Which means **function, $V = x^2 + y^2 - 2z^2$ satisfies Laplace's equation. Ans.**

Example 14. *Show that the potential function $V = q(x^2 + y^2 + z^2)^{-1/2}$ satisfies the Laplace's equation.*

The Laplace's equation in three dimension has the form

$$\nabla^2 V = \frac{\partial^2 V}{\partial x^2} + \frac{\partial^2 V}{\partial y^2} + \frac{\partial^2 V}{\partial z^2} = 0$$

$$\nabla^2 q(x^2 + y^2 + z^2)^{-1/2} = \frac{\partial^2}{\partial x^2}[q(x^2 + y^2 + z^2)^{-1/2}] + \frac{\partial^2}{\partial y^2}[q(x^2 + y^2 + z^2)^{-1/2}]$$

$$+ \frac{\partial^2}{\partial z^2}[q(x^2 + y^2 + z^2)^{-1/2}] \quad ...(1)$$

Let us false the first past of above equation

$$\frac{\partial^2}{\partial x^2}[q(x^2 + y^2 + z^2)^{-1/2} = q\frac{\partial}{\partial x}\left[\frac{\partial}{\partial x}(x^2 + y^2 + z^2)^{-1/2}\right]$$

$$= q\frac{\partial}{\partial x}\left[-\frac{1}{2}(x^2 + y^2 + z^2)^{-(1/2)-1} \cdot (2x)\right]$$

$$= -q\frac{\partial}{\partial x}[x(x^2 + y^2 + z^2)^{-3/2}]$$

$$= -q[(x^2 + y^2 + z^2)^{-3/2}\frac{\partial}{\partial x}(x) + \frac{\partial}{\partial x}(x^2 + y^2 + z^2)^{-3/2}]$$

$$= -q[x^2 + y^2 + z^2)^{-3/2} + x(-3/2)(x^2 + y^2 + z^2)^{-(3/2)-1}.(2x)]$$

$$= -q[(x^2 + y^2 + z^2)^{-3/2} - 3(x^2 + y^2 + z^2)^{-5/2}.x^2]$$

In the same way other parts may be written as

$$\frac{\partial^2}{\partial y^2}[q(x^2 + y^2 + z^2)^{-1/2}] = -q[(x^2 + y^2 + z^2)^{-3/2} - 3(x^2 + y^2 + z^2)^{-5/2}y^2]$$

and $\frac{\partial^2}{\partial z^2}[q(x^2 + y^2 + z^2)^{-1/2}] = -q[(x^2 + y^2 + z^2)^{-3/2} - 3(x^2 + y^2 + z^2)^{-5/2}y^2]$

So eq. (1) may be written as

$$\nabla^2 q(x^2 + y^2 + z^2)^{-1/2} = -q[(x^2 + y^2 + z^2)^{-3/2} - 3x^2(x^2 + y^2 + z^2)^{-5/2}]$$

$$- q[(x^2 + y^2 + z^2)^{-3/2} - 3y^2(x^2 + y^2 + z^2)^{-5/2}] - q[(x^2 + y^2 + z^2)^{-3/2}$$

$$- 3z^2(x^2 + y^2 + z^2)^{-5/2}]$$

$$= -q[3(x^2 + y^2 + z^2)^{-3/2} - 3(x^2 + y^2 + z^2)(x^2 + y^2 + z2)^{-5/2}]$$

$$= -q[3(x^2 + y^2 + z^2)^{-3/2} - 3(x^2 + y^2 + z^2)^{-3/2}] = 0$$

or $\nabla^2 V = 0$

Which shows that, **the function $V = q(x^2 + y^2 + z^2)^{-1/2}$ satisfies Laplace's equation. Ans.**

Example 15. *Is it possible for the electric potential in a charge free region to be given by (i) $V = x^2 + y^2 + z^2$ (ii) $v = 3x^2 + y^2 + 2z^2$? If not, find the charge density.*

Solution. In the charge free region, we have the Laplace's equation as

$$\nabla^2 V = \frac{\partial^2 V}{\partial x^2} + \frac{\partial^2 V}{\partial y^2} + \frac{\partial^2 V}{\partial z^2} = 0$$

This case $V = x^2 + y^2 + z^2$

$$\therefore \quad \nabla^2 V = \frac{\partial^2}{\partial x^2}(x^2 + y^2 + z^2) + \frac{\partial^2}{\partial y^2}(x^2 + y^2 + z^2) + \frac{\partial^2}{\partial z^2}(x^2 + y^2 + z^2)$$

$$= 2 + 2 + 2 = 6$$

Which is not zero, So the eq. $V = x^2 + y^2 + z^2$ can not represent electric potential in a charge free region . In the presence of free charge poisson's equation is applicable.

So $$\nabla^2 V = -\frac{\rho}{\varepsilon_0}$$

Thus, $$6 = -\frac{\rho}{\varepsilon_0} \quad \text{or} \quad \therefore \ \rho = -6 \in_0 \text{ coulomb}/m^2$$

(*ii*) here, $V = 3x^2 + y^2 + 2z^2$

$$\nabla^2 V = \frac{\partial^2}{\partial x^2}(3x^2 + y^2 + 2z^2) + \frac{\partial^2}{\partial y^2}(3x^2 + y^2 + 2z^2) + \frac{\partial^2}{\partial z^2}(3x^2 + y^2 + 2x^2)$$

$$= 3 \times + 2 + 2 \times 2 = 6 + 2 + 4 = 12$$

Which is non-zero. So, V cannot represent electric potential in charge free region. However in the presence of free charge Poisso*n's equation is valid. So we have,

$$\nabla^2 V = -\frac{\rho}{\varepsilon_0}$$

Thus,

Which gives $$12 = -\frac{\rho}{\varepsilon_0}$$

or $\rho = -12 \in_0$ coulomb/ m^2 **Ans.**

1.36. Dielectrics

While classifying solids on the basis of their band structure, we refer to the group of *solids as insulators which have energy gap of 3 e.v. or more.* The large magnitude of energy gap in an ideal insulator excludes the possibility of electrons being excited from valence band to conduction band by **thermal means** much less so by the application of an external electric field. It is due to this reason that insulators are poor conductors of electricity. **We can also think of it as the sum of the work needed when q_1** is and are also known as **dielectrics**. Dielectric materials find extensive use in electric industry and as capacitors.

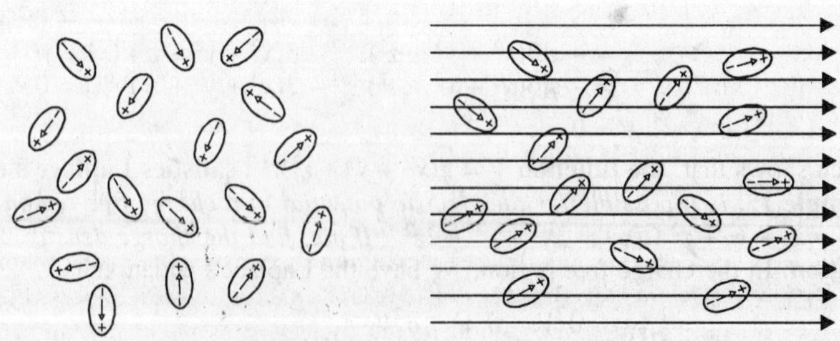

Fig. 1.38

Polar and non-polar molecules: In the case of polar molecules the *centre of gravity of the positive (i.e., protons) and negative charges (i.e., electrons) do not conicide.* Such molecules are called **permanent electric dipoles** as these have permanent dipole moments. Some examples of polar molecules are N_2O, H_2O and HCl. In the absence of an electric field, the *electric dipole moments* of these polar molecules are in random direction [Fig. 1.38 (a)] and cancel each other.

So even though each molecule has a dipole moment, the average moment per unit volume is zero: on the application of an electric field, the dipole moments of these molecules align themselves parallel to the direction of electric field as shown in Fig. 1.38 (b)]. However, this alignment is not complete because of thermal vibrations of the molecules. It is clear that the alignment of the molecules with applied field increase if

(a) Electric field intensity is increased,

(b) Temperature is lowered.

It may be noted that with the increase of electric field intensity, the dipole moment also increases as due to increase of electric field intensity the distance between positive and negative charges is increased and so the dipole moment is increased.

Non-polar molecules. In case of non-polar molecules the *centre of gravity of positive charges (protons) and negative charges (electrons) conicide,* so these molecules do not have **any permanent dipole moment.** Some common examples of non-polar molecules are oxygen (O_2) nitrogen (N_2) and hydrogen (H_2).

When a non-polar molecule is placed in an electric field, the centres of positive and negative charges get displaced and the molecules are said to have been *polarized* as shown in Fig. 1.33.

Fig. 1.39

Such a molecule is then called *induced electric dipole* and its electric dipole moment is called ***induced electric dipole moment.***

As soon as the electric field is removed, the induced dipole moment disappears. It may be noted that the induced dipole moment is proportional to the applied electric field but is *independent of temperature.* Also, the induced dipole moment is parallel to the electric field, right from the moment it is created.

Polarization. A dielectric may be made up of polar or non-polar molecules but the net effect of external field is almost the same *i.e.,* the external field will align the dipole moments along its own direction.

Let us consider the case of a dielectric slab in the electric field as shown in Fig. 1.40. The arrangement of negative and positive charges within the molecules of the dielectric in the electric field is shown in the figure. In this state the total dielectric and its molecules are said to be polarized.

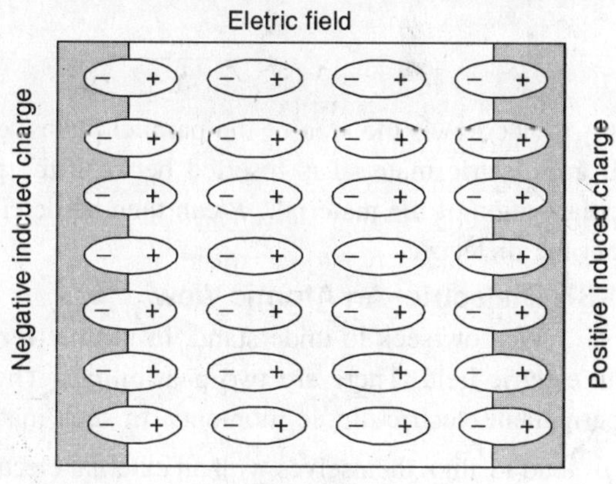

Fig. 1.40

The alignment of the dipole moments of the permanent or induced dipoles with the direction of applied electric field is called polarization, surface layers extremely thin shown as shaded part of the figure. Excess positive charge is shown in one thin layer and an equal excess negative charge is shown in the other layer.

The induced charges on the surfaces of dielectric are due to these layers. It may be noted that these charges are not free but are bound to the molecules lying in the surface or near the surface. It is because of this reason that these charges are called **bound charges or fictitious charges.** Also within the remaining dielectric, the *net charge per unit volume remains zero.* So, though the dielectric is polarized, yet it remains **electrically neutral**. Also in polarization, the internal state of the slab is characterized not by an excess charge but by the relative displacement of the charges within it.

So polarization can also be considered as a phenomenon in which an alignment of positive and negative charges takes place within the dielectric resulting no net increase in the charge of the dielectric.

The relation between the electric flux density D (charge per unit area in units of Cm^{-2}) at a point in a material and the electric field strength E (force charge in units of NC^{-1} or Vm^{-1}) at any point in space is given by

$$D = \varepsilon_r \varepsilon_0 E \qquad ...(1)$$

where ε_0 is dielectric constant or permittivity of free space (vaccum) ε_r is relative dielectric constant or relative permittivity.

It may be noted that ε_r is dimensionless and is a property of the material related to its atomic structure. Its value is 1 for free space and more than 1 for all materials. In S.I. units $\varepsilon_0 = 8.854 \times 10^{-12}$ farad/meter. As can be seen, ε_0 has this particular value from C.G.S. to S.I. units.

$$\varepsilon_0 = \frac{D}{\varepsilon_r E} = \frac{Cm^{-2}}{Vm^{-1}} = Fm^{-1}$$

As shown above, due to the application of electric field, displacement of local dipoles takes place. *The dipole moment/unit volume of the solid is the sum of all the individual dipole moments within the volume* (called polarization of the solid).

As polarization measures additional flux density arising due to the presence of the material as compared to free space, it has the same units as D and is related as under

$$D = \varepsilon_0 E + P \qquad ...(2)$$

From Eqn. (1) and (2), we get

$$P = \varepsilon_0 (\varepsilon_r - 1) E \qquad ...(3)$$

Macroscopically, the polarization or relative dielectric constant can be measured using a parallel plate capacitor, when a voltage V is applied to the capacitor, with vacuum between the plates, a charge develops on the plates. In this case capacitance is given by

$$C = \frac{\varepsilon_0 A}{d} \qquad ...(4)$$

where A is the area of the parallel plates, d is the distance of separation between the plates. If a dielectric material is inserted between the plates, the charge on the plates increases due to polarization of the material. P can then be determined by measuring capacitance, with and without the dielectric.

1.37. Dielectric–An Atomic View

We now seek to understand, **in atomic terms**, what happens when a dielectric is placed in an electric field. There are two possibilities. The molecules of some dielectrics, like water, have permanent electric dipole moments. In such materials (called *polar*) the electric dipole moments $\vec{P}$ tend to align themselves with an external electric field, as in Fig. 1.40 (a). Because the molecules are in constant thermal motion the degree of alignment will not be complete but will increase as the applied electric field is increased or as the temperature is reduced.

Whether or not the molecules have permanent electric dipole moments, they acquire them by induction when placed in an electric field. We know that the external electric field tends to separate the negative and the positive charge in the atom or molecule. This induced electric dipole moment is present only when the electric field is present. It is proportional to the electric field (for normal field strengths) and is created for already lined up with the electric field.

Let us use a parallel-plate capacitor, carrying a fixed charge q and not connected to a battery (see Fig. 1.41) to provide a uniform external electric field E_0 into which we place a dielectric slab.

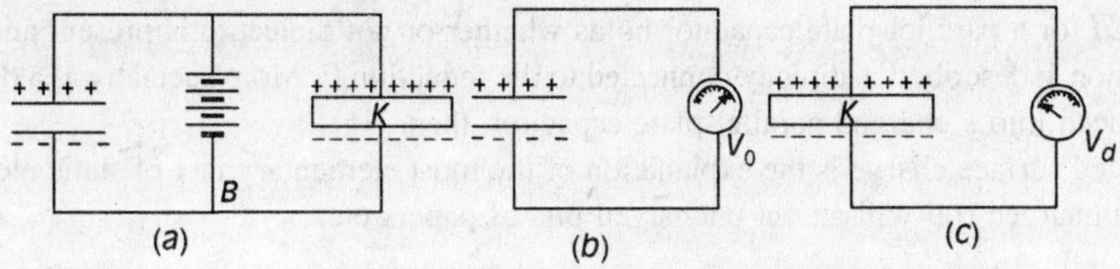

Fig. 1.41

The over-all effect of alignment and induction is to separate the centre of positive charge of the entire slab slightly from the centre of negative charge. The slab, as a whole, although remaining electrically neutral, becomes *polarized,* as Fig. 1.42 shows. The net effect is a pile-up of positive charge on the right face of the slab and of negative charge on the left face; within the slab no excess charge appears in any given volume element. Since the slab as a whole remains neutral, the

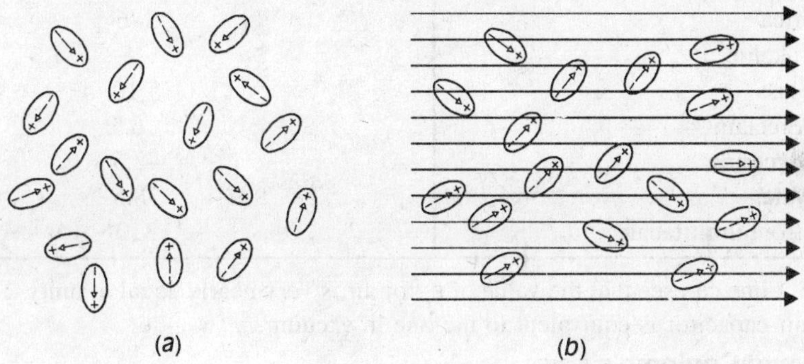

Fig. 1.42

positive *induced surface charge* must be equal in magnitude to the negative induced surface charge. It should be noted that in this case electrons in dielectric are displaced from their equilibrium positions by distances that are considerably less than an atomic diameter. There is no transfer of charge over macroscopic distance such as occur, when a current is set up in a conductor.

Figure 1.43 (c) shows that the induced surface charges will always appear in such a way that the electric field set up by them (E′) opposes the external electric field E_0. The *resultant* field in the dielectric E is the vector sum of E_0 and E′. It points in the same direction as ε_0 but is smaller. *If a dielectric is placed in an electric field, induced surface charges appear which tend to weaken original field within the dielectric.*

This weakening of the electric field reveals itself as a reduction in potential difference between the plates of a charged isolated capacitor when a dielectric is introduced between the plates. The

$$\frac{E_0}{E} = \frac{V_0}{Vd} = \varepsilon_r$$

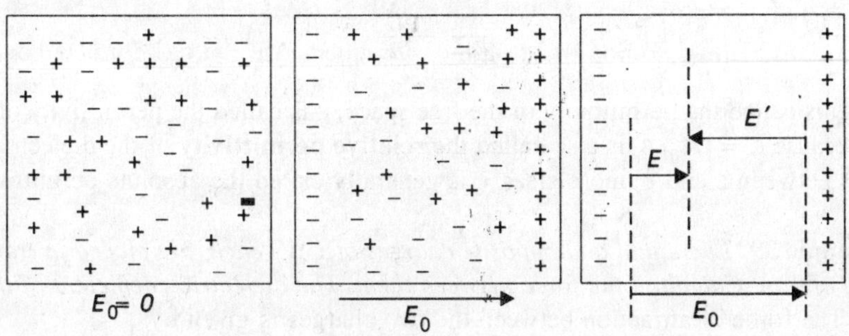

Fig. 1.43

relation $V = Ed$ for a parallel-plate capacitor holds whether or not dielectric is present and shows that the reduction V described is directly connected to the reduction E. More specially, if a dielectric slab is introduced into a charged parallel-plate capacitor, then.

Induced surface charge is the explanation of the most elementary fact of static electricity, namely, that a charged rod will attract uncharged bits of papers etc.

Table 3.1. Values Dielectric Coefficients of Some Materials.

Material	*Values of Dielectric Coefficients*
Vaccuum	1
Air (1 atm)	1.0006
Teflon	2.1
Paper	3.5
Mica	3–6
Bakelite	5
Glass	5.10
Porcelain	6.5
Glycerine	42.5
Water	78.5
Strontium titanate	310

From Table 3.1 one can see that the value of ε_r for air is very nearly equal to unity. Thus, for most purposes an air capacitor is equivalent to the one in vacuum.

1.44. Dielectrics and Coulomb's Law

For a capacitor, the force between the two charges on the plates is given by

$$F_0 \propto E_0 \quad \text{...(1)}$$

This force, after the introduction of the dielectric, reduces to

$$F \propto E \quad \text{...(2)}$$

So we have

$$\frac{F}{F_0} = \frac{E}{E_0} = \frac{1}{\varepsilon_r}$$

$$F = F_0 / \varepsilon_e \quad \text{...(3)}$$

So, the force between the two charges in the presence of a dielectric, reduces by a factor of ε_r. So which may be written as in dielectrics, the Coulomb's law becomes,

$$F = \frac{1}{4\pi\varepsilon_0\varepsilon_r} \frac{q_1 q_2}{r_{12}^2} \quad \text{... (4)}$$

$$F = \frac{1}{4\pi\varepsilon} \frac{q_1 q_2}{r_{12}^2} \quad \text{... (5)}$$

As ε_0 is called the permittivity of the free space, ε is called the permittivity of the dielectric. Further, because $\varepsilon_r = \varepsilon/\varepsilon_0$, ε_r is also called the **relative permittivity** of the dielectric. To make the distinction between ε and ε_r more clear, ε is generally called the absolute permittivity of the medium

Example 39. *Two equal and opposite charges of 1 μC each, are placed in transformer oil at a distance of 1 m. Calculate the force between them. The dielectric coefficient of the coil is 2.2.*

Solution. The force of attraction between the two charges is given by,

$$F = \frac{1}{4\pi\varepsilon_0\varepsilon_r}\frac{q_1q_1}{d^2}$$

$$= \frac{10^6 \times 10^{-6} C^2}{4\pi \times 8.85 \times 10^{-12} \text{ Fm}^{-1} \times 2.2 \times 1\text{m}^2}$$

$$= \frac{1}{4\pi \times 8.85 \times 2.2}\frac{C^2}{\text{Fm}}$$

$$= \mathbf{0.0046 \text{ N} \quad Ans.}$$

1.45. Gauss's Law in Dielectrics

Let us draw a Gaussian surfaces as shown in Fig. 1.43. From this we see that if ε is the resultant field inside the dielectric, then,

$$\varepsilon_0 EA = q_0 - q'$$

or
$$E = \frac{q_0}{\varepsilon_0 A} - \frac{q'}{\varepsilon_0 A} \qquad ...(1)$$

Here q', the charge induced at the surfaces of the dielectric, is called bound charge; and q_0, the charge at the capacitor plates, is called the free charge. In the absence of the dielectric, electric field between the capacitor plates is,

$$E_0 = q_0 / (\varepsilon_0 A) \qquad ...(2)$$

So we have

$$\frac{E_0}{\varepsilon_r} = \frac{q_0}{\varepsilon_0 A} - \frac{q'}{\varepsilon_0 A}$$

or
$$\frac{q_0}{\varepsilon_0 A \varepsilon_r} = \frac{q_0}{\varepsilon_0 A} - \frac{q'}{\varepsilon_0 A}$$

$$q' = q_0\left(1 - \frac{1}{\varepsilon_r}\right) \qquad ... 3$$

So, the Gauss's law becomes,

$$\varepsilon_0 \Sigma E.\,\Delta S = (q_0 - q')$$

$$= q_0 - q_0 + q_0/\varepsilon_r \qquad ... 4$$

or
$$\varepsilon_0 \sum \varepsilon,\, E.\Delta S = q_0$$

Hence, the effect of the polarization of the dielectric material on the Gauss's theorem is to multiply the ε-vector by ε_r.

Let us rewrite Eqn. (1) which applies to a parallel-plate capacitor containing a dielectric, as

$$\frac{q_0}{A} = \varepsilon_0\left(\frac{q}{K\varepsilon_0 A}\right) + \frac{q'}{A} \qquad ...(5)$$

The quantity in bracket is simply the electric field strength ε in the dielectric. The last term in Eqn. (5) is the *induced surface charge per unit area. We call it the electric polarization P,* or

$$p = \frac{q'}{1} \qquad ... (6)$$

1.40. Electric Displacement

Let us rewrite Eqn. (5) in the form

$$\frac{q_0}{\varepsilon_0 A} = E + \frac{q'}{\varepsilon_0 A}$$

which may be written as

$$\frac{q_0}{\varepsilon_0 A} = E + \frac{1}{\varepsilon_0} P$$

or
$$\frac{q_0}{A} = \varepsilon_0 E + P \quad \text{... (7)}$$

The quantity on the right of Eqn. (7) occurs quite often in theory. We call it electric displacement, and denote it by D. So we have

$$D = \varepsilon_0 E + P \quad \text{...(8)}$$

Also from Eqns. (7) and (8)

$$D = \frac{q_0}{A} = \frac{q_0}{\varepsilon_0 A} \times \varepsilon_0$$

$$= \varepsilon_0 E_0$$

$$= \varepsilon_0 E_0 = \varepsilon_0 \varepsilon_1 E$$

or
$$D = \varepsilon E \quad \text{... (9)}$$

Now
$$P = \frac{q'}{A} = \frac{q_0}{A}\left(\left[1 - \frac{1}{\varepsilon_r}\right]\right)$$

$$= \varepsilon_0 \varepsilon_r E - \varepsilon_0 E$$

or
$$P = \chi_e E \quad \text{... (10)}$$

The constant χ_e is called the electric susceptibility. Thus,

$$\varepsilon = \varepsilon_0 + \chi\varepsilon \quad \text{...(11)}$$

Combination of Eqns. (9) and (4) results in a simple form of Gauss's law,

$$\Sigma\Delta - \Delta\Sigma = q_0 \quad \text{...(12)}$$

One can note from above that E changes with medium D remains unchanged, and P vanishes in vaccuum.

1.46. Representation of Three Vectors

We have discussed above, the behaviour of dielectrics under special circumstances. But, the results obtained are general. These results hold good under all situation. We have introduced three vectors E, D and P where,

$$P = q'/A$$
$$D = q_0/A$$
and
$$\varepsilon E = (q_0 - q')/A$$

The expressions clearly show that D is connected with *free charges* only, P is connected with *bound charges* only, and E is connected *with all charges* that are actually present. Thus lines of D begin and end at free charges, lines of P begin and end at induced charges, and lines of E changes from one medium to another. It should be remembered that the aim of the representation D and P by lines is to visualise these, as we have done in representing E by lines of force. The pictorial

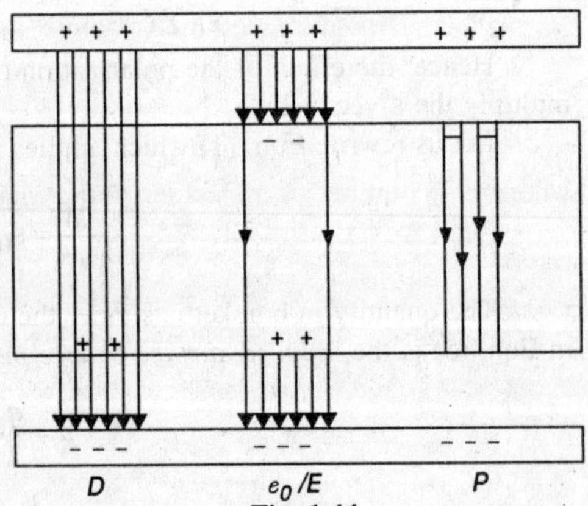

Fig. 1.44

representation of all these three vectors is given in Fig. 1.44. Out of these vector, E determines the force on a test charge. Therefore, this vector is of fundamental interest, while D and P are auxiliary vectors, helpful in more comoplex problems.

Table 1.2. Three Electric Vectors

Name	*Symbol*	*Associated with*	*Boundary Condition*
Electric field strength	E	All charges	Tangential comoponent continuous
Electric displacement	D	Free charges only	Normal comoponment continuous
Polarization (electric dipole moment per unit volume)	P	Polarization charges only	Vanishes in a vacuum

Defining equation for E $\quad F = qE$

General relation among the three vectors $\quad D = \varepsilon_0 E + P$

Gauss's law when dielectric media moment $\quad \oint D.ds = q$

(q = free charge only)

Emperial relations for certain dielectric materials $\quad D = \varepsilon_r \varepsilon_0 \mathrm{E}$

$P = (\varepsilon_r - 1)\varepsilon_0 \mathrm{E}$

1.47. Clausius Mosotti Equation

In Clasius Mosotti theory we consider the actual *non-polar molecules of a homogeneous isotropic dielectric* lying between the plates of a parallel plate condenser, size *of plates being very large as compared to the distance between them.*

C lausius Mosotti equation gives the *dielectric constant in terms of molecular properties.* When a non-polar molecule is placed is an electric field, the actual intensity of electric field *i.e.,* Lorentz field acting on any molecule is given by

$$F = E + \frac{P}{3\varepsilon_0} \qquad \text{...(1)}$$

Diipole moment induced in a molecule is given by

$$m = \alpha\, \varepsilon_0 F = \alpha\, \varepsilon_0 \left(E + \frac{P}{3\varepsilon_0} \right) \qquad \text{...(2)}$$

$\therefore$ Dipole moment/unit volume

$$P = n\varepsilon_0 \left(E + \frac{P}{3\varepsilon_0} \right) \qquad \text{....(3)}$$

where n $\rightarrow$ number of moleculers/unit volume

Also we have the relation

$$D = \varepsilon_0 + P$$

$$P = D - \varepsilon_0 E \qquad \text{...(4)}$$

From eq. (3) and (4) we get

$$D - \varepsilon_0 E = P = n\ \varepsilon_0 \left(E + \frac{P}{3\varepsilon_0} \right) \qquad \text{...(5)}$$

$$= n\varepsilon_0 \left(E + \frac{D - \varepsilon_0 E}{3\varepsilon_0} \right)$$

$$= na \left(\frac{D + \varepsilon_0 E}{3} \right) \quad \text{...(5)}$$

$$\therefore \quad \frac{D - \varepsilon_0 E}{D + 2\varepsilon_0 E} = \frac{n\alpha}{a}$$

Also $$D = \varepsilon E$$

$$\frac{\varepsilon - \varepsilon_0}{\varepsilon + 2\varepsilon_0} = \frac{n\alpha}{3} \quad \text{...(6)}$$

Put $\varepsilon = \varepsilon_0 . \varepsilon_r$, we get

$$\frac{\varepsilon_r - 1}{\varepsilon_r + 2} = \frac{n\alpha}{3} \quad \text{...(7)}$$

eqs. (5) and (6) give the simplest form of *Clausius-Mosotti equation* and gives the *permittivity or dielectric constant* in terms of the polarisability of the dielectric.

The function $\frac{\varepsilon - \varepsilon_0}{2 + 2\varepsilon_0}$ or $\frac{\varepsilon_r - 1}{\varepsilon_r + 2}$ is called *volume polarization* and depends only on the number of molecules/unit volume and their *polarizabilities.* It is dimensionless and is denoted by π.

$$\pi = \frac{\varepsilon - \varepsilon_0}{\varepsilon + 2\varepsilon_0} \quad \text{... (8)}$$

Put $\varepsilon_r = \chi + 1$, where χ is the susceptibility of the dielectric, we get

$$\frac{\chi}{\chi + 3} = \frac{n\alpha}{3}$$

Which gives us the *Clausius Mosotti* equation in terms of *susceptibility* of the dielectric. Multiplying sides of eq. (6) by specific volume v, which is reciprocal of density d, we get

$$P = \pi v = \frac{\varepsilon - \varepsilon_0}{\varepsilon + 2\varepsilon_0} . \frac{1}{d} = \frac{1}{3} \times n\, a\, v$$

$$= \frac{1}{3} za$$

where $z = n\nu$ is the number of molecules per unit mass, and p is called the *specific polarization.*

Again multiplying both sides of eq. (8) by molecular weight of he substance, we get

$$P_m = p.M = \frac{\varepsilon_0 - \varepsilon_0}{\varepsilon + 2\varepsilon_0} . \frac{M}{d} = \frac{1}{3} za$$

or $$P_m = \frac{\varepsilon - \varepsilon_0}{\varepsilon + 2\varepsilon_0} . Vm = \frac{1}{3} z\alpha \quad \text{... (9)}$$

where $V_m \rightarrow$ molecular volume

$N \rightarrow$ Avogadro's polarization

and $P_m \rightarrow$ molar polorization

it may be noted that quantities p, p_m and a have the dimensional of volume.
if we put $\varepsilon = \varepsilon\eta^2$ in eq (9) where η is the *refractive index of the medium we get*

$$\frac{\eta^2 - 1}{\eta^2 + 2} = V_m = \frac{1}{3} N\alpha \qquad ...(10)$$

This is known as **Lorentz** equation of light which was derived on e.m. theory considerations.

We can also write Clausius equation (eq. 6) as

$$\varepsilon_r = \frac{1 + \dfrac{2n\alpha}{3}}{1 - \dfrac{n\alpha}{3}} \qquad ...(11)$$

Which shows that ε_r must increase rapidly and approach infinity as the number of molecules per unit volume approach a *critical value* n_c i.e.,

$$\varepsilon_r = \infty \text{ if}$$

$$\frac{n_c\alpha}{3} = 1$$

Physically, $\varepsilon_r \to \infty$ corresponds to the dielectric volume filled with infinitely polarisable *i.e.,* conducting molecules. In case of gases and liquids this condition is never reached due to the fact that their polarizabilities are relatively smail. So the **Clausius Mosotti equation is valid for gases and liquids.** In case of solids, the interaction of dipoles is quite complicated and so the above simple model does not hold good. So the **Clausius Mosotti equations are not valid for solids.**

Electromagnetism

Some of the useful terms used in electromagnetism are given below:

(i) **Magnetic Forces** are due to the motion of electric charges, *i.e.,* due to electric currents. A magnetic force exists between two charges *if and only if both charges are moving relative to an observer.* Examples of moving charges are electric currents in wires, a beam of electrons in a cathode ray tube, and electrons rotating about the atomic nuclei and about their own axes.

(ii) **Magnetic Field** is the region surrounding a moving charge. A magnetic field exists at any point where a moving charge experiences a force exclusive of any electrostatic force.

(iii) **Direction of Magnetic Field of a Current** may be determined by a right-hand rule. *Point the thumb of the right hand in the direction of the current, then the curled fingers will point in the direction of the field.* Thus the magnetic field of a current is clockwise as viewed by an observer looking along the direction of the current. Refer to the Fig. 1.45.

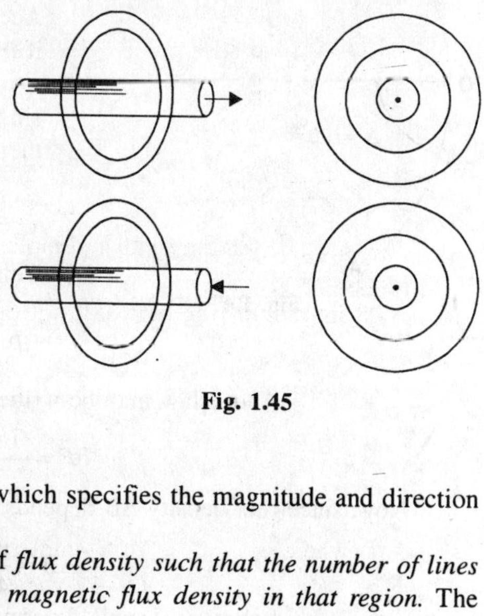

Fig. 1.45

(iv) **Magnetic Flux Density B,** also called **magnetic induction** of magnetic **B field,** is a vector which specifies the magnitude and direction of the magnetic field at a point.

A magnetic field may be represented by lines of *flux density such that the number of lines piercing perpendicularly a unit area is equal to the magnetic flux density in that region.* The direction of the lines at any point is that of B.

The magnetic flux Φ through a surface is the total number of lines of B extending through it. In a region where the magnetic flux density B is uniform and normal to a given area A,

$$\Phi = BA$$

If θ is the angle between the magnetic induction vector B and the normal to the reference area A,

$$\Phi = BA \cos \theta$$

The S.I. for B is the **tesla,** symbol T, and for Φ is the weber, symbol Wb. Thus the tesla is 1 Wb m^{-2} and

$$\Phi\ (\text{Wb}) = B(\text{T}) \times A(\text{m}^2) \times \cos \theta$$

1.48. Ampere's Law

Let XY be a current carrying conductor and let I e.m.u. current be flowing from X to Y.

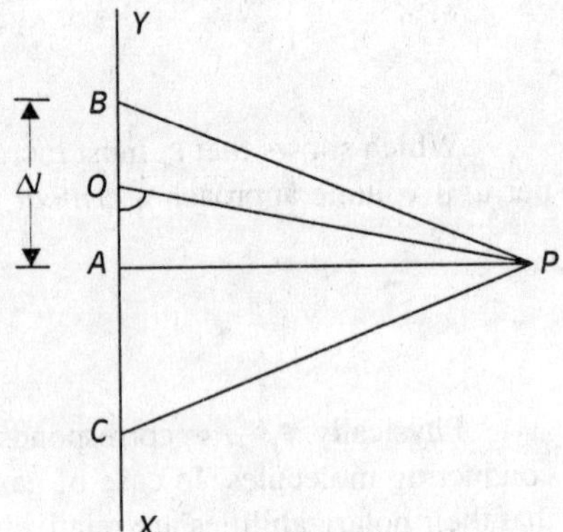

Fig. 1.46

The magnetic field at any point P at a distance r from the middle point O of AB is given by

$$\delta \overline{F} \propto \frac{i \Delta l \sin \theta}{r^2} \hat{r}$$

So the magnetic intensity at P is $\propto$ i, *i.e.,* strength of the current $\propto$ Δl, *i.e.,* length of the element $\propto$ $\sin \theta_r$ (where θ is angle between the direction of current and line joining middle point O of the element with P.)

$\propto \frac{1}{r^2}$, where r is distance OP.

$$\therefore \qquad \delta \vec{F} = \frac{\kappa .\ i\ \Delta l \sin \theta}{r^2} \hat{r}$$

where k is constant.

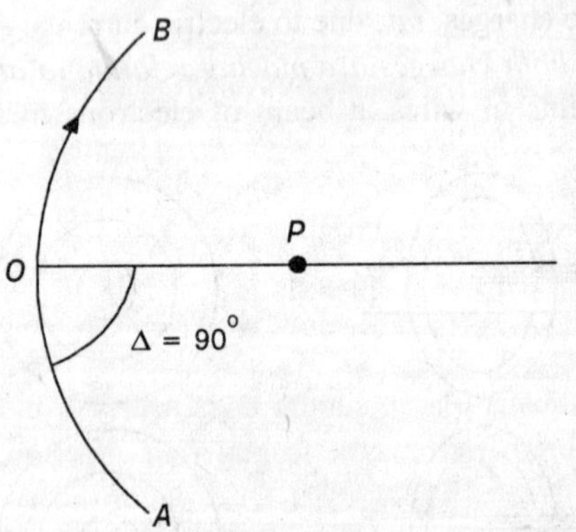

Fig. 1.47

(i) **e.m.u. of current.** One e.m.u. of current is that current which when flowing through a circular arc of length 1 cm and radius 1 cm produces a magnetic intensity of 1 oersted at the centre of the circular arc as shown in Fig. 1.47.

According to the above definition,

if $i = 1$

$r = 1$, $\Delta l = 1$,

$\sin 90° = 1$ and $F = 1$

then $k = 1$

$$\therefore = \delta \vec{F} = \frac{i\ \Delta l \sin \theta}{r^2} \hat{r} \qquad ...(1)$$

This is called **Ampere's** law or **Laplace's law.**

(*ii*) **Rationalised M.K.S.** units: In SI units $k = \frac{1}{4\pi}$

$\therefore$ Above law may be written as

$$\delta \vec{F} = \frac{i\ \Delta l \sin \theta}{r^2} \hat{r} \qquad ...(2)$$

Now, since flux density ΔB depends on the nature of the medium, we may write

$$\Delta \vec{B} = \frac{\mu_0}{4\pi} \cdot \frac{i . \Delta l \sin \theta}{r^2}$$

[where μ_0 is called permeability of the free space]

$\mu_0 = 1$ in C.G.S. units
$\neq 1$ in M.K.S units

For getting the force on a current carrying conductor placed in a magnetic field, we have to apply the Fleming's left had rule and it depends on the nature of the medium. So to get this force, magnetic flux density is used.

In this case $\mu_0 = 4\pi \times 10^{-7} = 12.57 \times 10^{-7}$ weber/amp-m or henry/m.

1.49. Magnetic Field of a Current

Consider that a current I eixsts in a wire element of length Δl. The magnetic flux density ΔB at point *P* due to the current element in a vacuum is

$$\delta\vec{B} = k\frac{I\Delta l \sin\theta}{r^2}$$

where r is the distance from Δl to P, θ is the angle between r and the current in the element δl, and K is a proportionality constant depending upon the units used. The direction of ΔB is perpendicular to the plane determined by Δl and r and is given by the right-hand rule.

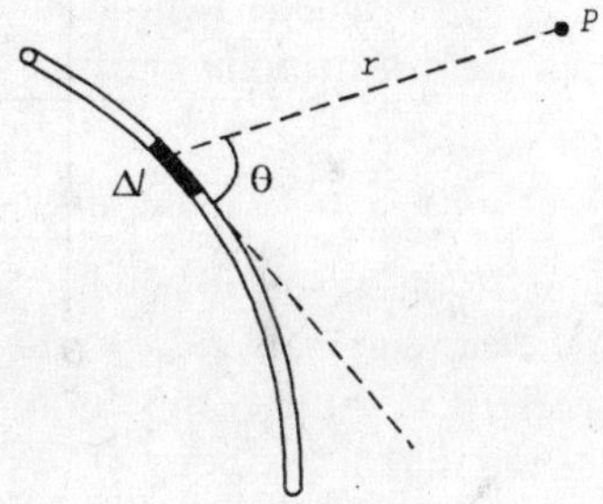

Fig. 1.48

The total magnetic flux density B at point P due to a current I in a wire of length l is the vector sum of the contributions of all the elements Δl. So we have

$$B = K\Sigma\frac{I\Delta l \sin\theta}{r^2} \quad \text{... (2)}$$

In SI units, K = 10^{-7} Wb $A^{-1}m^{-1}$ and it is written as $\frac{\mu_0}{4\pi}$ we may write eqn. (2) as

$$B = \Sigma\frac{\mu_0 I\Delta l \sin\theta}{4\pi^2} \quad \text{...(3)}$$

The constant μ_0 is called the permeability of free space or the permeability of vacuum and has the value

$$\mu_0 = 4\pi \times 10^{-7} \text{ Wb A}^{-1} \text{ m}^{-1}$$

Note. The unit of μ_0 can also be expressed in henry per metre (Hm^{-1}).

In infinite homogeneous media other than a vacuum, K takes the value $\frac{\mu_r\mu_0}{4\pi}$ where μ_r is called the relative permeability of the medium. For most substances, including μ_r is nearly 1 and only for ferromagnetics, it is substantially different. Common ferromagnetic materials are iron, steel, nickel and cobalt.

1.50. Flux Density near a Long Straight Current Carrying Conductor (Biot-Savart's Law)

In Fig. 1.50 an infinitely long, straight wire AB carries a current I. The flux density at a point *P* distant *S* from the wire may be found from the relation,

$$d\vec{B} = \Sigma K\frac{I\Delta l \sin\theta}{r^2} - \frac{\mu_0 I\Delta I \sin\theta}{r^2}\hat{r}$$

Since α and θ are complementary angles, we have

$$\sin\theta = \cos\alpha$$

$$\therefore \quad d\vec{B} = \frac{\mu_0}{4\pi}\frac{I\Delta l \cos\alpha}{r^2}\hat{r} \quad \text{... (i)}$$

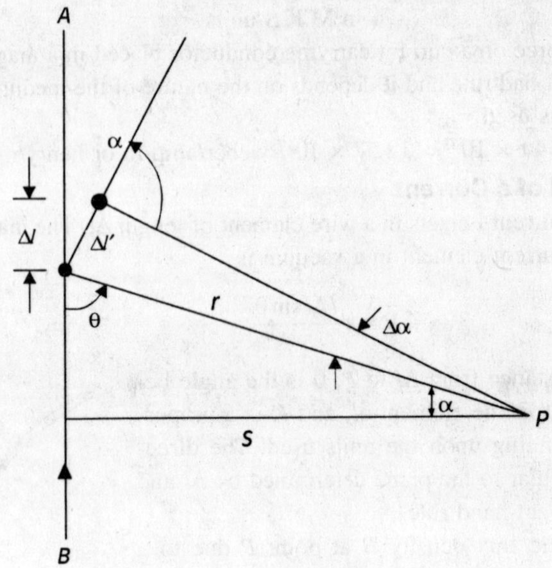

Fig. 1.49

By construction, $\Delta l = \Delta l \cos\alpha$ and $\Delta\alpha = \Delta l/r$.

Hence, $\Delta l \cos\alpha \propto \Delta l\, r\, \Delta\alpha$

Also, $r = S/\cos\alpha$

When these values are substituted in Eqn. (1), we have

$$d\vec{B} = \frac{\mu_0}{4\pi}\frac{I\cos\alpha\, d\alpha}{S}$$

The total flux density at P is the sum of the contributions from all the elements, that is, the integral when α varies from $-90°$ to $+90°$.

$$|\vec{B}| = \frac{\mu_0}{4\pi}\frac{I}{S}\int_{-90}^{+90°}\cos\alpha\, d\alpha$$

$$\therefore \quad = \frac{\mu_0}{4\pi}\cdot\frac{I}{S}\cdot|\sin\alpha|_{-90}^{90°}$$

$$= \frac{\mu_0}{4\pi}\cdot\frac{I}{S}\cdot 2 = \frac{\mu_0}{2\pi}\cdot\frac{I}{S} \qquad ...(2)$$

Eqn. (2) is sometimes called **Biot-Savart's law.** This Eqn. is valid for infinitely long currents. However, the equation gives satisfactory results for shorter wires if the distance of the point P is very small from the wire.

1.57. Ampere

An ampere is the strength of the current which when flowing through two parallel infinite long conductors placed in free space at a distance of 1 metre from each other, produces between them a force of 2×10^{-7} Newtons for 1 metre of their length.

The force between two parallel conductors

$$\vec{F} = \frac{\mu_0}{4\pi}\cdot\frac{2I_1I_2}{4\pi}\hat{r}$$

In this case $F = 2 \times 10^{-7}$ newton

$I_1 = I_2 = 1$ ampere

r = 1 metre

$$\therefore \quad 2 \times 10^{-7} = \frac{\mu_0 2\times 1\times 1}{4\pi}$$

$$\therefore \quad \mu_0 = 4p \times 10^{-7} \text{ weber A}^{-1} \text{ m}^{-1}.$$

1.52. Magnetic Field of Flat Circular Coil

To find the magnetic flux density at the centre of a circular loop of wire of radius r carrying a constant current I, use equation (3) of article 4.3. Here $\Sigma\, \Delta l$ = circumference of circle = $2\pi r$ and $\sin\theta = \sin 90° = 1$ for any element Δl. Thus

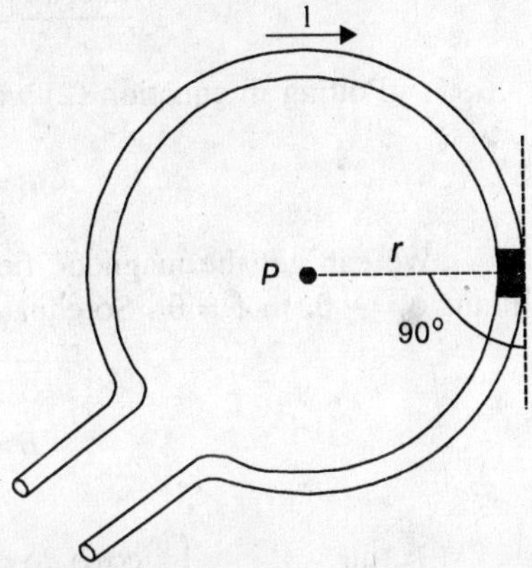

Fig. 1.50

$$B = \Sigma \frac{\mu_0 I \Delta l \sin\theta}{4\pi}$$

$$= \frac{\mu_0}{} \frac{I(2\pi r)}{r^2} = \frac{\mu_0 I}{2r}$$

It follows that for a closely wound flat circular coil of N turns, magnetic flux density at the centre is given by

$$B = \frac{\mu_0 NI}{2r}$$

1.53. Magnetic Field due to a Straight Current Carrying Conductor

Let ab be an elementary part of the conductor (Fig. 1.51) and P is observation point, such that $OP = R$. Let M be the middle point of the element ab and $\overrightarrow{MP}.\vec{r}$. Let the length ab be denoted vectorially as $\vec{L}$ such that the current element represented by ab is given by $l\ \overrightarrow{dl}$. If $OM = l$.

Then from Biot-Savart law, the magnetic field on the current element $I\overrightarrow{dl}$ at P is given by the relation.

$$d\vec{B} = \frac{\mu_0}{4\pi} \frac{I\overrightarrow{dl}\times\hat{r}}{r^2} \quad ...(1)$$

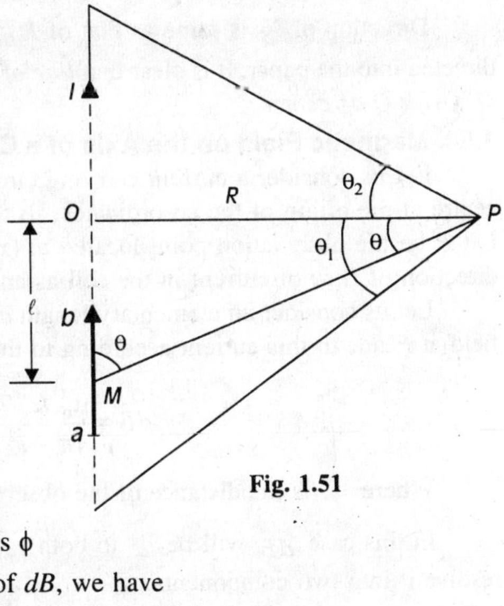

Fig. 1.51

But $\quad |\overrightarrow{dl}\times\hat{r}| = dl \sin\theta \cdot$

where $\quad \theta = \angle PMO.$

In ΔOMP, we have $\angle PMO = 90° - \angle MPO$

$$\theta = 90° - \phi$$

so $\quad |\overrightarrow{dl}\times\hat{r}| = dl \sin\theta = dl \cos\phi$

Putting in eqn. (1) and taking the magnitude of dB, we have

$$dB = \frac{\mu_0}{4\pi} \frac{I\, dl \cos\phi}{r^2} \quad ...(2)$$

In this case $l = R \tan \phi$

$\therefore \quad dl = R \sec^2 \phi \, d\phi$

Also we have $\frac{R}{l} = \cos\phi$ or $\frac{1}{r} = \frac{\cos \phi}{R}$

$$\therefore \quad \frac{dl \cos \theta}{r^2} = R \sec^2 \phi \, d\phi \cos \phi \frac{\cos^2 \phi}{R} = \frac{\cos \phi \, d\phi}{R}$$

Putting in equation (2) we have

$$dB = \frac{\mu_0}{4\pi} \frac{I \cos \phi \, d\phi}{R} \quad \text{... (3)}$$

We can get the magnetic field due to whole of the conductor by integrating between the limits $\phi = -\theta_1$ to $\phi = \theta_2$. So equation (3) becomes

$$B = \frac{\mu_0}{4\pi} \frac{I}{R} \int_{-\theta_1}^{\theta_2} \cos \phi \, d\phi \quad \text{... (4)}$$

But $\int_{-\theta_1}^{\theta_2} \cos \phi \, d\phi = \sin \theta_2 + \sin \theta_1$

Hence equation (4) can be written as

$$B = \frac{\mu_0}{4\pi} \frac{I}{R} [\sin \theta_2 + \sin \theta_1] \quad \text{...(5)}$$

For infinite conductor, we have $\theta_1 = \theta_2 = \pi/2$ and $\sin \theta_2 + \sin \theta_1 = 1 + 1 = 2$, therefore, magnetic field due to infinite conductor is

$$B = \frac{\mu_0}{4\pi} \frac{2I}{R} \quad \text{...(6)}$$

Direction of $\vec{B}$ is same as that of $I \, \vec{dl} \times r$, in the plane of the paper at P (Fig. 1.51) and is directed into the paper. It is clear that *line of magnetic force passing through P is a circle of radius OP, with O as centre.*

1.54. Magnetic Field on the Axis of a Circular Coil Carrying Current

Let us consider a current carrying circular coil of radius R, placed in the YZ plane with its centre at the origin of the co-ordinates, so that axis of the coil coincides with X-axis (Fig. 1.52). Let P be the observation point located at $(x, 0, 0)$ and let the current through the coil be I with direction of flow of current in the coil as anticlockwise.

Let us consider an elementary length ab of circular conductor and let $ab = dl$. The magnetic field at P due to this current according to the BiotSavart law is given as

$$\vec{dB} = \frac{\mu_0}{4\pi} \frac{I \vec{dl} \times \hat{r}}{r^2} \quad \text{...(1)}$$

Where $\vec{r}$ is the distance of the observation point P w.r.t. the current ba.

In this case $d\vec{B}$ will be $\perp$ to both $I \vec{dl}$ and $\hat{r}$, let it make an angle ϕ with X-axis. We can resolve it into two components

$$dB_2 = dB \cos \phi, \; dB_y = dB \sin \phi$$

$$\vec{dB} = \hat{i} \, dB \cos\phi + \hat{j} \, dB \sin \phi \quad \text{... (2)}$$

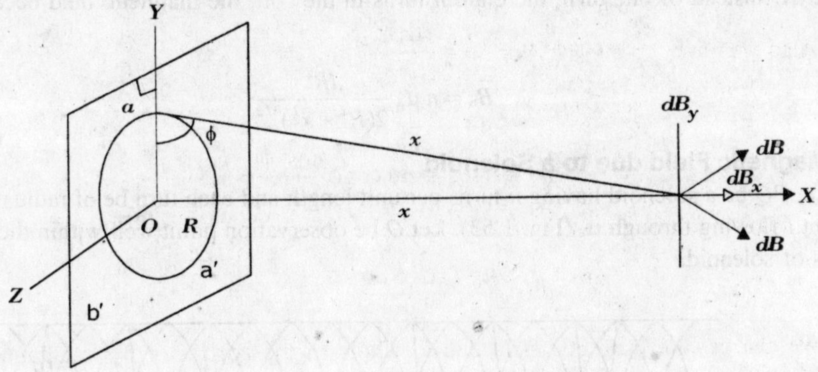

Fig. 1.52

Total magnetic field due to the coil will be vector sum of the magnetic field due to all the elements of the coil. So we may write

$$\vec{B} = \Sigma\, \overrightarrow{dB} = \hat{i}\, \Sigma\, dB\, \cos\phi + \hat{j}\, \Sigma\, dB \sin\phi \qquad ...(3)$$

Corresponding to every elementary length *ab*, there will be another element of length *a'b'* which will produce a magnetic field at *P* equal to *dB* but making angle – ϕ with the *X*-axis. Since cos (– ϕ) = cos ϕ and sin (– ϕ) = – sin ϕ, so the *Y*-components of *dB* in (3) will cancel out. Hence, net magnetic field will be

$$\vec{B} = \hat{i}\, \Sigma\, dB \cos\phi \qquad ...(4)$$

In the right angled triangle *OLP*, cos ϕ = *R*/*r*. Also the angle between, the elementary length al and *LP* is π/2. So we may write

$$B = \hat{i}\, \Sigma\, dB \frac{R}{r} \qquad ... (5)$$

$$\text{Also } I\, d\vec{l} \times \hat{r} = I\, dl \sin \pi/2 = I\, dl \qquad ...(6)$$

Putting in eqn. (5), we have

$$\vec{B} = \hat{i}\, \Sigma \frac{\mu_0}{4\pi} \frac{Idl}{r^2} \cdot \frac{R}{r}$$

$$= \hat{i} \frac{\mu_0}{4\pi} \frac{IR}{r^3} \Sigma\, dl$$

Now Σ *dl* = 2π*R*, so we may write

$$\vec{B} = \hat{i} \frac{\mu_0}{4\pi} \cdot \frac{IR}{r^3} \times 2\pi R = \hat{i} \frac{\mu_0}{2r^3} IR^3 \qquad ... (7)$$

Also from Fig. $r^2 = R^2 + x^2$, so we have

$$\vec{B} = \hat{i}\, \frac{\mu_0 IR^2}{2(p^2 + x^2)^{3/2}} \qquad ...(8)$$

Which gives expression for the magnetic field.

Special cases

(*i*) magnetic field at the centre of coil can be got by putting *x* = 0.

$$\vec{B}\,(\text{centre}) = \hat{i} \frac{\mu_0 I}{2R} \qquad ... (9)$$

(*ii*) If instead of one turn, there are n turns in the coil, the magnetic field becomes n times, this is

$$\vec{B}_n = \hat{n}\,\mu_0 \frac{IR^2}{2(R^2 + x^2)^{3/2}}$$

1.55. Magnetic Field due to a Solenoid

Let PQ be a solenoid having n turns per unit length and each turn be of radius R and having a current I flowing through it. (Fig. 1.53). Let O be observation point well within the solenoid and the axis of solenoid.

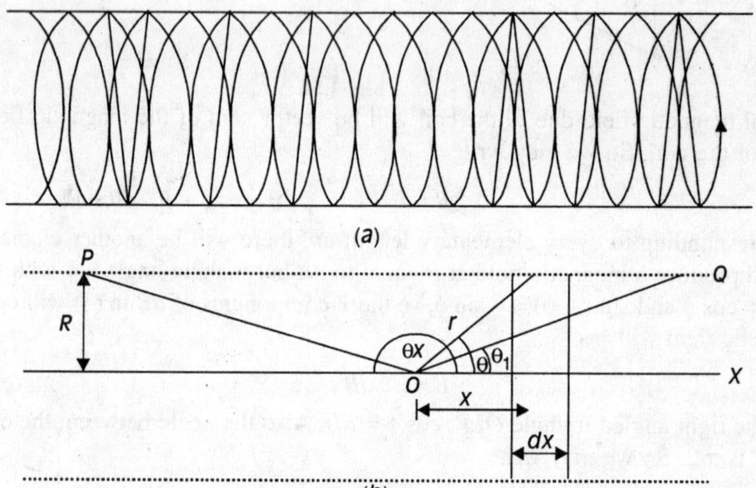

Fig. 1.53

Consider an elementary part of the solenoid having length dx to be located at a distance x from the observation point O.

Number of turns in the elementary length = $n\,dx$.

The magnetic field due to this elementary solenoid, treating it as coil '$n\,dx$' turn will be

$$d\vec{B} = \hat{i}\,\frac{\mu_0}{2r^3} IR^2 \times n\,dx \qquad ...(1)$$

We have assumed that the axis of the coil is along the X-axis so $\vec{B}$ is directed along $\hat{i}$. Let $\angle ROX = \theta$. This is the angle made by any point on the coil when joined to O with the X-axis.

From Fig. 1.53 we have $\cot\theta = x/R$

$\therefore$ $x = R\cot\theta$

and $dx = -R\,\text{cosec}^2\,\theta\,d\theta$

Substituting in (1), we get

$$d\vec{B} = \hat{i}\,n\frac{\mu_0}{2r^3} IR^2 \,\text{cosec}^2\,\theta\,d\theta \qquad ...(2)$$

Also $R/r = \sin\theta$. So eqn. (2) becomes

$$dB = -\hat{i}\,n\frac{\mu_0}{2} I\sin\theta\,d\theta \qquad ...(3)$$

Total magnetic field at O can be calculated by integrating equation (3) between the limits $\theta = \theta_1$ to $\theta = \theta_2$

$$\int d\vec{B} = -i\,n\frac{\mu_0}{2}I\int_{\theta_1}^{\theta_2}\sin\theta\,d\theta$$

So, the magnetic field at O, given by

$$\vec{B} = \hat{i}\,n\frac{\mu_0 I}{2}\,[\cos\theta_2 - \cos\theta_1] \quad ...(4)$$

Special cases

(i) If O lies on one end of the long solenoid say the end O, then $\theta_1 = \pi/2$, therefore, $\cos\theta_1 = 0$ and so we have

$$\vec{B} = \hat{i}\,n\frac{\mu_0 I}{2}\cos\theta_2 \quad ...(5)$$

(ii) If the solenoid is infinitely long and O lies in the middle then

$$\theta_1 \cong 0 \qquad \cos\theta_1 = 1$$
$$\theta_2 \cong \pi \qquad \cos\theta_2 = -1$$

and $\cos\theta_1 - \cos\theta_2 = 2$ we have

$$\vec{B} = \hat{i}\,n\,\mu_0 I \quad ...(6)$$

(iii) If the observation point is at one end of the infinite solenoid then

$$\theta_1 \cong 0 \qquad \cos\theta_1 = 1$$
$$\theta_2 \cong \pi/2 \qquad \cos\theta_2 = 0 \text{ and so} \quad ...(7)$$

(iv) **Magnetic field due to a toroid.** A solenoid bent into a circular form and whose ends are put together, is called endless solenoid or *toroid*. It behaves like an infinitely long solenoid, as regards magnetic field produced by the current flowing through it. So magnetic field due to a toroid will be given by the equation : $B = \hat{i}\,n\,\mu_0 I/2$.

1.56. Force between two Parallel Conductors

Consider that two wires, carrying currents I_1 and I_2 respectively, are parallel over a length l and separated by a distance r. The magnetic flux density at a distance r from the wire carrying current I_1 is $B = \mu_0 I/2\pi r$, and the force on the other wire carrying current I_2 in this field is $F = B' I_2 l = (\mu_0 I_1/2\pi r) I_2 l$. Hence the force between the parallel wires is

$$F = \frac{\mu_0 I_1 I_2 l}{2\pi r}$$

The force is attractive if I_1 and I_2 are in the same direction, the force is repulsive If I_1 and I_2 are in opposite directions.

The force per unit length is $F/l = \mu_0 I_1 I_2/2\pi r$. This explains the definition of the ampere *as that current which if existing in each of two infinitely long parallel wires separated by a distance of 1 meter in free space, cause force on each wire to be* 2×10^{-7} *newton per meter of length.*

1.57. Torque on Coil in Magnetic Field

The torque L on a coil of N turns carrying current in a field of magnetic flux density B is

$$L = BINA\cos\theta$$

where A is the area of the coil and θ is the angle the plane of the coil makes with the field. When the plane of the coil is parallel to the field, $\cos\theta = \cos 0 = 1$ and the torque is maximum. When the plane of the coil is normal to the field, the torque is zero.

1.58. Force on a Moving Charge in a Magnetic Field

Consider that a charge Q moves with velocity v in a field of magnetic flux density B and covers a distance l in time t. Then $l = vt$ and the moving charge represents a current $I = Q/t$. Substituting $Il = (Q/t)(vt) = Qv$ in $F = BIl\sin\theta$, the force on the charge is

$$F = BQ\,v\sin\theta$$

where θ is the angle between vectors *B* and *v*. *The direction of F is perpendicular to the vectors B and v.*

1.59. Path of Charged Particle in Magnetic Field

When a particle of mass *m*, charge *Q* and initial velocity *v* moves normal to a field of magnetic flux density *B*, the force acting on it as $F = BQv$ directed perpendicular both to *B* and *v*. Since *F* is constant in magnitude and directed always at right angles to *v*, the particle has uniform circular motion with radius *r* and centripetal acceleration $a = v^2/r$. Now combining $F = ma = mv^2/r$ and $F = BQv$, we have $mv^2/r = BQv$ from which

$$r = \frac{mv}{BQ}$$

A cyclotron is used to accelerate massive particles (*e.g.*, protons, deuterons, alpha particles) by applying small voltage many times in succession as the particles spiral outward in a magnetic field. This is possible because the angular velocity $\omega = v/r = BQ/m$ of a particle is independent of its speed and of the radius of its path, as long as *v* is small compared with the speed of light.

A mass spectrograph is used to determine the masses of positively charged ions and to measure the relative magnetic field *B* perpendicular to the vector *v* and thus are caused to move in a circular path of radius $r = mv/BQ$. Since v/BQ is the same for all ions, the mass *m* of the ion is proportional to *r*.

1.60. Lorentz Force

Total force expercuied by a changes particle when both electric and magnetic fields are present is called **Lorentz Force.** Let a particle of charge *Q* move with a velocity $\vec{v}$ in the presence of look electric field $\vec{E}$ and a magnetic field $\vec{B}$. The charge will experience an electric force $\overrightarrow{Fe}$ as well as magnetic force $\overrightarrow{Fm}$ depending upon the conditions of motion.

Electric and magnetic force on the charged particle are given by

$$\overrightarrow{Fe} = Q\vec{E} \text{ and } \overrightarrow{Fm} = Q\left(\vec{v} \times \vec{B}\right)$$

$$\therefore \text{ Lorentz force } \vec{F} = Fe + Fm$$

$$= Q\vec{E} + Q\left(\vec{v} \times \vec{B}\right)$$

or
$$F = Q\left[\vec{E} + \left(\vec{v} \times \vec{B}\right)\right]$$

This equation is called **Lorentz equation.** Form the alone equation, we may conclude

(i) For a given change $\overrightarrow{Fe}$ depends on the magnitude of $\vec{E}$ only while $\overrightarrow{Fm}$ depends on magnitude of $\vec{B}$ as well as that of $\vec{v}$.

(ii) if $v = 0$, *i.e.*, the particle is stationary $Fm = 0$ and only *Fe* acts.

(iii) *Fm* is perpendicular to both $\vec{v}$ and $\vec{B}$ so $\overrightarrow{Fm}$ does no work, it simply changes the directions of motion of charged particle.

(iv) *Fm* does not change kinetic energy of the charged particle while $\overrightarrow{Fe}$ changes the kinetic energy of the charged particle.

1.61. Ratio of Electronic Charge to Mass

Consider that an electron stream with velocity *v* passes through two mutually perpendicular electric and magnetic fields, that the electric field deflects each electron upward and the magnetic field deflects each electron downward, and that when both fields act simultaneously the resultant deflection is zero. The electrostatic force on each electron of mass *m* and charge *e* in an electric field E is $F = eE$, and the magnetic force $F = Bev$. Since the force balance $eE = Bev$ and

$$v = \frac{E}{B}$$

If the electric field is cut off and the magnetic field maintained, each electron will move in the magnetic field in a circular path of radius $r = mv$. Substituting $v = E/B$ and sloving we get

$$\frac{e}{m} = \frac{E}{B^2 r}$$

The ratio $e/m = 1.759 \times 10^{11}$ C kg_{-1}. Since $e = 1.602 \times 10^{-19}$C the rest mass of the electron is $m = (1.602 \times 10^{-19}/1.759 \times 10^{11}) = 9.107 \times 10^{-31}$ kg.

The electron-volt (eV) is the energy acquired by a body whose charge equals that of the electron when the body falls through a potential difference of one volt.

$$1 \text{ eV} = (1.602 \times 10^{-19}\text{C})$$
$$= 1.602 \times 10^{-19} \text{ joules}$$

1.62. Ampere's Circuital Law:

According to this law the line integral of the magnetic field along a closed hyprothetical path in vacuum is equal to μ_0 times the current enclosed by the closed path.

Mathematically, we may write

$$= \oint \vec{B} . d\vec{l} = \mu_0 I$$

Proof:

Let us consider an infinitely long conductor carrying a constant current I. Consider any arbitrary closed path around the conductor. Divide the closed path into large number of segments dl_1, dl_2, dl_5, ... dl_n at a mean distance of r_1, r_2, r_3 ...... r_n respectively. Let the angle subtented by these elements at O be dq_1, dq_2, dq_3 .... dq_n, respectively.

From the figure,

dq_1, + dq_2 + ... dq_n = 2π radian

i.e., $\frac{dl_1}{r_1} + \frac{dl_2}{r_2} + \dots \frac{dl_n}{r_n} = 2\pi$ radian

– (1)

We have considered the elements to be very small so that it may be assumed that each forms an arc of a circle with O as centre.

Using Right hand Thumb rule, we observe that the magnetic lines of field of the conductor have anticlockwise direction when seen from above.

Now, current enclosed by the closed path is I. so we have

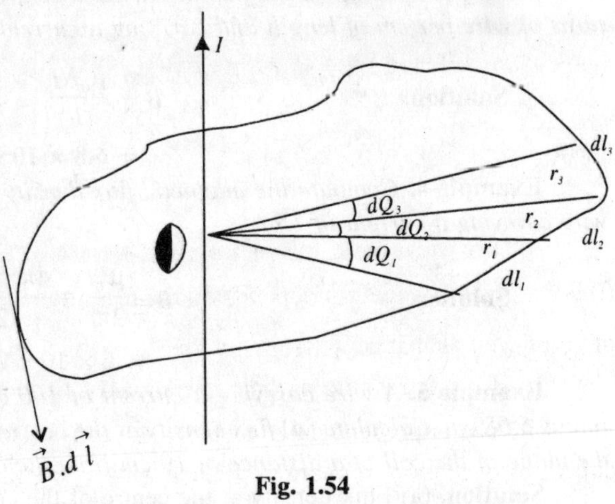

Fig. 1.54

$$\oint \vec{B}.d\vec{l} \;=\; B_1$$

$dl_1 \cos 0° + B_2\, dl_2 \cos 0° +$ .... $B_n\, dl_n \cos 0°$

$$= \frac{\mu_0}{4\pi}.\frac{2I}{r_1} dl_1 + \frac{\mu_0}{4\pi}.\frac{2I}{r_2} dl_2 \quad + \dots\dots \frac{\mu_0}{4\pi}.\frac{2I}{r_n}.dl_n$$

$$\left[\because \text{For a straight conductor } B = \frac{\mu_0}{4\pi}\cdot\frac{2I}{r}\right]$$

$$= \frac{\mu_0 I}{2\pi}\left[\frac{dl_1}{r_1} + \frac{dl_2}{r_2}\right] + \ldots\ldots\ldots \frac{dl_n}{r_n}$$

$$= \frac{\mu_0 I}{2\pi}.2\pi \quad \text{[using eq. (1)]}$$

$$= \mu_0 I$$

Hence the proof.

Example 1. *A charge of 4 × 10⁻⁴ C moves at 15 revolutions per second in a circle of radius 10 cm. What is the magnetic flux density at the centre of the circle?*

Solution. *Current in circle*

$$I = Q/t = (4 \times 10^{-6}\ \text{C}) \left[\frac{1}{15} s\right] = 6 \times 10^{-5}\ \text{A}$$

$$\frac{B - \mu_2 N_1}{2r} = 4\pi \times 10^{-2}\ \text{Wb A}^{-1}\ \text{m}^{-1} = \frac{(1)\,(6\times10^{-5}\ A)}{2\times0.2\ \text{m}}$$

$$= \mathbf{1.88 \times 10^{-10}\ T}$$

Example 2. *A circular coil of 40 turns of wire and negligible section has diameter has diameter 32 cm. What current must exist in the coil to produce a flux density of 3×10^{-4} T at its centre?*

Solution. $B = \dfrac{\mu_0 NI}{2r}$, $3 \times 10^{-4}\ \text{T} = \dfrac{4\pi\times10^{-7}\times Wb\ \text{A}^{-1}\text{m}\times40\times1}{2\times(0.16\ \text{m})}$

I = 1.9 *A* Ans.

Example 3. *Compute the flux density at a centre of the air cone of a long solenoid having 9 turns of wire per cm of length and carrying a current of 6 A.*

Solution. $B = \dfrac{\mu_0 NI}{l} = 4\pi \times 10^{-4}\ \text{Wb A}^{-1}\ \text{m}^{-1} \times \dfrac{9\ \text{turns}}{10^{-2} m} \times 6\text{A}$

$$= \mathbf{6.8 \times 10^{-3}\ T\ \ Ans.}$$

Example 4. *Compute the magnetic flux density in air at a point 5 cm from a long straight wire carrying a current of 15 A.*

Solution. $B = \dfrac{\mu_0 l}{2\pi r} = \dfrac{4\pi\times10^{-7}\ \text{WbA}^{-1}\text{m} - 15\text{A}}{2\pi r\times(5\times10^{-2}\ \text{m})}$

$$= 6 \times 10^{-5}\ \text{Wb m}^{-2} = \mathbf{6 \times 10^{-5}\ T\ \ Ans.}$$

Example 5. *A wire carrying a current of 100 amperes is bent into the form of a circle of radius 5.08 cm. Calculate (a) flux density at the centre of the coil (b) flux density perpendicular to the plane of the coil at a distance of 19 cm from the coil.*

Solution. (a) Flux density at the centre of the coil is given by

$$\text{B} = \frac{\mu_0 I}{2a}$$

a = 5.08 cm = 0.0508 m

I = 100 A, $\mu_0 = 4\pi \times 10^{-7}$ henry/m

$$\therefore \quad B = \frac{4\pi \times 10^{-7} \times 100}{2 \times 0.0508} = \mathbf{12.37 \times 10^{-4}\ Wb/m^2} \quad \textbf{Ans.}$$

(b) Flux density ⊥ to the plane of the coil at a distance r from the coil of radius a is given by

$$B = \frac{\mu_0 I r^2}{2(a^2 + r^2)^{3/2}}$$

a = 5.08 cm = 0.0508 m

r = 12 cm = 0.12 m

I = 100 A and $\mu_0 = 4\pi \times 10^{-7}$ henry/m

$$B = \frac{4\pi \times 10^{-17} \times 100 \times (0.0508)^2}{2 \times [(0.0508)^2 + (0.12)^2]^{3/2}}$$

$= \mathbf{7.33 \times 10^{-5}\ Wb/m^2}$ **Ans.**

Example 6. *Two long parallel wires, C and D, are 10 cm apart in air and carry currents of 6A and 4A respectively. Compute the force of attraction on each metre length of wire if the currents are (a) in the same direction, (b) in opposite direction.*

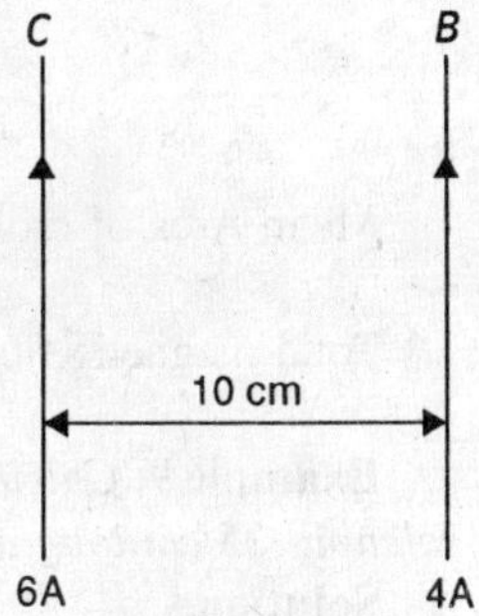

Fig. 1.55

Solution. (*a*) Flux density at *D* due to current in *C* is

$$B_0 = \frac{\mu_0 I_C}{2\pi r}$$

$$F = B_D I_D\, l = \left(\frac{\mu_0 I_C}{2\pi r}\right) I_D l$$

$$= \frac{\mu_0}{2\pi} \frac{I_C . l_D . l}{r}$$

$$= 2 \times 10^{-7} \text{ Wb A}^{-1}\text{m}^{-1} \times \frac{(6\text{A})\,(4\text{A})\,(1\text{m})}{0.1\text{m}}$$

$= \mathbf{4.8 \times 10^{-5}}$ **N attraction** **Ans.**

(*b*) $\mathbf{4.8 \times 10^{-5}}$ **N repulsion** **Ans.**

Example 7. *An electron of energy 1000 eV, describes a circle in a field of magnetic induction 0.02 Wb/m². Calculate the radius of the circle [e/m of electron = 1.76 × 11⁻¹¹ coulomb/kgm, 1 eV =* 1.6 × 10⁻¹⁹ *joules].*

Solution. K.E. of electron = 1000 eV

= 1000 × 1.6 × 10^{-19} joules

$$\therefore \quad \frac{1}{2} mv^2 = 1000 \times 1.6 \times 10^{-19} \text{ joules}$$

$$\therefore \quad v = \sqrt{\frac{2K.E}{m}}$$

$$= \sqrt{\frac{2 \times 1000 \times 1.6 \times 10^{-19}}{9.1 \times 10^{-13}}}$$

= 1.875 × 10^7 m/sec.

We can get the radius of the circle, by equating the deflecting force with centripetal force *i.e.*,

$$\frac{mv^2}{r} = q.v.B$$

$$r = \frac{mv}{qB}$$

where q is the charge on electron

$$\therefore \quad r = \frac{v}{(e/m).B} = \frac{1.875\times10^{7}}{1.76\times10^{-11}\times0.02}$$

$$= 0.00533 \text{ m}$$

$$= \mathbf{0.533 \text{ cm.}} \quad \textbf{Ans.}$$

Example 8. *A solenoid has a mean diameter of 0.03 m and length 2 metres. It has layers of 1000 turns each. Calculate the flux density at its centre when a current of 2.5 amperes flows through it. Also calculate the flux at the centre.*

Solution. In this case

$$l = 2\ m,\ I = 2.5 \text{ ampere}$$

$$N = 4 \times 1000 = 400 \text{ turns}$$

$$\mu_0 = 4\pi \times 10^{-7} \text{ henry/m}$$

Flux density

$$B = \frac{\mu_0 NI}{l}$$

$$= \frac{4\pi\times10^{-7}4000\times2.5}{2}$$

$$= 6.28 \times 10^{-3} \text{ weber/m2}$$

Mean Area of cross-section

$$A = \pi r^2 = \pi\ (0.025)^2 \text{ sq.m}$$

Total magnetic flux

$$= 6.28 \times 10^{-3} \times \pi\ (0.025)^2$$

$$= \mathbf{1.234 \times 10 \text{ weber.}} \quad \textbf{Ans.}$$

Example 9. *Calculate the flux density at the centre of a coil if there are 100 turns wound on a solenoid 25 cm long and carrying a current of 3 amperes.*

Solution. $N = 100,\ I = 3 \text{ A},\ l = 25\ cm = 0.24\ m$

$$\mu_0 = 4\ \pi \times 10^{-7} \text{ henry/m}$$

The flux density at the centre of the coil is given by the relation

$$B = \frac{\mu_0 NI}{l.}$$

$$= \frac{4\pi\times10^{-7}\times100\times3}{0.25}$$

$$= \mathbf{1.508 \times 10^{-3} \text{ wb/m}^2} \quad \textbf{Ans.}$$

Example 10. *A wire of square shape of each side 10 cm long is carrying a current of 2 A in the anticlockwise direction. Calculate the megnetic field at its centre.*

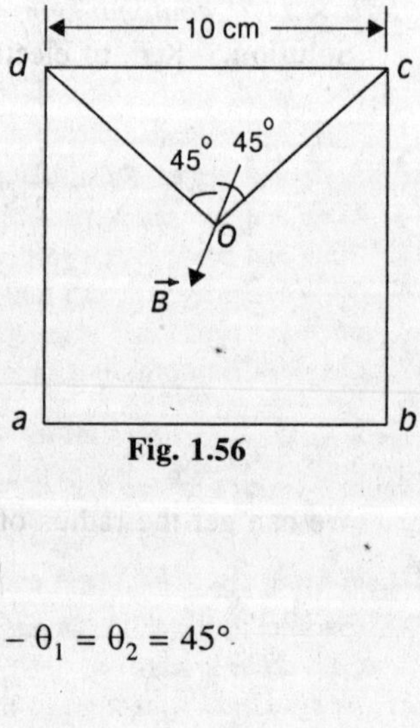

Fig. 1.56

Solution. Each side of the square will subtend an angle of 90° at the centres as shown in the Fig. 1.56. The point of observation is at a distance OL = 1/2 length of each side = 5 cm from each side.

The symmetry of the wire about θ allows us to calculate the magnetic field as the net magnetic field B = 4 × magnetic field due to each side.

Let us calculate the magnetic field due to one side say *'ab'*. Making use of the relation

$$B_{ab} = \frac{\mu_0 I}{4\pi R}\frac{1}{R}\ (\sin\theta_1 + \sin\theta_2)$$

In this case $I = 2\text{A},\ R = 5 \text{ cm} = 0.05 \text{ m},\ -\theta_1 = \theta_2 = 45°$

$\therefore \qquad \sin\theta_1 = \sin\theta_2 = \dfrac{1}{\sqrt{2}}$

Also $\qquad \mu_0 = 4\pi \times 10^{-7}$ T.m.A^{-1}

And $\qquad \sin\theta_1 + \sin\theta_2 = \dfrac{2}{\sqrt{2}} = \sqrt{2}$

So $\qquad B_{ab} = \dfrac{10^{-7}\ \text{Tm A}^{-1} \times 2\text{A}}{0.05\text{m.}}\sqrt{2}$

$= 10^{-7} \times 2 \times 20$ T

Total magnetic field $B = 4 \times B_{ab}$

$= 4 \times 2\sqrt{2} \times 20 \times 10^{-7}$ T

$= \mathbf{226.24 \times 10^{-7}}$ **T** **Ans.**

We can get the direction of the field, by right hand screw rule. According to which it points outwards perpendicular to the plane of the paper.

QUESTIONS

1. What are scalar and vector field? Give atleast on example.
2. What do you mean by gradient of scalar field? Find an expression for the gradient of a scalar field in terms of operator $\vec{\nabla}$.
3. Show that the gradient for a scalar field is a vector.
4. Explain clearly what do you mean by divergence and curl of a vector field F.
5. What is conservative field? Give examples of such field.
6. What do you understand by the field on any physical quantity? Give examples of scalar and vector fields.
7. Give the physical significance of grad ϕ. Express grad ϕ in cartesian Coordinates. Hence, define $\vec{\nabla}$ operator.
8. Discuss in brief line integral of a vector and conservative field on its basis.
9. What do you mean by divergence of a vector function? Obtain an expression for divergence of a vector field in Cartesian coordinates.
10. What are line and surface integrals?
11. What is the physical significance of (i) gradient of a scalar field (ii) divergence of a vector field and (iii) curl of a vector field? Explain by giving one example of each.
12. Define curl of a vector field. Give the physical significance of the curl of a vector field. Derive an expression for it. Prove, curl $\vec{A} = \vec{\nabla} \times \vec{A}$.
13. State and prove Gauss's divergence theorem. What is its importance?
14. State and prove Gauss theorem of divergence of Gauss-divergence theorem.
15. State and prove Stock's theorem.
16. State and prove Gauss's law in electrostatics. Use this law to determine the electric intensity inside and outside a uniformly charged sphere
17. Decuce Coulomb's law from Gauss's theorem.
18. What do you mean by the flux of an electric field? Calculate the flux passing over any closed surface.
19. State and prove Gauss's law in electrostatics.
20. Use Gauss's law to find the electric field intensity due to a uniformly charged sphere at an (i) external point (ii) internal point, and (iii) at the surface of a charged distribution.
21. What is electric flux? A cylinder of radius R is immersed in a uniform electric field $\vec{E}$ with its axis (parallel to the field (ii) perpendicular to the field. Prove that the electric flux through the cylinder in each cases in zero.
22. Apply Gauss's theorem to find electric field due to a spherical symmetric charge distribution.

23. State and prove Gauss's law in electrostatics. Derive Coulomb's law from the Gauss's law.
24. Use Gauss's law to find electric field due to a non-uniform spherical charge distribution at an external and internal points.
25. State and prove Gauss's law. Apply it to find the electric field due to an infinite line of charge.
26. Using Gauss's law, find the electric field intensity due to uniformly charged wire to infinite length.
27. State Gauss's law in electrostatics in the integral from and apply it to find the electric field near a uniformly changed long wire.
28. Apply Gauss's theorem to find electric intensity due to
(i) a charged spherical shell
(ii) two concentric spherical conductors.
(iii) an uniform charged wire.
(iv) an uniformly charged hollow cylinder.
(v) a charged insulating sheet.
(vi) a charged conductor.
(vii) Two parallel charged sheets.
29. Apply Gauss's theorem to show that the electric field strength near a charged conductor is twice of electric field strength in front of a non-conducting thin sheet of charge, with the same surface charge density as the conductor.
30. State Gauss's law in electrostatics and express it in the differential from.
31. Deduce Gauss's law in differential from, that is $\vec{\nabla}\vec{E} = \rho/\varepsilon_0$.
32. Derive an expression for the force per unit area on the surface of a charged conductor.
33. Derive Laplace's equation for electrostatics potential and hence prove that there can be no electric field in the interior of a hollow conductor.
34. Derive Poisson's and Laplace equations. What are the importance of these equations?
35. Starting from Gauss's law, establish Laplace's equation and discuss its significance.
36. State and prove Arlpere's's circular at law.
37. State and explain Biot Savat's law.
38. Explain Lorentz law of electromagnetism
39. Calculate the field due to a solenoid
40. Find the magnetic field on the axis of a circular coil.

NUMERICAL PROBLEMS

1. Prove that, $dV = (d\vec{r}.\vec{\nabla})V + \frac{\partial V}{\partial t} dt$
2. If $\vec{a} = a\,xi + \beta yj + \gamma\, zk$, show that $\vec{\nabla}(\vec{a}.\vec{r}) = 2\vec{a}$
3. Let $\vec{r}$ be the vector from some fixed point (x_0, y_0, z_0) to the point (x, y, z), and let r be its length. Show that, $\vec{\nabla} r^2 = 2\vec{r}$.
4. Find the gradient of f, where f is given as $yz + zx + xy$. **(Ans. 0)**
5. Find the gradient of f, where f is given as **(Ans.**(*i*)$2(y\,\hat{\mathbf{i}} + \hat{\mathbf{j}})$ (*ii*) $(y\,\hat{\mathbf{i}} - x\,\hat{\mathbf{j}})$**)**
(i) $f = 2xy$ (ii) $f = x/y$
6. Equipotential surfaces in some problems are spherical in shape : $V(x, y, z) = C(x^2 + y^2 + z^2)$. Show that the force is radial, that is, along the radius vector.
7. If $\vec{A} = 3xyz^2\ i + 2xy^2\ j - x^2yz\ k$ and $\phi = 3x^3 - yz$, find
(*a*) $\vec{\nabla}.\vec{A}$, (b) $\vec{A}.\vec{\nabla}\phi$ (c) $\vec{\nabla}.(\phi + \vec{A})$ (d) $\vec{\nabla}.\vec{\nabla}\phi$ at (1, –1). **(Ans.** (*a*) 4 (*b*) –15 (*c*) –1 (*d*) 6**)**
8. Evaluate $\vec{\nabla}.(r^3\vec{r})$ **(Ans 6r³)**
9. Prove that if $\vec{F}$ represents a vector field and curl $\vec{F} = 0$, but div $\vec{F} \neq 0$, then field has a vector potential.
10. If $\vec{F} = yi + (x^2 + y^2)j + (yz + zx)k$, calculate div $\vec{F}$. **(Ans.** $x + 3y$**)**

11. If $\vec{A} = 3x^2 i + 5xy^2 j + xyz^2 k$, then find div $\vec{A}$ at the point (1, 2, 3) (**Ans.** 38)

12. If $\vec{A} = \vec{r}/r$, find grad div $\vec{A}$. (**Ans** $- 2r^3.\hat{r}$)

13. Show that vector $\vec{A} = y^2\ i$ is solenoidal.

14. If $\vec{A} = kr^n \hat{r}$, prove that div $\vec{A} = k\ (n+2)\ r^{n-1}$

15. Find the work done W in moving a particle in a force field given by $\vec{A} = 3xy\ i - 5zj + 10x\ k$ along the curve $x = r^2 + 1, y = 2t^3, z = t^3$ from $t = 1$ to $t = 2$. (**Ans.** 303)

16. Prove that [grad $f(r)$] $\times \vec{x} = 0$

17. Evaluate curl $\vec{F}$ if $\vec{F} = yi + (x^2 + y^2)j + (yz + zx)\ k$.

18. Evaluate $\oint \vec{A}.d\vec{r}$ around the closed curve, $y^2 = x$, y goes from 1 to 0, where $\vec{A} = (x - y)i + (x + y)j$. (**Ans.** 0.667)

19. Find the div and curl of the following vector fields:

(a) $\vec{F} = 2x^2 i + (3y^2 + z^2) j + 2yzk$

(b) $\vec{F} = 2xi + (5y - 9z) j - 9jk$ [(a) **Ans.** 4x + 8y, 0 (b) **Ans.** 7.0]

20. If $\vec{A} = xz^3 i - 2x^2 yzj + 2yz^4 k$, find curl $\vec{A}$ at the point (1,–1,1). (**Ans.** $3j+4k$)

21. Calculate div and curl for the following electric field:

$$\vec{E} = \frac{1}{4\pi\varepsilon_0 r^3}$$

Explain wether this field is conservative of non-conservative? (**Ans.** 0, 0, conservative)

22. Prove that div grad $\phi_1 \times$ grad $\phi_2 = 0$

23. If S is any closed enclosing a volume V, then prove that $\iint_S (3xi + 5yj + 7zk).d\vec{S} = 15V$.

24. Prove $\iiint_V (\vec{\nabla} \times \vec{B})\ dV = \iint_S (\hat{n} \times \vec{B})\ ds$.

25. $\iiint_V \vec{\nabla}\phi dV = \iint_S \phi \hat{n} dS$

26. Evaluate $\oint_C \vec{F}.d\vec{r}$, $\vec{F} = (x-3y)i + (y-2x)j$ and C is the closed curve in the X–Y plane, $x = 2\cos t$, $y = 3\sin t$, from $t = 0$ to $t = 2\pi$. [**Ans.** 6π]

27. A solenoid of length 20 cm and radius 2 cm is closely wound with 200 turns. Calculate the field at the centre of one end of the solenoid. The current in the winding is 5 amperes. [**Ans.** 31. 43 oersted]

28. Two concentric circular coils x and y of radii 16 cm respectively lie in the same vertical plane containing north-south direction. Coil x has 20 turns and carries a current of 16 A; coil y has 25 turns and carries a current of 18 A. The sense of the current in x anticlockwise and in y clockwise, for a observer looking at the coils facing west. Give the magnitude and direction of the net magnetic field due to the coils at the centre. [**Ans.** 1.8×10^{-3} T, directed towards west]

29. Two straight wires each 1 m long are parallel to one another and are each carrying a current of a 40 amperes. What force in Newtons will be set up between the wire if the distance between the wires is 0.02 m ($\mu_0 = 4\pi \times 10^{-7}$ henry/m). [**Ans.** 16×10^{-2} N]

30. A short conductor of length 5.00 cm is placed parallel to a long conductor of length 1.5 m near its centre. The conductor carry currents 4.0 A and 3.0 A respectively in the same direction. What is the total force experienced by the long conductor when they are 3.0 cm apart? [**Ans.** 4×10^{-2} N]

31. A condenser charged to 2 volts in discharged through a ballistic galvanometer, when the corrected deflection is 9.6 cms and the current sensitively is 2.2×10^{-8} amp/cm and periodic time is 12 second. Calculate the capacity of the condenser.

2

MAGNETIC PROPERTIES OF MATERIALS

2.1. Introduction

Students are familiar with the various properties of solids and their importance in various fields. However, few more characteristics of solids will be discussed in this chapter. We will consider some important aspects of solids and know about their applications and importance.

Magnetic Characteristics

The best known magnets contain *metallic iron*. However many other elements, which are not metallic, still show magnetism. In modern days, in many fields, *metallic* as well as *ceramic magnets* are being put to use.

Before taking up the classification and various applications of magnetic materials, we will review some of the important definitions and properties of magnets.

2.2 Magnetic Dipole Moment

A small bar magnet when suspended in a uniform magnetic field of induction B, experiences a torque, which tends to align the magnet with the field. The force action on each pole of the bar magnet due to B is 'mB' where m corresponds to the strength of the magnet. The forces acting on north and south poles of the magnet act in opposite directions and constitute a couple (Fig. 2.1) having moment $mB \times d \sin\theta$, where d is the distance of separation of the poles and θ is the angle that the axis of the bari magnet makes with the magnetic induction vector B.

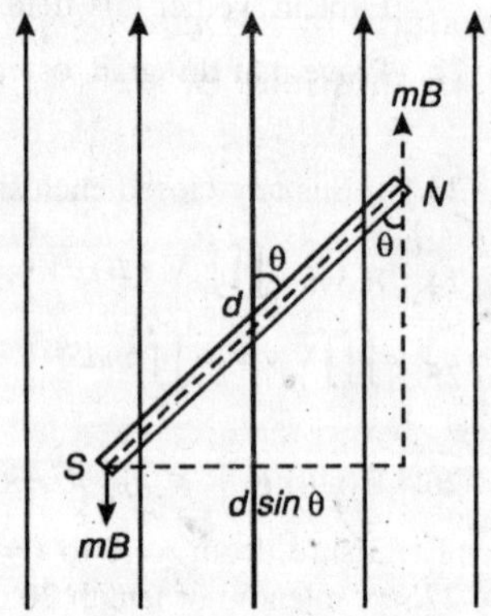

Fig. 2.1

$\therefore$ The torque $\tau = md\, B \sin\,\theta$

or $\vec{\tau} = \vec{\mu}_m \times \vec{B}$ ...(1)

where $\mu_m = md$ is known as the **magnetic dipole moment** μ_m *is a vector, directed from the south pole to the north pole.*

Also we know that a current loop freely suspended in a homogeneous magnetic field acts the same way as the bar magnet in Fig 2.2. It gets deflected due to the force acting on it (Fig.2.2*a*). The torque τ, acting on a loop, carrying a current I, is directly proportional to the area of the loop A to the current I and to the magnetic induction B and is given by

$$\tau = IAB\sin\theta; \quad \vec{\tau} = \vec{\mu}_m \times \vec{B}$$

where $\vec{\mu}_m = IA$.

μ_m characterizes the magnetic properties of the current loop and accounts for its behaviour in an external magnetic field. Again $\vec{\mu}_m$ is known as the magnetic dipole moment or simply **magnetic moment** of the current loop. The direction of vector $\vec{\mu}$ in this case, is determined with the aid of the **right hand screw rule**. *If the screw is rotated in the direction of the current flowing in the loop the translational motion of the screw will indicate the direction of the vector* μ_m [Fig. 2.2(*b*)]

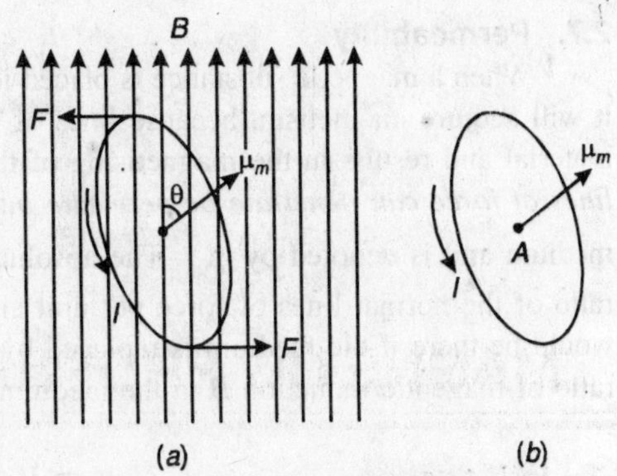

Fig. 2.2. (*a*) A current loop experiences a torque in a uniform magnetic field. (*b*) Direction of magnetic dipole moment of the current loop is determined by right hand screw rule.

The bar magnet of the current loop can be in a state of equilibrium in an external magnetic field, when the directions of **vectors** $\vec{B}$ and $\vec{\mu}_m$ coincide.

Examples of magnetic dipoles are a bar magnet, a compare or a current loop.

2.3. Magnetising Field (*H*)

This is the field in which a substance is placed and this field is usually uniform and is measured by the *number of unit tubes of force passing per unit area normally*,

2.4. Intensity of Magnetisation (*I*)

It gives the measure of the magnetisation of a magnetised specimen. It is defined as *magnetic moment per unit volume i.e.,*

$$I = \frac{\text{Magnetic Moment}}{\text{Volume}} = \frac{M}{V}$$

Since we normally take a small specimen, its magnetisation can be supposed to be uniform and also if the specimen has a uniform area of cross - section a and magnetic length 2*l*, and has pole strength *m*, then

$$M = m \times 2l \quad \text{and} \quad V = a \times 2l$$

$$\therefore \quad \frac{m \times 2l}{a \times 2l} = \frac{m}{a}$$

That is, the intensity of mangetisation for a specimen of specimen of uniform area of cross-section is given by **pole strength/unit area.** It is measured in ampere turn/metre.

2.5. Magnetic Induction or Flux Density

The total number of lines of force/unit area due to magnetising field and due to the field induced in the substance is called flux density (*B*) and is measured in weber/m^2 or Telsa.

2.6. Magnetic Susceptibility

Since the intensity of magnetisation for a material is proportional to the magnetic field, so the magnetic susceptibility (X) of a specimen measures the **ease with which the specimen** *can be magnetised and it is defined as the ratio of the intensity of magnetisation induced in it and magnetising field i.e.,*

$$\chi = \frac{1}{H}$$

2.7. Permeability

When a magnetic substance is placed in a magnetic field, due to the induction phenomenon, it will acquire magnetism, because lines of force of the magnetising field concentrate inside the material and results in the magnetising of the material. The measure of the *degree to which the lines of force can penetrate or permeate into the medium* is called absolute permeability of the medium and is denoted by μ_a. The absolute permeability of the medium can be defined as the ratio of the normal lines of force per unit area in the medium to the lines of force/unit area that would be there if the medium is replaced by air or vacuum so the permeability is defined as the ratio of magentic induction B in the medium to the magnetising field i.e.,

$$\mu_a = \mu_0\mu_r = \frac{B}{H} \qquad ...(1)$$

the term absolute permeability μ_a of material is expressed as the product of the permeability of the free space μ_0 (= $4\pi \times 10^{-7}$ henry/metre) and μ_r (the relative permeability).

i.e., $$\mu_a = \mu_0 \times \mu_r$$

where μ_r is given as the ratio of the *force between two poles* when placed at a fixed distance in vacuum to the force between them at the same distance in that medium. Since μ_r is ratio, it has no units and $\mu_0 = 1$ in air or vacuum.

When a magnetic material of A cross-sectional area and μ_r relative permeability is placed in a uniform field of intensity H, two types of lines of force pass through it, one due the magnetising field H and the other due to the material being magnetised by induction. So the total flux density will be given by

$$B = \mu_0 H + I \qquad ...(2)$$

where μ_0 corresponds to the permeability of free space. Now we have absolute permeability μ_a as

$$\mu_a = \mu_0\mu_r = \frac{B}{H} = \frac{\mu_0 H + I}{H}$$

$$= \mu_0 + \frac{I}{H} = \mu_0 + \chi$$

Dividing both sides by μ_0, we get the relation for relative permeability as :

$$\mu_r = 1 + \frac{\chi}{\mu_0} \qquad ...(3)$$

2.8. Magnetic Circuits

The path followed by magnetic flux is called **magnetic circuit**. As in the case of electric current which follows a closed circuit, magnetic flux follows a complete circuit or loop coming back to the starting point. Magnetic flux leaves from N pole and passing through air enters at S pole and finally reaches N pole after travelling through the magnet.

The flux ϕ is produced by the magnet due to some force, called **magnetomotive force**.

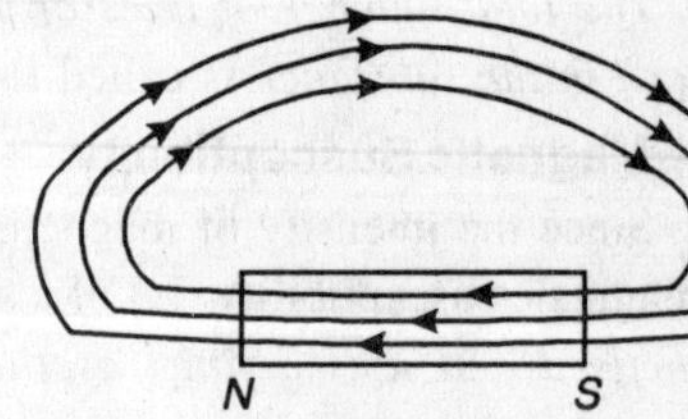

Fig. 2.3

2.9. Magnetomotive Force

Like electromotive force in electricity, we have magnetomative force (*m.m.f*) which may be defined as the *force that drives or tends to drive magnetic flux through a magnetic circuit.*

An m.m.f is produced in a solenoid when electric current flows in a coil of several turns and m.m.f. is measured by the product of current in amperes and number of turns. The unit of m.m.f. is **ampere turn** (*AT*) and denoted by M or F.So, we may write

$$M = I \times N\ (AT)$$

2.10. Reluctance

Similar to the resistance characteristic in electricity, we have reluctance in magnetism which is the property of the substance **that opposes the magnetic flux through it**.

Reluctance of any part of a magnetic circuit may be defined as the *ratio of the drop in magnetomotive force to the flux produced in that part of the circuit*. It is measured in ampere turns/weber and is denoted by R.

Reluctance is dependent on :

(*i*) Nature of the substance.

(*ii*) Area of cross-section of the material through which flux is passing (a).

(*iii*) Length of magnetic path. Mathematically, we may write

$$R = \frac{1}{a\mu_0\mu_r}\ AT/Wb$$

where μ_0 is absolute permeability

and $\mu_0 = 4\pi \times 10^{-7}$ henry/ m

2.11 Permeance

The reciprocal of reluctance is called permeance and is a measure of the readiness with which the magnetic flux is developed. It is analogous to conductance in an electric circuit and is measured by weber/ampere turn. So

$$\text{Permeance } P = \frac{\text{flux}}{\text{m.m.f.}}\ Wb/AT = \frac{1}{\text{reluctance}}\ (Wb/AT)$$

2.12 Classification of Magnetic Materials

We can classify the magnetic materials from the concept of *motion of an electron* in an atom. As discussed earlier, orbital motion of a charge carrying electron, around the central nucleus, is analogous to a *current carrying conductor*. With the application of a magnetic field, motion of electrons get modified in such a way that a weak magentic moment, opposing the field, is induced. **Dia-magnetism is the result of this interaction**.

(*a*) **Diamagnetism**. In a diamagnetic solid, the magnetic lines of force, due to an applied field, are repelled. Diamagnetic susceptibility is negative and very small in magnitude. (typically 10^{-5}). An important exception being a *super conductor,* though a perfect diamagnetic, yet having susceptibility value of –1. *All lines of force are repelled by a superconductor, thereby making them suitable for shielding of magnetic fields*.

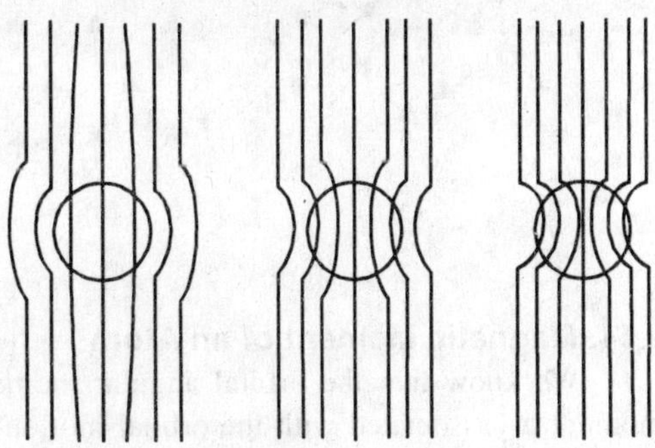

Fig. 2.4. (*a*) A diamagnetic solid repels slightly the magneticlines of force. (*b*) A paramagnetic solid weakly attracts the lines of force. (*c*) A frerromagnetic solid attracts them very strongly.

(*b*) **Paramagnetism**. Some atoms and molecules possess (intrinsic) permanent magnetic moments. In the absence of external applied field, the moments of atoms in a solid are randomly oriented *w.r.t.* one another and so the solid, on the whole, has no magnetic moment. However, if an external magnetic field is applied, the various magnetic moments tend to alien themselves **parallel to the applied field** thereby lowering the potential energy of the system. A typical example is in case of a suspended magnetic needle which aliens itself spontaneously with the **Earth's magnetic field**. Similarly the atoms of a solid try to alien themselves in the direction of the applied field, thereby intensifying the lines of force in the field direction. This phenomenon is called paramagnetism and is shown in Fig. 2.4 (*b*). The aligning force on the permanent moments of the atoms with ordinary magnetic fields is quite small, so that the **paramagnetic** effect is weak. Paramagnetic susceptibility is small and positive ~ 10^{-3}. It is important to note that this value decreases with rise in temperature as **thermal energy tends to change the alignement of atoms.**

(*c*) **Ferromagnetism**

Ferromagnetic solids are those solids in which the permanent magnetic moments (due to electron spins) are **already aligned due to bonding forces**. In case of ferromagnetic materials susceptibility is very large in the range of 10^2 to 10^5. They strongly attract the magnetic lines of force as shown in Fig. 2.4 (*c*)

2.13. Importance of Dipole Moments in Classification of Magnetic Materials

The magnetic materials may be re-classified, on the basis of permanent magnetic dipole materials into two broad groups, **as materials whose atoms carry permanent magnetic dipoles and those which do not have permanent magnetic dipoles**. Materials composed of atoms or molecules having zero magnetic moment, are called **diamagnetic**. Further, materials composed of atoms or molecules, having permanent magnetic moment are classified into four categories, depending on the interaction between the **individual magnetic dipoles**. If the interaction between the atomic magnetic dipoles is negligible, the material is called *paramagnetic* [Fig 2.5 (*a*)]. If the magnetic dipoles interact in such a way that they tend to orient in the same direction, the material is ferromagnetic [Fig 2.5 (*b*)]. If neighbouring dipoles orient in opposite directions and if the dipoles are of equal magnitude, the material is **antiferromagnetic** [Fig. 2.5(*c*)]. If the neighbouring dipoles are of different magnitude and orient antiparallel, the material is **ferri-magnetic**. [Fig. 2.5 (*d*)] (All the four types of orientations are given in the [Fig. 2.5]

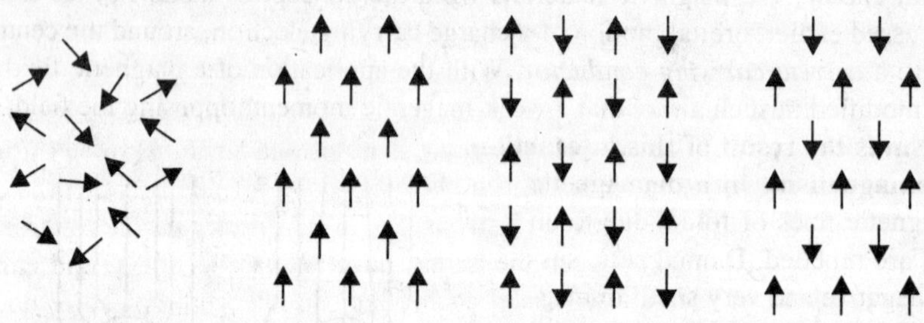

(a) (b) (c) (d)

Fig. 2.5

2.14. Magnetic Moment of an Atom

We know that the orbital angular momentum L of an orbiting electron and the magnetic moment μ_m associated with the orbital motion of the electron is given by

$$\mu_{morb} = -\frac{e}{2m}L \quad ...(1)$$

As the orbital angular momentum is space quantized and its projection L_z in a given direction z can take only $(2l + 1)$ specific values, so we have

$$L_z = m_e\hbar \quad ...(2)$$

Using the value of L_z into equation (1) we get

$$\mu_{morb} = -\frac{eh}{2m}m_e = -m_e\mu_B \quad ...(3)$$

The quantify $$\mu_B = \frac{e\hbar}{2m} = 9.274 \times 10^{-24}\ A.m^2 \qquad ...(4)$$
is known as the **Bohr magneton.**

The Bohr magneton is the elementary electron magnetic moment. It is elementary in the sense that no electron can have a magnetic moment below μ_B. So, it represents the minimum non-zero value of the projection of the magnetic moment of a electron in an arbitrary direction. In general, the component of the electron orbital **magnetic moment along an external field is equal to m_j Bohr magnetons.**

The spin angular momentum and the spin magnetic dipole moment of an electron are given by

$$\mu_{m\ spin} = -\frac{e}{m}S_z \qquad ...(5)$$

where S_z can have the values

$$S_z = \pm\frac{1}{2}\hbar \qquad ...(6)$$

$$\mu_{m\ spin} = \frac{e\hbar}{2m} \qquad ...(7)$$

Each orbit in an atom acommodates two electrons which orient themselves in opposite directions and revolve in opposite directions. **Because of this the spin and orbital magnetic moments of the pair cancel each other**. So, an atom, having completely filled electron subshells or shells does not posses permanent magnetic moment. So, the magnetic moment of an atom is determined **only by the electrons which occupy partially filled shells**. In majority of cases, the partially filled shells happen to be the outer valence shells. The electron spins and orbital magnetic momente in an outer shell tend to orient themselves in opposite directions so as to compensate each other to the maximum possible extent. So, the magnetic moment of a free atom is mainly determined by **uncompensated** spins of the outer electrons.

2.15. Magnetic Moments of Molecules

Although the magnetic moment of a free atom is mainly found by uncompensated spins of valence electrons, the magnetic moment of a molecule **is not equal to the sum of the magnetic moments of the atoms constituting it,** because a chemical bond formed between the atoms requires a certain rearrangement of the valence electrons. So, in a N_2 molecule the electrons participating in the covalent bond, formed between the atoms, have antiparallel spins, **and cancel each others effect**. Their orbital moments also cancel each other. So nitrogen molecules do not have permanent magnetic moment and hence N_2 is not **paramagnetic**. Such cancellation of magnetic moments takes place in molecules having an ionic bond. For example, *a* NaCl molecule has an ionic bond between Na^+ and Cl^-. Both ions have closed electron shells and hence their total magnetic moment is equal to zero. **They form a diamagnetic molecule.**

It means that the general tendency in the formation of molecules, is to ensure that the net magnetic moment is zero. However, oxygen molecule is an exception where the spins of collective electrons are compensated. In case of NO and NO_2, the total number of electrons is odd and so the spin of one electron is left uncompensated. As a result **oxygen, nitric and nitrous oxide gases exhibit paramagnetic properties.**

2.16 Origin of Magnetization

Ampere (1775-1836), postulated the molecular currents to explain the magnelixation of materials but he was not able to explain the origin of molecular currents. Rutherford's atomic model helps to explain the magnetic properties of the materials, which originate as a consequence of magnetic, moments associated with atoms. In general, magnetic moment of an atom may be due to the following parameters:

(a) **The orbital motion of electron :** The motion of an electron constitutes a current and the circular path of electron is identical to a current loop. Such a current loop behaves as an elementary magnet having a magnetic moment. The total orbital magnetic moment of an atom is the sum of orbital magnetic moments of individual electrons.

(b) **The electron spin.** The spinning electron may also contribute magnetic moment.

(c) **The nuclear spin.** The magnetic moment of the nucleus is about $\frac{1}{2000}$ as much as that of an electron. So, in studying the magnetic properties of materials, the magnetic moment due to nuclear spin is neglected.

The two factors, *i.e.*, orbital motion and spin of electrons, contribute atoms with a **permanent magnetic moment**. However, in many cases, the magnetic moments of electrons in an atom cancel each other and the atom may not possess a permanent magnetic moment. When such atoms are placed in a magnetic field, the orbital motion of electrons undergoes changes and the atoms acquire an **induced magnetic moment**.

2.17. Curie Law and Curie Point

The variations in the susceptibility of paramagnetic and diamagnetic materials was studied by Piere Curie in 1895. For paramagnetic materials, susceptibility is inversely proportional to the absolute temperature. This is known as Curie law *i.e.*,

$$\chi \propto \frac{1}{T} \text{ or } \chi = \frac{C}{T} \text{ where } C \text{ is curie constant.}$$

However, the later experimental data showed that Curie law is only approximately true. Another more accurate law called **Curie Wess law** is $\chi_m = \frac{C_m}{T - T_c}$, where χ_m is the molar susceptibility of the material, C_m is the Curie constant, T the absolute temperature of the specimen and T_C is the Curie temperature χ_m is the susceptibility for 1 *gm* equivalent of the material. When ferromagnetic materials are heated above certain critical temperature, their most of the magnetic properties are lost and they behave as paramagnetics. A data for approximate values of Curie temperature for various substances is given under.

Iron = 770° C, Cobalt = 1150°*C*, Nickel = 360°C *Ni-Fe* (*Fe* 30%) = 70°*C*, Permalloy (*Fe* 22% Ni 78%) = 550°*C*, Thermalloy (Nickel-Cu alloys) = 10°*C*.

2.18. Magnetic Domains :

Magnetic properties of the materials can be explained using the **Domain Theory**. Ferromagnetic property of a magnetic material is due to **regular or order arrangement** of neighbouring atoms in the same direction. Ferromagnetic materials have a magnetic moment even in the absence of an external magnetic field. This is called **Spontaneous magnetization.** Spontaneous magnetization has non zero value below certain temperature T_C called *Critical temperature.* Above critical temperature ferromagnetic national behaves like a **paramagnetic material**. Normally value of T_C is higher than the room temperature, and so these materials can be use as **permanent magnets** at room temperature.

In order to explain **Ferromagnetism** Weiss put toward. **Domain Theory** According to this theory a single, Ferromagnetic crystal may be considered to be constituted by a large number of small regions known as domains which are oriented in the same direction *i.e.*, each region has saturated value of magnetism. At room temperature, small thermal agitations do not disturb the dipoles in a particular domain. However, since each domain to the sample is randomly oriented so the net magnetization of the sample is zero. Even by the application of a weak external field, magnetization can be observed. This magnetization is due to the following two mechanisms.

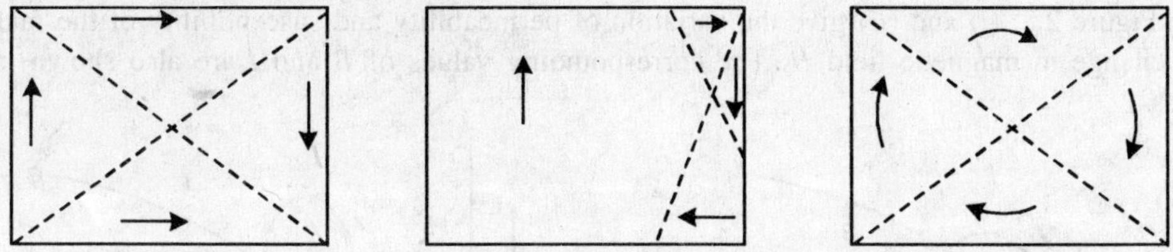

(a) Unmagnetized material (b) Magnetization due to domain growth (c) Magnetization due to domain rotation

Fig. 2.6

(*i*) In case a small external magnetic field is applied, then the magnetic domains in the material grow in the direction of the applied external field. In case external fields in raised to higher values, the domains intensity grows in the direction of applied external field, till it attains saturation value (Fig. 2.6b)

(*ii*) If the magnetic field is increased further than the **saturation value** as explained above then, magnetic domains rotate like a single entity (fig. 2.6 c). Finally, all domains orient themselves in the direction of the applied external field and maximum value of magnetization for the material is attained. This value becomes constant and does not increase even if more external magnetic field is applied.

In case the external field in remarked, the material loses some of the magnetic properties. However, in case of ferromagnetic materials, the magnetic properties still exist even if the external magnetic field in removed and so **a Ferromagnetic crystal** behaves like a permanent magnet in the absence of the external field.

Magnetic domains contributions can be observed in hysteresis phenomenon of Ferromagnetic materials.

2.19. Hysteresis

Introduction. If a ferromagnetic material, say iron is magnetised slowly by gradually increasing the intensity of the magnetic field *H*, the magnetic induction starts increasing. A plot of the *B-H* curve is given is Fig. 2.7. The portion *OA* corresponds to initial magnetisation when the magnetic field is increased from *O* to *G*. When the value of the magnetising field is changed from *G* to *G* and back to + *G*, the variation in magnetic induction is given by the curve *ABCDEFA*. This closed curve is called Hysteresis curve.

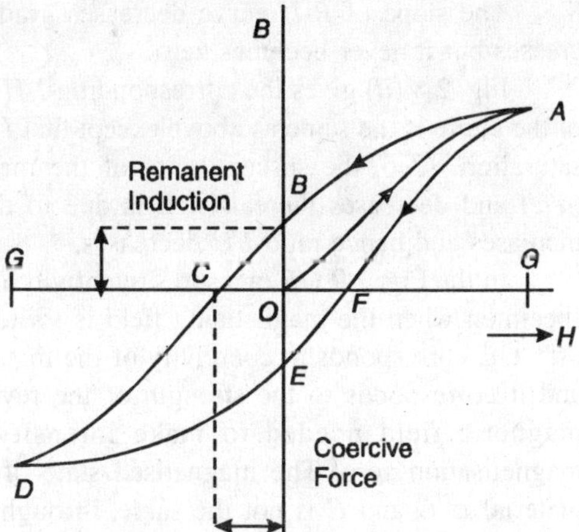

Fig. 2.7

2.20 Hysteresis Curve or Hysteresis Loop.

The permeability of the specimen μ (= *B*/*H*) is normally a function of *H* and the rate of increase of permeability at any point on the curve is given by $\frac{dB}{dH}$ *i.e.*, the slope of the curve at that point. For very small *H*, the initial permeability is independent of *H* and the slope of the curve corresponding to steepest portion measures the maximum permeability of the specimen. In the same way, we can plot an *I-H* curve and the initial and max value of susceptibilities χ of the specimen can be calculated. The slope $\frac{dI}{dH}$ at any point on *I-H* curve measures the incremental susceptibility.

Figure 2.8. (*i*) and (*ii*) give the variation of permeability and susceptibility of the material with change in magnetic field *H*. The correspoinding values of *B* and *I* are also shown in the curves.

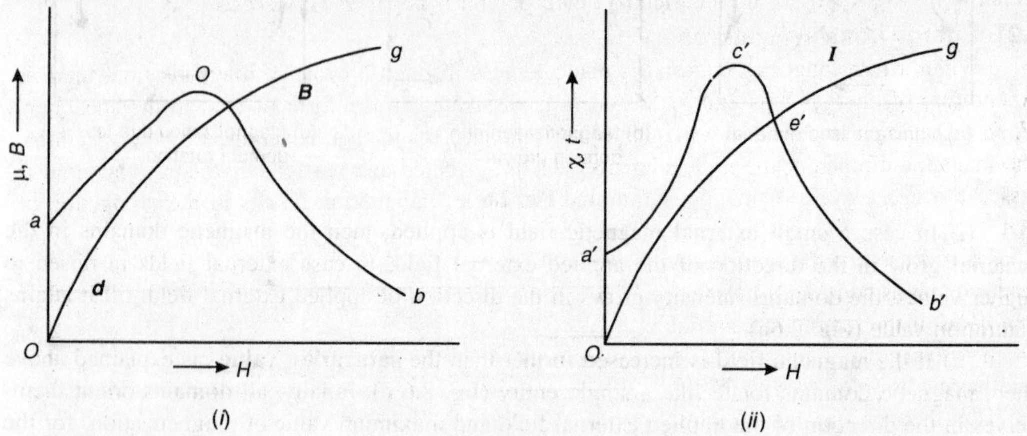

Fig. 2.8

From Fig. 2.8. (*i*), we can see that the value of *B* increases rapidly for small fields (portion Odg) and for higher values of *H*, the increase in value of *B* is not proportional to the increase in *H* (portion fg). This is due to the magnetic saturation of the magnetic substance. However, *B* does not become constant because $B = H + 4\pi I$ and so even if *I* is constant, the value of B increases with *H*. However for low fields, *I* also increases and so there is a steep rise in *B-H* curve.

It is clear that permeability of a material (*i*) increases at first with increase of *H* (*ii*) becomes maximum at the steepest portion of the curve (*iii*) and then it decreases.

The slope of *B-H* curve decreases gradually and so the incremental permeability also decreases but it never becomes zero.

Fig. 2.8 (*ii*) gives the corresponding *I-H* and χ-*H* curves. It is clear that the general variation of the curve is the same as above except that *I* attains constant value at high fields due to magnetic saturation. Also, the susceptibility of the material increases rapdily for low magnetising fields (*a′c′*) and decreases thereafter. It is due to the fact that the value of I remains constant but H increases and hence ratio *I*/*H* decreases.

In the Fig. 2.9 *OB* measures retentivity *i.e.*, the residual intensity of magnetisation left in the specimen when the magnetising field is switched off , *OC* corresponds to coercivity of the material and it corresponds to the strength of the reverse magnetic field needed to make intensity of magnetisation zero. The magnetised state of the material at *O* and *C* is not the same, through the value of *I* is zero at both these points. At *C*, a small change in *H* gives a large change in value of *I* but this is not the case at the point *O*. The curves (*a*) and (*b*) give *B-H* and *B-II* hysteresis loops. The values of retentivity, coercivity, saturation value of *I* and the area of *I-H* and *B-H* hysteresis loops give us an idea of the *nature of magnetic material* and these values determine the susceptibility for use in particular electrical *appliances like transformers, electromagnets, permanent magnets cores of dynamoes, motors etc.*

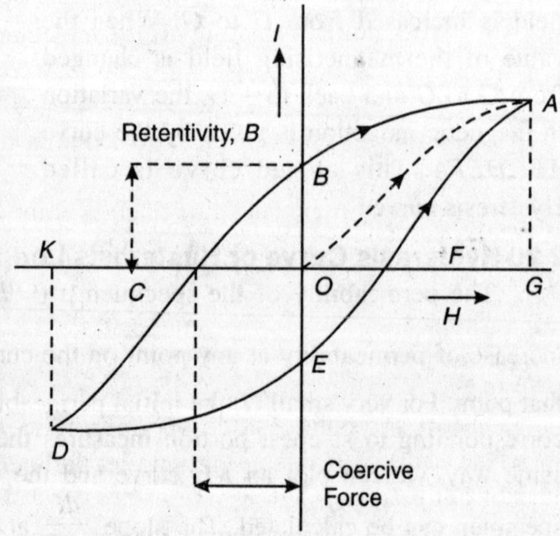

Fig. 2.9

Another important aspect shown from *I–H* and *B–H* curves is that the values of *l* and *B* do not become zero when the magnetising field is withdrawn. This characteristic of the material to retain its magnetisation or the lagging of *I* or *B* behind *H* is called **Hysteresis.**

2.21. Energy Loss in Hysteresis

When a ferromagnetic material is made to pass through a cycle of magnetisation, there is expenditure of energy. This energy is generally dissipated in the form of heat in the specimen. When a specimen is placed in a magnetic field, induced magnetism is acquired by the specimen. The magnetic dipoles or the molecular magnets are oriented and reoriented when the material is passed through a cycle of magnetisation and this molecular motion results in the production of heat.

In order to find the energy loss per cycle, let us consider a unit volume *abcd* of the magnetic material which is placed in a magenetising field having an intensity *H*.

Let there be a molecular magnet *NS* of magnetic moment *M* and inclined at an angle θ to the direction of magnetising field Me component of the maqgnetic moment along the direction of field = $M \cos\theta$ and that $\perp$ to the field in $M \sin\theta$. So, we have the total magnetic moment of all the molecular magnets/unit volume in the direction of field $\Sigma M \cos\theta$ and the component $\perp$ to the field $\Sigma M \sin\theta = 0$

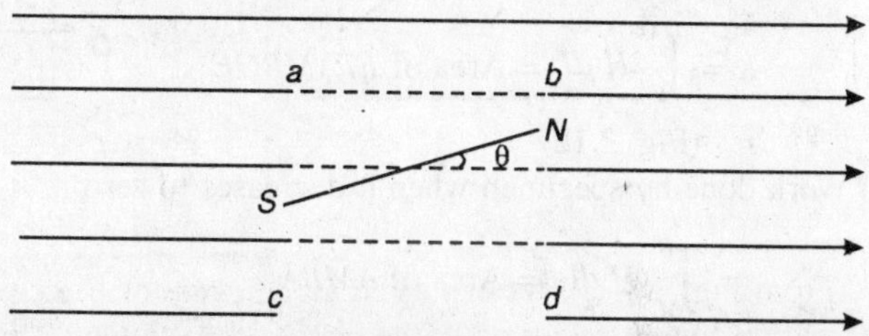

Fig. 2.10

Now from definition, the magnetic moment/unit volume of the specimen $\Sigma \cos\theta$ is equal to the intensity of magnetisation *I*.

$\therefore$ $$I = \Sigma M \cos\theta$$

Differentiating *w.r.t.* θ $$dI = \Sigma M(-\sin\theta)\, d\theta$$

$$= -\Sigma M \sin\theta\, d\theta \quad ...(1)$$

The couple acting on a molecular magnet inclined at an angle θ with the direction of the field – $MH \sin\theta$ and the work done in rotating this magnet by a small angle $d\theta = MH \sin\theta\,(-d\theta)$. The – *ve* sign indicates that $d\theta$ is the decrease in angle.

Total work done in rotating all the molecular magnets in unit volume

$$= -\Sigma MH \sin\theta d\theta \quad ...(2)$$

$$= -\Sigma\, [-\Sigma M \sin\theta d\theta]$$

From eqns. (1) and (2), the work done in increasing the intensity of magnetisation by $dI = HdI$.

If intensity of magnetisation changes from I_1 to I_2, we have

$$\text{Work done} = \int_{I_1}^{I_2} H\, dI \quad ...(3)$$

Let us consider the *I – H* curve of the material, and let it be having the shape *ABCDEFA* as shown in Fig. 2.10

Consider two points *r* and *s* very near to each other on the curve. Since these points, are very near so the value of *H* is constant and let the increase in value of *I* be *dI*, the difference between values of *l* at *p* and *q*.

From eqn. 3, the amount of work done on the specimen to increase the intensity of magnetisation by $dl = \int_{l_1}^{l_2} H\,dl$, which is corresponding to the area to the strip $pqsr$. Similarly we can imagine the whole cycle to be made up of small rectangular strips and the total work done on the specimen when H increases or by the specimen when H decreases to zero can be calculated.

Let us consider the complete cycle.

(*i*) Work done on the specimen when the value of H increases from zero to

$$K = \int_E^M H\,dl = \text{Area of } EFAMBOE$$

(Fig. 2.12).

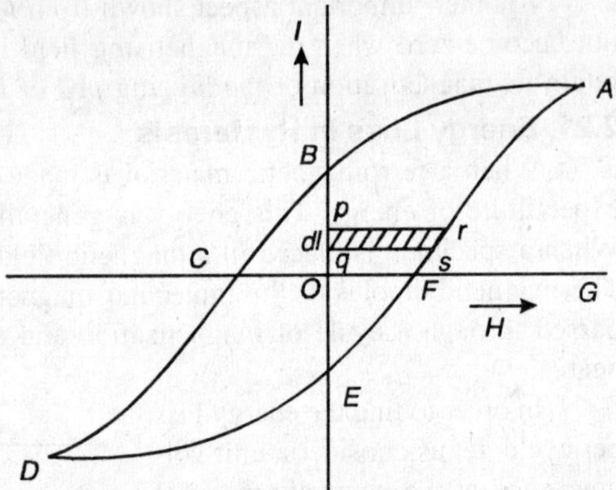

Fig. 2.11

(*ii*) Work done by specimen when it decreases to zero

$$= \int_M^B H\,dl = \text{Area of } AMBA$$

(*iii*) Work done on the specimen when it increases from O to L in the opposite direction

$$= \int_B^N Hdl = \text{Area of } BCDNEOB$$

(*iv*) Work done by specimen when H decreases to zero

$$= \int_N^E H\,dl = \text{area } DNED$$

So the total work done one the specimen/unit volume/cycle = Area *EFAMBOE* - Area *AMBA* + Area *BCDNEOB* – Area *DNED* = Area *ABCDEFA* = Area of $I - H$ curve.

So the work done on the specimen per unit volume when taken through a complete cycle of magnetisation is equal to the area enclosed by $I - H$ curve. The larger is the area of the curve, the larger is the loss of energy/ unit volume/cycle.

2.22. Area of $B - H$ Curve

We have the relation for magnetic Induction as

$$B = H + 4\pi I$$

Differentiating

$$dB = dH + 4\pi\,dI$$

Multiplying both sides by H

$$HdB = HdH + 4pHdI$$

Integrating for a complete cycle

$$\oint HdB = \oint HdH + \oint HdI$$

Now $\oint HdB$ = Area of $B - H$ curve

$\oint HdI$ = Area of $I - H$ curve

and $\oint HdH = 0$

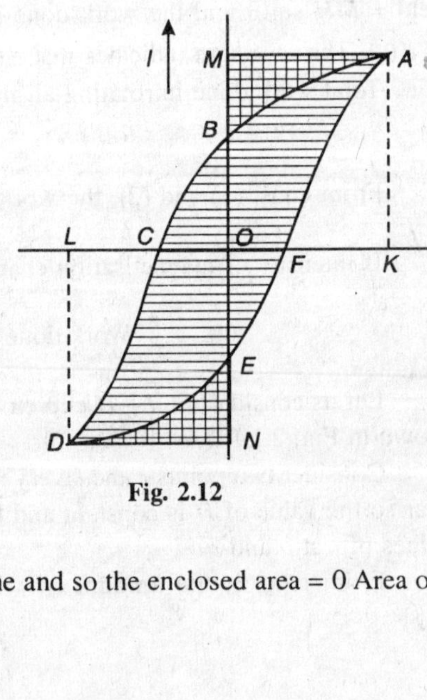

Fig. 2.12

Since a curve between H and H will give a straight line and so the enclosed area = 0 Area of $B - H$ curve = 4π × area of $I - H$ curve or Area of $I - H$ curve $= \frac{1}{4\pi}$ the area of B – H curve

So the loss in energy/cycle/unit volume of the specimen

$$= \frac{1}{4\pi} \text{ time the area of } B - H \text{ curve}$$

Now if I is expressed as magnetic moment/unit volume and H is in oersted, then $I - H$ curve gives the energy loss/cycle in ergs on the other hand if B is in Gauss and H in oersted, then $\frac{1}{4\pi}$ times the area of $B - H$ curve **gives energy loss/cycle in ergs.**

If rationalised *M.K.S.* or *S.I.* units are employed *i.e.*, B is in weber/metre^2 and H in ampere turn/metre, then the energy loss per cycle/cubic metre of the material will be equal to the area of $B - H$ curve and the **units of energy will be joule**.

2.23. Steinmetz Rule

We see that the area of *B-H* loop increases with increase in the strength of the magnetising field. Fig. 2.13, gives *B-H* curve for maximum magnetising fields. Where H_1, H_2, H_3... and A_1, A_2, A_3... correspond to the maximum values of B in each case. The locus of the points corresponding to A_1, A_2, A_3 ...corresonds to the initial magnetisation curve of the specimen after complete demagnetisation so the maximum flux density (B_{max}) is different for different magnetising values of H.

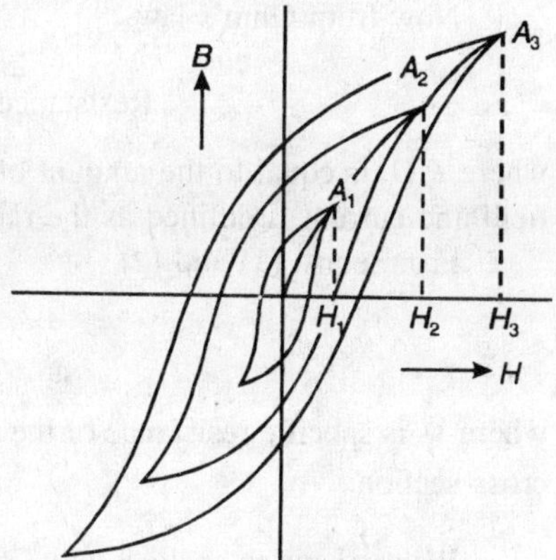

Fig. 2.13

Steinmetz gave an empirical equation for the calculation of energy loss per cycle/unit volume of the material as:

$$= \eta B_{\max}^{K}$$

where $\eta \longrightarrow$ coeff. of hysteresis loss

$K \longrightarrow$ is constant and varies from 1.4 to 1.8 for different ferromagnetic materials.

However, this equation is valid for low field strengths only.

2.24 Magnetic Circuit

If, we consider a ferromagnetic material in the form of a ring and having windings of N turns. If I be the strength of the current in amperes passing through the winding and l the mean circumference of the ring, then magnetic intensity within the toroidal winding is given by the relation

$$H = \frac{4\pi NI}{10l}$$

and magnetic induction $$B = \mu H = \frac{4\pi NI\mu}{10l}$$

where μ corresponds to the permeability of the material.

Magnetic flux across any section of area A of the ring is given by

$$\phi = B \times A = \frac{4\pi NI \mu A}{10l}$$

$$\therefore \quad \frac{4\pi NI}{10} = \frac{l}{A \cdot \mu}\phi$$

Now $\frac{4\pi NI}{10}$ corresponds to the work done in moving a unit north pole around the axis of the coil *i.e.*, along the mean circumference of length l against magnetic field intensity $\frac{4\pi NI}{10l}$

$$\therefore \quad \text{Work done} = \frac{4\pi NI}{10l} = \phi.\left(\frac{l}{A\mu}\right)$$

$$= \text{Flux} \times \left(\frac{l}{A\mu}\right) \qquad ...(1)$$

In rationalised *M.K.S.* units or *S.I.* units

$$= \frac{NI\mu A}{l} \text{ weber}$$

Now from Ohm's law

$$\text{Resistance} = \frac{P.D}{\text{current}} \qquad ...(2)$$

where *P.D.* is equal to the amount of work done in moving a unit positive charge against electric field and current is defined as the rate of flow of charge.

From eqns. (1) and (2)

$$R = \frac{\rho l}{A}$$

where ρ is specific resistance of the material and l is the length of the conductor and A the area of cross-section.

We can have an analogy with Ohm's law as (i) the work done $\frac{4\pi NI}{10}$ is called **magnetomotive** (*MMF*) or the line integral of the magnetic field. This is analogous to *e.m.f.* in an electrical circuit. The magnetic flux ϕ is analogous to the current and $\frac{l}{A\mu}$ is analogous to resistance $\frac{l}{A\mu}$ is called magnetic resistance or reluctance or $\frac{l}{\mu}$ corresponds to **specific resistance**.

Eqn. (1) may be written as

$$MMF = \text{Flux} \times \text{Reluctance}$$

or
$$\text{Flux } \phi = \frac{MMF}{\text{Reluctance}}$$

where NI is expressed in ampere turns, Flux in **Maxwells** and *MMF* in **Gilbert**.

However, the following dissimilarities in analogy of the above terms be kept in mind.

(1) $\frac{1}{\mu}$ corresponds to specific resistance ρ and so μ will correspond to specific conductivity. But, permeability in magnetism corresponds to K, the dielectric constant or permittivity in electricity.

(2) The variation of reluctance of a magnetic circuit is considerable, however, the permeability depends on intensity of magnetisation of the material, whereas the resistance of a conductor does not depend on the magnitude of the current in the circuit.

(3) In the magnetic circuit, some energy is required to establish the magnetic flux, whereas in an electrical circuit, the energy is spent so long as there is flow of current through the conductor.

Special cases

(*i*) In case of a closed magnetic circuit of different cross-section as shown in Fig. 2.14 the products B_1A_1, B_2A_2, B_3A_3... etc., at the planes P_1, P_2, P_3... etc. will be equal because in the absence of magnetic leakage, the total magnetic flux ϕ through any plane of the closed circuit must be the same because of the fact that flux is to have continuity .

(*ii*) If we have a magnetic circuit of different magnetic materials of different lengths and permeability, then if there is no magnetic leakage, the total reluctance in the circuit can be had as

$$\frac{l_1}{A_1\mu_1}+\frac{l_2}{A_2\mu_2}+\frac{l_3}{A_3\mu_3}+\frac{l_4}{A_4\mu_4}$$

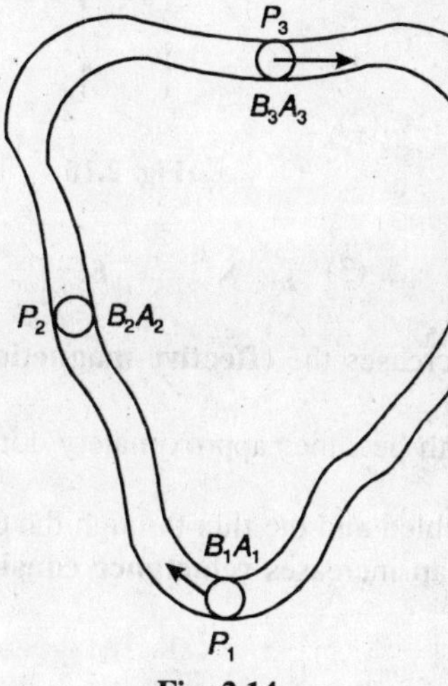

Fig. 2.14

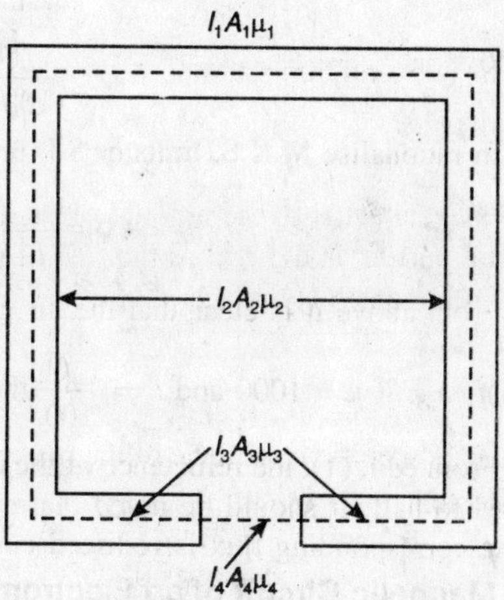

Fig. 2.15

In general reluctance of the magnetic circuit may be written as

$\Sigma\, A\mu$ and total flux is given by

$$\phi = \frac{F}{\frac{1}{A\mu}}$$

Where F is M.M.F. (magnetic motive force).

2.25. Magnetic Circuit of an Iron Ring with a Small Air Gap

Let the mean circumference of the ring be l and the air gap be of length l_1. respectively, then reluctance

$$= \frac{l-1_1}{\mu}+\frac{l_1}{A\mu_0}$$

$$= \frac{l-l_1}{A\mu}+\frac{l_1}{A} \quad \text{[Since } \mu_0 = 1 \text{ C.G.S. units for air]}$$

$$= \frac{Al-Al_1+A\mu l_1}{A\cdot A\mu} = -\frac{l_1+l+\mu l_1}{A\mu}$$

$$= \frac{l+(\mu-1)l_1}{A\mu} \qquad ...(1)$$

Now magnetomotive force $F = \dfrac{4\pi NI}{10}$

Where N is the number of turns and I the strength of current in amperes through windings.

Then, we have

$$\text{Flux } \phi = \frac{F}{\text{Reluctance}}$$

$$= \frac{4\pi NI}{10\left[\dfrac{l+(\mu-1)l_1}{A\mu}\right]}$$

$$\therefore \qquad \phi = \frac{4\pi NIA\mu}{10[l+(\mu-1)l_1]} \qquad ...(2)$$

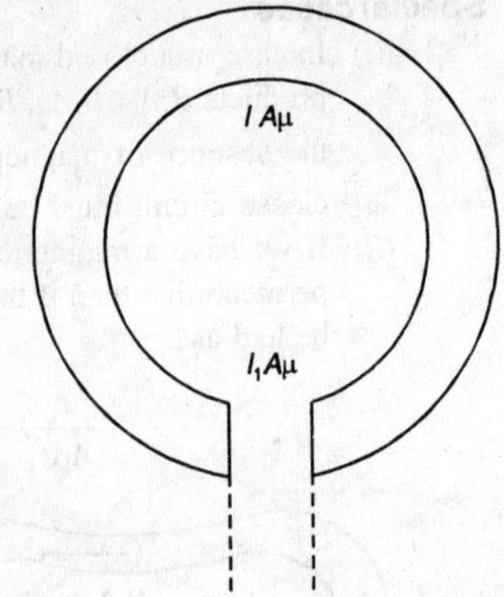

Fig. 2.16

In rationalise M.K.S. units or S.I. units,

$$\phi = \frac{NIA\mu}{l+(\mu-1)l_1} \text{ weber} \qquad ... (3)$$

From above it is clear that the air gap of length l increases the **effective magnetic length** $(\mu_1 - 1)l_1$ *e.g.* if $\mu = 1000$ and $l_1 = \dfrac{l}{100,}$ then magnetic length becomes approximately double.

From eqn. (1), the reluctance of the circuit is also doubled and the flux through the circuit is reduced to half. It should be noted that even a small air gap increases **reluctance considerably** and the corresponding flux is **reduced considerably.**

2.26. Magnetic Circuit of an Electromagnet

Figure 2.16 shows an electromagnet in which C,C correspond to the cores of electromagnet, A and B are pole pieces and G is the air gap while X is an iron yoke. Let l, l_1, l_2 and l_3 be the lengths of air gap, pole pieces core and yoke respectively as shown in Fig. 2.17 and if A_1, A_2, A_3 be the corresponding areas of cross-section of pole pieces, cores and yoke respectively, then the reluctance of the various parts of the magnetic circuit will be given as under:

Reluctance of air gap $= \dfrac{R}{\mu_0 A_1}$

Reluctance of pole pieces $= \dfrac{2l_1}{\mu_1 A_1}$

Reluctance of core $= \dfrac{2l_2}{\mu_2 A_2}$

Reluctance of yoke $= \dfrac{l_3}{\mu_3 A_3}$

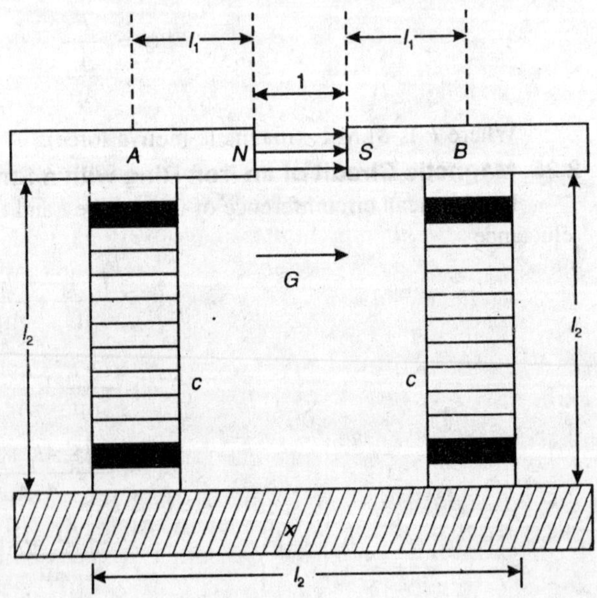

Fig. 2.17

where μ_0, μ_1, μ_2, μ_3 are permeabilities of air, pole pieces, cores and yoke respectively. So total reluctance of the circuit may be given as

$$\sum \frac{1}{A\mu} = \frac{1}{\mu_0 A_1} + \frac{2l_2}{\mu_1 A_1} + \frac{2l_2}{\mu_2 A_2} + \frac{l_3}{\mu_3 A_3}$$

Now magnetomoive force is given by

$$F = \frac{4\pi NI}{10}$$

$$\text{and flux } \phi = \frac{F}{\text{Reluctance}}$$

or

$$F = \phi \times \text{ reluctance}$$

$$= \phi\left[\frac{l}{\mu_0 A_1} + \frac{2l_1}{\mu_1 A_1} + \frac{2l_2}{\mu 2 A_2} + \frac{l_3}{\mu_3 A_3}\right]$$

Now for air $\mu_0 = 1$ and if pole pieces, yoke and core are of the same material, say, iron of permeability μ, we may have magnetomotive force as

$$F = \phi\left[\frac{l}{A_1} + \frac{2l_1}{\mu A_1} + \frac{2l_2}{\mu A_2} + \frac{l_3}{\mu A_3}\right]$$

It should be noted that the above calculations are approximately true because of leakage of flux in air gap. In case the air gap is very very small, the leakage of flux will be negligible and so the above equations becomes valid.

2.26 Magnetostriction

A ferromagnetic material when placed in an **external magnetic** field either expands or contracts and this phenomenon is called **magnetostriction**.

Longitudinal strain produced due to magnetization is given by $\frac{\Delta l}{l}$. Magnetostriction is due to the **notation of domains** which may result in **contraction or expansion** of the material.

Magnetostriction is revisable *i.e., there* can be alternating contraction and expansion as the external field is applied or removed. This phenomenon is quite useful in high frequency sound (Ultra sonics) production and detection in air and water or some other medium.

2.27. Magnetic Anisotropy

In some of the **single crystals** like iron, magnetic properties depend upon the direction in which these properties to be observed this phenomenan is called **magnetic anisotropy**. In case an external field is not present, the spontaneous magnetization takes place in some specific direction w.r.t. the axis of the crystal rather than any arbitrary direction.

In the case of iron, it is observed that there are six preferred directions for spontaneous magnetization. In the case of polycrystalline solids, since the domains are randomly oriented, so the properties observed are same in all the directions. However, on cold rolling, a polycrystalline solid may also have preferred directions of magnetization. This type of magnetization produced is called **Induced Anisotropy**.

The practical application of anisotropy is in case of computers and other storage devices where use is made by depositing a thin film of Ni-Fe alloy to a substrate by evaporation in vacuum in the presence of a suitable external magnetic field.

Example 1. *A solenoid consisting of* 500 *turns and carrying* 4 *ampere current is* 0.5 *m long Calculate* (*i*) *magnetomotive force* (*ii*) *Total flux if area of cross-section is* 0.0008 m^2.

Solution. In this case

Current flowing through the solenoid

$$I = 5\ A$$

Length of solenoid = 0.5m, No of turns of solenoid $N = 500$

Area of cross-section = $4 \times 10^{-4}\ m$

(*i*) Magnetomotive force (m.m.f.)

$$= \text{No. of turns} \times \text{current}$$
$$= 500 \times 5 = 2500 \text{ AT}$$

(*ii*) If air be the medium $\mu_r = 1$

Magnetising force H = ampere turn/length

$$= \frac{NI}{l} = \frac{2500}{0.5} = 5 \times 10^3\ AT/\text{m}$$

Now flux density is given by

$$B = \mu_0 \mu_r H = 4\pi \times 10^{-7} \times 5 \times 10^3$$
$$= 62.83 \times 10^{-4} \text{ Wb/m}^2$$

Total flux $\phi = B \times A$

$$= 62.83 \times 10^{-4} \times 4 \times 10^{-4}$$
$$= \mathbf{251.3}\ \boldsymbol{Wb}.\ \textbf{Ans.}$$

Example 2. *An iron ring of mean diameter* 100 *cm and* 10 *cm*^{2} *cross-section area has* 1000 *turns of insulated copper wire wound informly on if. If it produces a flux of* 2.5×10^{-3} *webers when a current of* 8 *amp. flows through it, calculate the permeability of the medium.*

Solution. We have

$$\text{Flux } \phi = \frac{\text{m.m.f}}{\text{Reluctance}}$$

In this case

$$\phi = 2.5 \times 10^{-3} \text{ Wb}$$
$$(m.m.f)\ M = N \times I = 1000 \times 8 = 8000\ AT$$

$$\text{Reluctance} = \frac{l}{\mu_0 \mu_r A} \text{ AT/Wb}$$

Now

$$l = 2\pi_r = 100 \times 10^{-2} \times \pi \text{ m}$$
$$\mu = 10 \times 10^{-4}\ m^2$$
$$\mu_0 = 4\pi \times 10^{-7} \text{ Henry}/m$$

Putting in eqn. (1)

$$2.5 \times 10^{-3} = \frac{8000}{\dfrac{\pi \times 100 \times 10^{-2}}{(4\pi \times 10^{-7} \mu_r \times 10 \times 10^{-4})}}$$

$$\therefore \quad \mu_r = \mathbf{781.25}\ \textbf{Ans.}$$

Example 3. *The mean length of an iron is* 50 *cm and its area of cross-section is* 4 *cm*^{2}. *An air gap* 1*mm in length is cut in the ring. If the ring is wound with* 200 *turns. What current is required to maintain a magnetic field of 10,000 gauss* (*μ for iron* = 500 μ_0).

Solution.

$$\text{Flux } \phi = \frac{4\pi N I A \mu}{10[l + (\mu - 1) l_1]}$$

or

$$I = \frac{10\phi[l + (\mu - 1) l_1]}{4\pi N A \mu}$$

In this case $\phi = 10{,}000$ gauss, $l = 50$ *cm*.

$$l_1 = 0.1\ cm,\ N = 200 \text{ turns}$$

$$A = 4\ \text{cm}^2,\ \mu = 500$$

$$\therefore \qquad I = \frac{10 \times 10000\,[50 + (500 - 1) \times 0.1]}{4\pi \times 200 \times 4 \times 500}$$

= **1.987 Amperes. Ans.**

2.28. Soft and Hard Magnets

Magnets are broadly divided into two categories :

(*i*) Soft magnets (*ii*) Hard magnets.

(*i*) A soft magnet can be magnetized and may be attracted by other magnets, but it retains **magnetism only in a magnetic field**. It can not be permanently magnetized.

(*ii*) **Hard magnet.** A hard magnet retains **magnetism more or less permanently** and it affects the presence of other magnetic materials in its field. A hard magnet retains its **magnetic characteristic** even in the absence of magnetic field.

(*i*) Soft magnetic materials are used in applications requiring frequent **reversal of direction of magnetization.** In many applications, the hysteresis *losses are to be kept as low as possible*. This can be achieved in case of soft magnets as in this case for large induction corresponding to a small applied field, loop area is small and so the hysteresis loss is reduced. In the design of a soft magnet, materials are chosen which have **easily moving domain walls**.

Another source of energy loss in soft magnets is due to *eddy currents*. With the change of magnetic flux, an *e.m.f.* is induced which according to Lenz's law is proportional to rate of change of flux and consequently to the *frequency of alternating current*. This induced current set up eddy currents in the medium and power loss due to eddy currents is equal to $\frac{V^2}{R'}$ where V is the induced *e.m.f.* and R is the resistance of the medium.

Eddy current losses can be minimized by increasing the **resistivity of the magnetic medium. Iron,** which was used as core of a transformer, is now almost completely being replaced by **Fe-Si** solid solution with about 4% of Si, which has quite high resistivity, as compared to pure iron. However, concentration of Si more than 4% is not preferred as beyond this concentration, **iron becomes brittle.** *In addition, the transformer core is laminated, as the resistance of laminations is much more than that of a one piece core.*

In these days, **metallic glasses,** produced from iron base allows, containing 15 to 25% *Si* + *B* + *C* offer substanital reduction in core-losses. Owing to larger concentration of impurity atoms, electrical resistivity is higher than that of Fe-Si (4%) allow; which substantially reduced the eddy current losses.

For frequencies higher than **MHz**, metals and alloys are generally not suitable as soft magnets because of the high eddy current losses. However thin films of **Permalloy** (45% Ni) of about 2000 Å in thickness, vapour deposited on a glass substrate, can be used. In this case, because of very low thickness, eddy current losses are reduced to bare minimum. **Ferri-magnetic oxides, ferrites and garnets** are quite suitable in high frequency ranges, as they being insulators will have much higher resistivity than alloys.

However, the choice of a particular soft magnet depends on its application. **Nickel-Zinc** Ferrites are used for *audio and TV transformers*, Magnetisem-Manganese ferrites with high resistivity are used in microwave isolators in kHZ to MHz range. With higher manganese to magnesium ratio, these are used in **memory core in computers.**

Recently, soft magnetic materials are used in the form of magnetic bubbles. Bubbles are magnetic domains of a few μm **diameter** which are embedded in a matrix of different spin orientation. All bubbles are arranged so that **they repel each other and maintain a constant distance between them.** These bubbles are the *electronic analogue* of the magnetic disc memories used in computers. Magnetic film in a disc memory is moved mechanically at high speed, while the bubbles move electronically in a bubble memory device.

2.29. Hard Magnetic Materials

For permanent magnets, Hard magnetic materials are used. In case of permanent magnets, hysteresis losses are of no significance, as no repeated reversals of magnetization is involved. Permanent magnets must have a high *residual induction* B_r and a *large corecive force* H_c. Area of hysteresis loop between B_r and H_c represents the energy required to demagnetize a permanent magnet. Maximum value of area is $B_r H_c$ and is called the *energy product*, which should be as large as possible in the case of permanent magnets.

High carbon-steels and low alloy tungston-chromium steels are used as permanent magnets. Some factors, which improve the mechanical hardness, provide better resistance to domain wall motion in permanent magnets. Among the metallic alloys, the best known are *Al-Ni-Co (alnico)* alloys.

Among the non-metallic oxides used as permanent magnets is Barrium Ferrite (BaO $6Fe_2O_3$) which has a high coercivity (more than 100 kAm-1). More recently, permanent magnets of cobal alloyed with rare earth elements have been developed with a very large energy product in the range of 200 k J m^{-3}. $\mathbf{Nd_2 F_{14} B}$, another alloy discovered in 1984 has energy product ~ 400 kJm^{-3}.

Tables 2.1 (*a*) and 2.1 (*b*) are given with some of the important properties of Soft and Hard magnetic materials.

Table 2.1 (*a*)
Properties is Soft Magnetic Materials

Material	*Initial relative permeability*	*Electrical resisitivity, (ohm m)*	*Hysteresis loss (J m^{-3})*	*Saturation induction, (Wbm^{-2})*
Commercial iron	250	0.1×10^{-6}	500	2.2
Fe-4% Si	500	0.6×10^{-6}	100	2.0
Fe-Si oriented	1500	0.6×10^{-6}	90	2.0
Permalloy (45% Ni)	2700	0.55×10^{-6}	120	1.6
Supermalloy (79% Ni, 5% Mo)	100000	0.65×10^{-6}	21	0.8
Ni-Zn ferrite	200-1000	$\sim 10^6$	35	0.4
Mn-Zn ferrite	200	$\sim 10^6$	40	0.3

Table 2.1 (*b*) Properties of Hard Magnetic Materials

Materials	B_r Wbm^{-2}	H_c kAm^{-1}	$B_r H_c$ kJm^{-3}
High carbon steel (martensitic)	0.90	3.98	3.58
Tungsten steel (5% W)	1.05	5.57	5.85
Chromium steel (4% Cr)	0.95	5.17	4.91
Cobalt steel (36% Co)	0.95	18.31	17.40
Al-Ni-Co alloys	0.8-1.2	60-120	48-144
Barium ferrite	0.21	140	29.4
Cobalt rare-earths Co_5 (Sm, Pr)	1.0	200	200

The differences between permanent, or hard, magnet and soft magnets, are best shown with the hysteresis loop (Fig. 2.16).

2.30. Types of Magnetic Materials

We know that the magnetic flux produced by an electric current is proportional to the relative permeability of the materials through which the flux passes. As explained earlier, in case of

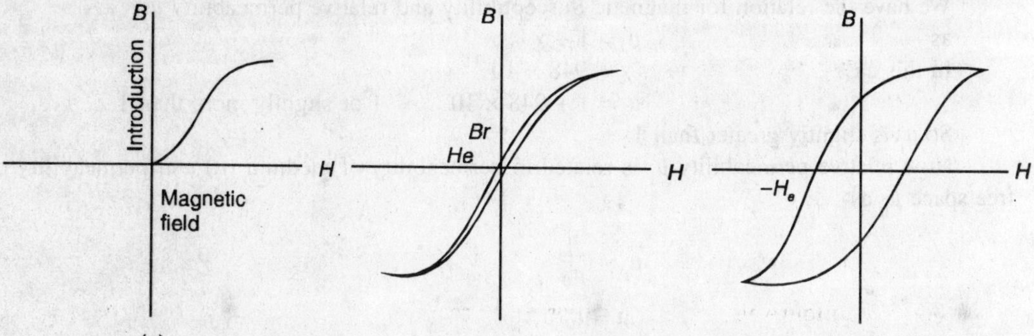

(a) (b) (c)

Fig. 2.17. Magnetization Curves. (a) Initial induction B versus magnetic field H. (b) Hysteresis loop (soft magnet). (c) Hysteresis loop (hard magnet). Both the remanent induction (fluxdensity) B_c, and coercivefield, -He, are large for hard magnets.

Ferromagnetic and Ferrimagnetic materials, since they possess high permeabilities, so they are capable of producing high flux densities with lower currents. So these materials are widely used as **flux paths and flux multipliers** in electrical machines. Transformers, motors, generators etc. *i.e.* the rotating electrical machines use magnetic materials as **core materials.**

Ferromagnetic materials are called *metallic magnets* while the **Ferrimagnetic** materials are called **ceramic magnets**. The most common ceramic magnets are Ferrites with general composition of $[R\ Fe_2\ O_4]_8$; where indicated iron is Fe^{3+} and R may be Fe^{2+}, Ni^{2+}, Zn^{2+}, Mn^{2+} etc. *i.e.*, each ferrous ion has lost two electrons.

Nature's original magnet is magnelite (Fe_3O_4) formerly known is Lodestone Magnetite is natural ceramic phase with an fcc lattice of O^{2-} ions.

Table 2.2. Units, Constants and Conversion Factors used in Magnetism

Quantity		*SI units*	*Other units*
Magnetic induction B	$Wb\ m^{-2}$	(weber per square	gauss
Saturation induction B_s	or	metre) (tesla)	
Residual induction B_r	T		
Magnetic field strength H	$A\ m^{-1}$	(ampere per metre)	oersted
Coercive field H_c			
Magnetization M			
Magnetic permeability μ	$H\ m^{-1}$	(henry per metre)	–
Relative permeability μ_r	–	–	– –
Magnetic susceptibility χ	–	–	–
Magnetic moment m	Am^2	(ampere metre^2)	–
Energy product B_rH_c	$J\ m^{-3}$	(joule per cubic metre)	–
Hysteresis loss	W	watt	–
Eddy current-loss			
Frequency of alternating current	Hz	hertz	sec^{-1}
Permeability of free space	μ_0	$H\ m^{-1}$	
Conversion factors			

1 gauss = 10^{-4} *wb* m^{-2}

1 oersted = 79.6 Am^{-1}

Example 5. *Magnetic susceptibility of a medium is 948 × 10^{-11}. Calculate the permeability and relative permeability.*

Solution.

We have the relation for magnetic Susceptibility and relative permeability μ_r as

$$\mu_r = 1 + \chi$$

In this case $\chi = 948 \times 10^{-11}$

$\therefore$ $\mu_r = 1 + 948 \times 10^{-11} \approx$ 1 or slightly more than 1

So μ_r is slightly greater than 1.

Now relative permeability μ_r is related to permeability of medium (μ) and permeability of free space μ_0 as

$$\mu_r = \frac{\mu}{\mu_0}$$

$$\therefore \quad \mu = \mu_0 \cdot \mu_r$$

Now $\mu_0 = 4\pi \times 10^{-7}\ H/m$

$$\therefore \quad \mu = 1 \times 4\pi \times 10^{-7}$$

$$= \mathbf{4\pi \times 10^{-7}\ H/m} \quad \textbf{Ans.}$$

Example 6. *Estimate the number of ampere-turns necessary to produce a flux of* 0.001 *Wb around an iron of* 6 *cm*^{2} *cross-section and* 20 *cm mean diameter, having an air gap of* 2 *mm. The permeability of iron is* 1200 μ_0.

Solution. Let l_i be the length of the iron ring, and *la* that of air gap, So by ampere-turns necessary to produce a flux ϕ_B around a ring of area of cross section *A* is given by

Number of amp-turn, $$Ni = \phi_B\left(\frac{l_a}{\mu_0 A} + \frac{l_i}{\mu A}\right)$$

But $\mu = 1200\,\mu_0$ (given) $\quad \therefore \quad \text{Ni} = \dfrac{\phi_B}{\mu_0 A}\left(l_a + \dfrac{l_i}{1200}\right)$

Now $\mu_0 = 4\pi \times 10^{-7} H/m$, $A = 6\ cm^2 = 6 \times 10^{-4}\ m^2$, $\phi_B = 0.001$ Weber

$r = 20/2 = 10$ cm $= 0.1$ m

$l_a = 2\text{mm} = 2 \times 10^{-3}$ m and $l_i = 2\pi r - l_a = 2 \times 3.14 \times 0.1 - 2 \times 10^{-3} = 0.626$

$$\therefore \quad \text{Ni} = \frac{0.001}{4 \times 3.14 \times 10^{-7} \times 6 \times 10^{-4}}\left(2 \times 10^{-3} + \frac{0.626a}{1200}\right) = 1.327 \times 10^6\ (2.5217 \times 10^{-3})$$

$$= \mathbf{3346\ Ans.}$$

Example 7. *Find the total reluctance and permeance between the ends of a parallel connected rectangular iron blocks of lengths $l_1 = l_2 = 4$ cm, width $w_1 = w_2 = 5$ cm and thickness $h_1 = h_2 = 4$ cm. Given that $\mu_1 = 1000\,\mu_0$ and $\mu_2 = 2000\,\mu_0$.*

Solution. Two rectangular blocks have identical dimensions. Let *l* is the length and *w* the width of each slab and μ_1 and μ_2 their permeabilities then their reluctance will be given by

$$R_1 = \frac{l}{\mu_1 A} \text{ and } R_2 = \frac{l}{\mu_2 A}, \text{ where } A = l \times w$$

In this case $l_1 = l_2 = l = 20\ cm = 20 \times 10^{-2}\ m$, $w_1 = w_2 = w = 5\ cm = 5 \times 10^{-2}\ m$

So, $A = 20 \times 10^{-2} \times 5 \times 10^{-2} = 100 \times 10^{-4}\ m^2$ and $\mu_0 = 4\pi \times 10^{-7}\ H/m$

$$R_1 = \frac{20\times10^{-2}}{1000\mu_0 A} = \frac{20\times10^{-2}}{1000\times4\times3.14\times10^{-7}\times100\times10^{-4}} = \mathbf{7961.78}$$

$$R_2 = \frac{20\times10^{-2}}{2000\times4\times3.14\times10^{-7}\times100\times10^{-4}} - 7961.78$$

As the blocks are connected in parallel, So, their resultant reluctance R is given by

$$\frac{1}{R} = \frac{1}{R_1} + \frac{1}{R_2} = \frac{1}{15923.57} + \frac{1}{7961.78} = 1.8844 \times 10^{-4}$$

$\therefore$ R = **5307.85 ampere-turns/weber Ans.**

The permeance $= \frac{1}{\text{reluctance}} = \frac{1}{5307.85}$ = **1.884 × 10^{-4} weber/amp-turns Ans.**

Example 8. *A solenoid consisting of* 100 *turns and carrying* 10 *amp current is* 0.5 *m long. Calculate* (*a*) *magnetomotive force and* (*b*) *total flux, if area of cross-section is* 0.0004 *m*^{2}.

Solution. (*a*) Magnetomotive force is given by, $mmf = Ni$

In this case $N = 100$ and $i = 10$ amp

$\therefore$ $mmf = Ni = 100 \times 10 = 1000$ **Amp.-turns**

(*b*) Total flux, $\phi_B = B.A$, but $B = \mu_0 H$ (With air as medium *i.e.* $\mu_r = 1$)

Also $Hl = Ni$, So $H = Ni/l$, put $l = 0.5$ *m*.

$\therefore$ $H = \frac{1000}{0.5} = 2000$ amp-turns/m

and $B = \mu_0 H = 4 \times 3.14 \times 10^{-7} \times 2000 = 2.512 \times 10^{-3}$ weber/ m^2

$\therefore$ $\phi_B = 2.512 \times 10^{-3} \times 0.0004 =$ **1.005 × 10^{-6} weber Ans.**

Example. 9. *Calculate the number of ampere-turns required to produce to produce a flux of* 10^{-3} *weber round an iron ring of* 5 *cm*^{2} *cross-section and* 20 *mm mean diameter having an air gap* 2 *mm wide across it. The relative permeability of the iron may be taken to be* 1000. *Neglect the leakage flux.*

Solution. As the number of ampere-turns per unit length is equal to the magnetising force H,

$\therefore$ $H = N_i$

If l is the length of the ring, then ampere-turn, $AT = H \times l$

In this case, $\phi_B = 10^{-3}$ weber and $A = 5\text{ cm}^2 = 5 \times 10^{-4}\ m^2$

$$\therefore \text{ Flux density, } B = \frac{\phi_B}{A} = \frac{10^{-3}}{5\times10^{-4}} = 2 \text{ weber/m}^2$$

$$\therefore \text{ Ampere-turn} = Hl = \frac{Bl}{\mu} = \frac{Bl}{\mu_0\mu_r}$$

In case of air gap, $\mu_r = 1$ and $l = 2$ mm $= 2 \times 10^{-3}$ m

$$\therefore \text{ Ampere-turn for air gap} = \frac{Bl}{\mu_0} = \frac{2\times2\times10^{-3}}{4\pi\times10^{-7}} = 3184.7$$

In the same say, ampere-turn for iron path :

$$\frac{Bl}{\mu_0\mu_r} = \frac{2\times2\pi r}{4\pi\times10^{-7}\times1000} = \frac{2\times2\pi\times10\times10^{-3}}{4\pi\times10^{-7}\times1000} = 100$$

So, total ampere-turn = AT for air gap + AT for iron path = 3184.7 + 100 = **3284.7 Ans.**

Example 10. *An iron ring of mean radius 10 cm and area of cross-section 5 sq. cm has an air gap 1 cm wide. The ring is wound unifomly with coil of 900 turns. A flux of 10^{-4} Wb is required in the air gap. Find the strength of current required to produce the flux. Given $\mu_{iron} = 65 \times 10^{-4}$ H/m.*

Solution. An iron ring is a combination of iron and air gap. Let l_i and l_a be the length of iron of magnetic circuit and air gap respectively. If μ is the permeability of iron and μ_0 that of air and A the area of cross-section of air gap as well as of iron ring, then from magnetic circuit equation, the ampere turn is given by the relation.

Ampere-turn, $$\text{Ni} = \phi_B\left(\frac{l_a}{\mu_0 A}+\frac{l_i}{\mu A}\right) = \frac{\phi_B}{A}\left(\frac{l_a}{\mu_0}+\frac{l_i}{\mu}\right)$$

where $i \longrightarrow$ current through the core $\phi_B \longrightarrow$ flux required in air gap or iron sing $N \longrightarrow$ No of turns

Now $\phi_B = 10^{-4}$ weber, $A = 5$ sq. cm $= 5 \times 10^{-4}$ sq. m, $l_a = 1$ cm $= 10^{-2}$ m,
$l_i = 2\pi r = 2 \times 3.14 \times 10 \times 10^{-2}$, $\mu_0 = 4\pi \times 10^{-7}$ and $\mu = 65 \times 10^{-9}$ H/m

So $$Ni = \frac{10^{-4}}{5\times10^{-4}}\left(\frac{10^{-2}}{4\times3.14\times10^{-7}}+\frac{2\times3.14\times10\times10^{-2}}{65\times10^{-4}}\right)$$

or $$Ni = 0.2(7961.78 + 96.615)$$

Hence $$i = \frac{0.2\times8058.4}{900} = \mathbf{1.79\ amp}\quad \textbf{Ans.}$$

Example. 11. *An iron rod of length 50 cm and cross-sectional area 4 cm² is in the form of a closed circular ring. If the permeability of iron is 65×10^{-4} H/m, calculate the number of ampere-turns required to produce a flux of 4×10^{-4} Wb.*

Solution. Number of ampere-turns (*Ni*) required to produce a flux ϕ_B by a circular ring of length l and area of cross-section A is given by the relation.

$\text{Ni} = \phi_B \dfrac{l}{\mu A}$, where μ is the pemeability of iron.

Now $\phi_B = 4 \times 10^{-4}$ Weber, $l = 50$ cm $= 50 \times 10^{-2}$ m, $A = 4$ cm² $= 4 \times 10^{-4}$ m
and $\mu = 65 \times 10^{-4}$ H/m

$$\text{Ni} = \frac{4\times10^{-4}\times50\times10^{-2}}{65\times10^{-4}\times4\times10^{-4}} \cong \mathbf{77\ Ampere\ turns}\quad \textbf{Ans.}$$

Example. 12. *An iron ring has cross-sectional area 10 cm² and an air gap of width 2 mm and mean length 60 m including the air gap. Find the number of ampere-turn required to produce the flux density 1 Wb/m². Given H = 10^{-3} H/m.*

Solution : Ampere turn *Ni* is given by

$$\text{Ampere-turn } Ni = \phi_B\left(\frac{l_a}{\mu_0 A}+\frac{l_i}{\mu A}\right)$$

where l_a is the width of air gap, l_1 the length of iron ring excluding air gap. Area of cross section, μ_0 the permeability of air gap and μ that of iron.

In this case $l_a = 2$ mm $= 2 \times 10^{-3}$m, $l_i = 60 - 2 \times 10^{-3} = 59.998$m, $B = 1$ Wb/m²
$A = 10$ cm² $= 10 \times 10^{-4}$ m², $\mu_0 = 4\pi \times 10^{-7}$ and $\mu = 10^{-3}$ H/m

$\therefore$ Ampere-turn, $$Ni = B\left(\frac{l_a}{\mu_0}+\frac{l_i}{\mu}\right) \quad (\text{As } B = q/A)$$

$$= 1\left(\frac{2\times10^{-3}}{4\times3.14\times10^{-7}} + \frac{59.998}{10^{-3}}\right)$$

$= 1592.357 + 59.998 \times 10^3$

$=$ **61590 Ans.**

QUESTIONS

1. Define the following :
   (*a*) magnetic induction, (*b*) intensity of magnetisation, (*c*) magnetic susceptibility, (*d*) magnetic permeability, and (*e*) relative permeability.
   Establish $\mu_r = 1 + \chi$.
2. Define the three magnetic vectors namely, magnetic induction $\vec{B}$, Magnetisation $\vec{I}$ and magnetic intensity $\vec{H}$. Establish the relation $\vec{B} = \mu_0(\vec{H} + \vec{I})$
3. Establish the relation (*a*) $B = \mu_0(H + 1)$, and (*b*) $\mu = \mu_0 (1 + \chi)$
4. Define permeability and magnetic susceptibility.
   show that $\mu = \mu_0(1 + \chi)$
5. Give a short account cf dia-, para and ferromagnetism.
6. Define $\vec{B}, \vec{M}$ and $\vec{H}$. Establish the relation $\vec{B} = \mu_0 \vec{H} + \mu_0 \vec{M}$.
7. Prove that the magnetic moment due to orbital motion of an electron must be an integral multiple $eh/4\pi m$
8. What are ferromagnetic domains ? How are their existence explained ? Distinguish between ferromagnetism, anti-ferromagnetism and ferrimagnetism.
9. What is anti-ferromangetism ? Explain the difference between ferro-mangetism and anti-ferromagnetism.
10. Discuss domain theory of ferro-magnetism ?
11. Distinguish between paramagnetic, diamagnetic, and ferromagnetic substances and illustrate their behaviour by simple experiments. How can you explain the behaviour of these substances ?
12. What is magnetisation curve ? Explain residual magnetism, coercive force and hysteresis.
13. Explain formation of domains in a ferromagnetic material and show how the hysteresis curve is explained on the basis of the domain theory.
14. What is meant by Hysteresis ? Explain hysteresis loss. How would you use the hysteresis curves select material for the constructive of permanent magnets ?
15. What is Weiss theory for ferromagnetic materials ? Establish the relation $\chi = C/(T - \theta)$.
16. Explain the magnetic circuit, magnetomotive force and reactance. How does a magnetic circuit differ from an electric circuit.
17. Define (*a*) magnetomotive force, (*b*) reluctance, and (*c*) magnetic flux. Establish a relation between them.
18. What is magnetic circuit ? Compare and contrast the magnetic circuit with electric circuit between expression for magnetic flux in a circuit with an air gap.
19. Describe the importance of hysteresis curve. How would you use the hysteresis curve for selecting the material for the construction of a permanent ?

NUMERICAL PROBLEMS

1. A soft iron ring has a mean circumference of 250 cm and a uniform circular cross-section of 1.0 cm diameter. The relative permeability of the iron is 2500 . A flux of 500 micro-weber is required in the iron. How many turns of wire carrying a current of 2 amp must be wound around the ring? **(Ans.** 2536)
2. An iron rod 20 cm long, 1 cm in diameter and of a permeability 1000, is placed inside a solenoid, 1 metre long would uniformly with 600 turns. If the current of 0.5 ampere is passed through the solenoid find the magnetic moment of the rod. **(Ans** .4.70 ampere –m^2)
3. A paramagnetic salt contains 10^{28} ions per m^3 with magnetic moment of 1 Bohr magneton (9.28 ×10^{-24} amp-m^2). Calculate the paramagnetic susceptibility and the magnetisation produced in a uniform magnetic field of 0.1 weber/m^2 when the temperature is 27°C. Given Boltzmann constant K = 1.38 × 10^{-23} joule/K. **(Ans.** $\chi = 8.6 \times 10^{-5}$ **MKS** unit, I = 69 A/m)
4. A material core has 10 turns per cm of wire wound uniformly upon it, which carries a current of 2.0 ampere. The flux density in the material is 1.0 weber/metre^2. Calculate the magnetising force H and the magnetisation I of the material. What would be the relative permeability of the core ? ($\mu_0 = 4\pi \times 10^{-7}$ weber/amp-metre). **(Ans.** 2000 Ampere turns/m, 7.9 × 10^5 A turns/m, 397)
5. The mean length of an iron ring having 200 turns of wire wound upon it is 0.5 metre and its cross-section is 4 × 10^{-4} metre^2. What current in the winding should be sent to produce a flux of 4 × 10^{-4} weber in the ring ? Take the permeability of iron at this flux as 65 × 10^{-4} Weber/amp-metre. **(Ans.** 0.384A)
6. A solenoid consisting of 500 turns and carrying 5 ampere current is 0.5m long. Calculate (*i*) magneto-motive force and (*ii*) total flux if the area of cross-section is 0.0004m^2. **(Ans.** 2500 A-turns,251wb)
7. For soft iron, hysteresis loss per unit volume per cycle is 10^3 joule. take the density of iron 7.5gm/cm^3 and specific heat 100 calories/kg-°C to calculate the temperature increase per minute whin an alternating mangetic field of 50 cycle per second is applied on the iron specimen. **(Ans.** 0.95°C)
8. A mild steel ring having a cross-sectional area of 5 cm^2 and mean circumference of 40 cm has a coil of 200 turns would uniformly around it. Calculate (*i*) the reluctance of the ring, (*ii*) the current required to produce a flux 800 micro-webers in this ring. (Assuming relative permeability of mild steel to be 380 as the flux density developed in the core). **(Ans.** 1.675 × 10^6 A turns/ wb, 6.7A)
9. An iron rod of 0.2 cm^2 cross- sectional area is subjected to a magnetising field of 1200 amp-m^{-1}. The susceptibility of iron is 599. Find the permeability and the magnetic flux produced. **(Ans.** 7.54 × 10^4 JmA^{-1})
10. An iron rod 0.2 m long, 10 mm in diameter and of permeability 100 is placed inside a long solenoid wound with 300 turns/metre. If a current of 0.5 ampere is passed through the rod, find the magnetic moment of the rod. **(Ans.** 0.2325 Am^2)

3

LASERS

3.1 Introduction

Laser is the acronym for "Light Amplification by Stimulated Emission of Radiation" Laser light differs from conventional light in several aspects. Even though both are electromagnetic in nature, Laser light has certain properties that make it highly desirable in a multitude of fields and applications. Unlike conventional light, Laser light is

(a) highly monochromatic i.e. light of almost single wavelength
(b) temporally and spatially coherent (i.e. same phase)
(c) highly polarised in one plane.

3.2 Milestones in the history of lasers :

1917 Einstein's treatment of stimulated emission
1951 Development of Maser by C.H. Townes
1958 Proposal of CH Townes and AL Schawlow that the Maser concept could be extended to optical frequencies
1960 The Mainman at Hughes Laboratories reports the first Laser, the Ruby pulsed Laser.
1961 The continuous wave Laser (He-Ne Laser) was reported.
1964 Nicolay Basov, Charlie Townes and Aleksandr Pokhorov get Novel prize for "fundamental work in the field of quantum electronics, which has led to the construction of oscillators and amplifiers based on maser - laser principle"
1981 Art Schalow and Nicholas Bloembergen get the Nobel Prize for "their contribution to the development of Laser spectroscopy.
1997 Steven Chu, Claude Cohen Tannoudji and William D Phillips get the Nobel prize for " development of methods to cool and trap atoms with Laser light. The present state of art :
Peak Powers > 10^{12} Watt
Laser pulses shorter than 10^{-15} second
Blue light emitting Lasers.

Before the description of lasers and their applications, we may underline some of the basic concepts of the various energy states.

3.3. Atomic Excitation

An atom can be excited to a state higher than its ground state by the following two principal methods.

(*i*) By collision with another atom during which part of the kinetic energy is absorbed by the atom and an atom in this way returns to ground state in time ~ 10^{-8} sec by emitting one or more photons (Fig. 3.1)

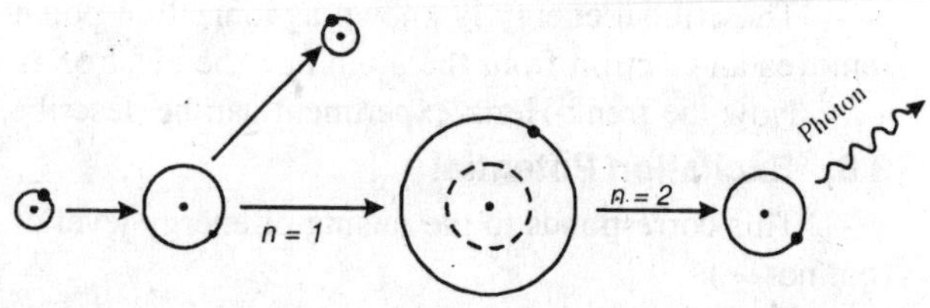

Fig. 3.1

(*ii*) By establishing an electric field that accelerates **electrons and atomic ions** till kinetic energies are sufficient to excite atoms they collide with. Since the energy transfer is maximum when the colliding particles are of the same mass, so electrons in luminous discharge in a rarefied gas are more effective than the ions in providing *energy to atomic electrons*. **Neon signs and mercury vapour lamps** are some of the familiar examples of the characteristic colours in spectral emission when strong electric fields are applied between electrodes in a gas filled tube. Neon gives reddish light while mercury vapour give bluish light.

3.4. Excitation by Absorption

A different kind of excitation is involved when an atom absorbs a photon of light whose energy is just the right amount of energy to higher energy level. For example, a photon of $\lambda = 121.7$ *nm* is emitted when a hydrogen atom in $n = 2$ state comes to $n = 1$ state. The absorption of a photon of $\lambda = 121.7$ *nm* by hydrogen at ground state $n = 1$ will bring it to $n = 2$ state. Fig. 3.2 explains the absorption spectrum.

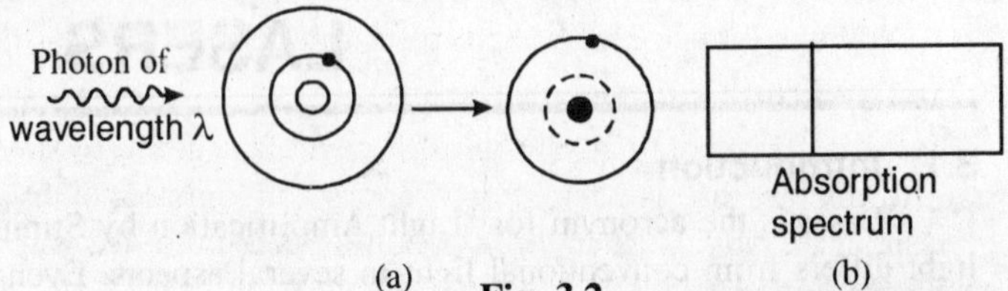

Fig. 3.2

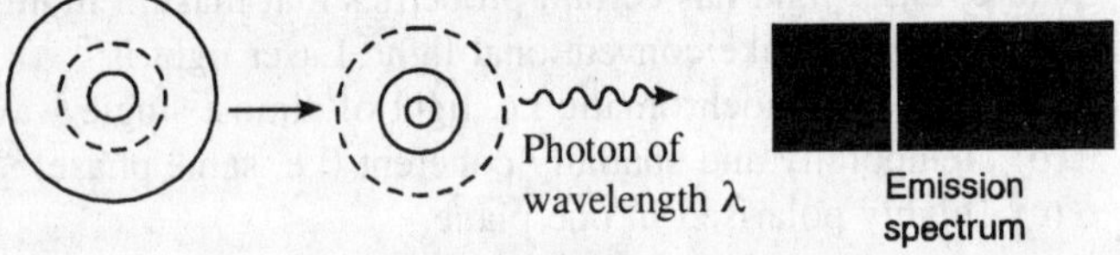

Fig. 3.3

We can explain the absorption and emission spectrum as under :

White light consisting of all the colours (wavelengths) when passes through hydrogen gas, then photons of those wavelengths that correspond to the transitions between energy levels are absorbed *resulting in the excitation of the hydrogen atom.* The excited hydrogen atom radiates photons almost (instantly at once) but these photons come in random directions and only few of them in the same direction as the original beam of white light. So the dark lines in an absorption spectrum are never completely dark but only appear to be *black because of bright background.*

The dark lines in the absorption spectrum are at the same position as the bright lines in the emission spectrum.

To explain the discrete energy levels of the atoms, a series of experiments based on the excitation phenomenon by collision were performed by **James Franck and Gustav Hertz.** These experiments provide a direct demonstration of the *existence of discrete atomic energy levels* and these levels agree with those suggested by the line spectra.

Before the description of the experiment few more terms may also be clearly understood.

3.5. Critical Potential

This corresponds to the minimum energy in electron volts required to excite an electron from ground state to a higher state.

Ionization Potential

The critical energy is known as ionization potential when the absorption of the energy just removes an electron from the atom *i.e.*, the electron is raised from level $n = 1$ to $n = \infty$.

Now the frenk-Hertz experiment can be describe, to illustrate about these energy states.

3.6. Excitation Potential

This corresponds to the raising of energy level of an unexcited atom to a higher energy state (but not ∞).

From Bohr's theory, we have the energy of hydrogen atom given by

$$E_n = \frac{13.6 \text{ eV}}{n^2}$$

when $n = 1, E_1 = 13.6$ eV
$n = 2, E_2 = -3.4$ eV
$n = 3, E_3 = -1.51$ eV
$n = \infty, E_\infty = 0$ eV

So we have the energy required for hydrogen as
For first excited state = 13.6 – 3.4 = 10.2 eV
2nd excited state = 13.6 – 1.51 = 12.09 eV

So the discrete amounts of **energy are required to excite hydrogen atom to various energy states which are thereby quantized.**

Now the Franck-Hertz experiment can be described to illustrate about these energy states.

3.7. The Frank-Hertz Experiment

As is clear from above, the excitation of an atom, by any of the two mechanisms from one state to another involve transfer of **energy in quanta**. This quanta either absorbed or emitted corresponds to some line in the spectrum of that atom. It was with these considerations that corresponds to some line in the spectrum of that atom. It was with these considerations that Frank and Hertz in 1914 performed the experiment using **inelastic collision** of accelerated electrons with the atoms as the source of **excitation**. This experiment provides a direct demonstration that *atomic energy levels are present in the atom.*

Experiment Arrangement. A tube filled with the gas or vapours of the atoms under study is fitted with an indirectly heated cathode *C*, grid *G* and anode *A*.

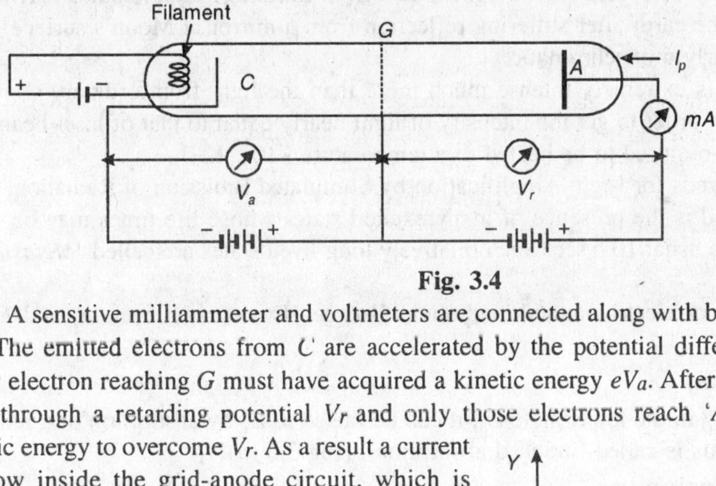

Fig. 3.4

A sensitive milliammeter and voltmeters are connected along with batteries, as shown in Fig. 3.4. The emitted electrons from *C* are accelerated by the potential difference Va towards *G*, so every electron reaching *G* must have acquired a kinetic energy eVa. After crossing *G* the electrons pass through a retarding potential Vr and only those electrons reach '*A*' which have sufficient kinetic energy to overcome Vr. As a result a current Ip flow inside the grid-anode circuit, which is measured by the milliammeter *mA*. A typical plot of Ip and Va is shown in fig. 3.5

Results of the Experiments. As Va is increased from zero, the kinetic energy eVa of electron is increased so that more of them cross *G* and reach *A*, increasing Ip. The rise from *O* to *M* in the curve in figure 7.5, indicating that upto the corresponding value of the accelerating voltage, the collision of the electrons with the atoms are elastic and number of electrons reaching *A* increases with Va. The abrupt fall in Ip form *M* to *N* or $Va = Va_1$ indicates that when the kinetic energy of the electron is eVa these **undergo inelastic**

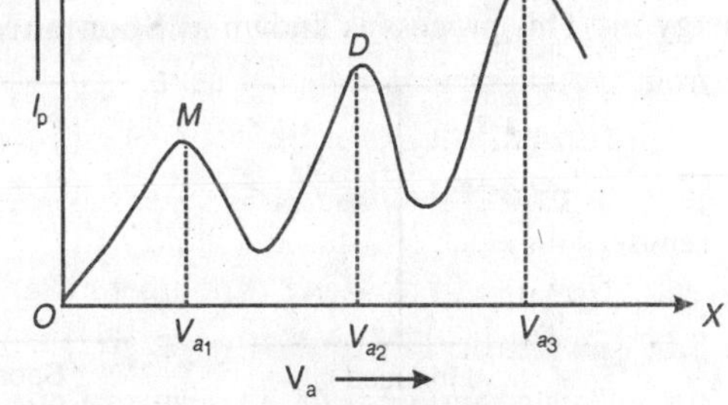

Fig. 3.5

collisions with the atom close to G so that their kinetic energy after crossing G is smaller than eV_r. Further rise from N to O shows that the collisions of the electrons are against elastic. If they undergo an inelastic collision it involves energy eV_{a_2} and after that these have gained kinetic energy sufficient to overcome I_r. Thus the collisions continue to be elastic upto V_{a_2} which again most of the collisions are inelastic as there is sudden drop in I_p from O to P.

From the above discussion it is clear that the collisions of the electrons with the atoms are inelastic only for some critical potentials *i.e.*, $V_{a_1}, V_{a_2}, V_{a_3}$ ...... The direct evidence from this is that atoms absorb discrete energies $eV_{a_1}, eV_{a_2}, eV_{a_3}$ ......to get excited. The energies other than these are not transferred. To check the interpretation of critical potential as being due to discrete atomic energy levels. Franck and Hertz observed the emission spectra of vapour during electron bombardment. In the case of mercury vapours, they found that a minimum electron energy of 4.9 eV was required to excite 2536 eV spectral line of mercury. A photon of 2536 eV has an energy of just 4.9 eV. It was concluded that energy levels in the atoms are quantized and the **absorption and emission of radiation by these are in quanta** and not continuous.

It is observed that mercury accepts only and *exactly* 4.86 eV, 6.67 eV and 8.84 eV which means that mercury atoms have a set of *discrete energy* states above ground state and continuous states above 10.4 eV.

Some of the following terms must be clearly understood to know the properties laser beams:

1. The light is coherent with all the waves exactly in phase with each other.
2. Laser beam hardly diverges. Such a light if sent from Earth to Moon remains narrow to be detected on the earth after suffering reflection from a mirror at Moon's surface.
3. The beam is nearly monochromatic.
4. The laser beam is extremely intense much more than the light from ordinary conventional sources. In order to get the intensity of light nearly equal to that of laser beam an object would be required to be heated to a temperature $\sim 10^{30}$ K.

The term LASER stands for Light Amplification by Stimulated Emission of Radiation. The basic phenomenon involved is the presence of many excited states whose life times may be 10^{-3} sec. or more, instead of the usual 10^{-8} sec. Such relatively long lived states are called *'Metastable States'*.

We have three kinds of electromagnetic radiations between two energy levels E_0 and E_1 in an atom.

3.8. Induced Absorption

If the atom is initially in the lower state E_0, it can be raised to E_1 by absorption of a photon of energy $E_1 - E_0 = hv$. This is called 'induced absorption'. [Fig 3.6 (*a*)]

3.9. Spontaneous Emission

If the atom is initially in the upper state E_1, it can drop to E_0 by emission of a photon of energy hv. This process is known as **Spontaneous Emission** [Fig. 3.6 (*b*)].

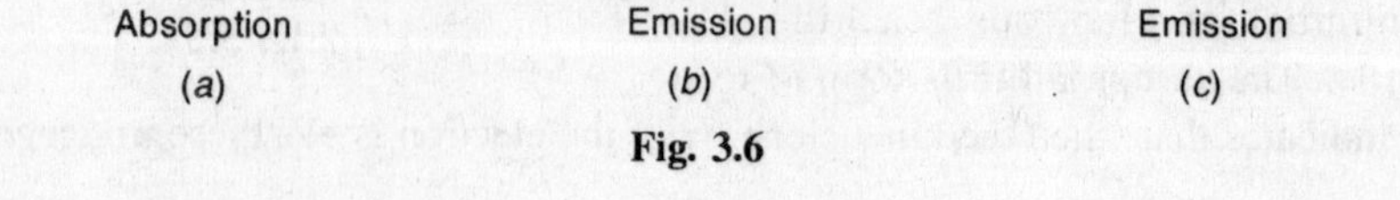

Induced Absorption (*a*) Spontaneous Emission (*b*) Induced Emission (*c*)

Fig. 3.6

3.10. Induced Emission

In this case an incident photon of energy hv causes a transition from E_1 to E_0. In the induced emission, the radiated light waves are exactly in phase with the incident waves, so that we get an enhanced beam of coherent light. In 1917 Einstein showed that *induced emission and induced absorption have the same probabilities.* Which means that a photon of energy hv when incident on an atom has the same *likelihood of induced emission as that of induced absorption.*

The concept of induced absorption forms the basis of LASERS.

3.11. Spontaneous and Stimulated Emission

Spontaneous emission is the phenomenon by which an atom or molecule in an excited state drops to ground state, resulting in the creation of a photon.

It the atom is in the excited state E_2, it may spontaneously decay to ground state E_1, releasing the difference in energies between the two states as photon.

(a) Energy Level of Atom in excited state **(b) Atom dexcites and photon in emitted**

Fig. 3.7

Stimulated emission is the process by which an atom or molecule in an excited state drops to ground state when it interact with external photon having energy equal to excited state energy

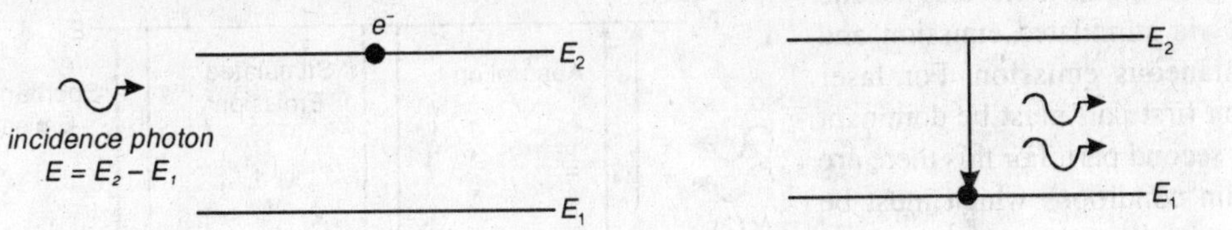

(a) Excited atom **(b) Atom dexcites and photon in emitted.**

Fig. 3.8

The external photon is not destroyed in the process, and the second photon is created with same phase, polarisation and frequency. It is quantum mechanical process. Stimulated emission is responsible for laser operations stimulated emission is responsible for Laser operation.

3.12 Population Inversion

Laser action requires very large no. of stimulated emissions. This is possible when there are large no. of atoms/molecules in excited state. At thermal equilibrium there are very large no. of atoms/molecules in lower state than in upper state. By creating non-equilibrium state this condition can be inverted.

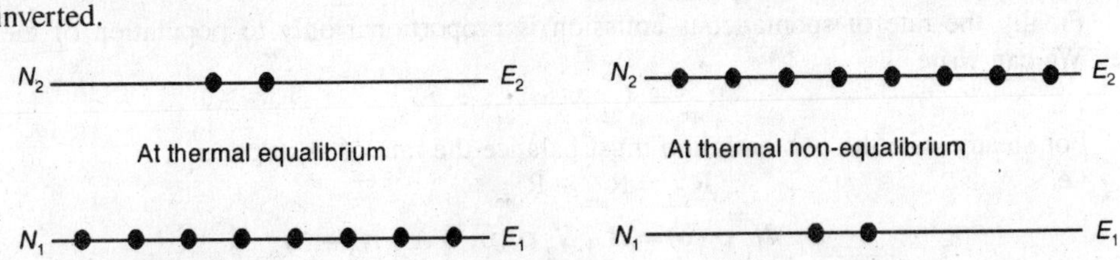

(a) Population power state is large **(b) Population of higher state is large**

Fig. 3.9

When there are large no. of atoms/molecules in upper energy level than in lower energy level, this is called population inversion.

3.13. Pumping

For mainting the condition of population inversion, the atoms/molecules have to be raised continuously to excited state. In requiters energy. The process of supplying energy to create the state of population inversion in called pumping.

There are no. of methods of pumping

(a) **Optical pumping :** Light energy from appropriately selected sources such as gaseous discharge flashtubes, or continuously burning tables is used.

(b) **Electrical Pumping :** The intense electrical discharge raises the atoms/molecules to excited state. This is particularly suited to gas media.

(c) **Chemical Pumping :** Suitable exothermic reaction can create the condition of population inversion.

(d) **Heat Pumping :** The medium is at first brought to high temperature and then rapidly cooled.

3.14. Active Medium

The heart of the laser is a certain medium-solid; liquid, or gaseous - called an active medium in which population inversion exists. In active medium only small fraction of particular atoms/molecules undergo stimulated emission. **They are called active centres.** The remaining bulk of the medium may plays the role of host and support active centres or may help in creating population inversion.

3.15. Conditions for Lasing Action (Einstein's Coefficients)

The atoms/molecules dexcite both via stimulated emission and spontaneous emission. For laser action first part must be dominant over second part. For this there are certain conditions which must be maintained.

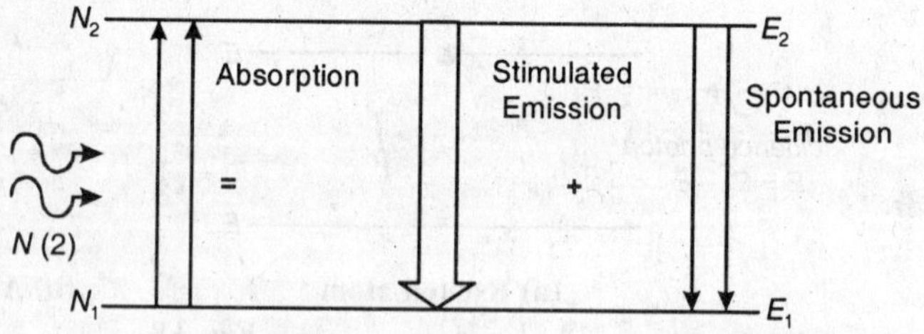

Fig. 3.10 absorption and emission of radiation

The rate of stimulated absorption in proportional to the population of lower level and to the energy density of radiation.

$$\text{Rate of stimulated absorption} \propto N_1 \, \rho(\upsilon)$$

$$R_{sta} = B_{12} N_1 \, \rho(\upsilon)$$

Similarly the rate of stimulated emission is also proportional to radiation energy density and population of upper level. Thus we can write

$$R_{ste} = B_{21} N_2 \, \rho(\upsilon)$$

Finally the rate of spontaneous emission is proportional only to population of the upper level. We can write

$$R_{spe} = A_{21} N_2$$

For steady state the two emission must balance the rate of absorption

i.e.
$$R_{sta} = R_{ste} + R_{spe}$$

$$B_{12} N_1 \, \rho(\upsilon) = B_{12} N_2 \, \rho(\upsilon) + A_{21} N_2 \quad - \text{(i)}$$

This relation was described by Einstein, and the coefficients B_{12}, B_{21} and A_{21} are called Einstein coefficients.

At thermal equilibrium, the ratio of the stimulated to spontaneous emission rates in negligible.

$$\frac{R_{ste}}{R_{spe}} = \frac{B_{21}\,N_2\,\rho(\upsilon)}{A_{21} N_2} = \frac{B_{21}}{A_{21}} \rho(\upsilon).$$

Above ratio can be increased by increasing the concentration of photons. This can be achieved by providing an optical resonant cavity in which the photon density can build up to a large value through multiple internal reflections

Similarly to obtain more stimulated emission than absorption we must have $N_2 > N_1$

$$\frac{R_{ste}}{R_{ste}} = \frac{B_{21} N_2 \rho(\upsilon)}{B_{12} N_1 \rho(\upsilon)} = \frac{B_{21} N_2}{B_{12} N_1}$$

i.e. ratio can be increased by creating population inversion.

Mare about Einstein Coefficients

From Statistical Physics it can be shown that radiation density $\rho(\upsilon)$, and relative ratio of population of lower and upper level are given by

$$\rho(\upsilon) = \frac{8\,\pi h \upsilon^3}{C^3} \frac{1}{e^{h\upsilon/KT} - 1} \quad \text{— (2)}$$

and $$\frac{N_1}{N_2} = e^{(E_2 - E_1)/KT} = e^{h\upsilon/KT} \quad \text{(Maxwell - Boltzman relation) — (3)}$$

Recalling, $$B_{12} N_1 \rho(\upsilon) = B_{21} N_2 \rho(\upsilon) + A_{21} N_2$$

$$\rho(\upsilon) = \frac{A_{21} N_2}{(B_{12} N_1 - B_{21} N_2)}$$

$$\rho(\upsilon) = \frac{\dfrac{A_{21}}{B_{21}}}{\left(\dfrac{B_{12}}{B_{21}} \dfrac{N_1}{N_2} - 1\right)},$$

Comparing it with equations (2) and (3)

$$\rho(\upsilon) = \frac{\dfrac{A_{21}}{B_{21}}}{\dfrac{B_{12}}{B_{21}} e^{h\upsilon/KT} - 1} = \frac{8\,\pi\, h\upsilon^3 / c^3}{e^{h\upsilon/KT} - 1} \quad \text{(From Planlk's radiation law)}$$

$$\boxed{\frac{B_{21}}{A_{21}} = \frac{c^3}{8\,\pi h \upsilon^3}} \quad \text{— (4)} \quad \text{and} \quad \frac{B_{12}}{B_{21}} = 1 \quad \boxed{B_{12} = B_{21}} \quad \text{— (5)}$$

Units of Einstein's coefficients

$$\frac{B_{21}}{A_{21}} = \frac{m^3 s^{-3}}{J - s \times s^{-3}} = m^3 J^{-1} s^{-1} \qquad \therefore A_{21}/B_{21} = J - s/m^3.$$

From equation (4) it is clear that for laser action the frequency of photon density must be small, and from equation (5) probability of stimulated absorption is same as that of absorption.

3.16. The Principal Pumping Schemes :

Atoms, ions or molecules used as active centres often exhibit large and complicated system of energy levels. Among them only three or four levels are of importance.

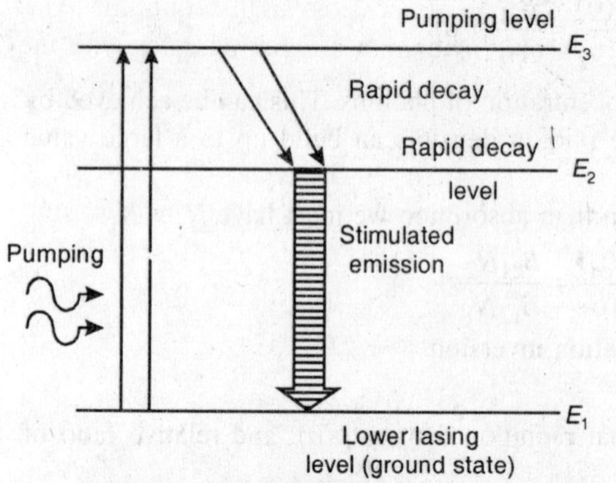

Fig. 3.11 Three level laser

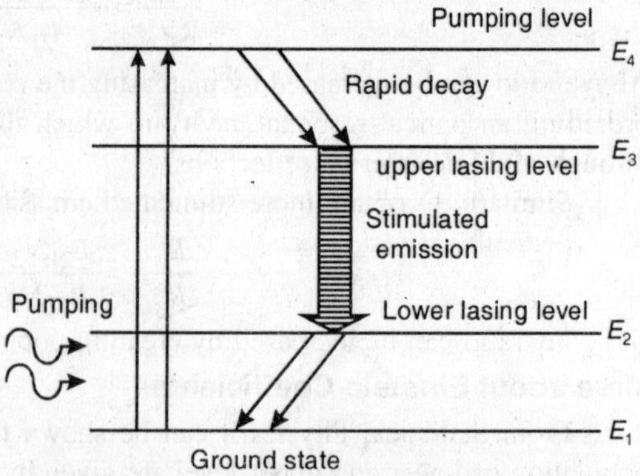

Fig. 3.12 Four level laser

In three level laser, the pumping raises the population of state E_3 called pumping level, and is unstable. So Atoms immediately dexcited to power level E_2 called upper lasing level, which is metastable. When the process of pumping continuous population of state E_2 build up. When population of this state becomes more than the half of the population of ground state. The population inversion is achieved between state E_2 and E_1. Now photon of energy $h\upsilon = E_2 - E_1$ can initiate the stimulated emission.

In this scheme, the lower lasing level is simultaneously the ground state. To maintain population of upper level more than half as that of ground level, a very high pump power is required.

In four level laser the lower lasing level is not the ground level. When stimulated emission occurs between lasing levels, atoms rapidly decays to ground level from lower lasing level thereby keeping lower lasing level virtually empty. So there always exists population inversion. But we must remember that each extra transition implies additional energy losses.

3.17. Optical Resonator

We have learned already that a light beam incident on an active medium can be amplified because stimulated emission dominates over absorption in an inverted medium. The question in hour to initiate the stimulated emission and too in only one direction. It is the spontaneously emitted photons in active medium, which initiate the stimulated emission in various directions. The stimulated emission can be made to exist for only one direction by restricting other directions. Optical resonator serves this purpose.

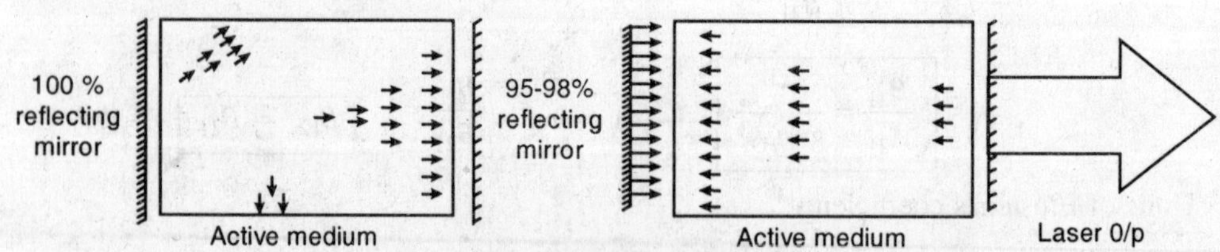

Fig. 3.13 action of optical Resonator

An optical resonator generally consists of two opposing plane mirrors, with the active materials placed in between them. One of the mirror is semitransparent while the other is 100% reflecting. The mirrors are set normal to optic axis of material.

This structure is known as **Fabry-Perot resonator.**

Just after achieving population inversion, active centres dexcite and emit spontaneous photons in all the directions. Each of these photons initiate stimulated emissions in their own directions.

The directions which are not parallel to optic axis are finally escaped out of the active medium. The direction which is parallel is selected and to attain maximum possible amplification of light, the stimulated photons are to be made to pass through the medium a number of times with the help of mirror. Finally an intense light called lasser light comes out through semi transparent mirror.

3.18. Ruby Laser

It was the first solid state working laser built by T.H. Mainman. It employs three level laser and operates in discontinuous wave mode.

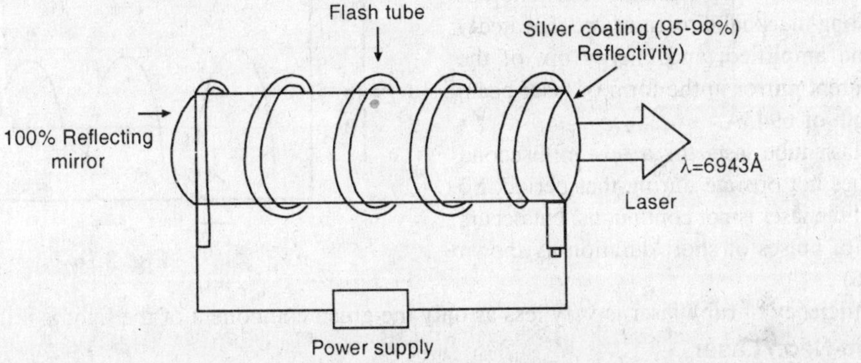

Fig. 3.14 Schematic description of first ruby laser.

Construction :

(a) **Active Medium :** It is a Ruby crystal, family of gems consisting of $Al_2 O_3$ with 0.05% of chromium atoms by weight as impurity. Al^{3+} ions in the crystal are substituted by Cr^{3+} ions. These ions act as **active centres**.

(b) **Optical Resonator :** The cross sectional faces are cut parallel and polished with a high reflectivity materials such as Al or Ag. One face act as 100% reflecting mirror and other's reflectivity in about 95-98%.

(c) **Pump :** The flash tube filled with Xe gas in wounded helically over the ruby rod. The burst of white light from tube serves as optical pumping.

Energy Level Diagram :

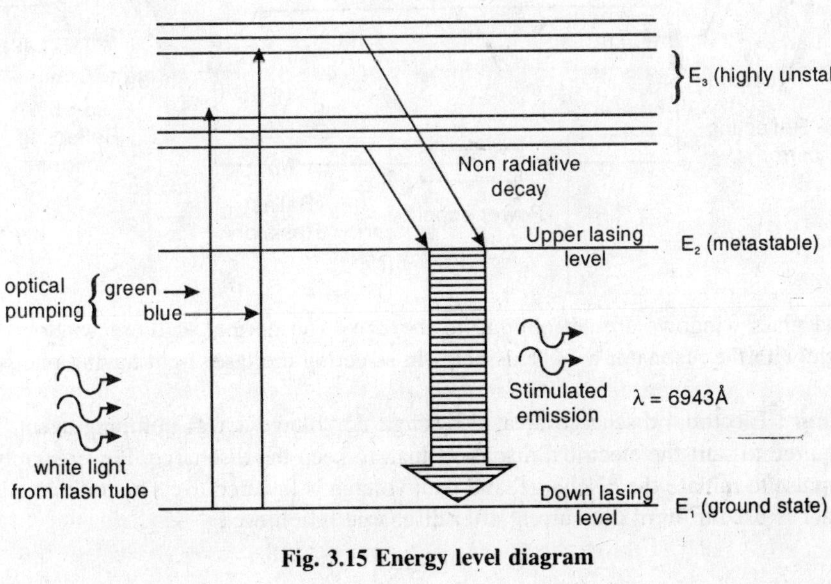

Fig. 3.15 Energy level diagram

Working

The energy level diagram of Cr^{3+} ions is shown in fig. 3.15. The green and blue component of light from flash tube is absorbed (former is dominant) by Cr^{3+} ions raising them to the band of level E_3. This level is highly unstable. Cr^{3+} ions rapidly lose part of their energy (E_3-E_2) to the crystal and undergoes non radiative transition to level E_2. This level is metastable. If pumping continuous the population of this level exceeds that of the ground state. This is the state of population inversion. A spontaneously emitted photon by Cr^{3+} ion at E_2 level initiates the stimulated emission by the other Cr^{3+} ions in the metastable state. Photons travelling along the axial directions are repeatedly reflected and amplified, and energy out of the semi-transparent mirror in the form of laser beam of wavelength of 6943Å.

The flash tube lasts for a few millisecond and laser does not operate during that period. So the output of the laser is not continuous, but occurs in the form of pulses of short duration as shown in Fig. (3.16) .

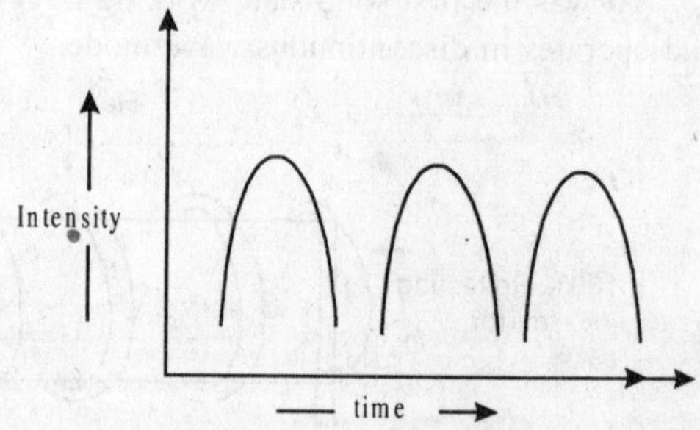

Fig. 3.16

The efficiency of ruby laser is very less as only the green component of the light is utilized.

3.19 Helium-Neon Laser

The He-Ne laser was the first gas laser and was most common untill the spread of semiconductors laser. It was first built by Ali Javan in 1961. It employs four level laser and operates in continuous wave (CW) mode.

Construction

(a) **Active Medium :** It is the mixture of Helium and Neon in the ratio of about 10:1 at a pressure of about 1 torr. It is enclosed in long discharge tube. The tube is hermetically sealed. The Ne atoms act as active centres.

(b) **Optical Resonator :** Two confocal type mirror are placed across the ends of the discharge tube. One is 100% reflecting mirror and other is semitransparent from where laser is emitted.

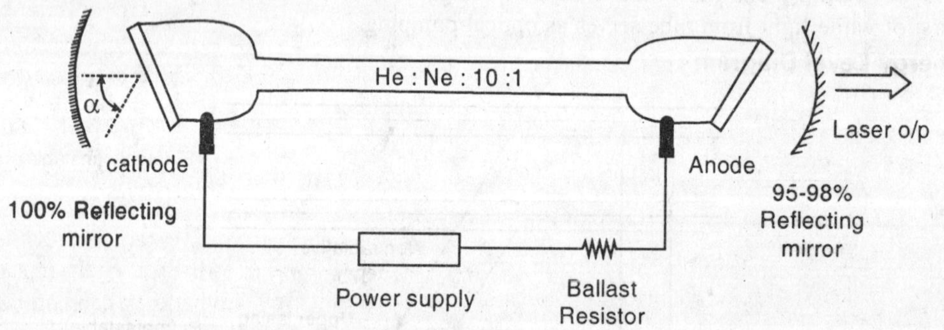

Fig. 3.17

The end glass windows are set oblique to the axis. The normal to these windows makes Brewster angle with the resonator axis. This helps in selecting the laser light having one plane of polarization.

(c9 **Pump :** Electrical discharge created by high dc voltage act as pumping agent. Higher voltage is required to start the electrical discharge than to keep the discharge. Power supply gives high voltage pulse to initiate the discharge, and then voltage is lowered to its operating value. The Ballast Resistor is used to limit the current after discharge is achieved.

Energy Level Diagram

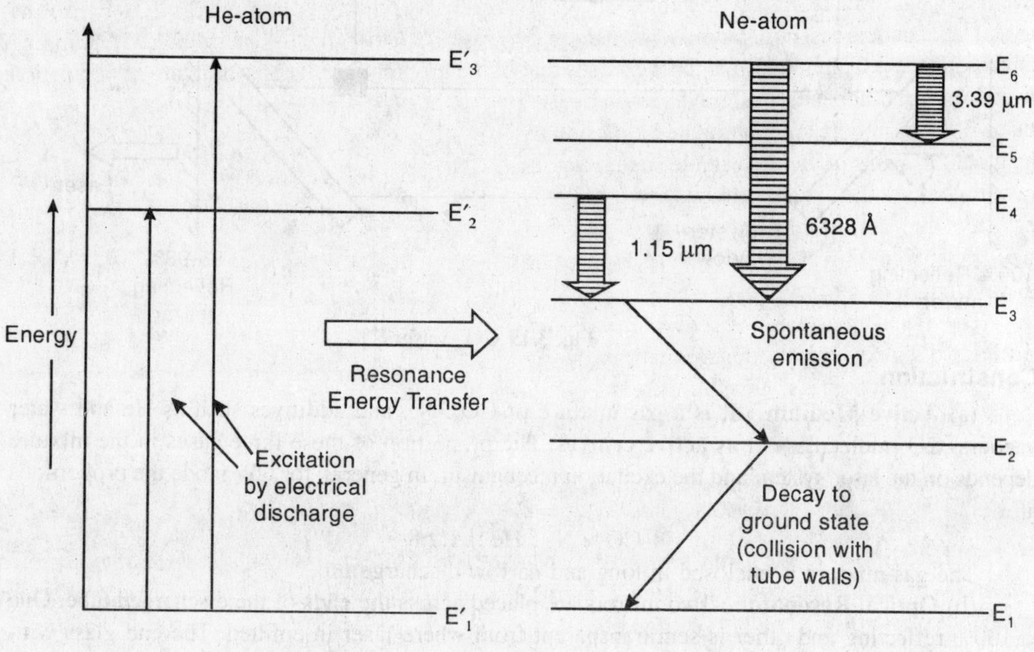

Fig. 3.18 Energy level diagram of He-Ne laser

Working : The electrical discharge accelerates the electrons. These electrons collide with He atoms and these atoms are raised to level E'_2 and E'_3. These excited He atoms can not dexcite spontaneously rather they collide with Ne atoms and transfer their excited energies through the process of resonance energy transfer. Ne-atoms are excited to level E_6 and E_4. Reverse energy transfer also takes place, but due to small concentration of Ne-atoms, it is neglegible. As collisions continues population of Ne-atoms in these two excited states increases. Three stimulated transitions occur which produce laser light of three wavelengths

(i) $E_6 \rightarrow E_3$ transition λ = 6328Å Red colour

(ii) $E_4 \rightarrow E_3$ transition λ = 1.15 μm IR

(iii) $E_6 \rightarrow E_5$ transition λ = 3.39 μm for IR

After these transitions Ne-atoms are rapidly decay to level E_2 which is again metastable level. In order to dexcite them to ground state, tube is made narrow so that they can collide with inner wall of the tube frequently.

3.20. Carbon-Dioxide Laser

Lasing action in a CO_2 molecule was first demostrated by C Patel in 1964. It is molecular laser and employs four level laser from the standpoint of potential industrial applications, this laser unquestionably ranks first. It is capable of continuously generating as high powers as 10KW at a relatively high efficiency (upto 40%).

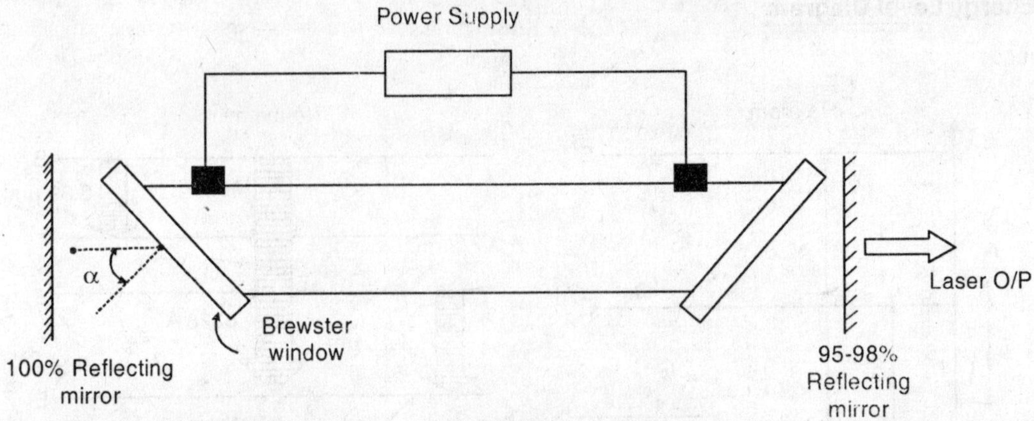

Fig. 3.19 CO_2 Laser

Construction

(a) **Active Medium :** It is a gas mixture of CO_2, N_2, and additives such as He and water vapours. CO_2 molecules act as **active centres**. The proportion of these three gases in the mixture depends on the laser system and the excitation mechanism. In general, for CW mode the proportions are:

$$CO_2 : N_2 : He :: 1:2:8$$

The gas mixture is enclosed in long and narrow discharge tube.

(b) **Optical Resonator :** Two mirrors are placed across the ends of the discharged tube. One is 100% reflecting and other is semitransparent from where laser in emitted. The end glass windows are set oblique. The normal to these windows makes angle equal to Brewster angle with the resonator axis. This helps in selecting the laser light having one plane of polarization.

Pump : Electrical discharge created by dc voltage act as pumping agent.

Energy Level Diagram :

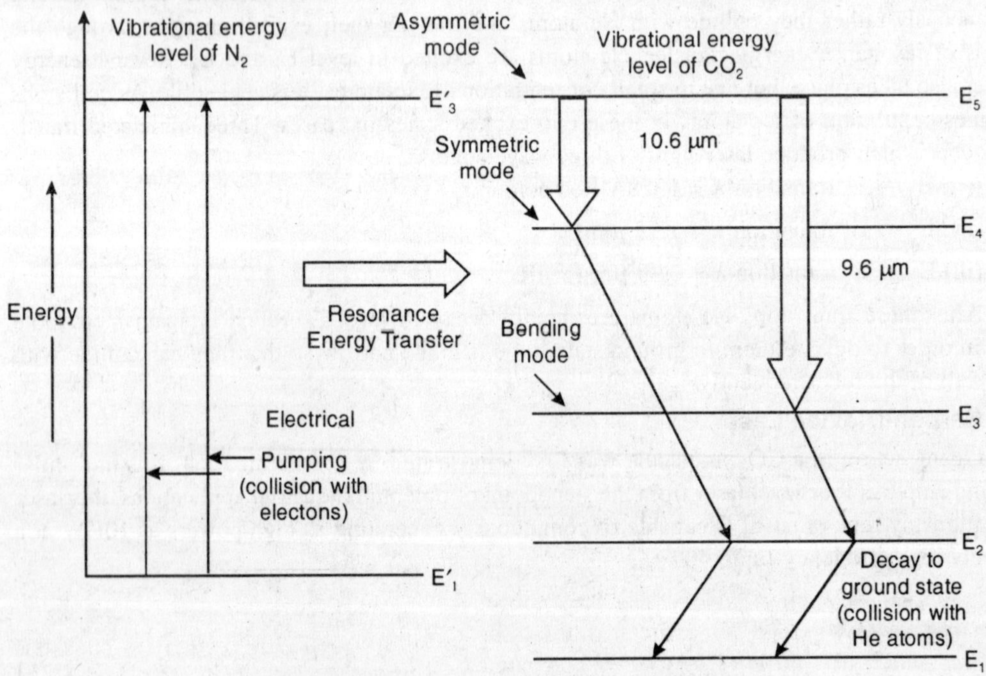

Fig. 3.20 Vibrational energy level of N_2 and CO_2

Vibrational Modes of CO_2 : CO_2 is a linear molecule which can possess three vibrating mode

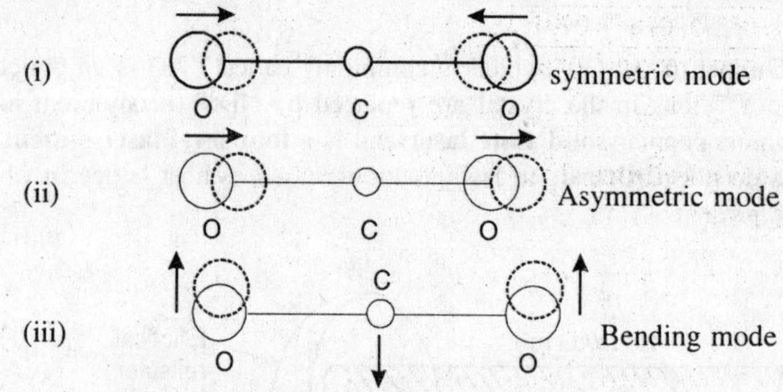

Fig. 3.21 Vibrational modes of CO_2 molecule

If the forward current is large enough, the region about the junction contains a large concentration of electrons within the conduction band and hales within valance band, which is definitely a conduction of population inversion. This inversion exists between the bottom of the conduction band and the top of the valence band. If the front and back faces of junction are cleaved, they can act as resonator. Then spontaneously emitted photons initiate stimulated emission and laser is emitted through the cleaved junction. Various techniques are used to provide adequate heat sinking of the device.

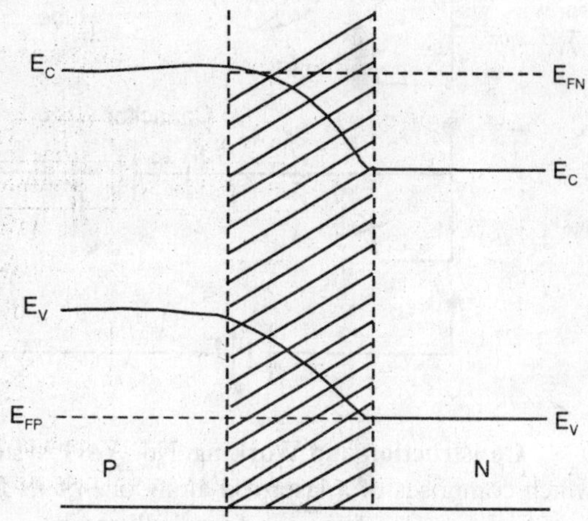

PN junction at forward bias. Shaded region indication the invertion region at the juntion.

Working : The electrical discharge raises the vibrational energy level of N_2 to upper state F'_2. This state is metastable state, so N_2 can not dexcite spontaneously. These excited N_2 molecules transfer this energies to CO_2 molecules via resonance energy transfer. The CO_2 molecules are excited to level E_5 where they possess asymmetric mode of vibrations. As the population of CO_2 molecules builds up at E_5 level, population inversion is achieved between E_5 and E_3, and E_5 and E_3 levels. The stimulated transition between $E_5 \rightarrow E_4$ (prom asymmetric mode to symmetric mode) level produces for IR laser at wavelength 10.6 μm and transition between $E_5 \rightarrow E_3$ (from asymmetric mode to bending mode) level produces for IR laser at wavelength 9.6 μm. The E_4 and E_3 levels are also metastable, CO_2 molecules at these levels falls to lower level E_2 through collisions with normal CO_2 molecules. The presence of He-atoms further dexcite the CO_2 molecules to their ground state. **It also serves as coolent.** This whole process repeats.

3.21. Semiconductor Laser

This laser was became an important part of semiconductor device technology in 1962 when the first PN junction laser were built in Ga As (IR region) and Ga As P (visible region). Junction lasers are remarkably small, they exhibit high efficiency, and laser output is easily modulated by controlling the forward current. Semiconductor lasers operate at low power compared to, e.g. ruby or CO_2 lasers; on the other hand, these junction

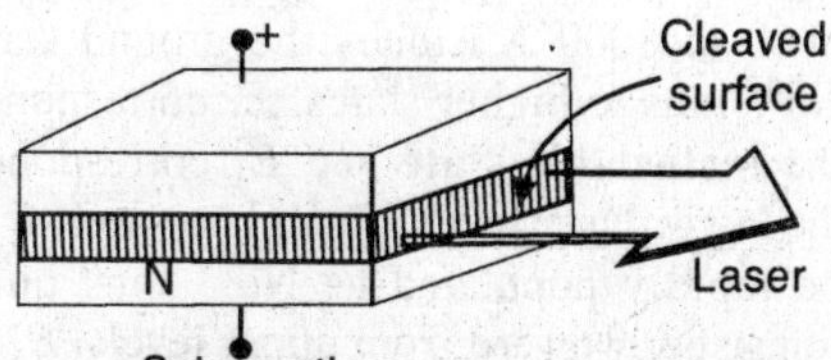

Fig. 3.22 diagram of PN junction laser

lasers compare with He-Ne lasers in power output. They are particularly suitable for fiber optic communication.

3.22. Nd : YAG Laser

Yttrium Aluminium Garnet ($Y_3Al_5O_{12}$) which is commonly called YAG is an optically isotropic crystal. Some of the Y^{3+} ions in the crystal are replaced by ND^{3+} (neodymium ions) Nd : YAG laser is one of the most popular **solid state laser** and is a **four level laser system.** In Nd : YAG laser, crystal atoms do not participate in lasing action but act as host lattice in which Nd^{3+} ions (active centres) are present.

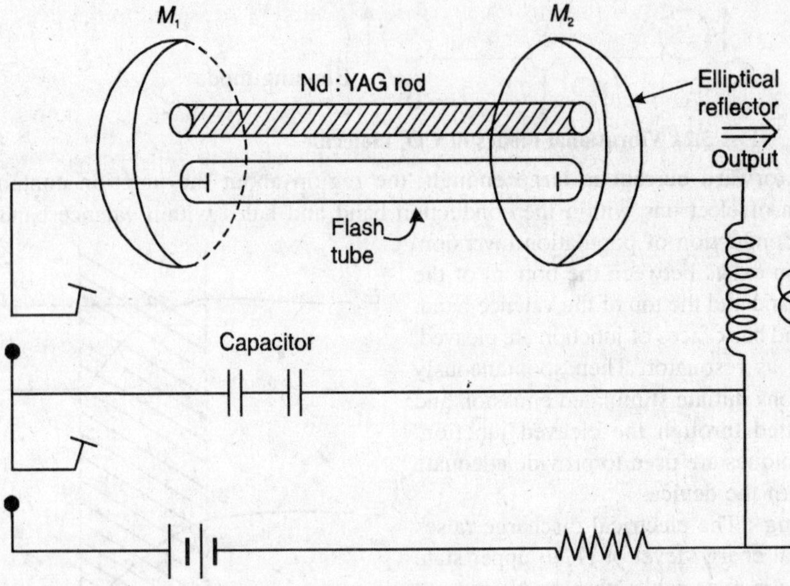

Fig. 3.23

Construction and Working. Nd : YAG system consists of an **elliptically cylindrical** reflector which comprises of a laser rod along one of its focus lines and flash lamp along the other focus line. Light coming from one focus will reach the other focus at free suffering multiple reflections from the silvered surfaces of the reflector. So the total light is received by the Nd : YAG rod, the two ends of which are polished and silvered so as to constitute and **optical resonator.**

As shown in the energy level diagram, the energy levels of this laser system are complex. The pumping of Nd^{3+} ion to the upper states is done using **krypton arc lamp.** Optical pumping with light of wavelength 5000 to 8000 Å excites the ground state of Nd^{3+} ions to higher states. E_3 corresponds to the **metastable state** and E_2 corresponds to the **lower laser level.** Upper laser level E_3 will be rapidly populated as Nd^{3+} ions quickly transist downward from upper levels. E_2 level is quite above the ground level and so it cannot be populated by Nd^{3+} ions through

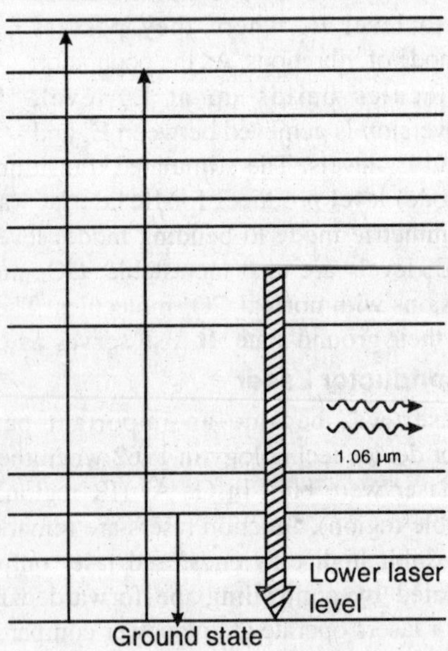

Fig. 3.24 Energy levels of Nd : YAG laser.

normal transitions from the ground level. So the pupulation inversion is achieved be tween E_3 **and** E_2 **level** and laser emission occurs in the infrared (IR) range of spectrum, *i.e.*, about 10600 Å (1.06 μ m).

Nd : YAG laser finds applications in many machine operations like drilling, welding, etc. Also they are being used in medical surgery and many other fields.

Energy and momentum of e.m. waves (lasers). Electromagnetic waves also **transport momentum**. An object experiences pressure when a beam of light shines on it and this is called **radiation pressure. Radiation pressure is usually very small.**

Consider a plane e.m. wave or laser incident on an object for a very small time *t*. Let this wave be completely absorbed during this time so the momentum *P* delivered to the object due to absorption of energy (say *E*) may be written as:

$$p = \frac{E}{c} \qquad ...(1)$$

Now if the energy *E* is entirely reflected, the momentum imparted to the perfectly reflecting object will be twice the above value, *i.e.*,

$$p = \frac{2E}{c} \qquad ... (2)$$

Example 1. *A laser beam of intensity 50 w/m² falls on a perfectly reflecting plane mirror for an hour. The area of the mirror is 0.5 cm² Calculate:*

(i) Momentum imparted to the mirror in this time.

(ii) Average force acting on this mirror.

Solution. Energy falling on the mirror in one hour is given by

$$E = 50 \times 0.5 \times 3600 = 9 \times 10^4 \text{ J}$$

∴ (*i*) Momentum imparted to the mirror

$$P = \frac{2E}{c} = \frac{2\times9\times10^4}{3\times10^8} = \mathbf{6\times10^{-4}\,kg\,ms^{-1}}$$

(*ii*) Average force $F = \dfrac{\text{change of momentum}}{\text{time}} = \dfrac{p}{t}$

$$= \frac{6\times10^{-4}}{3600} = \mathbf{1.67\times10^{-7}\,N\ Ans.}$$

Example 2. *A laser beam having a wavelength of 8000 Å and aperture 0.5 cm is sent to moon. Calculate (1) angular spread of the beam (ii) areal spread of the beam when it reached to moon. (Given: distance of moon from earth = 4 × 10⁸ m).*

Solution. In this case $\lambda = 8000\text{ Å} = 8\times10^{-7}\text{ m}$

Aperture $d = 0.5 \text{ cm} = 5 \times 10^{-3} \text{ m}$

Distance of moon $D = 4 \times 10^8 \text{ m}$

Now angular spread in given by :

(*i*) $$d\theta = \frac{\lambda}{d} = \frac{8\times10^{-7}\text{ m}}{5\times10^{-3}\text{ m}} = \mathbf{1.6\times10^{-4}}\text{ radian}$$

(*ii*) Areal spread = $(D.d\theta)^2 = (4 \times 10^8 \times 1.6 \times 10^{-4})^2 = \mathbf{4.096 \times 10^9\ m^2}$. **Ans.**

Example 3. *A laser beam λ = 6000 Å on earth is focussed by a lens of diameter 2 m on to the surface of moon. How big is the spot on the moon. (Given distance of moon from earth = 4 × 10⁵ km)*

Solution. In this case, $\lambda = 6000 \text{ Å} = 6 \times 10^{-7} \text{ m}$

$d = 2 \text{ m}$

$D = 4 \times 10^5 \text{ km} = 4 \times 10^8 \text{ m}$

Angular spread, $$d\theta = \frac{\lambda}{d} = \frac{6\times10^{-7}}{2} = 3\times10^{-7} \text{ radian}$$

Areal spread, *i,e,.* area of the spot on the moon's surface

$$= (D \times d\theta)^2 = (4 \times 10^8 \times 3 \times 10^{-7})^2$$

$$= \mathbf{1.44 \times 10^4\ m^2}$$

Example 4. *A laser beam has a power of 50 mW. It has an aperture of 5 × 10⁻³ m and wavelength 7200 Å. A beam is focussed with a lens of focal length 0.1 m. Calculate the area and the intensity of the image.*

Solution. In this case $\lambda = 7200 \text{ Å} = 7.2 \times 10^{-7}$ m

$$d = 5 \times 10^{-3} \text{ m}$$

Focal length of the lens = 0.1 m

$$\therefore \quad d\theta = \frac{7.2\times10^{-7}}{5\times10^{-3}} = 1.44\times10^{-4} \text{ radian}$$

(*i*) $$\text{Areal spread} = (d\theta.f)^2 = (1.44\times10^{-8}\times0.1)^2 = 2.074\times10^{10}\,\text{m}^2$$

(*ii*) $$\text{Intensity} = \frac{\text{Power}}{\text{Area}} = \frac{\text{Power}}{\text{Areal spread}}$$

$$\text{Power} = 50 \times 10^{-3} \text{ watt}$$

$$\text{Areal spread} = 2.074 \times 10^{-10}\ \text{m}^2$$

$$\therefore \quad I = \frac{50\times10^{-3}}{2.074\times10^{-10}} = \mathbf{2.411 \times 10^8\ watt/m^2}$$

3.23. Applications of Lasers

Lasers have found use and applications in almost all the fields of science and technology. Some of the important applications may be described as under :

1. Because of the intense energy, these have been used in drilling holes in hard metals and diamonds.

2. In surgery, these are used because due to the production of localized heat, they cut and seal all the blood vessels simultaneously. They are infact used in 'blood' less surgery'. These are also used for welding the detached retina of the eye and also in the treatment of maligned tumours.

3. These are used in radios and televisions because being coherent, these can be modulated to send hundred of messages simultaneously. Also since these are not absorbed, by water, so for underwaters transmissions, these are highly useful.

4. These have also been used for automatic control of rockets and satellites which may be used for determination of distant stars and planets.

5. These have also found use in warfare for detecting and destroying aeroplanes, missiles and tanks.

6. In scientific research, these are being used for chemical and crystalline structures of various molecules. In Raman spectroscopy and various other related phenomenon these are being extensively used.

7. In high speed photography, Lasers are useful because these can detect the fast moving bullets and missiles. Holography, *i.e.*, the three dimensional photography has been possible because of lasers. *Also in fibre optics lasers are being extensively used.*

7.24. Masers

The term Maser stands for "*molecular* (previously microwave) *amplification by simulated emission of radiation*" It is extremely useful as a source for carrier of microwaves.

The working of maser is similar to that of laser. Actually the principle of stimulated emission on which laser is based was first employed in microwave region (in frequency range ~ 10^{10} Hz).

In the same way, the principle of maser can be stated as under :

The maser action is based on the principle of *'population inversion'* followed by *'stimulated emission.'* For *stimulated emission* to predominate, most of the atoms should be in the excited state. So population of atoms in higher energy level is increased by pumping the matter by **photons of appropriate size.** If during the excited state of the atom, it is struck by a matching photon having precisely the energy as that emitted by the atom in spontaneous emission, then the excited atoms emits photons that fall precisely in phase with the stimulating photons. So the excited atoms emits waves in step with incoming wave and increases its intensity. Finally resulting emergent wave is extremely intense divergence.

3.25. Wave Packet

Light is emitted by an excited atom when it comes to lower state and difference of energy is given in the form of a photons. So a photon corresponds to a light wave. As the process of tranisition from higher to lower state takes about 10^{-8} s, the atom will emit light wave in about 10^{-8} second. It means that the light emitted from an atom is not a continuous sine wave, **but a train of wave oscillations, having a finite length.**

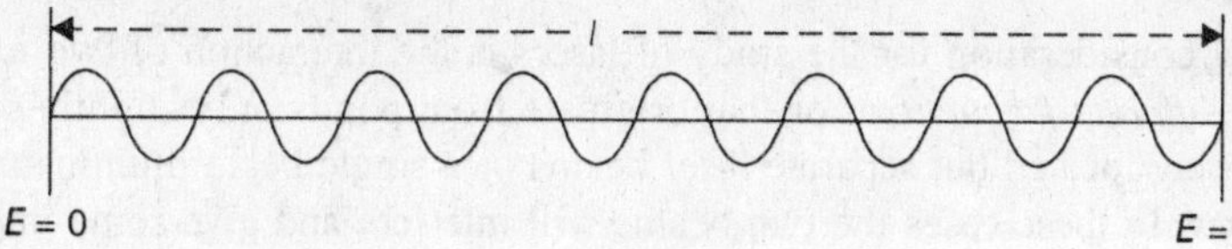

Fig. 3. 25

A wave train is given by $E = E_0 \sin (kx - \omega t)$

Such a wave representation is called a **wave packet** or a **wave train**.

Let us length of the wave train in vacuum be l, then $l = x_2 - x_1 = ct$.

If $t = 10^{-8}$ *s*, and $c = 3 \times 10^8$ ms^{-1}

$l = 3 \times 10^8$ ms$^{-1} \times 10^{-8}$ s = 3 m

∴ No. of wave oscillation present in a wave train of mean wavelength 5000 Å (say) will be given by

$$N = \frac{l}{\lambda} = \frac{3m}{5 \times 10^{-7}} = 6 \times 10^6$$

which means that a wave train comprising of a single photon contains about a million wave oscillations.

3.26. Band Width

It may be clearly understood that a wave packet is not a **harmonic** wave and so can not be represented by simple sine functions. A mathematical representation of a wave packet is done in terms of **Fourier Integrals** and is a **complex identity**. This representation we will not take up here but will describe some important features of wave packets.

If light given out by a source is analysed by a spectrograph, we observe it

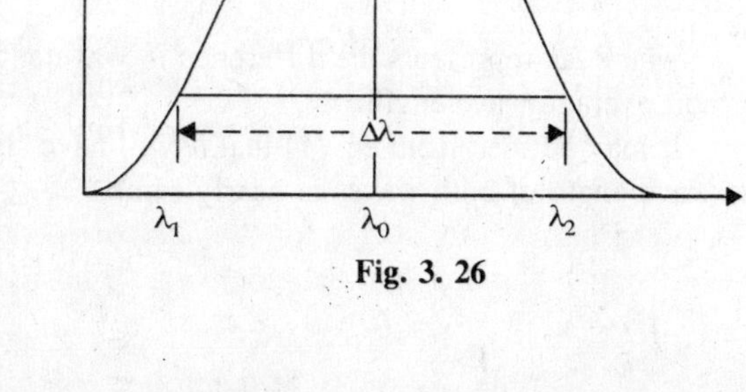

Fig. 3. 26

to be made up of **discrete spectral line** which are formed by wave packets emitted by atoms. So a **spectral line and a wave packet corresponds to the same phenomenon** referred in different ways.

As stated, wave length of a **wave packet of a spectral line is not precisely** defined. It means that there is a continuous spread of wavelengths over a finite range $\Delta\lambda$, centred around a wavelength λ_0. A wave packet may be considered as comprising of a number of harmonic waves which differ only by very **very small wavelengths**. Maximum intensity of the wave packet will occur at λ_0 and the intensity falls off rapidly on either side of l0 as shown in Fig. 3.26.

The spread of wave length is called line width or band width. As shown in the figure, the band width is the wavelength interval from $\lambda_0 - \frac{\Delta\lambda}{2}$ to $\lambda_0 + \frac{\Delta\lambda}{2}$ that contains the major part of the energy possessed by the wave packets.

Fourier analysis of the **wave packet** gives a relation between **band width and frequency of oscillation** as

$$\Delta\nu = \frac{1}{\Delta t}$$

where Δt corresponds to the life time of the light emitting atom into its excited state.

3.27. Coherence

An important consideration for the study of lasers is the interaction of two e.m. waves that have only *slightly different frequencies* or that originate from points only slightly separated. Spatially, *e.g.*, two closely located but separate laser beams or a single beam illuminating two *closely positioned apertures.* In these cases the two beams will interfere and give some important effects like longitudinal modes, mode locking, phase matching and frequency multiplication etc.

When complete coherence is not achieved but still some interference takes place, the condition is called *partial interference.* Partial interference condition can be grouped into two categories.

(*i*) Temporal wherence associated with difference in frequencies of two waves. It is also referred to as *longitudinal coherence*.

(*ii*) Spatial coherence associated with location of two waves. It is also referred to the *transverse coherence*.

3.28. Temporal Coherence

It refers to the relative phase or coherence of the two waves at two separate locations along the propagation direction of the two beams.

If we assume that the two beams are exactly in phase at *initial* (say first) location then they will be at least partially in phase in some other (say 2nd) location upto some length say *lc* which is called coherence length and may be found as

Fig. 3.27

$$l_c = \lambda. = \left(\frac{\lambda}{\Delta\lambda}\right) = \frac{\lambda^2}{\Delta\lambda} \quad \text{...(1)}$$

where $\Delta\lambda$ represents the difference in wavelength between the two waves and λ corresponds to their average wavelength.

It may be seen from eq. (1) that *lc* will have significant value only when $\Delta\lambda << \lambda$ *i.e.*, when the wavelength of both waves is nearly same.

3.29. Spatial Coherence

Spatial coherence is also referred to as *transverse coherence* and it describes, how far apart two sources or two portion of the same source can be located in a direction transverse to the direction of observation and still exhibit coherent properties over a range of observation points. This is some times called lateral coherence.

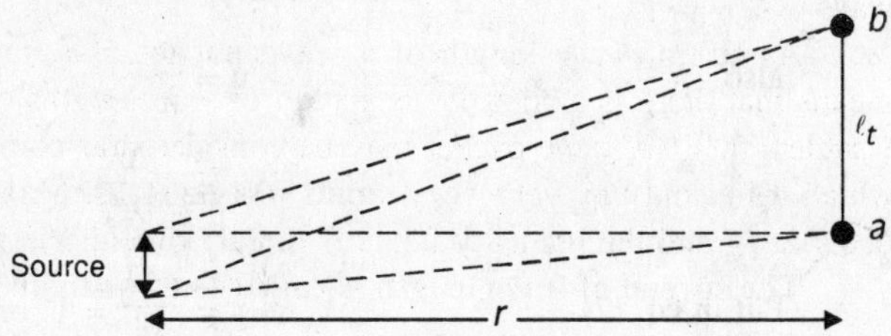

Fig. 3.28

In other words, it means that the distance l_t by which two points are separated in transverse direction and still interference effects are observed at *Specific distance from the source.*

Let two sources be separated by a distance S in *transverse direction* to the direction of observation and distance r from point of observation. If the two sources exhibit *interference phenomenon* at point a then transverse length l_t is the transverse distance from a to b, where two sources show interference effects. This distance may be calculated from

$$l_t = \frac{\lambda}{\theta_s} = \frac{r\lambda}{S} \qquad [\because \theta_s = S/r]$$

Temporal coherence can be studied from the *Michelson's interferometer* by moving one of the mirrors. In the case of an interferometer, the visibility V of the fringes is a measure of *degree of coherence*. In the case of Michelson interferometer, path difference is adjusted to be zero to get *maximum visibility*. From this position, one of the mirror is moved to a position where no fingers are observed. Then distance between these positions give us the *maximum range over which fringes can be observed*. The time $\tau = \frac{2d}{c}$ is the time taken by the wave to travel a distance $2d$ and V_1, V_2 are the frequencies of the two waves. The distance l for which the field remains sinusoidal is called coherence length and is given by

$$l = \tau.c$$

where c is the velocity of light.

Now if $2d < l$, there will be a definite phase relationship between two interfering beams and if $2d <$ l, no definite phase relationship between the beams will be observed, so starting with equal separation of mirrors from the silvered glass plate, (*i.e., zero* path difference between interfering beams), if now the path difference is continuously increased, the dask and bright fringe contrast goes on reducing, till the fringes disappear. The path difference at disappearence of fringes gives us an estimate of *coherence length.*

For sodium light, coherence length is about 3 cm, so coherence time is $\frac{l}{c} = \frac{3}{3\times10^{10}}$ and for Cd red light, coherence length is about 30 cm. *i.e.,* 10 times more than that of sodium light.

3.30. Purity of a Spectral Line

Finite purity Q of a spectral line is defined on the basis of quantum theory as

$$Q = \frac{\lambda}{\Delta\lambda} \qquad ...(1)$$

The purity of Cd red light $\simeq 10^6$ and that of sodium yellow light $\simeq 10^5$.

If the radiations emitted from a source of frequency V has a frequency spread $\Delta\nu$ and its coherence time τ, then the radius of amplitude is explained by putting the difference of number of oscillations of frequency ν and $\nu + \Delta\nu$ in time by unity.

i.e., $\tau(\nu + \Delta\nu) - \tau\nu = 1$

$\therefore$ $\tau\,\Delta\nu = 1$ ...(2)

also $\nu = \dfrac{c}{\lambda}$

$\therefore$ $\Delta\nu = c\,\dfrac{\Delta\lambda}{\lambda^2}$ (in magnitude)

Put in eq. (2) $\tau = \dfrac{c\Delta\lambda}{\lambda^2} = 1$ or $\Delta\lambda = \dfrac{\lambda^2}{c\tau}$ ...(3)

also coherence length l is given by

$$l = \tau\, c$$

$\therefore$ eq. (3) may be written as $\Delta\,\lambda = \dfrac{\lambda^2}{l}$

$\therefore$ coherence length $l = \dfrac{\lambda^2}{\Delta\lambda}$ ...(4)

which may be written in terms of purity Q of spectral line as

$$1 = \lambda.\ \frac{\lambda}{\Delta\lambda} = \lambda Q \qquad ...(5)$$

i.e., coherence length = wavelength of a spectral line × finite purity

3.31. Relation between Coherence Length and Line Width

We have $c = \nu\,\lambda$

$\therefore$ $\lambda = \dfrac{c}{\nu}$

Differentiating $\Delta\lambda = -\,c.\ \dfrac{\Delta\nu}{\nu^2} = -\,c.\ \dfrac{\Delta\nu}{c^2}\lambda^2$

$$= -\frac{\lambda^2\Delta\nu}{c} \qquad ...(1)$$

Negative sign shows that change in $\Delta\lambda$ is opposite to change in frequency.

Writing the above expression in magnitude

$$\Delta\lambda = \lambda^2\frac{\Delta\nu}{c} = \frac{\lambda^2}{c}\frac{1}{\Delta t} \qquad \left[\therefore \Delta\nu = \frac{1}{\Delta t}\right]$$

$\therefore$ $\Delta t = \dfrac{\lambda^2}{c\,.\,\Delta\lambda}$ ...(2)

and coherence length is given by

$$\Delta\,l = c.\ \Delta t$$

$$= c.\ \frac{\lambda^2}{c\,.\,\Delta\lambda}$$

or $\Delta\,l = \dfrac{\lambda^2}{\Delta\lambda}$ ...(3)

which gives us a relation between coherence **length Δl and line width $\Delta\lambda$.**

Example 5. *An optical filter has a line width of 1.5 nm and mean wavelength 550 nm with white light incident on the filter, calculate:*

(i) coherence length (ii) number of wavelengths in wave train.

Solution.

In this case $\Delta\lambda = 1.5 \text{ nm} = 1.5 \times 10^{-9}$ m

$\lambda = 550 \text{ nm} = 550 \times 10^{-9}$ m

(*i*) coherence length

$$\Delta l = \frac{\lambda^2}{\Delta\lambda} = \frac{(550\times10^{-9})^2}{1.5\times10^{-9}}$$

$$= \mathbf{2.017 \times 10^{-4}\ m}$$

(*ii*) number of waves in wave train

$$n = \frac{\Delta l}{\lambda} = \frac{2.017\times10^{-4}}{550\times10^{-9}}$$

$$= \mathbf{367\ (approx.)\quad Ans.}$$

Example 6. *Cadmium red light of mean wavelength 6438 Å is used in Michelson interferometer. Line width of the source is 0.013 Å for the initial setting of the interferometer for zero optical path difference. One of the mirrors is slowly moved until the fringes first disappear. Calculate the distance by which the mirror is moved, also calculate the number of wavelength corresponding to the distance.*

Solutions. In this case

$\lambda = 6428 \text{ Å} = 6438 \times 10^{-10}$

$\Delta\lambda = 0.013 \text{ Å} = 0.013 \times 10^{-10}$

(*i*) coherence length

$$\Delta l = \frac{\lambda^2}{\Delta\lambda} = \frac{\left(6438\times10^{-10}\right)^2}{0.013\times10^{-10}}$$

$$= \mathbf{0.319\ m\quad Ans.}$$

Now the condition for first disappearance of feringes is that the optical path must exceed the coherence length $\Delta\lambda$.

$\therefore$ $2d = 0.319$ m

or $d = 0.1595 \text{ m} = 15.95$ cm.

which means that the mirror has to be moved through a distance 15.95 cm.

also $d = n\lambda$

$\therefore$ $0.1595 = n \times 6438 \times 10^{-10}$

$\therefore$ $n = \mathbf{24770\ waves\quad Ans.}$

Example 7. *Quasi monochromatic source emits radiations of mean wavelength $\lambda = 5461$ Å and has a band width $\Delta v = 10^9$ Hz. Calculate:*

(*i*) *coherence time* (*ii*) *coherence length* (*iii*) *frequency stability.*

Solution. In this case

$\lambda = 5461 \text{ Å} = 5461 \times 10^{-10}$ m

$\Delta v = 10^9$ Hz

$\therefore$ (*i*) $$\Delta t = \frac{1}{\Delta v} = \frac{1}{10^9} = 10^{-9} \text{ sec.}$$

(*ii*) $\Delta l = \text{c.}\ \Delta t$

$= 3 \times 10^8 \times 10^{-9}$

$= 0.3$ m.

(*iii*) we have

$$\nu = \frac{c}{\lambda} = \frac{3\times10^8}{5461\times10^{-10}}$$
$$= 5.5 \times 10^{14} \text{ Hz.}$$

∴ Frequency stability

$$\frac{\Delta\nu}{\nu} = \frac{10^9}{5.5\times10^{14}} = \mathbf{1.82\times10^{-6}}$$

Example 8. *Calculate temporal coherence length lc for*

(i) mercury vapour lamp emitting in green portion of spectrum at wavelength of 546.1 nm with emission band width of $\Delta\nu = 6 \times 10^8$ *Hz.*

(ii) A helium neon laser operating at wavelength of 632.8 nm.

Solution. We have the relation

$$c = \nu\lambda$$

or $$\nu = \frac{c}{\lambda} \qquad \therefore \quad \Delta\nu = -\frac{c}{\lambda^2}\Delta\lambda.$$

∴ (*i*) For mercury lamp, we have

$$\Delta\lambda = \frac{\lambda^2\Delta\nu}{c} \text{ (in magnitude)}$$

$$= \frac{(546.1\times10^{-9}\text{ m})^2\times(6\times10^8\text{ s}^{-1})}{3\times10^8\text{ m s}^{-1}} = 5.96\times10^{13}\, m$$

and temporal length is given

$$l_c = \frac{\lambda^2}{\Delta\lambda} = \frac{(546.1\times10^{-9}\text{ m})^2}{5.96\times10^{-13}\text{ m}} = \mathbf{0.50\ m}$$

(*ii*) For helium-neon lasers, we have

$$\Delta\lambda = \frac{\lambda^2\Delta\nu}{c} = \frac{(632.8\times10^{-9}\text{ m})^2\ (1\times10^6 2^{-1})}{3\times10^8\text{ ms}^{-1}}$$

$$= 1.33 \times 10^{-15} \text{ m} = 1.33 \times 10^{-6} \text{ nm}$$

Temporal length is thus given by

$$l_c = \frac{\lambda^2}{\Delta\lambda} = \frac{(632.8\times10^{-9}\text{ m})^2}{1.33\times10^{15}\text{ m}} = \mathbf{30m}$$

So temporal coherence length of laser beam is much longer than that of mercury lamp.

Example 9. *A laser produced plasma consisting of a 100 μm diameter ball radiates very strongly at a wavelength of 10 nm. Calculate spatial coherence at a distance 0.5 m from the source.*

Solution. In this case

$$r = 0.5 \text{ m}$$
$$\lambda = 10 \text{ nm}$$
$$S = 100\ \mu\text{m} = 10^{-4} \text{ m.}$$

We have the relation for spatial coherence as

$$l_t = \frac{r\lambda}{S} = \frac{(0.5\text{m})\ (1\times10^{-8}\text{ m})}{1\times10^{-4}\text{ m}} = \mathbf{5\times10^{-5}\ m} \quad \textbf{Ans.}$$

Example 10 . *Find the average number of photons/m^3 in a monochromatic beam of radiation having frequency 10 kHz and intensity 10^4 W/m^2.*

Solution. Intensity of a beam of radiation is the total energy of the no. of photons passing normally through unit area a cross-section of beam/sec.

Let x be the no. of photons/unit volume each of energy hv. Let us consider a length 1 of the beam having cross-section area A.

So in volumne Al, the total photon energy = $nAl\ hv$ and time taken by this energy to pass through the length 1 is given by

$$\bar{t} = l/c$$

where c is the velocity of light in free space.

∴ Intensity I of the beam is given by

$$I = \frac{nAlhv}{A.t} = \frac{hl.hv}{t}$$

$$= \frac{nl\ hv}{l/c} = nhvc$$

$$\therefore \qquad n = \frac{I}{hvc}$$

Putting the various value, we get :

$$n = \frac{10^4}{6.62\times10^{-34}\times10^4\times3\times10^8}$$

$$= \frac{10^{26}}{19.88} = 5.03\times10^{24} \text{ photons/m}^3$$

Example 11 . *A certain ruby laser units 1.0J pulses of light having wavelength 6940Å. Find the minimum number of Cr^{3+} ions in the ruby*

Solution. We have the relation for power of light pulses;

$$P = \frac{nhc}{\lambda}$$

$$\therefore \qquad 1.0J = \frac{n\times6.62\times10^{-34}J-S\times3\times10^8 ms^{-1}}{6940\times10^{-10}m}$$

$$\therefore \qquad n = \frac{1.0\,J\times6940\times10^{-10}m}{6.62\times10^{-3}J-S\times3\times10^8 ms^{-1}}$$

$$= \mathbf{3.4945\times10^{18}\ ions} \qquad \textbf{Ans.}$$

Example 12 . *Find the intensity of laser beam of 20 mw and diameter 1.5 mm. Assume the intensity to be uniform throughout the beam.*

Solution. Power P = 20 mw = 20 × 10^{-3} w diameter d = 1.5 m. m = 1.5×10^{-3} m.

$$\therefore \text{ Intensity} \qquad I = \frac{Power}{Area} = \frac{P}{\pi d^2/4} = \frac{4P}{\pi d^2}$$

$$I = \frac{4\times20\times10^{-3}w}{3.142\times(1.5\times10^{-3}m)^2}$$

$$= \mathbf{11.32\ kw/m^2} \qquad \textbf{\textit{Ans.}}$$

3.32. Holography

In ordinary photography, only the intensity distribution of light coming from the object is recorded on the photographic film and two dimensional image is obtained i.e. solidity of the object is lost.

In 1947, Dennis Gabor invented new technique of photography in which three dimensional image is formed. This can be obtained if in addition to recording of intensity distribution, the phase distribution of light waves coming from different parts of the object is also recorded. To record phase conditions of interference is required. This technique is called **Holography**.

This technique comprise of two steps. One is recording of Hologram and other is the reconstruction process.

Step I : Recording of hologram: The light from the object and the reference light is made

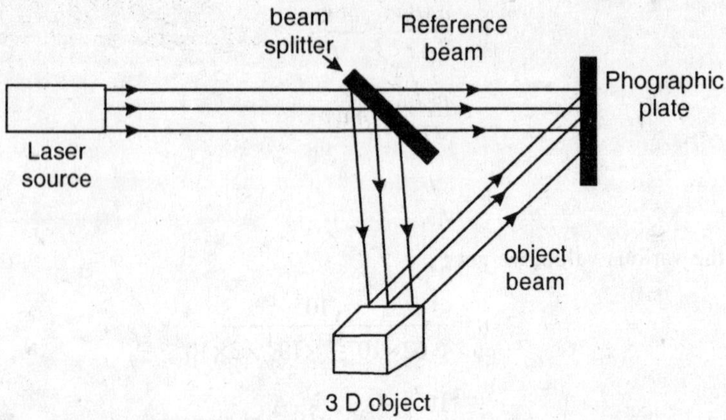

Fig. 3.29

to interfere on the photographic plate, where they interfere according the phase distribution. This photographic plate now becomes hologram. Hologram contains interference pattern.

Step II : Reconstruction of Image : In the reproduction of image, the hologram is illuminated by the laser, which was used as reference beam for the recording of hologram. Two images are

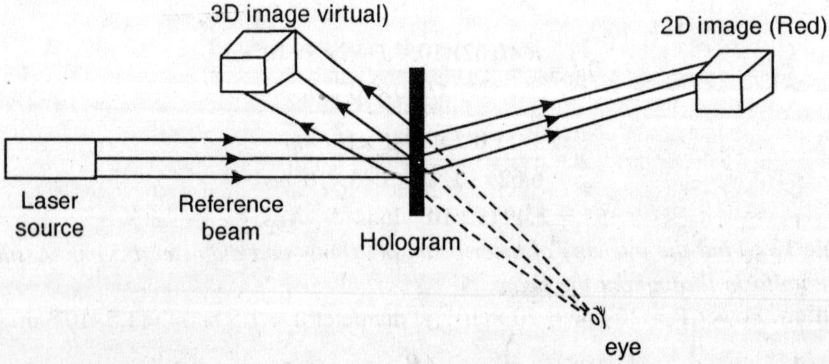

Fig. 3.230

formed by the diffracted waves coming out of hologram. One of them, projected towards the illuminating source, seem to originale object located at position where the original object was present. Other is real image, that can be photographed directly, without using a lens i.e. by placing a photographic plate at the position of the image.

QUESTIONS

1. When a suitable system of atom is exposed to a radiation flux of absorbable energy quanta, the process of absorption and emission of energy occur simultaneously. Emission, however, occurs by two distinctive mechanisms. Distinguish between them and explain the condition under which one of them predominates over the other.
Explain how the condition favourable to lasing action are realised in a ruby laser. Why are the end faces of the ruby rod silvered?

2. (a) Distinguish between the mechanism and characteristics of spontaneous emission and stimulated emission. Which one of the two emission processes in maximized in laser operation? And how?
(b) Explain tne construction and working of a semiconductor laser.

3. Distinguish between.
Spontaneous emission and stimulated emission.

4. (a) Explain the terms:
(*i*) Stimulated emission,
(*ii*) Population inversion.
(*iii*) Pumping and
(*iv*) Metastable states.
(*v*) What is the role played by them in the working of the laser?
(b) In He-Ne laser, what is the function of He atoms? Explain the answer with the help of energy level diagrams for He and Ne. In He-Ne laser why is it necessary to use narrow tubes?

5. (a) Explain the terms:
(*i*) Spontaneous emission
(*ii*) Optical pumping.
(b) Why laser action cannot occur without population inversion between atomic energy levels?
(c) Discuss in brief two engineering applications of laser.

6. (a) Explain the terms:
(*i*) Optical pumping and
(*ii*) Population inversion.
(b) Describe with a neat sketch the working of a Roby Laser.

7. (a) What is population inversion? Why is it necessary for lasing?
(b) Explain in brief how four level laser is more efficient than three level laser.
(c) In helium-neon laser lasing in through neon gas. What is then the role of helium gas?

8. Explain working of Ruby or Helium-Neon laser. Explain the role of end mirrors in laser. What frequencies are amplified by this mirror system?

9. What is an optical resonator cavity? What is its role in a laser?

10. (a) Explain with diagram absorption, spontaneous emission and stimulated emission of radiation
(b) Why population inversion in necessary to achieve lasing? In laser, active media should have preferably broad absorption band. Explain why?
(c) What is the role of metastable state in laser media? Is it must for occurrence of lasing?
(d) What is the role of resonant cavity length in supporting different frequency modes?
(e) Increase in diameter of He-Ne laser may reduce lasing efficiency. Explain why?

11. (a) What is LASER? Explain how the basic lasing action is achieved?
(b) What are the important properties of laser beam? Mention a few of its applications based on these properties?

12. (a) What is a LASER? How it differs from an ordinary sources of light? Mention a few engineering applications of a laser?
(b) With the help of a neat energy level diagram explain how the population inversion is achieved in He-Ne laser.

13. Explain clearly spontaneous and stimulated emission of radiation with the help of neat diagrams. In laser how stimulated emission is made to dominate over spontaneous emission?

14. Describe how a hologram is produced and viewed.

15. What is a maser? Explain how is it different than the laser.

4

FIBRE OPTICS

4.1. Fibre Optics (General Introduction)

In case of electronic communication, the information is normally carried over in the form of **radio waves and microwaves** through copper wires and co-axial cables. However, the information carrying capacity of these wires is restricted due to their limited band width and is not sufficient as per the needs of the modern communication techniques.

If **light waves** are used instead of radio waves, or microwaves, the number of signals transmitted can be increased enormously. Technically, **Grahm Bell** in 1880 invented **Photophone,** which transmitted speech on a beam of light. By 1960, it was established that light could be guided by glass fibre. However, the glass available at that time *heavily attenuated light propagating through them.*

It was due to the invention of solid state lasers in 1970 and fabrication of **low-loss glass fibres,** that made the optical communication practical, with this the era of **photonics** started.

In addition to the use of optical fibre in communication, these have been extensively used as **sensors** to detect **electrical, mechanical and thermal energies.** Also they are being used in copying machines and in medical diagnostics (Endoscopy) **optical fibres.**

These are made up of **glass or plastic conducts** as thin as a **human hair** which is designed to guide light waves along the length of the fibre with the help of successive **total internal reflections** from side walls of the fibre. The movement of light is in a *zig-zag* path and some fraction of light may escape through side walls but major fraction comes out from the other end of the fibre. Fig. 4.1.

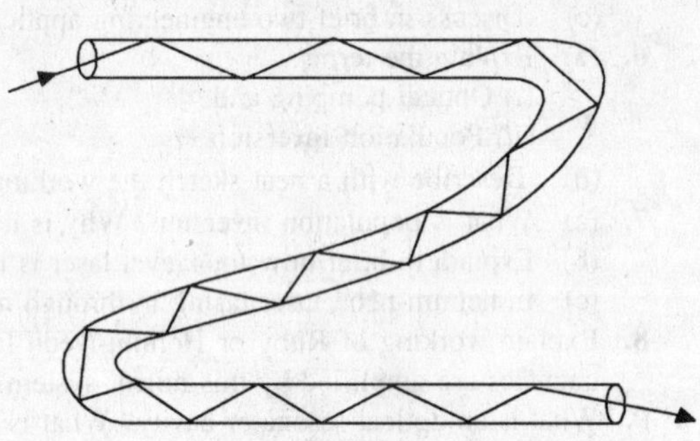

Fig. 4.1

Construction

An optical fibre consists in general the following three regions:

(i) Inner most region which is the light guiding region is called **core.**

(ii) Core region is surrounded by a middle region called **cladding**. The refractive index of cladding is always lower than that of the core.

(iii) The outermost region is called sheath. The sheath protects the core and cladding from moisture, abrasion contamination and to give mechanical strength to fibre.

Basic purpose of cladding is to confine the light to the core as the light falling on core and striking core-cladding interface at angle greater than the critical angle will be reflected back to the core.

Optical fibres are either made as a single fibre or a flexible bundle or cable. **A bundle fibre is a number of fibres in a single jacket.**

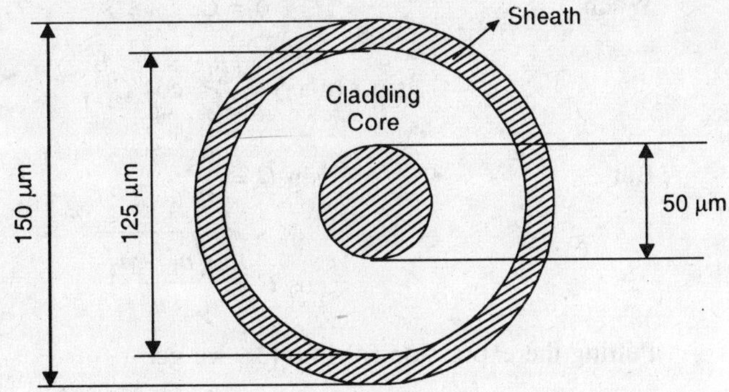

Fig. 4.2

4.2. Propagation of Light Through Fibre

Main function of the optical fibre is to accept maximum light and transmit the same with minimum **attenuation**.

Light gathering ability of a fibre depends on two factors:"

(i) Core size

(ii) Numerical aperture

Numerical aperture of fibre is determined by **acceptance angle** and **fractional refraction index change.**

4.3 Acceptance Angle and Acceptance Cone

Consider an optical fibre through which light is being sent. The end at which light enters is called **launching end**. Let the refractive indices of the core and cladding be n_1 and n_2 respectively; $n_2 < n_1$. Let the refractive index of the medium k through which light is launched be n_0.

Let the light beam enter at an angle i to the axis of the fibre. The ray gets refracted at an angle r and strikes the core cladding interface at an angle ϕ. For angle ϕ more than the critical angle C, light will undergo **total internal reflection.** The ray will also undergo total internal reflection at interface as $n_1 > n_2$. It means that so long as ϕ is greater than critical angle C, light will **stay within the fibre.**

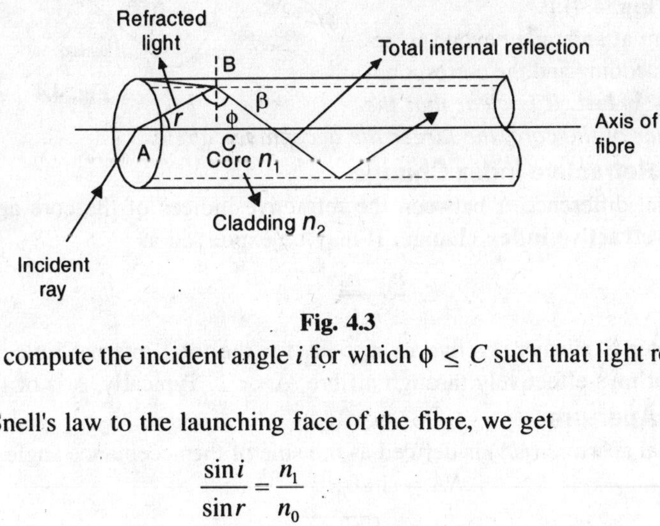

Fig. 4.3

Let us now compute the incident angle i for which $\phi \leq C$ such that light rebounds within the fibre.

Applying Snell's law to the launching face of the fibre, we get

$$\frac{\sin i}{\sin r} = \frac{n_1}{n_0} \quad ...(1)$$

If i is increased beyond a light, ϕ will drop below the critical value C and the ray escapes from the side walls of the fibre. The largest value of i occurs when $\phi = C$.

For the $\Delta\, ABC$, we have

$$\sin r = \sin (90° - \phi) = \cos \phi \quad ...(2)$$

From equation (1) into equation (2), we get

$$\sin i = \frac{n_1}{n_0} \cos \phi$$

When $\phi = C$

$$\sin i\,(\max) = \frac{n_1}{n_0}\cos C \qquad ...(3)$$

But

$$\sin C = \frac{n_2}{n_1}$$

$$\therefore \quad \cos C = \frac{\sqrt{n_1^2 - n_2^2}}{n_1} \qquad ...(4)$$

Putting the expression (4) into (3), we get

$$\sin i\,(\max) = \frac{\sqrt{n_1^2 - n_2^2}}{n_0} \qquad ...(5)$$

When incident ray is launched from air medium, we have $n_0 = 1$
Designing i (max) = θ_0, equation (5) may be simplified to

$$\sin\theta_0 = \sqrt{n_1^2 - n_2^2} \qquad ...(6)$$

$$\therefore \quad \theta_0 = \sin^{-1}[\sqrt{n_1^2 - n_2^2}\,] \qquad ...(7)$$

The angle θ_0 is called the *acceptance angle* of the fibre. Acceptance angle may be defined as the *maximum angle that a light ray can have relative to the axis of the fibre and propagate down the fibre.*

The light rays contained within the cone having a full angle $2\theta_0$ are accepted and transmitted along the fibre. The cone is called the acceptance cone. (Fig. 4.4).

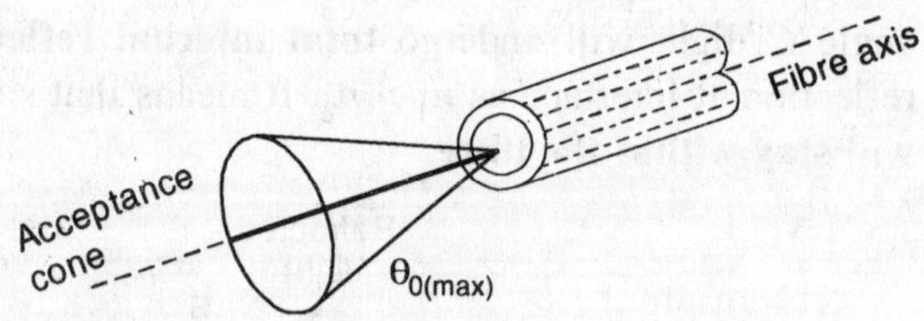

Fig. 4.4

Light incident at an angle beyond θ_0 refracts through the cladding and the corresponding **optical energy is lost**. *It is clear that the greater the diameter of the core, the larger the acceptance angle.*

4.4 Fractional Refractive Index Change

The fractional difference Δ between the refractive indices of the core and the cladding is called **fractional refractive index change.** It may be expressed as

$$\Delta = \frac{n_1 - n_2}{n_1} \qquad ...(8)$$

This parameter is always positive as $n_1 > n_2$ for the total internal reflection condition. In order to guide light rays effectively through a fibre, $\Delta \ll 1$. Typically, Δ is of the order of 0.01.

4.5. Numerical Aperture

The numerical aperture (*NA*) is defined as the sine of the acceptance angle. So we may write

$$NA = \sin\theta_0$$

$$\therefore \quad NA = \sqrt{n_1^2 - n_2^2} \qquad ...(9)$$

$$n_1^2 - n_2^2 = (n_1 + n_2)(n_1 - n_2) = \left(\frac{n_1 + n_2}{2}\right)\left(\frac{n_1 - n_2}{n_1}\right)2n_1$$

Approximating $\frac{n_1 + n_2}{2} \approx n_1$, we can express the above may be written as

$$(n_1^2 - n_2^2) = 2n_1^2\,\Delta$$

$$\therefore \quad NA = n_1\sqrt{2\Delta} \qquad ...(10)$$

Numerical aperture accounts for the light gathering ability of the fibre and it measures the amount of light accepted by the fibre. As is clear from eq. (9). Numerical aperture is depends only on the refractive indices of the core and cladding materials.

Numerical apertures ranges between 0.13 to 0.50 and a larger value of numerical aperture will mean that the *fibre can accepted more light from the source.*

Example 1. *Calculate the refractive indices of the core and cladding material of a fibre from following data : NA = 0.22 and Δ = 0.012.*

Solution. Given, $$\Delta = \frac{n_1 - n_2}{n} = 0.012 = 0.012 \quad ...(1)$$

and $$NA = n_1 \sqrt{2\Delta}$$

∴ $$n_1 = \frac{NA}{\sqrt{2\Delta}} = \frac{0.22}{2 \times .012} = \mathbf{1.42}$$

Put in eq. (1), $$\frac{1.42 - n_2}{1.42} = 0.012$$

∴ $$\mathbf{n_2 = 1.40} \quad \textbf{Ans.}$$

Example 2. *Calculate the numerical aperture, acceptance angle and the critical angle of a fibre having core refractive index = 1.50 and the cladding refractive index = 1.45.*

Solution. We have the relation, $$\Delta = \frac{n_1 - n_2}{n_1} = \frac{1.5 - 1.45}{1.5} = 0.033$$

Numerical aperture $$(NA) = n_1 \sqrt{2\Delta} = 1.5 \times \sqrt{2 \times 0.033} = 0.387$$

Acceptane angle $$\theta_0 = \sin^{-1} NA = 22.78°$$

and critical angle $$C = \sin^{-1} \frac{n_2}{n_1} = \sin^{-1} \frac{1.45}{1.50} = \mathbf{75.2°} \quad \textbf{Ans.}$$

Example 3. *An optical fibre has a NA of 0.20 and a cladding refractive index of 1.59. Determine the acceptance angle for the fibre in water which has a refractive index of 1.33.*

Solution. $$NA = \frac{\sqrt{n_1^2 - n_2^2}}{n_0}$$

When the fibre is in air $$n_0 = 1 \text{ and } NA = \sqrt{n_1^2 - n_2^2} = 0.20$$

∴ $$n_1 = \sqrt{(NA)^2 + n_2^2} = \sqrt{(0.20)^2 + (1.59)^2} = \mathbf{1.6025}$$

When the fibre is in water $$n_0 = 1.33$$

∴ $$NA = \frac{\sqrt{n_1^2 - n_2^2}}{n_0} = \frac{\sqrt{(1.6025)^2 - (1.59)^2}}{1.33} = \mathbf{0.15}$$

∴ $$\theta_0 (\max) = \sin^{-1} (NA) = \sin^{-1} (0.15) = \mathbf{8.6°} \quad \textbf{Ans.}$$

Example 4. *A glass clad fibre is made with core glass of refractive index 1.5 and the cladding is doped to give a fractional index difference of 0.0005. Determine (i) the cladding index, (ii) the critical internal reflection index, (iii) the external critical acceptance angle, (iv) the numerical aperture.*

Solution. Give, $$n_1 = 1.5, \Delta = 0.0005$$

(i) Let the refractive index of cladding be n_2. So we have

$$\Delta = \frac{n_1 - n_2}{n_1}$$

$\therefore \quad 0.0005 = \dfrac{1.5 - n_2}{1.5}$

or $\quad n_2 = 1.5 - 1.5 \times 0.005 = \mathbf{1.49925}$ **Ans.**

(ii) Let the critical internal reflection angle be C

$$\sin C = \frac{n_2}{n_1}$$

$$\therefore \quad C = \sin^{-1}\left[\frac{n_2}{n_1}\right] = \sin^{-1}\left[\frac{1.4925}{1.5}\right] = \sin^{-1}(0.9995) = \mathbf{88.2°} \text{ **Ans.**}$$

(iii) Let the external critical acceptance angle be θ_0, so we have,

$$\sin C = \frac{\sqrt{n_1^2 - n_2^2}}{n_0} \text{; where } n_0 = 1$$

$$\therefore \quad \theta_0 = \sin^{-1}\left[\sqrt{n_1^2 - n_2^2}\right]$$
$$= \sin^{-1}(1.5^2 - 1.49925^2)^{1/2}$$
$$= \sin^{-1}(2.25 - 2.24775)^{1/2}$$
$$= \sin^{-1}(0.0474) = \mathbf{2.72°}$$

(iv) Numerical aperture $NA = n_1\sqrt{2\Delta}$

$\therefore \quad NA = 1.5\sqrt{2 \times 0.005} = 1.5\,(0.03162) = \mathbf{0.0474}$ **Ans.**

4.6 Modes of Propagation

In an optical fibre light travels as an e.m. wave and all the waves moving in directions **above the critical angle** will be trapped in the fibre due to total internal reflections. However, all such waves do not propagate through the fibre, and only certain ray directions are allowed for propagations. These allowed directions corresponds to **modes of the fibre.**

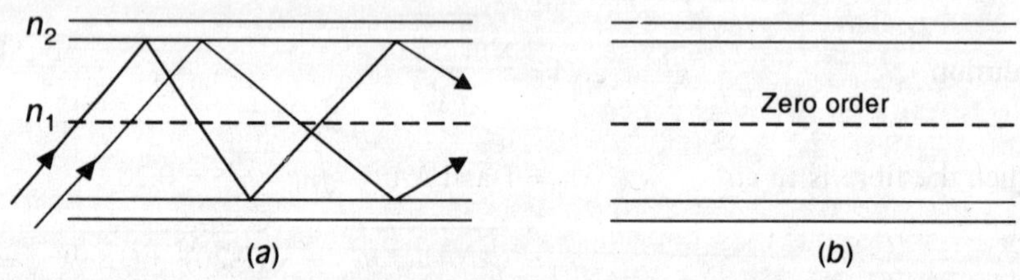

Fig. 4.5

In simple form, we can understand the modes as the **number of paths of light** in an optical fibre. The paths are all **zig-zag** with the exception of light moving along axial directions. So the light rays moving through a fibre may be classified as (*i*) axial rays (*ii*) zig-zag rays.

Since the rays get repeatedly reflected at the walls of the fibre, there occurs a phase shift. So waves along certain zig-zag path be in phase and get intensified while some other waves moving in other paths will be out of phase and may fade out due to **destructive interference.**

The light ray paths along which the waves are in phase inside the fibre are called modes. Number of modes, a fibre can support depends on the ratio $\frac{d}{\lambda}$ where ***d* is the diameter of the core and λ is the wavelength of the wave transmitted.**

In general, modes are represented by an order number *m*. In a fibre of fixed thickness, the higher order propagates at smaller angles than the lower order modes.

Axial ray that travels along the axis of the fibre is called **zero order ray**.

4.7. Types of Optical Fibre

Optical fibres are in general of two types:

(i) Single Mode Fibre (SMF)

(ii) Multimode Fibre (MMF)

A single fibre has a smaller core diameter and can support only one mode of propagation while a multimode fibre has a large core diameter and can support a large number of modes.

Multimode fibres are further classified on the basis of index profile. An index profile is a graph of refractive index (along X-axis) and distance from the core (along Y-axis). Index profile of a MMF can be either step index (SI) type or graded index (GRIN) type. Index profile of SMF is usually a step index (SI) type.

4.8 Single Mode Step Index Fibre

A single mode step index fibre consists of a very fine thin core of uniform refractive index surrounded by a cladding of refractive index lower than that of the core. Since the refractive index

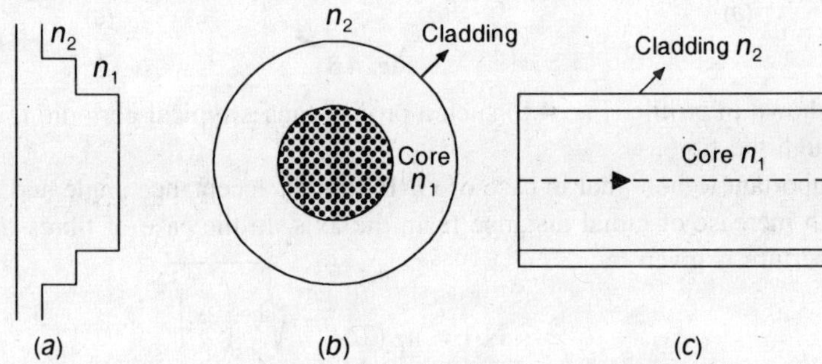

Fig. 4.6

abruptly changes at the core cladding boundary, it is known as step index fibre. A typical SMF has a core diameters of 4μm which corresponds to some of the wavelengths of light waves. Light travels along a single path, *i.e.*, along the axis only and so zero order mode is supported by SMF. Usually SMF is characterised by a very small value of Δ (~ 0.002).

The fibre is surrounded by some **opaque protective sheath.**

4.9. Multimode Step Index Fibre

It is similar to the single mode step index fibre with the exception that it has a large core diameter (~ 100 μm). Core diameter is very large as compared to the wavelength of transmitted light. Light moves along zig-zig paths along MMF. A typical structure along with profile of step index MMF are shown in Fig. 4.7.

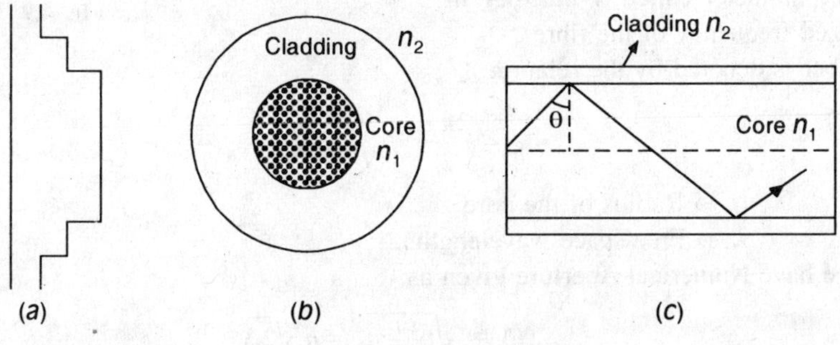

Fig. 4.7

NA in case of a MMF is quite large as core diameter is larger and is ~ 0.3.

4.10. Grain Index Fibre (GRIN)

A GRIN is a multimode fibre which has **concentric layers of refractive** indices. Which means that the refractive index of the core varies with distance from the fibres axis, i.e., it has high value at the centre and falls of rapidly as the radial distance increases from the axis.

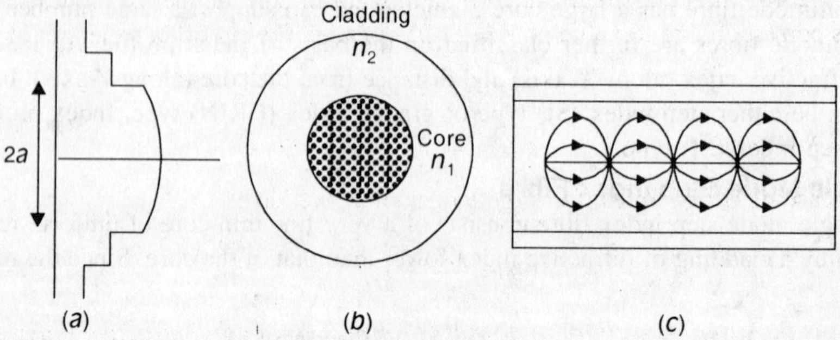

Fig. 4.8

As is shown in profile (Fig. 4.8) such a profile causes typical periodic focussing of the ligth moving through the fibre.

It is important to note that in case of GRIN fibres, acceptance angle and numerical aperture diminish with increase of radial distance from the axis. In the case of fibres of parabolic profile, numerical aperture is given by:

$$\text{NA} = n_1 \, (2\Delta)^{1/2} \sqrt{1-\left(\frac{r}{a}\right)^2}$$

The convention of writing the size of optical fibre is to given the core/cladding diameter.

Sizes of SI fibre are $\frac{50}{125}, \frac{100}{140}, \frac{200}{230}$ etc.

Size of GRIN fibres are $\frac{50}{125}, \frac{62.5}{125}, \frac{85}{125}$

etc. For example, a $\frac{50}{125}$ fibre has 50 μm core and 125 μ m cladding.

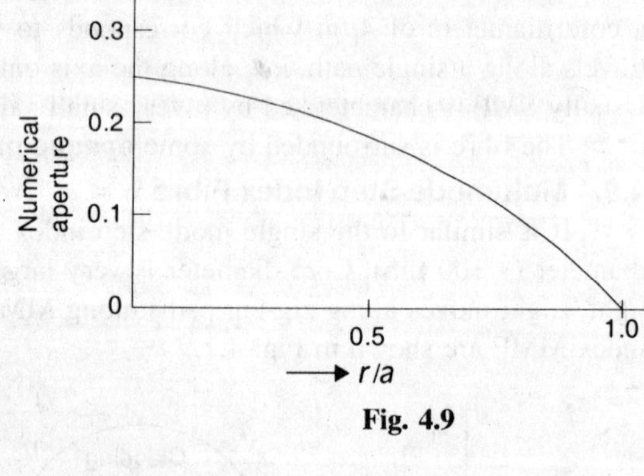

Fig. 4.9

4.11. Normalized Frequency (V-number) of a Fibre

An optical fibre may be **characterised** by one more parameter called **V-number or the normalized** frequency of the fibre.

V-number is denoted by the relation

$$\text{V} = \frac{2\pi \, a}{\lambda} \cdot \sqrt{n_1^2 - n_2^2} \qquad ...(1)$$

where $\alpha \rightarrow$ Radius of the core

and $\lambda \rightarrow$ Free space wavelength.

Also we have Numerical Aperture given as:

$$NA = \sqrt{n_1^2 - n_2^2} = n_1\sqrt{2\Delta} \qquad ...(2)$$

$$\text{V} = \frac{2\pi \, a}{\lambda} \, (NA) = \frac{2\pi \, a}{\lambda} n_1 \sqrt{2\Delta} \qquad ...(3)$$

Miximum number of modes N_m supported by an SI fibre may be got using the relation

$$N_m = \frac{1}{2} V^2$$

e.g., for $V = 10$, $Nm = 50$

In case V < 2.405, fibre can support only one mode and is called SMF.

For V > 2.405, it is classified as MMF and it can support number of modes simultaneously. Wavelengths corresponding to value of $V = 2.405$ is called the cut off wavelength of the fibre and is given as

Cut off wavelength $$\lambda_C = \frac{\lambda V}{2.405} \qquad ...(4)$$

In the case of GRIN fibres, for large values of V, we have

$$Nm = \frac{V^2}{4} \qquad ...(5)$$

Example 5. *A step index fibre in air has a NA of 0.16, a core refractive index of 1.45 and a core diameter of 60 cm. Determine the normalized frequency for the fibre when light at a wavelenth of 0.9 mm is transmitted.*

Solution. Given,

$$NA = 0.16$$
$$n_1 = 1.45$$
$$d = 60 \text{ cm} = 0.6 \text{ m}$$
$$\lambda_0 = 0.9 \text{ µm} = 9 \times 10^{-7} \text{ m}$$

The normalized frequency (V-number) is given by

$$V = \frac{\pi d}{\lambda_0}\sqrt{n_1^2 - n_2^2} = \frac{3.143 \times 0.60 \text{ m}}{9 \times 10^{-7} \text{ m}} \times 0.16$$

$$\therefore \qquad V = 335103.22 = \mathbf{3.35 \times 10^5} \textbf{ Ans.}$$

Example 6. *Find the core diameter necessary for single mode operation at 850 µm in S.I. fibre with n_1 = 1.480 and n2 = 1.47. What is the numerical aperture and maximum acceptance angle of this fibre?*

Solution. (*i*) For single mode propagation, *V*-number is 2.405, so we have,

$$V = \frac{\pi d}{\lambda_0}\sqrt{n_1^2 - n_2^2}$$

$$\therefore \qquad 2.405 = \frac{\pi d}{450 \times 10^{-9} \text{ m}} \times 0.1717$$

$$d = \frac{2.405 \times 850 \times 10^{-9}}{3.143 \times 0.1717} \text{ m} = \mathbf{3.79 \text{ µm}}$$

(ii) $$NA = \sqrt{n_1^2 - n_2^2} = \sqrt{1.48^2 - 1.47^2} = 0.1717$$

(iii $$\sin \theta_0 = NA$$

$$\theta_0 = \sin^{-1}(NA) = \sin^{-1}\left[n_1^2 - n_2^2\right]^{1/2}$$

$$= \sin^{-1}(0.1717) = \mathbf{9°53'12''} \textbf{ Ans.}$$

Example 7. *A step index fibre is made with a core of index 1.52, a diameter of 29 µm and a fractional difference index of 0.007. It is operated at a wavelength of 1.3 µm : (i) the fibre V-number, and (ii) the number of modes the fibre will support.*

Solution. Care refractive index $n = 1.52$

$$d = 29\ \mu m = 29 \times 10^{-6}\ m$$
$$\Delta = 0.007$$
$$\lambda_0 = 1.3\ \mu m = 1.3 \times 10^{-6}\ m$$

Using the relation, $$\Delta = \frac{n_1 - n_2}{n_1}$$

We get, $$0.007 = \frac{1.52 - n_2}{n_1}$$

$\therefore$ $$n_2 = 1.52 - 1.52 - 0.007 = 1.5189$$

(*i*) $$V = \frac{\pi d}{\lambda_0}\sqrt{n_1^2 - n_2^2}$$

$\therefore$ $$V = \frac{3.143 \times 29 \times 10^{-6}}{1.3 \times 10^{-6}}[1.52^2 - 1.5189^2]^{1/2} = \mathbf{4.049}$$

(*ii*) Number of nodes $N = \frac{1}{2}V^2 = \frac{1}{2}(4.049)^2 = 8.19 =$ **8 modes Ans.**

Example 8. *Calculate the maximum radius allowed for a fibre having core refractive index = 1.47 and cladding refractive index = 1.46. Fibre has to support only one mode at a wavelength of 1300 nm.*

Solution. Condition for single mode is $V < 2.405$

or $$\frac{\pi d}{\lambda}\sqrt{n_1^2 - n_2^2} < 2.405$$

or $$d < \frac{2.405.\lambda}{\pi\sqrt{n_1^2 - n_2^2}} < \frac{2.405 \times 1300 \times 10^{-9}}{3.142 \times \sqrt{(1.47)^2 - (1.46)^2}} < 5.82\ \mu m$$

$\therefore$ Maximum diameter = **5.82 μm Ans.**

Example 9. *Calculate the maximum value of Δ in the case of a single mode fibre core of diameter 10 μm and core refractive index 1.5. Fibre is coupled to a light source of wavelength 1.3 μm. Also calculate the cladding refractive index and the acceptance angle. Given V cut off for a single mode propagation = 2.405.*

Solution. We have the relations,

$$V_{\text{cut-off}} = \frac{\pi d}{\lambda} NA\ (\text{cut off})$$

$\therefore$ $$NA\ (\text{cut off}) = \frac{V \times \lambda}{\pi d} = \frac{2.405 \times 1.3 \times 10^{-6}}{3.142 \times 10} = 0.09957$$

$\therefore$ Acceptance angle $= \sin^{-1}(0.09957) = 5.71°$

Also numerical apreture, $NA = n_1 = \sqrt{2\Delta}$

$\therefore$ $$\Delta = \frac{(NA)^2}{n_1^2} \times \frac{1}{2} = \frac{(0.09957)^2}{(1.5)^2} \times \frac{1}{2} = 2.2 \times 10^{-3}$$

Also $$\Delta = \frac{n_1 - n_2}{n_1}$$

$\therefore$ $$n_2 = n_1 - \Delta n_1 = n_1(1 - \Delta) = 1.5(1 - 2.2 \times 10^{-3})$$
$$= \mathbf{1.496\ Ans.}$$

4.12. Fabrication of Optical Fibres

Materials which are used for optical fibres are transparent to optical frequencies. Normally, step index fibres are produced in the following three forms.

(*i*) Glass core claded with glass which has lower refractive index than the glass-core.
(*ii*) Silica glass core cladded with plastic.
(*iii*) Plastic core cladded with another plastic.

In case of all glass fibres, **refractive step index** is smallest, slightly more for plastic cladded silica (PCS) and largest for plastic core and plastic cladded fibre.

(*i*) **Glass Fibre**

Most abundant and inexpensive optical fibres are made from silica (SiO2) which has refractive index = 1.458 for λ = 8500 Å.

We get material for different refractive index by dopping the basic material. If basic material is doped with Germenia (GeO_2) or Phosphorus pentoxide (P_2O_5), value of refractive index increases **and such materials can be used as core while pure silica is used as cladding**.

In pure silica is doped with B_2O_3 or flourine, its refractive index reduces and **so it be used as cladding while pure silica is used as core material.**

Glass optical fibres exhibit very low losses and are used for long distance communication.

(*ii*) **PCS Fibres**

Core is normally of pure quartz while cladding and sheath are of some transparent polymer such as silicon resin of low refractive index. These fibres are less expensive but transmission losses are very high. So these are used for short distances.

(*iii*) **Plastic Fibres**

In this case both core and cladding are of plastics. Using appropriate polymers, a high refractive index difference can be achieved between core and cladding. So this type of fibre will have large **Numerical Aperture**. The main advantage of all plastic fibres is low cost and high mechanical flexibility. However, since those fibres are temperature sensitive. So they are limited in application between **80°C to 100°C temperatures.** Also they have high transmission losses and so have limited use because of this limitation.

4.13. Attenuation

An optical signal passing through a fibre will get progressively reduced. This reduction or attenuation of signal may be defined as the **ratio of the optical output power from a fibre of length *L* to the input optical power.** It is expressed in terms of decibel/km (dB/km)

$$\alpha = \frac{10 \log P_1 / P_2}{L}$$

where $P_1 \longrightarrow$ Power of optical signal at launching end

$P_2 \longrightarrow$ Power of optical signal at the receiving end.

In an ideal fibre $P_1 = P_2$ and attenuation will be 0 dB/km. However an attenuation upto 3 dB is within permissible limit. It is important to note that attenuation is wavelength dependent and so the wavelength must also be specified.

4.14. Fibre Losses

The following three losses may occur in optical fibres

(*i*) Absorption
(*ii*) Geometric effects
(*iii*) Rayleigh scattering

(*i*) **Absorption**

Even very pure glass absorbs **light of** a specific wavelength. Strong **electronic absorption** occurs in UV region and **vibrational absorption** occurs in IR region of wavelength 7 μm to 12 μm. These losses are attributed due to inherent property of the glass and is called intrinsic absorption.

However, this loss is insignificant.

Impurities are major extrinsic source of losses in fibre. Hydroxyl radical ions (OH) and transition metals like Nickel, **Chromium**, **Copper**, Manganese etc. have **electronic** losses in **near visible range of spectrum.** These **impurities** should be kept away as far as **possible** from the fibre. Intrinsic as well as extrinsic losses are found to be minimum at about 1.3 μm.

(*ii*) **Geometric Effects**

These may occur due to manufacturing defects like irregularities in fibre dimensions during drawing process or during coating, cabling or insulation processes.

(*iii*) **Rayleigh Scattering**

As glass has disordered structure having local microscopic variation in density which may also cause variation in refractive indices. So light travelling through these structures may suffer scattering losses due to Rayleigh, *i.e.*, scattering $\propto \frac{1}{\lambda^4}$. It means Rayleigh scattering sets a lower limit on wavelength that can be transmitted by a glass fibre at 0.8 μm below which scattering loss is appreciably high.

Example 10. *A certain optical fibre has an attenuation of 3.5 dB/km at 850 nm. If 0.5 mw of optical power is initially launched with fibre, calculate the power level after 4 km.*

$$\text{Attenuation } \alpha = 3.5 \text{ dB/km}$$
$$\text{Initial power level } P_1 = 0.5 \text{ mW}$$
$$\text{Final power level } P_2 = ?$$
$$\text{length of cable} = 4 \text{ km}$$

We have the relation

$$a = \frac{10}{L} \log \frac{P_1}{P_2}$$

$$\therefore \qquad 10 \log_{10} \frac{P_1}{P_2} = a.L. = 35 \times 4 = {}^{14}\text{dB}.$$

$$\log_{10} \frac{P_1}{P_2} = \frac{14}{10} = 1.4 \text{ or } \log_{10} \frac{0.5}{P_2} = 1.4$$

Taking antilog on both sides

$$\frac{0.5}{P_2} = 25.11 \text{ or } P_2 = \frac{0.5}{25.11} \, mw = \mathbf{19.9 \ \mu w} \quad \textbf{Ans.}$$

Example 11. *A 15 km optical fibre link uses fibre with a loss of 1.5 dB/km. The fibre is joined every km with connectors, which give attenuation of 0.80 dB each. Find the minimum mean optical power which must be launched with the fibre to maintain a mean optical power level of 0.3 μω at the detector.*

Solution. (*i*) Connector loss = 0.8 dB/km

∴Connector loss for 15 km length = 0.8 × 15 = 12 dB.

(*ii*) Fibre loss = 1.5 dB/km

∴Fibre loss of 1.5 km = 1.5 × 15 = 22.5 dB

Total loss = 12 dB + 22.5 dB = 34.5 dB

Optical power level at detector P_2 = 0.3 μW

Optical power at launch position = P_1

Now we have the relation

$$a = 10 \log \frac{P_1}{P_2}$$

$$34.5 = 10 \log \frac{P_1}{0.3\mu W} \text{ or } \log \frac{P_1}{0.3\mu W}$$

Taking antilogs on both sides

$$\frac{P_1}{0.3\mu m} = 2818.3$$

$$P_1 = 8.455 \times 10^{-4} \text{ W} = \mathbf{0.86 \text{ mW Ans.}}$$

4. 15 Optical Windows

A typical curve of fibre attenuation v/s wave length for a silica based optical fibre is given 4.10

As may be observed from the above curve, there are minima at a particular optical wave lengths. The band of wavelengths at which attenuation is minimum is called an optical window or transmission window. Optical signals having wavelengths in this region travel through the fibre with least attenuation. Because of the said property such windows are quite suitable for transmission of information.

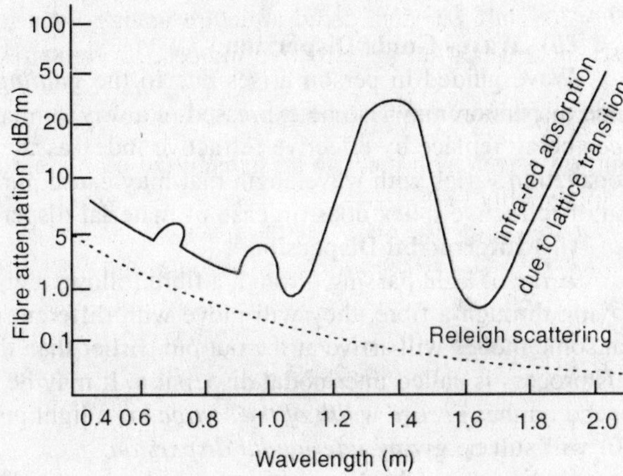

Fig. 4.10

In fibre optical communication, two such windows are chosen:

(*i*) First window lies between 0.8μm to 0.9μm

(ii) Second window lies between 1.3 μm to 1.6 μm

4. 16 Dispersion

As stated earlier a pulse launched with a fibre gets attenuated doe to losses in fibre. More over the incoming pulse also spreads during the transit through the fibre. So a pulse at the output is wider than the pulse at the in put *i.e.,* the pulse gets distorted as it moves through the fibre (Fig. 4.11). This distortions of pulse is due to dispersion effects which is measured in terms of nanoseconds per km).

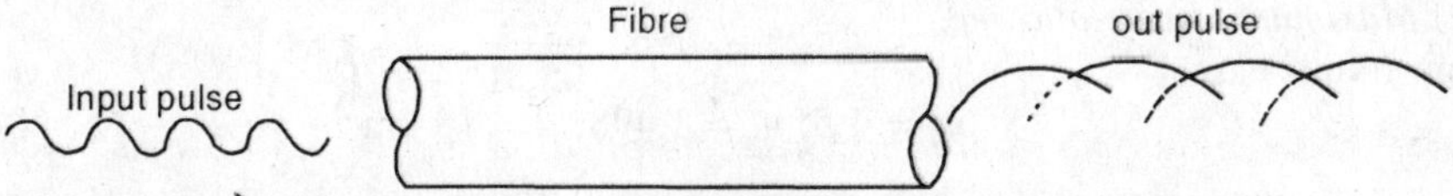

Fig. 4.11

There are three phenomenon that may contribute towards the distorting effects.

(*i*) Material dispersion

(*ii*) Wave guide dispersion.

(*iii*) Intermodal dispersion.

(*i*) Material dispersion

Light waves having different wavelength will travel with different speeds through a medium. Light waves having *shorter wavelength* will move slower than those of longer wavelengths. So narrow pulses of light tend to broaden as they move through an optical fibre. This process is called material distortion. As may be evident, the spectral width of the source will determine the extent of dispersion and the material dispersion is given by the equation.

$$D_m = \frac{\lambda(\Delta\lambda)}{C} . L \frac{d^2 n}{d\lambda^2} \quad \text{... (1)}$$

Where λ = corresponds to peak wavelength

$\lambda\,\Delta$ = is the spectral width

L = is the angle; of the fibre

and n = corresponds to core refractive index

It may be mentioned that the material dispersion can be very much reduced by using a pure monochromatic source.

To give an example of material dispersion we observe that a LED operating at 820 nm and having a spectral spread of 38 nm gives dispersion of 3 ns/km. This dispersion can be reduced of to about 0.3 ns/km with the help of a laser diode that operates at 1140 nm with a spectral width 3 nm.

(*ii*) **Wave- Guide Dispersion**

Wave guided in person arises due to the *guiding properties of the fiber*. Amount of wave guide dispersion may also be expressed in a way similar to the eq. (1) with the *material refractive index* being replace by effective refractive index as the *effective refractive index* for any mode of propagation varies with wavelength that may cause purse spreading in the same way as the variation of refractive index does, in case of material dispersion.

(*iii*) **Intermodal Dispersion**

A ray of light passing through a fibre follows a zig-zag pth and when a number of modes are moving through a fibre, they will move with different net velocities w.r.t the fibre axis. It means that some modes will arrive at the out put earlier than the others *i.e.* there is spread of input pulse. This process is called intermodal dispersion. It may be noted that this type of dispersion does not depend on the *spectral width of the source i.e.* a light pulse from a pure monochromatic source ($\Delta\gamma = 0$) will still be giving *intermodal dispersion*.

In the case of MMF, all there spreading mechanisms are observed simultaneously while in case of SMF, *only material and wave guide* dispersion.

In the case of fibres with low numerical aperture (NA), dispersion in observed while for fibre with high NA, large dispersion in observed. Dispersion may be restricted by the use of a *low NA fibre and a narrow spectral width source.*

A solution to the above problem may be got by the use of GRIN fibres which produce less distortion than the SI fibres and they have pulse spread of only few ns/km, which is guide small as compared to the spread in case of SI fibres. However, GRIN fibres are more expensive. Also in the case of a short length of fibre, dispersion in proportional to the length of the fibre and so may ...

Example 12. *A step index multimode fibre has core index of 1.5 and cladding index of 1.498. Calculate (i) Intermodal factor for the cable; (ii) Total dispersion in a 18 km length and (iii) Maximum bitrate allowed.*

Solution. In this case:

$$n_1 = 1.5;\ n_2 = 1.498;\ l = 18 \text{ km}.$$

$$\therefore \qquad \Delta = \frac{n_1 - n_2}{n_1} = \frac{1.5 - 1.498}{1.5} = 0.00133$$

(*i*) For 1 km length fibre

$$\Delta t = \frac{n_1 . L}{c}\left(\frac{\Delta}{1-\Delta}\right) = \frac{1.5 \times 1000 m}{3 \times 10^8 ms^{-1}} \times \frac{.00133}{1 - .00133}$$

$$= 0.6659 \times 10^{-8} \text{ sec} = \mathbf{6.66\ ns/km.\ \ Ans.}$$

(*ii*) Total dispersion for 18 km length

$$\Delta t \text{ (total)} = (\Delta t)\ 1 \text{ km} \times 10 \text{ km}$$

$$= 6.66 \text{ ns/km} \times 18 \text{ km} = 119.9 \text{ ns}$$

(*iii*)

$$B_{max} = \frac{1}{5\Delta,\text{(total)}} = \frac{1}{5 \times 119.9 ns} = 0.00166 \times 10^9 \text{ bits/sec.}$$

$$= 1.66 \times 10^6 \text{ bit/sec} = \mathbf{1.66\ m\ bits/\ sec.\ \ Ans.}$$

Example 13. *compute the dispersion/km of length and total dispersion in a 10 km length of SI fibre from the following data:*

refractive index of core = 1.558 and Δ = 0.026.

Solution Maximum dispersion $= \frac{n_1}{c} \cdot \frac{\Delta}{1-\Delta}$

$\therefore$ max. dispersion/km $= \frac{1.558}{3\times10^8} \times \frac{0.26}{1-0.026} \times 1000\text{m}$ [∴ L = 1 km = 10^3m]

13.8×10^{-9} sec = 13.8 *n* sec.

∴ dispersion in 10 km = 13.8 × 10 = **138 *n* sec Ans.**

Example 14. *Compute the maximum dispersion for an optically graded fibre from the following data refractive index at the centre of core = 1.5, Δ = 0.0260 and length of fibre = 1 km.*

Solution. In case of GRIN fibre,

We have the relation,

$$\Delta t = \frac{n_1 . L . \Delta^2}{8c} = \frac{1.5\times1000m\times(0.026)^2}{8\times3\times10^8 ms^{-1}} = \mathbf{0.42\ ns/km\ Ans.}$$

4.17 Coupling Components of an Optical Fibre

(Splices, Connectors and Couplers

1. **Fibre Splices:** *Fibre splices are permanent joints in fibres and are analogous to soldered joints in electrical system.* A splice in a fibre which produces a localised optical distortion that may increase the signal attenuation and some of the signal may get scattered at splices. *Mode conversion generally* from low of high order modes may occur thereby increasing the attenuation beyond splices. Combination of all these effects may *attenuate* a signal significantly.

So, for less distortion and attenuation of the signal, proper matching of the front ends of the fibres to be jointed is required. These losses may be due to the *misalignment* of core and cladding of the fibres to be joined. The joining process generally introduces a loss of 0.2 to 0,3 db for a 50 micron *core index fibre.* Fibre mismatches of core diameter, cladding diameter, index profile and numerical aperture between the adjoining fibres increase this loss to about 05 db.

In view of the above concession, it is important that the fibre splices must have low *insertion loss* and sufficient mechanical strength. A number of techniques have been developed, *e.g.*, V-groove splices, fusion splices and sleeve splices etc. It should be noted that in designing a splice system, simplicity of the operation must have high priority because of the following reasons.

1. Working conditions in the fields to not allow because of the following reasons.
2. Operating companies do not want to increase cost factor by employing more persons to maintain the fibres.

4.18 V-Groove Technique

The technique is usually employed during testing of optical fibres in the laboratory or in field or for quick restoration of fibre fault in the communication link. The V-grooved system or *jig* consists of a central pillar and two sides alignment blocks called magnetic clamps.

On the central pillar V-grooved impressed by mandrel into copper substrate measuring about 6 × 2 mm are mounted. Fig. 4.12 shows the V-groove arrangement in which the fibres to be joined are held in alignment blocks by magnetic clamps. Central pillar is raised so that the fibres are forced down into V-grooves under there own spring force. Some of epoxy resin or glue in applied over the fibre ends which are brought near by moving the right hand fibre clamp towards *V-groove jig*. As soon as the fibre ends touch, a small heater is switched on in the central pillar so that the *epoxy resin* fixes the fibre in place. The fibres with V-groove substrates are then taken out and mounted in a joint box. Joints made with this technique for identical fibres show losses of the order of 0.1 db or even lesser than this value.

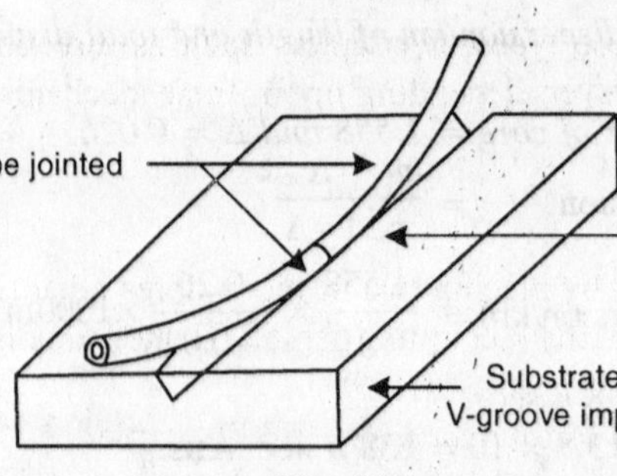

Fig. 4.12

In *sleeve splice* technique, a glass sleeve is used which has lower softening temperatures than the fibres to be joined. The basic principle in this method is the when the glass tube in heated to its softening point, surface tension forces tend to collapse the tube into a solid rod.

Fusion splices give lowest losses, however, they involve difficult techniques and are not suitable to be used in the fields. Fibre ends and are first cleansed and then butted and aligned using a *binocular microscope* which requires a considerable skill. Once aligned the ends are heated either by a microtorch or using *electric arc* till the ends are fused. This is a very sophisticated step as more heat melts the glass and so the fibre may be deformed. It properly used, this technique reduced the fibre losses to less 0.1 db.

4.18 Fibre Connectors

These are essential components of an optical fibre which allow the direct transition of optical power from one core to another. Fibre connectors are used at points where fibres terminate in transmitters and receivers either at the end of line or at the points where repeaters are put connectors must be dismountable so that the equipment may be removed from servicing connectors must not have any angular or misalignment effects as these may increase the fibre losses.

A variety of fibre connectors have been developed but all these comprise three main parts: *central alignment section and two normally identical insertion members* attached to the two fibres to be connected.

Ferrule type connectors are quite in use which are made by precisely centering the uncoated fibres into ferrules and fixing them with some *epoxy resin.* The ferrules and fibre surfaces and polished. Then the two ferrules are brought to touch each other inside a precisely guiding cylinder sleeve. Ferrule type connectors are made in different dimensions and have fibre losses in the range of 0.5 to 1 db.

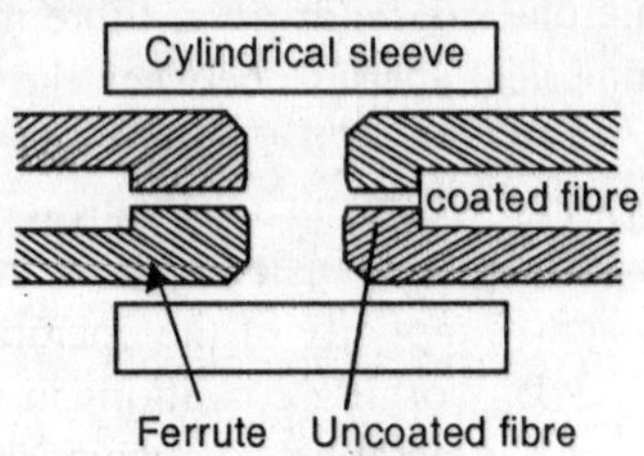

Fig. 4.13

4.19 Couplers

Couplers are needed for signal distributions while permanent splices and demountable connectors are used for connecting the fibres.

Main significance of these coupling components is that they do not require optical equipment like *lenses, mirrors, reflectors, etc.*

1. **Coupling ratio is Defined** as the optical power of a simple output divided by the total output power.
2. **Excess loss:** Defined as the told output power divided by total input power. Excess loss must be quite small and in modern couplers; losses ~ 2 db have been realized for couplers with more than 20 fibres.

For coupling ratio, it is essential that it should be very less and also independent of the fibre materials.

Functionally, there are three types of couplers such as directional, distributive and wavelength dependent couplers. However, depending upon their mechanism they are put as

(1) Diffusion type

(2) Arca-splitting type

(3) Beam-spliting type

A coupler is manufactured by placing two or more fibres adjacent to each other, then fusing and stretching them, thereby creating a coupling region. Heated area is then stretched till required coupling properties are attained.

This device is called *fused biconical taper* (FBT) coupler.

When heated fibre is elongated, the diameter of the core is decreased which means V number is decreased a V- number is given by.

$$V = \left[\frac{2\pi an}{\lambda}\right]\sqrt{2\Delta}$$

Since $2a = d$ corresponds to diameter of core and when it decreases, V decreases. 1 is the operating wavelength, n and D are average and relative refractive indices respectively.

Now, V number determines how tightly light power is confined within a core. It means that smaller the V-number, more the mode field diameter (MFD) exceeds the fibre core diameter. So by decreasing the core diameter, we allow a greater portion of an optical mode to propagate in the cladding where this mode will couple to the core of another fibre. This is illustrated in the Fig. 4.14.

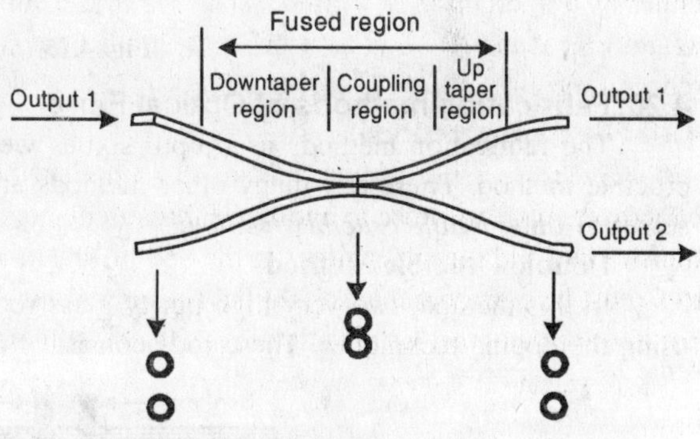

Fig. 4.14

What portion of input light will be coupled to be second fibre depends on the distance between two cores, core diameters within the coupling region and the operating wavelengths. So, by carefully choosing the size of the cores, output power ratios called *coupling ratio* can be controlled. A coupling ratio of 50 : 50 is quite popular, however, any coupling ratio can be attained by controlling the size of the cores and there distances.

Various type of couplers are show in the Fig. 4.15

A 2 ×1 coupler is used to combine two lights into single fibre Fig. 4.15(a). When the direction of propagation of lights is changed, tis will split the light signal into two as shown in Fig. 4.15(b). In this case, a coupler is called a splitter due to the function if performs.

There are couplers that couple or split $1 \times N$ or $N \times 1$ ports as shown in Fig. (4.15(c) and they are called *tree couplers* and may have $N \times m$ configuration.

An important coupler type for wavelength-division multiplexing (WDM) network is a star coupler shown in Fig. 4.15(d). In this case, the same number of ports serve as input and output.

In a sense, a star coupler is an $N \times m$ bidirectional coupler. However, a star coupler can also be built as an $N \times m$ *unidirectional coupler.*

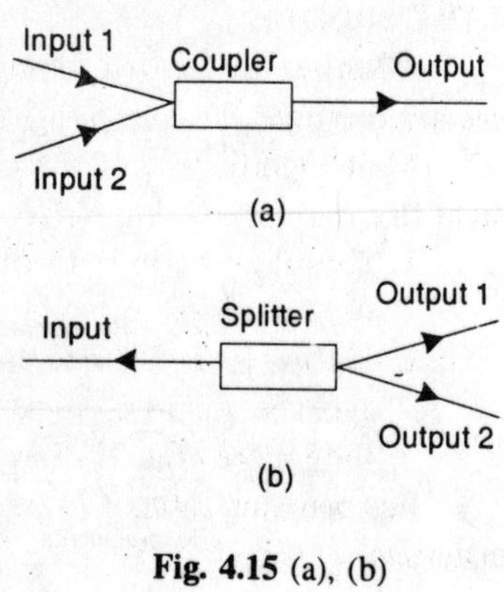

Fig. 4.15 (a), (b)

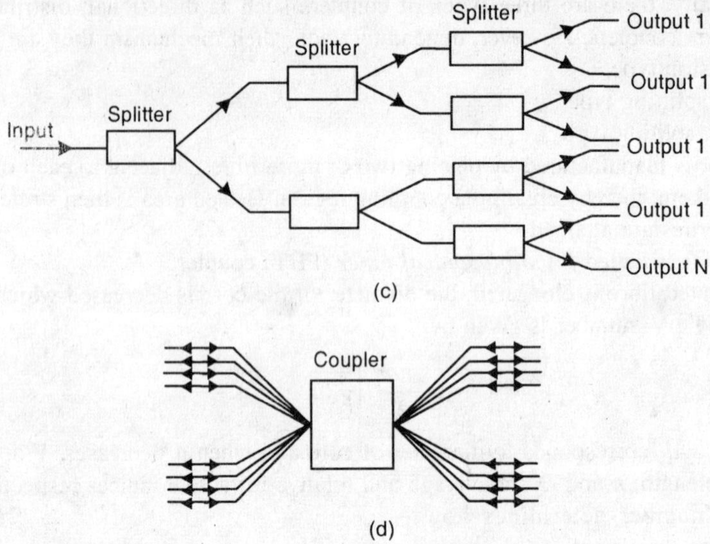

Fig. 4.15 (c), (d).

4.20. Fabrication methods of Optical Fibre

The fabrication methods used upto sixties were only two; *rod in tube method and double* crucible method. These day many other methods are being used, however, we will restrict ourselves to only *double crucible method.*

Double Crucible Method

In this method, two very high purity glass rods with different refractive indices *r* prepared using the doping techniques. These rods consist of different compositions for *cor and cladding.*

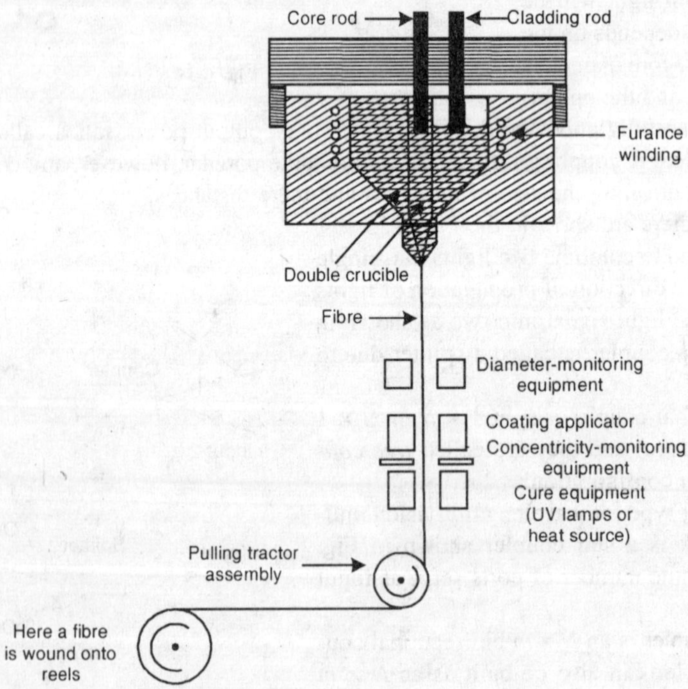

Fig. 4.16

In order to draw cladded core fibres from these glass rods, two cylindrical crucibles at arranged concentrally as shown in Fig. 4.16. The rods are allowed to move slowly down where they melt due to the heated Crucibles, thus filling up the crucibles with glass n.

The height of the melt in kept constant and the crucible temperature is maintained between 850 °C to 1100° C depending upon the composition of the glass.

At the base, the crucibles are tapered to concentric nozzle which mold the glass in fibre. Pulling speed of the fibre is controlled by the *drum* on which the fibre is wound. So in method fabrication of the fibre is achieved only in one step and fibres of unlimited lengths can manufactured.

This method has been improved by many British and Japanese companies producing fibre having attenuation vales below 5 db/km.

The method can also be used to fabricate graded index fibre (GRIN fibre). The achieved km can be attained in this case.

Coatings

As we know that an optical fibres has to be coated, the Fig. 4.35 shows how the coating line can be achieved. A liquid polymer is applied by the coating applicator as the fibre through the coating line. The liquid gets solidified by heat or using ultraviolet light.

Main function of the coating is to protect the fibre from external damage. However, look reveals the following important functions of the coating.

1. **Adhesion:** It firmly sticks to the glass due to adhesion property.

2. **Toughness:** This property is important to provide enhanced abrasion protection an handling and cabling without the loss of strength. Toughness is judged by the elastic (Modulus) testing which shows whether the coating is hard or soft.

3. **Moisture resistance:** This coating protects the glass from moisture. Moisture resistance can be judged from the aging of fibre when it is exposed to water.

4. **Glass transition temperature:** Low limit of temperature at which attenuation is within a stipulated margin can be judged from the characteristic curves of the coating if *i.e.*, variation of the property with temperature. From the curves variation of the thermal on fibre thereby resulting attenuation of the signals.

It should be noted that coating is critical component of the fibre. It determines I sensitivity, abrasion resistance, static fatigue protection and many other properties fibre.

4.21. Applications of Optical Fibres

1. **Military Applications.** Optical fibres are finding lot of applications in various warfares and military operations.

It conventional copper cables are to be transported by an aircraft, a ship or a tank for control mechanisms or communications of enormously reduced and also communication network is greatly improved as with fibre-optics secracy is also maintained.

During recent wars, fibre guided missiles have been extensively used. Sensores are mounted on missiles to collect the video information. This information is passed on to the ground control van which sends further commands to the missiles through sensors. Control van constantly monitors the course of missiles and corrects the direction of the missiles to hit the target precisely.

2. **Optical Fibre Sensors.** In this case the basic characteristics of the optical fibres are used, *i.e.*, the variation of refractice index of the fibre under the action of the external forces may be utilized for using the optical fibre as **transducer**.

In case the fibre is subjected to heat, the refractive index changes because of change of temperature. As temperature increases, the difference of refractive index between the core and cladding is reduced, **thereby some light leaks into cladding.** This characteristics can be used for measurement of temperature using a LED as light source. Such thermometer can be used for measuring temperatures in the range of 80°C to 700°C.

Also fibre optics may be use to measure **pollution and foreign suspended particles** in the air. A beam of light is send from one end of fibre and in case foreign particles are present, they scatter the light and measurement of variation of intensity of light will account of the extent of the presence of foreign particles.

Fibre optics may also be used to measure the level of a liquid in a container. A part of fibre is suspended in the liquid and light is directed to pass through it and its intensity is measured. A bare core loses more light when it is immersed in liquid than when it is placed in air. A sudden change in intensity of light will indicate the liquid level. A LED source, a photodector and a MMF is used to monitor the liquid level and in finding many applications in the filling of petroleum tanks.

In addition, fibre optics can be used in the fabrication of various sensors using the phenomenon of **double refraction and birefringes.** Due to the application of external forces, the optical properties of the fibre are varied. For example, we can have a **pressure sensor** using the induced birefringes, **current sensor** using the Faradays rotation, **mag. sensor** using the variation of magnetic fields and **voltage sensors** to account for the variation of voltages.

3. **Entertainment Applications.** A coherent optical fibre bundle can greatly enhance the size of the image on a TV screen. In case we have to use the conventional method of getting larger images then the projection system becomes quick bulky and unmanageable.

4. **Medical Applications.** Fibre optics are finding large number of applications in medical fields. Many parts of the human body which are otherwise inaccessible can be studied by illumination with fibre optics. A bundle of MMF is used to illuminate a part of human organ and the other part to collect the reflected light. This fibrescopic technique is employed for endoscopic applications. Fibre optics can be employed to attach a detached retina or rectify other eye defects using lasers.

In heart problems also **lasers angioplasty** is replacing the **balloon angioplasty** and **bypass surgery.** A special catheter is used which consists of three channels, viz., (i) to visualize the various viens (ii) to supply power for lasers (iii) an open tube to collect and suck the debris and other blocking material.

5. **In Communications.** A fibre optic communication system is similar to the **traditional communication** system with the exception that this system is highly efficient.

A block diagram of the typical communication system is shown in Fig. 4.17. A transducer changes a non-electrical message to an electrical signal and in case of fibre communication **light waves act as carrier waves.** Light waves are generally generated by **a laser diode or a light emitting diode (LED)** and the intensity of light given by these sources is **proportional to the current passing through them.**

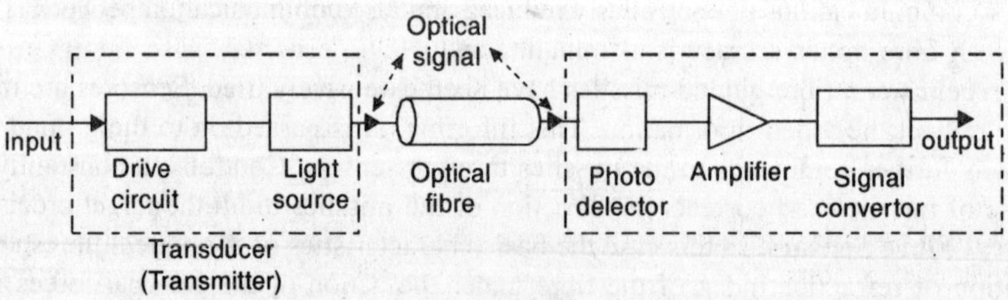

Fig. 4.17

As a message in the form of an electrical signal is received by the optical source, light ouput will follow the path of variation of signal with the message digital modulation is preferred in optical communication over the **analogue modulation** as it can pass through a larger distance with the **same power**.

However, one basic difficulty arises due to the **attenuation** of optical signal over large distance because of various losses and dispersion in the optical fibre. So to overcome it various **repeaters** are used in the transmission line to amplify and reshape the signal.

At the receiver end an optical coupler is used to direct the light coming from the fibre into a **photo-detector**, which converts light into **electric current**, which will contain the message fed into the optical fibre. In the next step, the message is extracted from this current and is amplified if the nearby. Then this message is put into a transducer to get it either in **video or radio** form depending upon the nature of the **original message**.

4.22. Use of Fibre Optics and Laser in Medical (Endoscopy)

Every doctor wants to see through the body of a patient without cutting any part of the body and then perform the surgery. This dream of the doctors has come true with the lasers and fibre optics.

For diagnostic purpose, an endoscopy (meaning 'to see within') is extensively used. An endoscope consists of a bunch of **optical fibres** which carries light to the inside of body and then transfer an image of the minor parts of the body to be viewed on the screen by the doctor. Endoscopy can better be done using **laser** beams because of the coherence characteristics.

Endoscope can be inserted in the body through the *mouth* or *rectum* and moved along any part of the body, where a particular defects has to be studied by a doctor.

Pulsed ruby lasers have been extensively used to the detached retina as the laser beam passes through transparent outer parts of the eye without causing any damage and it is then focussed on the defective part of the retina.

A burst of photons then generate sufficient energy to 'weld' the retina to the choroid from which it has become detached.

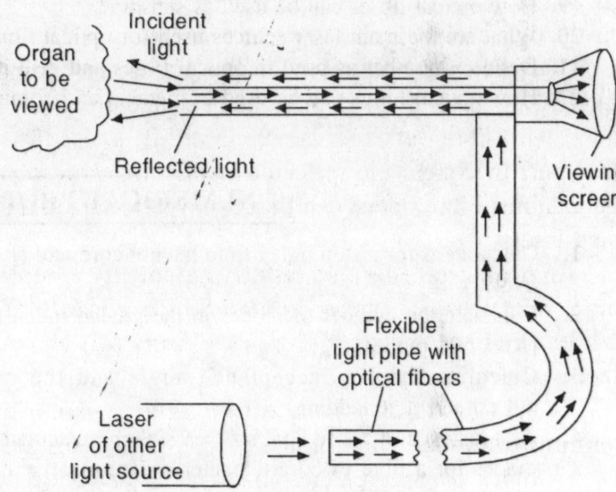

Fig. 4.18

High energy carbon dioxide (CO2) lasers are now being used to destroy cancer tissues. The 10,600 mm infrared light from a (CO2) laser is ideal for surgery as it is strongly asborbed by water which makes up 80% to 90%, of the body tissues that also includes the tumour tissue. Laser beam is used to sufficiently heat the defective tissue so that it is completely destroyed. Also the intense heat burns the defective tissue so that the disease does not spread to normal tissues."

QUESTIONS

1. What do you mean by excitation and ionization potentials? How these are used to explain the production of laser beams.
2. Define the terms :
   (i) Induced absorption (ii) Spontaneous emission (iii) Induced emission
3. How induced emission explain the lasers. Give Frank-Hertz experiment to explain the various energy states of atoms.
4. Explain the terms:
   (i) Life time of an energy states (ii) Metastable state
   (iii) Optical pumping (iv) Population inversion

5. Explain the production of lasers by Ruby crystal.
6. Explain how lasers can be produced by Helium-Neon gas.
7. Give the various uses of lasers in medical, engineering and scientific fields.
8. What does masers stand for? Give their field of applications.
9. Give the salient features of optical fibres.
10. What is numerical apreture? How is it useful in optical fibers?
11. Give the principle of optical fibres. How the terms angle of acceptance and cone of acceptance are used in optical fibres.
12. How optical fibres and lasers are useful in medical line?
13. What is index fibre? How you define the modes in an optical fibre? Give a relation between core diameter and numerical apreture.
14. Give the basic structure of an optical fibre. How the various refractive indices have to be related to get better working of an optical fibre?
15. How the light gets attenuated in an optical fibre? Derive an expression for fibre attenuation in terms of dB/km.
16. What are various modes of an optical fibre? Give their importance and applications.
17. Give the important components of a fibre optic system with specific use of each component.
18. Give the reasons of attenuation and distortion of light through the optical fibre.
19. How optical fibres can be used as sensors?
20. What are the main laser sources used for optical fibre system.
21. Define V-number as used in optical fibres and give its importance.
22. How optical fibres can be used in medical and communication fields?

NUMERICAL PROBLEMS

1. Calculate Δ for a step index fibre having core and cladding refractive indices 1.48 and 1.46 respectively. **[Ans.** 1.351 %]
2. Calculate the number of modes in the graded index fibre at an operating wavelength of 850 nm. [**Ans.** 504 modes]
3. Calculate the NA, acceptance angle and the critical angle of a fibre having μ1 (core) = 1.50 and μ_2 (cladding) = 1.45. [**Ans.** (i) 0.387; (ii) 22.78°; *(iii)* 75.2°]
4. At a distance of 1 km, spot size = 9800 cm; calculate the cut-off parameter and also the number of modes for a fibre of core diameter = 50 μm.μ1 = 1.47 (core), μ = 1.45 (cladding) and operating wavelength = 8500 Å. [**Ans.** 996 modes]
5. The core of a glass fibre has a refractive index of 1.5 while its cladding is doped to give a fractional refractive index change of 0.005.
   Find. (a) Reflective index of cladding
   (b) Critical internal reflecting angle
   (c) Acceptance angle
   (d) Numerical Aperture
   [**Ans.** (a) 1.4925 (b) 84.26 (c) 8.62° (d) 1.498]
6. Find the N.A. and acceptance angle for an optical fibre having core and cladding refractive indices 1.45 and 1.40 respectively [**Ans.** 0.377; 22.15°]
7. Find the core radius necessary for a single mode fibre operating at 850 nm is o step index fibre. The core and cladding refractive indices being 1.48 and 1.47 respectively. [**Ans.** 1.91 μm]
8. A step index single mode fibre has a core radius of 30 μm ; n_1 = 1.5 and n_2 = 1.47. If the operating wavelength is 800 nm, find the number of modes propagating through the fibre. [**Ans.** 2474]
9. A step index single mode fibre has a length of 5 km. Ratio of output to input power is 0.95. Find the alternation coefficient. [**Ans.** 0.045 dB/Km]
10. A step index fibre has a core of refractive index 1.3725 and cladding of refractive index 1.3274. It the cut off parameter is 2.405, find the N.A. and radius of core [Ans. 0.349; 1.43 μm]

5

THEORY OF RELATIVITY

5.1. Introduction

The first theory in Physics was *mechanics* developed by Newton and so is called *classical or Newtonian mechanics*. Actually, the Physics developed before the advent of 20th century is called *classical Physics.*

The development of 'theory of relativity' by Einstein in 1905 revolutionazed the old concepts. It discards the absolute motion through space and deals with objects or observer moving with relative velocities w.r.t. each other. This theory is divided into two parts.

(*i*) *Special theory* (*ii*) *General theory*

The special theory of relativity deals with objects and systems which are either moving at a *constant speed with respect to one another or are at rest.* **The general theory deals with objects or systems which are speeding up or slowing down with respect to one another.**

5.2. Frame of Reference

We know that the motion of the body has no meaning unless it is described with respect to some well defined coordinate system. The coordinate system is known as ***frame of reference***. The velocity is now measured with respect to this ***frame of reference***. The frame of reference is selected in such a way that the laws of nature may become fundamentally simpler in that frame of reference. There are two types of frames of reference:

(*i*) Inertial or unaccelerated frames.

(*ii*) Non-inertial or accelerated frames.

(*i*) **Inertial frames:** A frame of reference is said to be inertial when bodies in this frame obey Newton's law of inertia and other laws of Newtonian mechanics. In this frame, a body not acted upon by external force, is at rest or moves with a constant velocity. Experiments show that all inertial reference frames are equivalent for the measurement of physical phenomenon. Observers in different frames may obtain different numerical values for measured physical quantities but the relationship between the measured quantities *i.e.,* the laws of physics will be same for all observers.

(*ii*) **Non-inertial frames.** A frame of reference is said to be noninertial frame when a body, not acted upon by an external force, is accelerated. In this frame, **the Newton's laws are not valid.**

As the motion of Earth around the Sun is accelerated motion, ideally it should be **non inertial frame**. However since the acceleration of the Earth during the experiment is negligibly small, so for all practical purposes Earth can be considered as **inertial frame of reference**.

5.3. Galilean Transformations

Events and Coordinates. We begin by considering the concept of a physical event. The event might be the striking of a tree by a lightening bolt or the collision of two particles, and happens at a point in space assigning to it four coordinates: the three position coordinates x, y, z that measure the distance from the origin of a coordinate system where the observer is located, and the time coordinate t that the observer records with his clock.

Consider now two observers, O and O' where O' travels with a constant velocity v with respect to O along their common X–X' axis (Fig. 5.1). Both observers are equipped with meter-sticks and clocks so that they can measure coordinates of events. Further, suppose both observers adjust their clocks so that when they pass each other at $x = x' = 0$, the clocks read $t = t' = 0$. Any given event P will have eight numbers associated with it, the four coordinates (x, y, z, t) assigned by O and the four coordinates (x', y', z', t') assigned (to the same event) by O'.

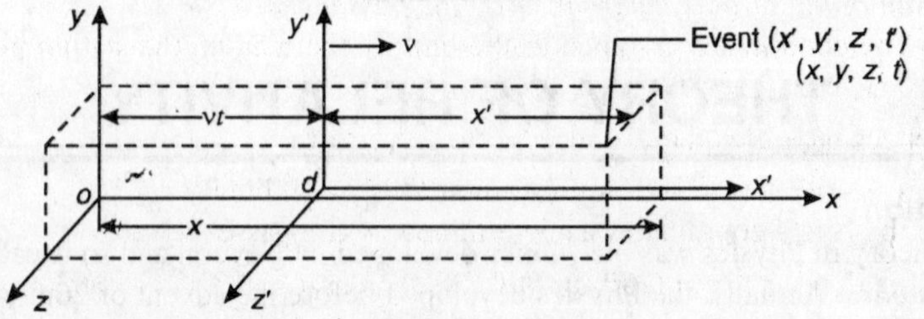

Fig. 5.1

5.4. Galilean Coordinate Transformations

The relationship between the measurements (x, y, z, t) of O and the measurements (x', y', z', t') of O for a particular event is obtained by examining Fig. 5.1.

$$x' = x - vt; \; y' = y, \; z' = z$$

In addition, in classical physics it is implicitly assumed that

$$t' = t$$

These four equations are called the Galilean coordinates transformations.

5.5. Galilean Velocity Transformations

In addition to the coordinates of an event, the velocity of particle is of interest. Observers O and O' will describe the particles' velocity by assigning three components to it, with (u'_x, u'_y, u'_z) being the velocity components as measured by O'.

The relationship between (u_x, u_y, u_z) and (u'_x, u'_y, u'_z) is obtained from the time differentiation of the Galilean coordinate transformations. Thus, from $x' = x - vt$

$$u'_x = \frac{dx'}{dt'} = \frac{d}{dt}(x - vt)\frac{dt}{dt'} = \left(\frac{dx}{dt} - v\right) \times (1) = u_x - v$$

Altogether, the Galilean velocity transformations are

$$u'_x = u_x - v \qquad u'_y = u_y \qquad u'_z = u_z$$

5.6. Galilean Acceleration Transformations

The acceleration of particle is the time derivative of its velocity, *i.e.*, $a_x = du_x/dt$. To find the *Galilean acceleration transformations* we differentiate the velocity transformations and use the facts that $t' = t$ and v = constant to obtain

$$a'_x = a_x \quad . \; a'_y = a_y \qquad a'_z = a_z$$

Thus the measured acceleration components are the same for all observers moving with uniform relative velocity.

5.7. Invariance of an Equation

By *invariance* of an equation it is mean that the equation will have the same form when determined by two observers. In classical theory it is assumed that space and time measurements of two observers are related by the Galilean transformations. Thus, when a particular form of an equation is determined by one observer, the Galilean transformations can be applied to this form to determine the form for the other observer. If both forms are the same, the equation is invariant under the Galilean transformations.

Example 1. *A passenger in a train moving at 30 m/s passes a man standing on a station platform at t = t′ = 0. Twenty seconds after the train passes him, the man on the platform determines that a bird flying along the tracks in the same directions as the train is 800 m away. What are the coordinate of the bird as determined by the passenger?*

Solution. The coordinates assigned to the bird by the man on the station platform are

$$(x, y, z, t) = (800 \text{ m}, 0, 0, 20 \text{ s})$$

The passenger measures the distance x' to the bird as

$$x' = x - vt = 800 \text{ m} - (30 \text{ m/s})(20\text{s}) = 200 \text{ m}$$

Therefore the bird's coordinates as determined by the passenger are

$$(x', y', z', t) = \mathbf{(200\ m, 0, 0, 20\ s)}\ \ \mathbf{Ans.}$$

Example 2. *A sample of radioactive material, at rest in the laboratory, ejects two electrons in opposite directions. One of the electrons has a speed of 0.6 c and the other has a speed of 0.7 c, as measured by a laboratory observer. According to classical velocity transformations, what will be the speed of one electron as measured from the other?*

Solution. Let observer O be at rest with respect to the laboratory and let observer O' be at rest with respect to the particle moving with speed 0.6 c (taken in the positive direction). Then, from the Galilean velocity transformation.

$$u_x' = u_x - v = -0.7c - 0.6c = \mathbf{-1.3c}\ \ \mathbf{Ans.}$$

This problem demonstrates that velocities greater than the speed of light are possible with the Galilean transformation, a result that is inconsistent with Special Relativity.

Example 3. *A train moving with a velocity of 60 km/hr passes through a railroad station at 12.00. Twenty seconds later a bolt of lightning strikes the railroad tracks one km from the station in the same direction that the train is moving. Find the coordinates of the lightning flash as measured by an observer at the station and by the engineer of the train.*

Solution. Both observers measure the time coordinate as

$$t = t' = \left(\frac{1 \text{ hr}}{3600\text{s}}\right) = \frac{1}{180} \text{hr}$$

The observer at the station measures the spatial coordinate to be $x = 1$ km. The spatial coordinate as determined by the engineer of the train is

$$x' = x - vt = 1 \text{ km} - (60 \text{ km/hr})\left(\frac{1}{180}\text{hr}\right) = \mathbf{2/3\ km.}\ \ \mathbf{Ans.}$$

Example 4. *A hunter on the ground fires a bullet in the north east direction which strikes a deer 0.25 km from the hunter. The bullet travels with a speed of 1800 km/hr. At the instant when the bullet is fired, an airplane is directly over the hunter at an altitude of one km and is travelling due east with a velocity of 600 km/hr. When the bullet strikes the deer, what are the coordinates as determined by an observant in the airpane?*

Solution. Using the Galilean transformations,

$$t' = t = \frac{0.25 \text{ km}}{1800 \text{ km/hr}} = 1.39 \times 10^{-4} \text{ hr.}$$

$$x' = x - vt = (0.25 \text{ km}) \cos 45° - (600 \text{ km/hr})(1.39 \times 10^{-4} \text{ hr})$$

$$y' = y = (0.25 \text{ km}) \sin 45° = 0.177 \text{ km}$$

$$z' = z - h = 0 - 1 \text{ km} = \mathbf{-1\ km}\ \ \mathbf{Ans.}$$

5.8. Concept of Ether

Newton considered the fixed stars as frame of reference for the explanation of his laws. However in later years a new search for a frame of reference was carried out. Later on Maxwell proved the electromagnetic nature of light waves and also it was proved that unlike sound waves (longitudinal waves), **light waves have transverse character** which was proved by the polarization phenomenon.

Again it was considered that a material medium is necessary for propagation of transverse light waves and Luminiferous ether was supposed to fill the whole space.

Since transverse waves require shearing forces and these forces can occur in solids only, so ether must be a rigid solid filling the whole space and the velocity of propagation of light will depend on the elasticity of the medium. Also the velocity of light being very high so the ether must also have a high elasticity.

Now if the light is propagated the ether medium, then we must expect a change in velocity (drift) depending on the direction of propagation of light w.r.t. the motion of the aparatus.

If we consider earth's motion around the sun in an orbit with a velocity of 3×10^4 m/s and velocity of light is 3×10^8 m/s. The ratio between the two is ~ 10^{-4}. We may consider the motion of the earth through stationary orbit and this led to the experiment of Michelson and Morley in 1881 – 1887 to verify the concept of ether filling the whole space.

5.9. Michelson Morley Experiment

Light from a monochromatic source S is rendered parallel with the lens L and is divided into two portions by the half silver plate A (Fig. 5.2). One portion of the beam travels to the mirror M_1 and is reflected back and the other portion of the beam travels towards M_2 and is reflected back. The two reflected beams interfere and the interference fringes are viewed through the telescope.

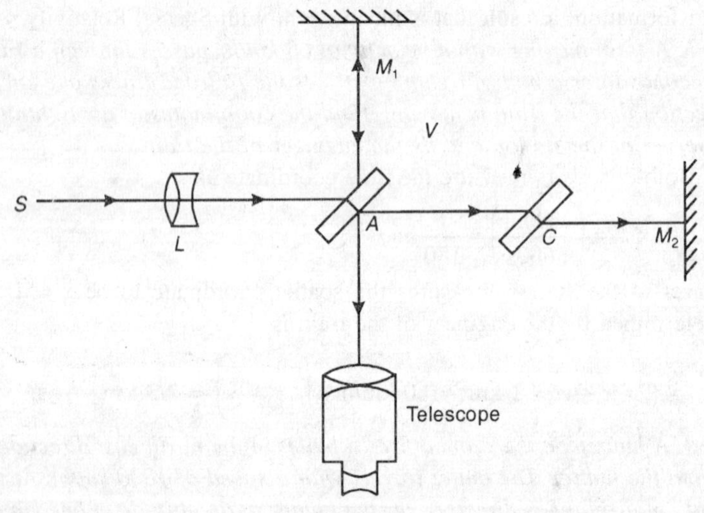

Fig. 5.2

The apparatus is arranged to move along the direction of the earths's orbit round the sun and the speed of movement of the apparatus is equal to v, *i.e.*, speed of the earth in its orbit. The whole apparatus is floated on mercury and *can be adjusted such that it is always moving along the direction of the earth's orbit round the sun.* The earth is considered to be stationary.

If we consider the Galilean frame of reference, F is a fixed frame of reference coresponding to the ether medium (Fig. 5.3) and F' is the frame of reference moving with a velocity v in the direction of the movement of the apparatus *i.e.*, in the direction of the movement of the earth in its orbit round the sun. The velocity of light in the direction of the movement of the frame $F' = c - v$

and in the opposite direction it is $= c + v$. Let T_1 be the time taken by light to travel from A to M_2' and T_2 that taken that M_2' to A' (Fig. 5.4).

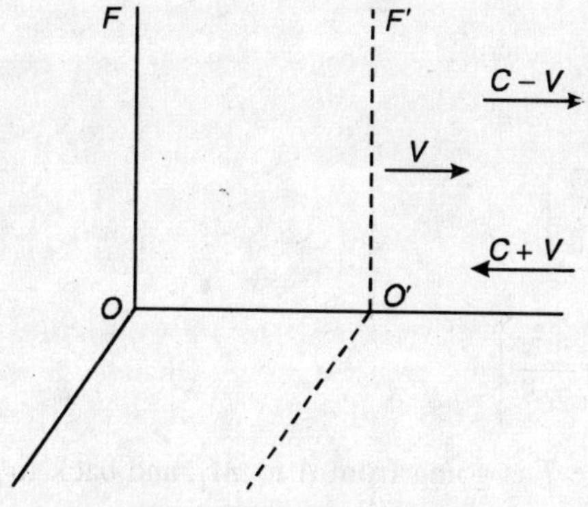

Fig. 5.3

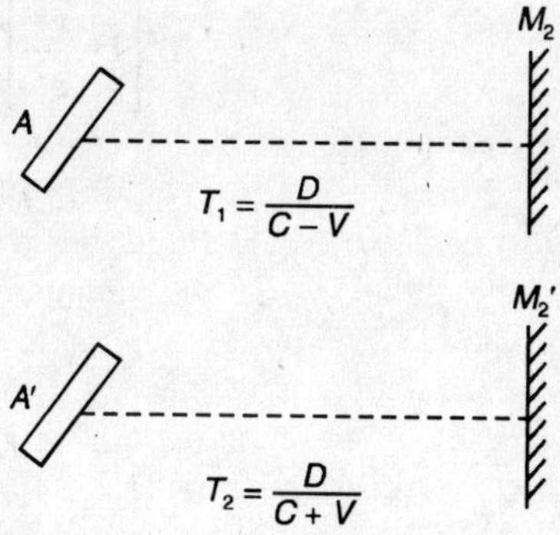

Fig. 5.4

In this case $D = AM_1 = AM_2$

So total to and fro times is given by

$$T = T_1 + T_2 = \frac{D}{c-v} + \frac{D}{c+v}$$

$$T = \frac{2Dc}{c^2 - v^2} \quad \text{.... (1)}$$

The total distance travelled by light is given by

$$x_1 = T \times c = \frac{2Dc^2}{c^2 - v^2}$$

$$x_1 \cong 2D\left[1 + \frac{v^2}{c^2}\right] \text{ [we neglect higher powers]} \quad \text{.... (2)}$$

Let the time taken by light to travel from A to M_1' be T'. In this case M_1 is shifted to M_1' in the time A is shifted to R (Fig. 5.5)

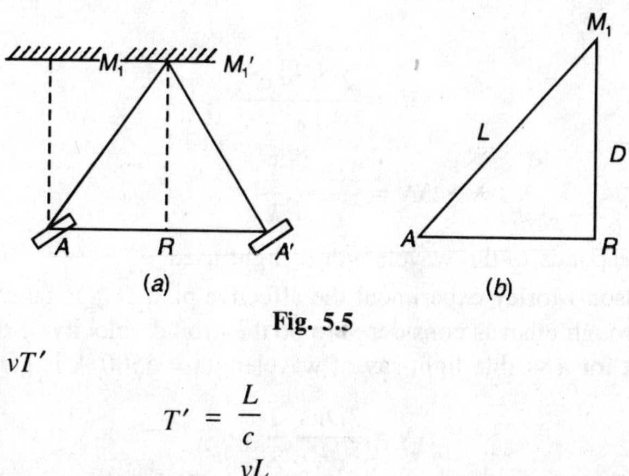

Fig. 5.5

The distance $AR = vT'$

Also $$T' = \frac{L}{c}$$

$\therefore$ $$AR = \frac{vL}{c}$$

$$L^2 = D^2 + \left[\frac{vL}{c}\right]^2 \quad \text{From Fig. 1.5 } (b)$$

$$L^2\left[1-\frac{v^2}{c^2}\right] = D^2$$

$$L^2 = \frac{D^2}{\left[1-\frac{v^2}{c^2}\right]}$$

$$L = D\left[1+\frac{v^2}{2c^2}\right]$$

So the total distance travelled by light in time T is going from A to M_1' and back to A'

$$= 2L = 2D\left[1+\frac{v^2}{2c^2}\right]$$

$$x_2 = 2D\left[1+\frac{v^2}{2c^2}\right]$$

From equations (2) and (3), we have the path difference ... (3)

$$x_1 - x_2 = 2D\left[1+\frac{v^2}{c^2}\right] - 2D\left[1+\frac{v^2}{2c^2}\right]$$

$$= \frac{Dv^2}{c^2} \quad \text{... (4)}$$

If the apparatus is turned through 90°, the path difference will be $\frac{-Dv^2}{c^2}$ *i.e.*, the displacement in the interference fringes is $\frac{2Dv^2}{c^2}$ which will shift the fringes in the telescope T through a number N, given by

$$\Delta N = \frac{2Dv^2/c^2}{\lambda}$$

or

$$\Delta N = \frac{2Dv^2}{c^2}.\frac{1}{\lambda}$$

where λ corresponds to the wavelength of light used.

In the Michelson-Morley experiment the effective path D was taken = 11 metres and if the velocity of earth through ether is considered to be the orbital velocity of the earth *i.e.*, $v = 3 \times 10^4$ m/s, the fringe shift for a visible light say of wavelength = 5500 Å is given by

$$\Delta N = \frac{2Dv^2}{c^2}.\frac{1}{\lambda}$$

or

$$\Delta N = \frac{2\times 11\times(3\times 10^4)^2}{(3\times 10^8)^2}\times\frac{1}{5.5\times 10^{-7}} = \mathbf{0.404}$$

This displacement could be very easily detected by Michelson-Morley apparatus. But actually no fringe shift was observed. The experiment was repeated at different times and different places, but all the times no change was observed. **Which concludes that the relative velocity between ether and earth is zero.**

Explanation of the negative results. Following separate interpretations were given to explain the negative results of Michelson-Morley experiment:

(*a*) **Ether-drag hypothesis.** The moving earth drags the ether along with it and hence there is no relative motion between the earth and the ether. This explanation of Michelson himself, however, is not reasonable because it goes against the observed phenomenon of aberration of starlight.

(*b*) **Lorentz-Fitzgerald contraction hypothesis.** Fitzgerald and Lorentz independently offered a more radical suggestion that a material body moving through ether is contracted in the direction of motion by a factor $\sqrt{[1-(v^2/c^2)]}$. Thus a length l in the direction of motion is shortened to $l\sqrt{[1-(v^2/c^2)]}$. It would equalize the times t_1 nd t_2 in Michelson-Morley experiment and hence no fringe shift can be expected. This contraction phenomenon is purely mathematical without any experimental confirmation.

(*c*) **Constancy of speed of light.** The proper explanation of zero results of Michelson-Morley experiment was given by Einstein. He proposed that the *speed of light is an invariant*, *i.e.*, the speed of light is constant and does not depend upon the motion of the source, observer or the medium. Then light would take equal times to travel the two paths in Michelson-Morele apparatus *i.e.*,

$$t_1 = t_2 = \frac{2D}{c}$$

and hence, no fringe shift is to be expected.

This negative result suggests that the *velocity of light is invariable and remains constant in all directions.* Also the effects of ether are undetectable. This has been verified by Doppler's principle *i.e.*, the speed of light is independent of the motion of the source or the observer.

So, all attempts to make ether as a fixed frame of reference have failed.

5.10. Einstein's Special Theory of Relativity

Einstein in 1905 announced the special theory of relativity and in 1915 proposed the general theory of relativity. The special theory deals with the problems in which one frame of reference moves with a *constant linear velocity* relative to another frame of reference. The general theory of relativity deals with problems in which one frame of reference is accelerated with respect to another frame of reference. Einstein assumed that all observers will notice that their motion in space will make no difference in the velocity of light with respect to them. He assumed that a fixed frame of reference cannot be located. All laws must be stated and they are applicable in any frame of reference.

According to Einstein's special theory of relativity:

(*i*) Observation on any particular frame of reference are not preferred to those of the other frames of reference. It means that the laws of physics are applicable equally well for an observer in the frame of reference A moving with a constant linear velocity with respect to an observer in anotehr frame of reference *B* and *vice versa* and

(*ii*) Velocity of light in free space is constant and is independent of the motion of the observer in any frame of reference.

Einstein proved the following facts based on his theory of relativity. Let v be the velocity of the spaceship with respect to a given frame of reference where an observer makes the observations.

(*a*) All clocks on the spaceship will go slow by a factor

$$\sqrt{1-\frac{v^2}{c^2}}$$

(*b*) The mass of the spaceship increases by a factor

$$\left[\frac{1}{\sqrt{1-\frac{v^2}{c^2}}}\right]$$

(*c*) All objects on the spaceship will have contracted in length by a factor

$$\sqrt{1-\frac{v^2}{c^2}}$$

(*d*) The speed of a material object can never exceed the velocity of light.

(*e*) Mass and energy are interconvertible

$$E = mc^2$$

(*f*) If two objects *A* and *B* are moving with velocities *u* and *v* with respect to each other along the *X*-axis, the relative velocity of *A* with respect to *B* is given by

$$V_x = \frac{u-v}{1-\frac{uv}{c^2}}$$

Here *u* and *v* are comparable with the value of *c*.

Example 5. *An electron is moving with a speed of 0.85 c in a direction opposite to that of a moving photon. Calculate the relative velocity of the electron with respect to the photon.*

Solution. Here, $u = 0.85\ c$, $v = -c$

$$V_x = \frac{u-v}{1-\frac{uv}{c^2}}$$

$$V_x = \frac{0.85c+c}{1+\frac{0.85c\times c}{c^2}} = \frac{1.85c}{1.85} = c \textbf{ Ans.}$$

From this example it is clear that the velocity of any object cannot exceed the velocity of light.

5.11. Lorentz Transformation of Space and Time

Without making any particular postulates, Lorentz algebraically tried transformation equations which may replace the Galilean transformations to give results in agreement with Michelson-Morley experiment.

Let us consider two inertial system *S* and S' as shown in Fig. 5.6. The system S' is moving with a velocity *v* with respect to *S* in the positive direction of *X*. For convenience, let us suppose that the axes of two coordinate systems coincide at $t = t' = 0$. Let a pulse of light is generated at time $t = 0$ at the origin which grows in the space. Now consider the situation when the pulse reaches at point *P*. Let (x, y, z, t) and (x', y', z', t') be the positions and times coordinates of *P* measured by observers O and O' in frames S and S' respectively. When the pulse is observed from S, then we have

$$x' = x - vt \quad \text{... (1)}$$

$$y = y \quad \text{... (2)}$$

and $$z' = z \quad \text{... (3)}$$

$$t = t' \quad \text{... (4)}$$

If a velocity is measured in system S and we want to get the components of velocity in system in terms of those in system S, we can write

$$v'_x = \frac{dx'}{dt} = v_x - v \quad \text{... (5)}$$

$$v'_y = \frac{dy'}{dt} = v_y \quad \text{... (6)}$$

and $$v'_z = \frac{dz'}{dt} = v_z \quad \text{... (7)}$$

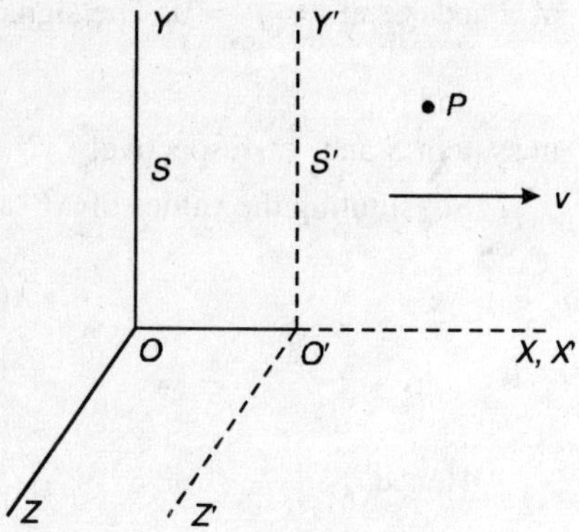

Fig. 5.6.

It is clear that the *Galilean transformations and the velocity transformations obtained in equations (5), (6) and (7) violate the postulates* of the special theory of relativity. **The equations of physics obtained by these transformations assume totally different forms in frames S and S'.** Moreover, if the velocity of light in frame S is c, that in system S' would be $c' = c - v$. It is, therefore, necessary that different transformation equations must be obtained if the postulates of the special theory of relativity are to be satisfied. These transformation equations – called the **Lorentz transformations** – are developed as under:

Let us assume that possible relation x and x' is of the type

$$x' = \lambda(x - vt) \quad \text{... (8)}$$

where λ is a constant of proportionality. It does not depend upon x and t but it may be a function of v. In choosing the relation between x and x' in the form expressed in equation (8) the following points are considered: (*i*) The relation is linear in x and x'. Hence, a single event in frame S corresponds to a single event in S'. (*ii*) It can be easily reduced to the form $x' = x - vt$ which is known to be correct in Newtonian mechanics if $\lambda = 1$ when $v << c$.

The inverse relation will be written as

$$x = \lambda'(x' + vt') \quad \text{... (9)}$$

in which primed quantities are replaced by unprimed quantities and *vice versa*. The other relations will be as before

$$y' = y \quad \text{... (10)}$$

and $$z' = z \quad \text{... (11)}$$

Time t and t' are not equal. To verify this, let us substitute the value of x' from equation (8) in equation (9). We get,

$$x = \lambda\lambda'(x - vt) + \lambda' vt'$$

$$t' = \lambda t + \left(\frac{1 - \lambda\lambda'}{\lambda' v}\right) x \quad \text{... (12)}$$

To determine the value of λ and λ' we use the second postulate of the special theory of relativity.

Let at $t = 0$, the origins of the two system S and S' coincide with each other. Let this instant corresponds to $t' = 0$ as well. Suppose that a signal of light is given from the common or origin of S and S' at $t = t' = 0$. The signal propagates in the two systems satisfying the equations

$$x = ct \quad \text{... (13)}$$

and

$$x' = ct' \quad \text{... (14)}$$

in system S and S' respectively.

Substituting the values of x' and t' from equations (8) and (12) into equation (14), we get

$$\lambda(x - vt) = c\lambda t + \left(\frac{1 - \lambda\lambda'}{\lambda' v}\right) cx$$

Hence,

$$x = ct\left[\frac{1 + v/c}{1 - \left\{\dfrac{1}{\lambda\lambda'} - 1\right\} c/v}\right] \quad \text{... (15)}$$

Comparing equation (15) with equation (13), we have

$$\frac{1 + v/c}{1 - \left\{\dfrac{1}{\lambda\lambda'} - 1\right\} c/v} = 1$$

or

$$\sqrt{\lambda\lambda'} = \frac{1}{\sqrt{1 - v^2/c^2}} \quad \text{... (16)}$$

Now, $\lambda\lambda'$ depends on v^2 and not on v, the relative velocity of S' with respect to S. In fact, we cannot choose between the two frames S' and S' except for the sign of v which does not affect the dependence of $\lambda\lambda'$. Therefore, we choose

$$\lambda = \lambda' = \frac{1}{\sqrt{1 - v^2/c^2}} \quad \text{... (17)}$$

Substituting these values in equations (8) and (12), we get

$$x' = \frac{x - vt}{\sqrt{1 - v^2/c^2}}$$

$$y' = y$$

$$z' = z$$

and

$$t' = \frac{t - \dfrac{vx}{c^2}}{\sqrt{1 - v^2/c^2}} \quad \text{... (18)}$$

Equations (18) are called the Lorentz transformations.

The inverse Lorentz transformations can now be written as

$$x = \frac{x' + vt'}{\sqrt{1 - v^2/c^2}}$$

$$y = y'$$

$$z = z'$$

and

$$t = \frac{t - \dfrac{vx'}{c^2}}{\sqrt{1 - v^2/c^2}}$$

So, the measurements of position and time are found to depend upon the frame of reference of the observer. It should be noted that Lorentz equations reduce to the Galilean transformations when relative velocity v is very small in comparison with velocity c of light. So, the consequences of special theory of relativity will be observable only when velocity v is practically comparable with the velocity of light.

Example 6. *An event occurs at x = 100 m, y = 10 m, z = 5 m and t = 1 × 10⁻⁴ second in a frame S. Find the coordinates of this event in a frame S′ which is moving with velocity 2.7 × 10⁸ metre per second with respect to the frame S along the common XX′ axes using (i) Galilean transformation and (iii) Lorentz transformation.*

Solution. (*i*) According to Galilean transformation, we get

$$x' = x - vt, \quad y' = y, \quad z' = z \quad \text{and} \quad t' = t$$

Substituting the given values, we get

$$x' = 100 - 2.7 \times 10^6 \times 1 \times 10^{-4} = -26900 \text{ m}$$

$$y' = y = 10 \text{ m}, \quad z' = z = 5 \text{ m}$$

and $$t' = t = 1 \times 10^{-4} \text{ second}$$

Thus the coordinates in frame S' are

$$\mathbf{x' = -26900 \text{ m}, \; y' = 10 \text{ m}, \; z' = 5 \text{ m} \text{ and } t' = 10^{-4} \text{ second} \;\; Ans.}$$

(*ii*) According to Lorentz transformation, we have

$$x' = \frac{x - vt}{\sqrt{1 - \frac{v^2}{c^2}}}$$

and $$t' = \frac{t - vx/c^2}{\sqrt{1 - \frac{v^2}{c^2}}}$$

Substituting the given values, we get

$$x' = \frac{100 - 2.7 \times 10^8 \times 1 \times 10^{-4}}{\sqrt{\left\{1 - \left(\frac{2.7 \times 20^8}{3 \times 10^8}\right)^2\right\}}} = -61720 \text{ m.}$$

$$y' = y = 10 \text{ m}, \quad z' = z = 5 \text{ m}$$

$$t' = \frac{1 \times 10^{-4} - \{(2.7 \times 10^3 \times 100)/(3 \times 10^8)^2\}}{\sqrt{\left\{1 - \left(\frac{2.7 \times 20^8}{3 \times 10^8}\right)^2\right\}}}$$

$$= 2.288 \times 10^{-4} \text{ sec.}$$

Thus the coordinates in S' are

$$x' = -61720 \text{ m}, \; y' = 10 \text{ m}, \; z' = 5 \text{ m and } t' = \mathbf{2.288 \times 10^{-4} \text{ sec.} \;\; Ans.}$$

Example 7. *Use Lorentz transformation to show that*

$$x^2 + y^2 + z^2 - c^2t^2 = x'^2 + y'^2 + z'^2 - c^2t'^2$$

or show that the quantity $x^2 + y^2 z^2 - c^2t^2$ is invariant under Lorentz transformation.

Solution. According to Lorentz transformation, we have

$$x' = \frac{x - vt}{\sqrt{1 - v^2/c^2}}, \; y' = y, \; z' = z$$

and $$t' = \frac{t - vx/c^2}{\sqrt{1 - v^2/c^2}}$$

$$\therefore \quad x'^2 + y'^2 + z'^2 - c^2t'^2$$

$$= \left(\frac{x - vt}{\sqrt{1 - v^2/c^2}}\right)^2 + y^2 + z^2 - c^2\left(\frac{t - vx/c^2}{\sqrt{1 - v^2/c^2}}\right)^2$$

$$= \frac{(x - vt)^2 - c^2(t - vx/c^2)^2}{1 - \frac{v^2}{c^2}} + y^2 + z^2$$

$$= \frac{(x^2 + v^2t^2 - 2xvt) - c^2\{t^2 + (v^2x^2/c^4) - (2xvt/c^2)\}}{1 - v^2/c^2 + y^2 + z^2}$$

$$= \frac{x^2(1 - v^2/c^2) - c^2t^2(1 - v^2/c^2)}{1 - v^2/c^2} + y^2 + z^2$$

$$= \frac{(x^2 - c^2t^2)(1 - v^2/c^2)}{(1 - v^2/c^2)} + y^2 + z^2$$

$$= x^2 - c^2t^2 + y^2 + z^2 = x^2 + y^2 + z^2 - c^2t^2$$

This shows that the quantity $x^2 + y^2 + z^2 - c^2t^2$ is same in both frames of reference. So, this is invariant under Lorentz transformation.

Example 8. *Show that the motion of one projectile as seen from another projectile will always be a straight line motion.*

Solution. Let the two projectiles be projected from the origin O with initial velocities u_1 and u_2 at angles of projections α_1 and α_2 respectively as shown in Fig. 5.7. After t sec the position coordinates of points P_1 (x_1, y_1) and P_2 (x_2, y_2) along X- and Y-axes may be written as

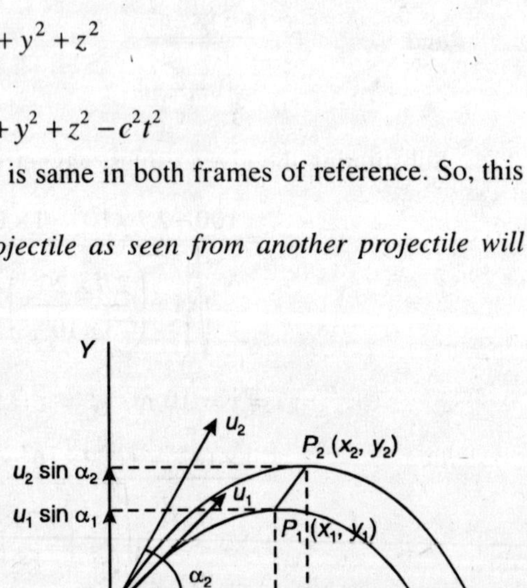

Fig. 5.7

$$x_1 = (u_1 \cos \alpha_1)t, \; y_1 = (u_1 \sin \alpha_1)t - \frac{1}{2}gt^2$$

$$x_2 = (u_2 \cos \alpha_2)t, \; y_2 = (u_2 \sin \alpha_2)t - \frac{1}{2}gt^2$$

$$\therefore \quad x_2 - x_1 = (u_2 \cos \alpha_2 - u_1 \cos \alpha_1)t \qquad \text{... (1)}$$

$$\text{and } y_2 - y_1 = (u_2 \sin \alpha_2 - u_1 \sin \alpha_1)t \qquad \text{... (2)}$$

Dividing (2) by (1), we get

$$\frac{y_2 - y_1}{x_2 - x_1} = \frac{u_2 \sin\alpha_2 - u_1 \sin\alpha_1}{u_2 \cos\alpha_2 - u_1 \cos\alpha_1} = m \text{ (Constant)}$$

($\because u_1, u_2, \alpha_1$ and α_2 are constants)

$\therefore \quad (y_2 - y_1) = m\,(x_2 - x_1)$

Let $\quad y_2 - y_1 = Y$ and $x_2 - x_1 = X$

$\therefore \quad Y = mX$

Which is the equation of a straight line. So, **the motion of one projectile as seen from another projectile will always be a straight line motion. Ans.**

Example 9. *A ball has velocity (4i – 5j + 10k) m/sec w.r.t to a train moving with velocity (3i + 4j) m/sec relative to an observer on the ground. Find the velocity of the ball relative to the ground.*

Solution. Let the train be equivalent to a moving frame of reference S' and let the ground be stationary frame of reference S. So we have

$$\vec{V}\ (3i + 4j) \text{ m/sec and } \vec{u'} = (4i - 5j + 10k) \text{ m/sec}$$

So the velocity of the ball w.r.t. ground will be

$$\vec{u} = \vec{u'} + \vec{V}$$
$$= (4i - 5j + 10k) + (3i + 4j)$$

or $\quad \vec{u} = 7i - j + \mathbf{10k}$ **m/sec. Ans.**

Example 10. *If at the time t = t′ = 0, the origins of systems S and S′ just coincide and a spherical pulse of light is produced at the common origin, show that the speed of propagation of the spherical wavefront is the same in both systems-namely c. (where system S′ is moving relative to stationary system S)*

Solution. As per problem the origin of two systems S and S' coincide at $t = t' = 0$. The system S is at rest while S' is moving with fixed velocity v relative to S. Let spherical pulse start from a point where coordinates w.r.t. S and S' are (x, y, z, t) and (x', y', z', t') respectively. We have from invariance of Lorentz transformation equations

$$x^2 + y^2 + z^2 = c^2t^2 \text{ and } x'^2 + y'^2 + z'^2 = c^2t'^2 \qquad \text{... (1)}$$

Now the Lorentz transformation equations are

$$x' = \frac{x - vt}{\sqrt{(1 - v^2/c^2)}},\ y' = y,\ z' = z \text{ and } t' = \frac{t - (vx/c^2)}{\sqrt{(1 - v^2/c^2)}} \qquad \text{... (2)}$$

From equation (1), we have

$$c = \frac{(x^2 + y^2 + z^2)^{1/2}}{t} \text{ and } c' = \frac{(x'^2 + y'^2 + z'^2)^{1/2}}{t'} \qquad \text{... (3)}$$

Putting the values of x', y', z' and t' in equation (3)

Now from eq. (2) we have

$$c' = \frac{\{[(x - vt)^2/(1 - v^2/c^2)] + y^2 + z^2\}^{1/2}}{[(t - vx/c^2)/\sqrt{(1 - v^2/c^2)}]}$$

or

$$c' = \frac{[(x - vt)^2 + (y^2 + z^2)(1 - v^2/c^2)]^{1/2}}{(t - vx/c^2)} \qquad \text{... (4)}$$

Adding and subtracting a factor of $x^2\left(1-\frac{v^2}{c^2}\right)$ in the numerator of equation (4) and writing in form

$$c' = \frac{[(x^2+y^2+z^2)(1-v^2/c^2)+\{v^2t^2-2xvt+(v^2/c^2).x^2\}]^{1/2}}{(t-vx/c^2)}$$

But, $x^2 + y^2 + z^2 = c^2 t^2$

$$\therefore \quad c' = \frac{\left[c^2t^2\left(1-\frac{v^2}{c^2}\right)+v^2t^2-2xvt+\frac{v^2}{c^2}x^2\right]^{1/2}}{\left(t-\frac{vx}{c^2}\right)}$$

$$= \frac{\left(c^2t^2-2xvt+\frac{v^2}{c^2}x^2\right)^{1/2}}{\left(t-\frac{vx}{c^2}\right)} = \frac{\left[c^2\left(t-\frac{vx}{c^2}\right)^2\right]^{1/2}}{\left(t-\frac{vx}{c^2}\right)} = c$$

$\therefore \quad c' = c$

Which means that the velocity of light is same in both systems or the speed of propagation of the spherical wavefront is same in both systems. Ans.

Example 11. *A light pulse is emitted at the origin of a frame of reference S' at time $t' = 0$. Its distance x' from the origin after a time t' is given by $x'^2 = c^2t'^2$. Use the Lorentz transformation to transform this equation to an equation in x and t and show that this is $x^2 = c^2t^2$. Discuss the implication of this result.*

Solution. From given problem, the distance of a light pulse x' from the origin after a time t' in the movin; frame S' may be given as

$$x'^2 = c^2t'^2$$

Now Lorentz transformation equations, have the form

$$x' = \frac{x-vt}{\sqrt{1-v^2/c^2}} \text{ and } t' = \frac{t-xv/c^2}{\sqrt{1-v^2/c^2}}$$

Putting these values of x' and t' in equation (2) so we have

$$\frac{(x-vt)^2}{1-v^2/c^2} = \frac{c^2[t-(xv/c^2)]^2}{1-v^2/c^2}$$

or $$(x-vt)^2 = c^2\left(t-\frac{xv}{c^2}\right)^2$$

or $$x^2+v^2t^2-2xvt = c^2t^2+\frac{x^2v^2}{c^2}-2txv$$

$$x^2+v^2t^2 = c^2t^2+\frac{x^2v^2}{c^2}$$

or $$x^2 - c^2t^2 + v^2t^2 - \frac{x^2v^2}{c^2} = 0$$

or $$x^2 - c^2t^2 - \frac{v^2}{c^2}(x^2 - c^2t^2) = 0$$

or $$(x^2 - c^2t^2)\left(1 - \frac{v^2}{c^2}\right) = 0$$

Now since $v \neq c$, $\therefore$ $\left(1 - \frac{v^2}{c^2}\right) \neq 0$

$\therefore$ $$x^2 - c^2t^2 = 0$$

or $$x^2 = c^2t^2$$

Which shows that the velocity of light is an absolute constant independent of the frame of reference or the equation is invariant under Lorentz transformation. Ans.

Example 12. *Find the time [as measured by a clock at rest (on the rocket)] taken by a rocket to reach a distant star and return to earth with a constant velocity* $v = \sqrt{(0.9999)}c$, *if the distance to the star is 4 light years (A light year is defined as the distance travelled by a light beam in vacuum in one year).*

Solution. Lorentz transformation equation for time, may be written as

$$t' = \frac{t - vx/c^2}{\sqrt{(1 - v^2/c^2)}} \quad \text{... (i)}$$

In this case $x = 2 \times 4$ light year $= 2 \times 4 \times 365 \times 24 \times 60 \times 60 \times c$ cm. $= 2.523 \times 10^8$ *cm*, where t is the time taken by the rocket to reach the star and back to the earth as measured by an observer in a stationary frame at earth and t' is the time measured by the clock in the moving rocket for the complete journey time is given by $t = x/v$, where $v = c\sqrt{(0.9999)}$

$\therefore$ $$t' = \frac{[(x/v) - (vx/c^2)]}{\sqrt{(1 - v^2/c^2)}} = \frac{\frac{x}{v}\left(1 - \frac{v^2}{c^2}\right)}{\left(1 - \frac{v^2}{c^2}\right)^{1/2}}$$

or $$t' = \frac{x}{v}\sqrt{(1 - v^2/c^2)} \quad \text{... (ii)}$$

Putting the values of x and v in equation (*ii*) we have

$$t' = \frac{2.523 \times 10^8 c}{c\sqrt{0.9999}}\sqrt{1 - 0.9999} = \frac{2.523 \times 10^8 \times (0.0001)^{1/2}}{\sqrt{0.9999}}$$

or $$t' = \frac{2.523 \times 10^8}{(9999)^{1/2}} = \frac{2.523 \times 10^8}{99.995} = \mathbf{2.523 \times 10^6}$$ **sec Ans.**

5.12. Length Contraction

Consider two coordinate systems S and S', the latter moving with velocity v relative to the former along the positive x-direction. Let a rod be placed in S' along X-axis. If x_1' and x_2' are coordinates of the abscissae of the ends of the rod, then its length l' in system S' is given by

$$l' = x_2' - x_1' \qquad \text{... (i)}$$

Now let an observer in S measure the length of the same rod. If x_1 and x_2 are the abscissae of the ends of the rod at the same time t, then the length l of the rod is given by

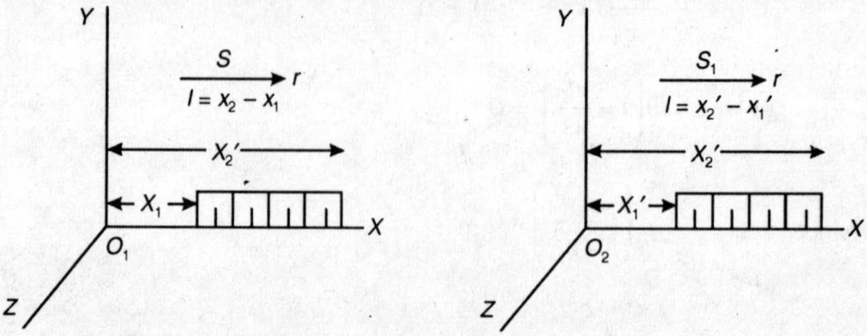

Fig. 5.8

$$l = x_2 - x_1 \qquad \text{... (ii)}$$

From Lorentz transformation equations we may write

$$x_2' = \frac{(x_2 - vt)}{\sqrt{1-\dfrac{v^2}{c^2}}}$$

and
$$x_1' = \frac{(x_1 - vt)}{\sqrt{1-\dfrac{v^2}{c^2}}}$$

Putting these values in eq. (*i*) we have

$$l' = \frac{(x_2 - vt)}{\sqrt{1-\dfrac{v^2}{c^2}}} - \frac{(x_1 - vt)}{\sqrt{1-\dfrac{v^2}{c^2}}}$$

$$= \frac{(x_2 - x_1)}{\sqrt{1-\dfrac{v^2}{c^2}}} = \frac{l}{\sqrt{1-\dfrac{v^2}{c^2}}}$$

or
$$l = l'\sqrt{1-\frac{v^2}{c^2}}$$

Thus the length of the rod moving with velocity v relative to the observer is contracted by a factor $\sqrt{1-\dfrac{v^2}{c^2}}$ *in the direction of motion. This is known as Lorentz-Fitgerald contraction.*

From above **we may** conclude that

(*i*) When v is very small compared to c, then $\dfrac{v^2}{c^2}$ will be negligible in comparison to unity, therefore

$$l = l'$$

i.e., the length of the moving rod or object is the same as that, when it was at rest.

(*ii*) When v is comparable to c, $\sqrt{1-\frac{v^2}{c^2}}$ will be less than unity, therefore $l < l'$ *i.e.,* the length of the moving object or rod appears to be less than the length when it was at rest.

(*iii*) When v is equal to c or greater than $c, \frac{v^2}{c^2}$ *will be equal to unity or greater than unity, therefore*

$$l = 0 \text{ or imaginary.}$$

This is impossible. *So no material body can ever attain the velocity of light.*

(*iv*) The contraction takes place only along the direction of motion and remains unchanged in a perpendicular direction.

(*v*) The contraction cannot be visualized, though it really exists.

5.13. Time Dilation

According to the special theory of relativity, time intervals are also affected by relative motion between two frames of reference and reference frame B is moving with a velocity v along the X-axis.

Let at any instant, the two reference frames coincide at $t = t' = 0$. The observer in B notes the time at any instant in his clock at t_1' whereas an observer in A notes the time as t_1.

According to inverse Lorentz transformation,

$$t_1 = \frac{t_1' + \frac{vx'}{c^2}}{\sqrt{1-\frac{v^2}{c^2}}} \qquad \text{... (1)}$$

Let t_2' and t_2 be the times measured by the observers in B and A respectively at the same instant,

Here
$$t_2 = \frac{t_2' + \frac{vx'}{c^2}}{\sqrt{1-\frac{v^2}{c^2}}} \qquad \text{... (2)}$$

Let t_0 be the interval of time as assumed by an observer in B and t the interval of time measured by an observer in A.

$$t_0 = t_2' - t_1' \qquad \text{... (3)}$$

$$t = t_2 - t_1 \qquad \text{... (4)}$$

$$t = \frac{t_2' + \frac{vx'}{c^2}}{\sqrt{1-\frac{v^2}{c^2}}} - \frac{t_1' + \frac{vx'}{c^2}}{\sqrt{1-\frac{v^2}{c^2}}}$$

$$t = \frac{t_2' - t_1'}{\sqrt{1-\frac{v^2}{c^2}}} = \frac{t_0}{\sqrt{1-\frac{v^2}{c^2}}} \qquad \text{... (5)}$$

Which show that an interval of time observed in a moving frame of reference will be less than the same interval of time observed in a stationary frame of reference. It also show that the clocks in the moving spaceships will appear to go slower than the clocks on the surface of the earth. **It is called time dilation.**

5.14. Experimental Verification of Time Dilation

Time dilation has been verified in case of a nuclear particle meson. There are π meson and μ mesons. The π^+ mesons decay (convert into μ mesons and neutrons) in such a way that in every 1.8×10^{-8} second half of them die away. In the laboratory the π^+ mesons are produced with speed of 0.99 c when high energy particles generated by Synchrotron strike a target. The flux of π^+ mesons was measured at two places separated by 30 metres. The laboratory time interval $\Delta t'$ for this distance is given by

$$\Delta t' = \frac{30\,\text{m}}{0.99 \times 3 \times 10^8\,\text{m/sec}} = \mathbf{10 \times 10^{-8}\ sec.}$$

This is nearly 5.6 times of 1.8×10^{-8} sec. Thus the flux of π^+ mesons should decrease to $(2)^{-5.6}$ or or less than 2% of the original flux in travelling a distance 30 metres. It was observed that the flux at the second place was nearly 60% of that of the first place.

This discrepancy can be explained by calculating t given by the relation

$$\Delta t' = \frac{\Delta t}{\sqrt{1 - \frac{v^2}{c^2}}}$$

or

$$\Delta t = \Delta t' \times \sqrt{1 - \frac{v^2}{c^2}}$$

$$= 10 \times 10^{-8} \sqrt{1 - \left\{\frac{.99c}{c}\right\}^2} = \mathbf{1.4 \times 10^{-8}\ sec.}$$

This is 0.78 times the half life 1.8×10^{-8} sec. Hence in so much time the flux should fall $(2)^{-0.7} \approx 60\%$.

This is exactly the same, what is observed experimetally. From this experiment, it is obvious that π^+ mesons take a time 10×10^{-8} sec to cover a distance 30 metres in laboratory measurements while a time 1.4×10^{-8} sec in their own frame of reference. Thus a 7-fold time dilation has occurred in this case.

Example 13. *A rocket ship is 100 metre long on the ground. When it is in flight, its length is 99 metres to an observer on the ground. What is its speed?*

Solution. We know that

$$l = l' \sqrt{1 - \frac{v^2}{c^2}}$$

Here $l'\ 100 \times 100 = 10000$ cm

and $l = 99 \times 100 = 9900$ cm

$$\therefore \quad 9900 = 10000 \sqrt{1 - \frac{v^2}{c^2}}$$

or $$\frac{99}{100} = \sqrt{1-\frac{v^2}{c^2}}$$

or $$\frac{99\times 99}{100\times 100} = 1-\frac{v^2}{c^2}$$

or $$\frac{v^2}{c^2} = 1-\frac{99\times 99}{100\times 100} = \frac{199}{100\times 100}$$

or $$v = c\frac{\sqrt{199}}{100}$$

$$v = 3\times 10^{10}\frac{\sqrt{199}}{100}$$

$$= 4.23 + 10^9 \text{ cm/sec.}$$

$$= \mathbf{4.23 \times 10^7\ ms^{-1}} \quad \textbf{Ans.}$$

Example 14. *What will be the length of a meter rod appear to be for a person travelling parallel to the length of the rod at a speed of 0.8 c relative to the rod?*

Solution. From length contraction relation, $L = L_0\sqrt{(1-v^2/c^2)}$

In this case $v = 0.8\ c$ and $L_0 = 1\text{m}$

$\therefore\ L = 1\sqrt{[1-(0.8c/c)^2]} = 1\sqrt{(0.36)}$ or $L = 1 \times 0.6 =$ **0.6 m or 60 cm. Ans.**

Example 15. *A certain process requires 10^{-6} sec to occur in an atom at rest in laboratory. How much time will this process require to an observer in the laboratory, when the atom is moving with a speed of 5×10^7 metres/sec.*

Solution. We have the relation

$$\Delta t' = \frac{\Delta t}{\sqrt{1-\left(\frac{v^2}{c^2}\right)}}$$

Here $$\Delta t = 10^{-6} \text{ sec and } v = 5\times 10^9 \text{ cm/sec.}$$

$\therefore$ $$\Delta t' = \frac{10^{-6}}{\sqrt{1-\left(\frac{5\times 10^9}{3\times 10^{10}}\right)^2}}$$

$$= \frac{10^{-6}}{\sqrt{1-\left(\frac{25}{900}\right)}} = \mathbf{1.013\times 10^{-6}\ sec.}\ \textbf{Ans.}$$

Example 16. *A train whose length is 150 m when at rest, has to pass through a tunnel of length 125 m. The train is moving with uniform speed of 2.4×10^8 m/sec towards the tunnel. Find the length of the train and that of the tunnel as observed by an observer (i) at the tunnel, (ii) at the train.*

Solution. (*i*) Observer at the tunnel, at rest, observes the train coming towards him with velocity 2.4×10^8 m/sec. So, the length L_1' of the train as it appears to an observer at the tunnel may be got by using length contraction relation as

$$L = L_0\sqrt{\left(1-\frac{v^2}{c^2}\right)}$$

In this case $L = L_1'$, $L_0 = 150$ m and $v = 2.4 \times 10^8$ m/sec.

$$\therefore \qquad L_1' = 150\sqrt{\left[1-\left(\frac{2.4\times10^8}{3\times10^8}\right)^2\right]} = 150 \times 0.6 = 90 \text{ m}$$

As the tunnel is at rest so the length of the tunnel as it appears to the observer at the tunnel will remain the same, that is, **125 m. Ans.**

(*ii*) Observer in the train will observe the tunnel moving towards him with velocity 2.4×10^8 m/sec. So, the length L_2' of the tunnel as it appears to the observer in train may be written as

$$L_2' = L_1'\sqrt{(1-v^2/c^2)}$$

$$\therefore \qquad L_2' = 125\sqrt{\left[1-\left(\frac{2.4\times10^8}{3\times10^8}\right)^2\right]} = \mathbf{125 \times 0.6 = 75}$$

Since the train is at rest with respect to the observer in the train, the length of the train appears to the observer in train will remain the same, that is, **150 m. Ans.**

Example 17. *With what velocity should a space ship fly so that every day spent on it may correspond to three days on earth's surface.*

Solution. We know that

$$\Delta t' = \frac{\Delta t}{\sqrt{1-\dfrac{v^2}{c^2}}}$$

According to the given problem, $t = 1$ day and $t' = 3$ days

$$\therefore \qquad 3 = \frac{1}{\sqrt{1-\dfrac{v^2}{c^2}}}$$

$$\text{or} \qquad \left(1-\frac{v^2}{9\times10^{16}}\right) = \frac{1}{9} \text{ or } v^2 = \frac{9\times10^{16}\times8}{9}$$

$$\therefore \qquad v = \sqrt{8}\times10^8 = \mathbf{2.83 \times 10^5 \text{ m/s.}}$$

Example 18. *A* μ *meson has a life time* 2×10^{-6} *sec.*

(*a*) *What is the mean life time when the particle is travelling with a speed of* 2.994×10^{10} *cm/sec?*

(*b*) *How far does it go during one mean life?*

(*c*) *What distance would be travelled without relativistic effects?*

Solution. (*a*) We have the relation

$$\Delta t' = \frac{\Delta t}{\sqrt{1-\dfrac{v^2}{c^2}}}$$

Here $\Delta t = 2 \times 10^{-6}$ sec

$v = 2.994 \times 10^{10}$ cm/sec

$$\therefore \quad \Delta t' = \frac{2\times10^{-6}}{\sqrt{1-\left(\frac{2.994\times10^{10}}{3\times10^{10}}\right)}} = \mathbf{31.7 \times 10^{-6}\ sec.}$$

(*b*) The distance travelled by the particle in one mean life

$$s = v \times \Delta t'$$

$$= 2.994 \times 10^{10} \times 31.7 \times 10^{-6} \text{ cm} = \mathbf{9510\ metres.}$$

$$s = v \times \Delta t$$

$$= 2.994\times10^{10} \times 2\times10^{-6} \text{ cm} = \mathbf{598.8\ metres\ Ans.}$$

Example 19. *Show that the circle, $x^2 + y^2 = a^2$ in frame S appears to be an ellipse in frame S' which is moving with velocity v relative to S.*

Solution. We have the equation of a circle in stationary frame S: $x^2 + y^2 = a^2$, where a is constant.

From Lorentz-Fitzgerald length contraction

$$x = x'\left(1-\frac{v^2}{c^2}\right)^{1/2} \quad \text{and} \quad y = y'$$

Putting these values of x and y in the equation of the circle, we have

$$\therefore \quad x'^2\left(1-\frac{v^2}{c^2}\right) + y'^2 = a^2 \text{ or } \frac{x'^2}{a^2/(1-v^2/c^2)} + \frac{y'^2}{a^2} = 1$$

Let $a^2/(1 - v^2/c^2) = b^2$,

$$\therefore \quad \frac{x'^2}{b^2} + \frac{y'^2}{a^2} = 1$$

So, in the moving frame S' the circle will look like an ellipse. **Ans.**

5.15. Addition of Velocities

Consider two systems S and S', S' moving with a velocity v relative to system S along the positive direction of X. Let us express the velocity of a body in them. Suppose the body moves a distance dx in time dt in system S and through a distance dx' in time dt' in system S'. then

$$\frac{dx}{dt} = u \text{ and } \frac{dx'}{dt'} = u' \quad \text{... (1)}$$

From Lorentz transformation equations, we have the relations

$$x = \lambda(x' + vt') \quad \text{... (2)}$$

and

$$t = \lambda\left(t' + \frac{vx'}{c^2}\right) \quad \text{... (3)}$$

Differentiating equations (2) and (3), we get

$$dx = \lambda(dx' + vdt') \quad \text{... (4)}$$

$$dt = \lambda\left(dt' + \frac{vdx'}{c^2}\right) \quad \text{... (5)}$$

Dividing eq. (4) by eq. (5)

$$\frac{dx}{dt} = \frac{\lambda(dx' + vdt')}{\lambda(dt' + vdx'/c^2)}$$

$$\frac{dx}{dt} = \frac{\frac{dx'}{dt'} + v}{1 + \frac{v}{c^2}\frac{dx'}{dt'}}$$

or $$v = \frac{u' + v}{1 + \frac{u'v}{c^2}} \qquad \text{... (6)}$$

The eq. (6) represents the relativistic law of addition of velocities whereas in classical mechanics it is simply $u = u' + v$. Following points are evident from here

(*i*) When u' and v are smaller as compared to c, $\frac{u'v}{c^2}$ can be neglected in comparison to unity, therefore

$$u = u' + v \qquad \text{... (7)}$$

which is classical formula.

(*ii*) When v or $u' = c$, then

$$u = \frac{u + c}{1 + \frac{u'c}{c^2}} = \frac{u' + c}{c + u'} \times c = c \qquad \text{... (8)}$$

i.e., if one object moves with velocity c with respect to other then, their relative velocity is always c, whatever may be the velocity of the other.

(*iii*) When $u' = c = v$, then

$$u = \frac{c + c}{1 + \frac{c^2}{c^2}} = \frac{2c}{2} = c$$

This shows that the addition of velocity of light to the velocity of light merely reproduces the velocity of light.

Example 20. *Two particles came towards each other with speed 0.8 c with respect to laboratory. What is their relative speed?*

Solution. We know that

$$u = \frac{u' + v}{1 + \frac{u'v}{c^2}}$$

Here $$u' = v = 0.8\,c$$

$\therefore$ $$u = \frac{0.8c + 0.8c}{1 + \frac{0.8c \times 0.8c}{c^2}} = \frac{0.8c + 0.8c}{1 + 0.8 \times 0.8}$$

$$= \frac{1.6c}{1.64} = \mathbf{0.975\,c}$$

Thus the relative speed is **0.975 *c*. Ans.**

5.16. Variation of Mass with Velocity

Einstein's concepts of relativistic change of space and time given an idea that mass also varies with velocity. In fact the mass has to be considered as a function of velocity with which a body moves.

Let us consider two inertial frames S and S' and let S' be moving with velocity V along the positive X-direction.

Consider two identical particles A and B of same masses in frame S' approaching each other with a velocity μ' in frames' (taking positive velocity of particle A and negative velocity of particle B as it is moving in opposite direction.

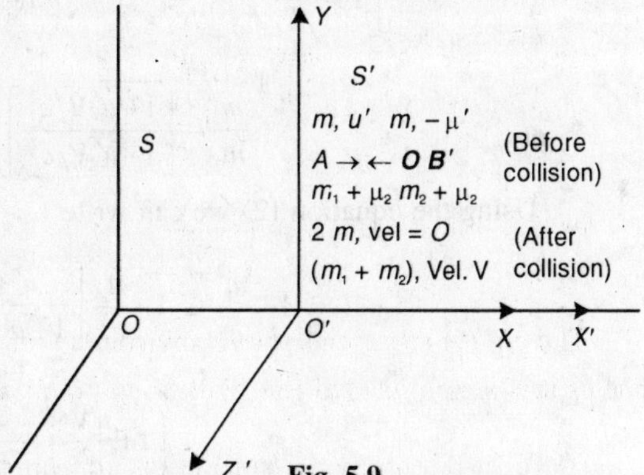

Fig. 5.9

Let, to an observer in fram S, the bodies be moving with velocities u_1 and u_2 before the collision. Since mass is a function of velocity, so to an observer in S; particles A and B will appear to have different masses. Let these masses be m_1 and m_2 as observed in from S.

After Collision, the two masses stick to each other as the collision in elastic. The combined mass in frame S' will be 2m and it will be at rest in frame S′.

In frame S, the two bodies will have mass $m_1 + m_2$ and will appear to move with a velocity v, the velocity of frame S′.

From Lorentz inverse velocity transformation equations.

$$u_x = \frac{u'_x + V}{1 + \frac{V}{c^2} u'_x} \qquad \text{... (1)}$$

For body A, but $u_x = u_1 \quad u'_x = u'$, so

We have from eq. (1)

$$u_1 = \frac{u' + V}{1 + \frac{u'V}{c^2}} \qquad \text{... (2)}$$

For body B, put and $u_x = u_2$ and $u'_x = -u'$

$$\therefore \qquad u_2 = \frac{-u' + V}{1 - \frac{u'V}{c^2}} \qquad \text{... (3)}$$

From the conservation law of linear momentum in frame S, we get

$$m_1 u_1 + m_2 u_2 = (m_1 + m_2)V \qquad \text{... (4)}$$

Putting the values of u_1 and u_2 from equations (2) and (3) is eq. (4), we get

$$m_1 \left[\frac{u' + V}{1 + \frac{u'V}{c^2}} \right] + m_2 \left[\frac{-u' + V}{1 - \frac{u'V}{c^2}} \right] = (m_1 + m_2)V$$

or $$m_1\left[\frac{u'+V}{1+\frac{u'V}{c^2}}-V\right] = m_2\left[V-\left(\frac{-u'+V}{1-\frac{u'V}{c^2}}\right)\right]$$

or $$\frac{m_1}{m_2}=\left[\frac{1+u'V/c^2}{1-u'V/c^2}\right] \quad \text{... (5)}$$

Using the equation (2) we can write

$$1-\frac{u_1{}^2}{c^2} = 1-\frac{1}{c^2}\left[\frac{u'+V}{1+u'V/c^2}\right]^2$$

$$= \frac{\left(1+\frac{u'V}{c^2}\right)^2-\frac{1}{c^2}(u'+V)^2}{\left(1+\frac{u'V}{c^2}\right)^2}$$

$$= \frac{\left(1-\frac{u'^2}{c^2}\right)\left(1-V^2/c^2\right)}{(1+Vu'/c^2)^2} \quad \text{... (6)}$$

Similary using eq. (3) we get

$$1-\frac{u_2^2}{c^2} = \frac{\left(1-u'^2/c^2\right)\left(1-V^2/c^2\right)}{\left(1-u'V/c^2\right)^2} \quad \text{... (7)}$$

Dividing eq. (7) by eq. (6), we get

$$\frac{\left(1-u_2^2/c^2\right)}{\left(1-u_1^2/c^2\right)}=\left[\frac{1+u'V}{1-\frac{u'V}{c^2}}\right]^2$$

$$=\left(\frac{m_1}{m_2}\right)^2 \quad \text{[From eq. (5)]}$$

$$\therefore \quad \frac{m_1}{m_2}=\frac{\left(1-u_2^2/c^2\right)^{1/2}}{\left(1-u_1^2/c^2\right)^{1/2}} \quad \text{...(7)}$$

Relation (8) is valid for all values of u_1 and u_2. If we select $u' = 0$ for particle *B* in frame *S* [This is possible if $u' = V$]

In that case $u_2 = 0$

and $m_2 = m_0$ = rest mass of particle B.

Then eq. (8) reduces to

$$\frac{m_1}{m_0}=\frac{1}{\left(1-u_1{}^2/c^2\right)^{1/2}}$$

or $$m_1 = \frac{m_0}{\left(a - u_1^2/c^2\right)^{1/2}}$$

As we have taken the two masses exactly identically so the rest mass of other body may also be considered as m_0. This in other words mean that the above equation is valid as if applied to one and same particle.

Replacing m_1 by m and u_1 by v, we can write the general relation

$$\text{m} = \frac{m_0}{\left(1 - \text{v}^2/c^2\right)^{1/2}}$$

Where m_0 is the rest mass and m is the mass when the body is making with a velocity v.

So from relativity theory, the measurements of an object or an event will be different for different observers who are in relative motion. If the rest mass of a body be m_0, then the mass of the body in motion is given by

$$m = \frac{m_0}{\left(1 - \frac{v^2}{c^2}\right)^{1/2}} \quad \text{... (9)}$$

where $v \rightarrow$ velocity of motion of the body and $c \rightarrow$ velocity of light

From (9) it is clear that $m > m_0$ *i.e.*, an object will appear to have more mass when in motion than at rest.

Relative mass of electron in motion moving with different speeds is shown in graph (Fig. 5.10). The relative mass of the electron increases enormously when its velocity approaches the velocity of light.

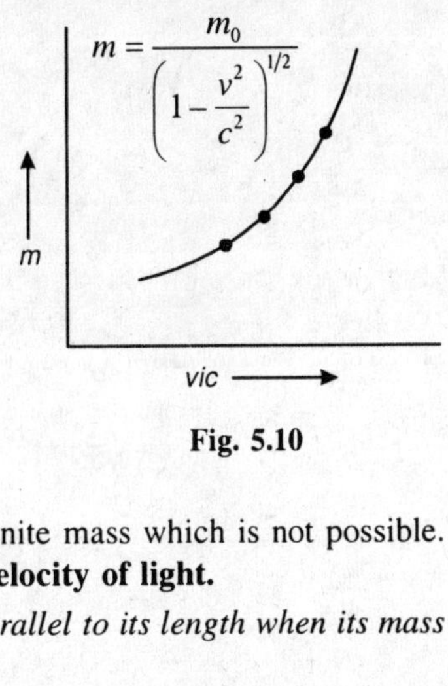

Fig. 5.10

If $$v = c$$

$$m = \frac{m_0}{\sqrt{1 - \frac{v^2}{c^2}}}$$

$$m = \frac{m_0}{0} = \infty$$

i.e., an electron moving with the speed of light will have an infinite mass which is not possible. **Hence no material particle can have a velocity equal to the velocity of light.**

Example 21. *What is the length of a metre stick moving parallel to its length when its mass is 3/2 of its rest mass.*

Solution. Here $$m = \frac{m_0}{\sqrt{1 - \frac{v^2}{c^2}}} \quad \text{... }(i)$$

and $$\frac{m}{m_0} = \frac{1}{\sqrt{1 - \frac{v^2}{c^2}}}$$

and $$L = L_0\sqrt{1 - \frac{v^2}{c^2}}$$

Dividing eq. (*ii*) by eq. (*i*)

$$\frac{L}{m} = \left(\frac{L_0}{m_0}\right)\left(1 - \frac{v^2}{c^2}\right)$$

$$L = \left(\frac{m_0}{m_0}\right) L_0 \left(1 - \frac{v^2}{c^2}\right)$$

$$\therefore \quad L = \left(\frac{m}{m_0}\right) L_0 \left(\frac{m_0}{m}\right)^2 = \left(\frac{m_0}{m}\right) L_0$$

Here $\frac{m_0}{m} = \frac{2}{3}$ and $L_0 = 1$

$$\therefore \quad L = \frac{2}{3} \times 1 = \mathbf{0.667\ m} \quad \textbf{Ans.}$$

5.17. Energy-Mass Relation

The increase in kinetic energy E_k when a force F displaces the body through a distance dS is given by

$$dE_k = F \cdot dS = F\left(\frac{dS}{dt}\right) dt = F\, v\, dt \quad \text{... (i)}$$

Also $$F = \frac{d}{dt}(mv)$$

or $$F dt = d\,(mv) \quad \text{... (ii)}$$

$\therefore$ $$dE_k = vd\,(mv) \quad \text{... (iii)}$$

or $$dE_k = vd\left[\frac{m_0 v}{\left(1 - \frac{v^2}{c^2}\right)^{1/2}}\right]$$

$$dE_k = vm_0\left[\left(1 - \frac{v^2}{c^2}\right)^{1/2} + \frac{v^2 / c^2}{\left(1 - \frac{v^2}{c^2}\right)^{1/2}}\right]\frac{dv}{\left(1 - \frac{v^2}{c^2}\right)}$$

or $$dE_k = \frac{m_0 v dv}{\left(1 - \frac{v^2}{c^2}\right)^{3/2}} \quad \text{... (iv)}$$

Also $$m = \frac{m_0}{\left(1 - \frac{v^2}{c^2}\right)^{1/2}}$$

Differentiating, we get

$$dm = \frac{m_0 v dv}{c^2\left(1-\frac{v^2}{c^2}\right)^{3/2}} \qquad \text{... (v)}$$

From equations (*iii*) and (*v*)

$$dm = \frac{dE_k}{c^2}$$

or $$dE_k = c^2\, dm \qquad \text{... (vi)}$$

Now consider that the mass of body at rest is m_0 and when it is moving with a velocity v it is m.

Integrating equation (*vi*)

$$\int dE_k = c^2 \int_{m_0}^{m} dm$$

$$E_k = c^2\,(m - m_0)$$

Total energy = kinetic energy of moving body + energy at rest

$$E = E_k + m_0\, c^2$$

$$E = c^2\,(m - m_0) + m_0\, c^2$$

$$E = mc^2 \qquad \text{... (viii)}$$

which gives us a relation between mass and energy.

5.18. Energy-Momentum Relation

If a body of mass m is moving with a velocity v, its momentum is given by

$$p = mv \qquad \text{... (i)}$$

Also energy is given by the relation ... (*ii*)

$$E = mc^2$$

and $$m = \frac{m_0}{\sqrt{1-\frac{v^2}{c^2}}} \qquad \text{... (iii)}$$

$$\therefore \quad E^2 - c^2p^2 = m^2c^4 - c^2m^2v^2$$

$$E^2 - c^2p^2 = \frac{m_0^2c^4}{\left(1-\frac{v^2}{c^2}\right)} - \frac{c^2m_0^2v^2}{\left(1-\frac{v^2}{c^2}\right)} \qquad \text{... (iv)}$$

$$E^2 - c^2p^2 = \frac{m_0^2c^4\left[1-\frac{v^2}{c^2}\right]}{\left(1-\frac{v^2}{c^2}\right)}$$

$$E^2 - c^2p^2 = m_0^{\,2}c^4$$

$$\mathbf{E^2 = m_0^{\,2}c^4 + c^2p^2} \quad \textbf{Ans.}$$

Example 22. *Calculate the percentage contraction of a rod moving with a velocity of 0.8c in a direction inclined at 60° to its own length.*

Solution. Let L_0 be the length of the rod in the frame S in which it is at rest. Let S' be the frame moving with a speed of 0.8 c in a direction making an angle 60° with X-axis. The components of L along and perpendicular to the direction of motion are L_0 cos 60° respectively.

We have the apparent length of the rod along the direction of motion as

$$= L_0 \cos 60° \sqrt{(1-(0.8)^2)} \quad [\because \; L = L_0\sqrt{(1-v^2/c^2)}]$$

$$= \frac{L_0}{2}\sqrt{(1-0.64)} = \frac{L_0}{2}\times 0.6 = 0.3\ L_0$$

and apparent length in a direction perpendicular to the direction of motion as

$$= L_0 \sin 60° = \frac{\sqrt{3}}{2}L_0 \text{ (as no change of length occurs in perpendicular direction)}$$

So, length of the moving rod, $L = \left[(0.3L_0)^2 + \left(\frac{L_0\sqrt{3}}{2}\right)^2\right]^{1/2} = \mathbf{0.916\ L_0}$ **Ans.**

Percentage contraction in Length $= \dfrac{L_0 - L}{L_0}\times 100 = \dfrac{L_0 - 0.916L_0}{L_0}\times 100$

$\therefore$ % contraction = **8.4% Ans.**

Example 23. *A vector in system S' is represented by 8i + 6j. How can the vector be represented in system S while S' is moving with velocity 0.8 ci with respect to S. i and j are unit vectors along the respective directions.*

Solution. As per problem system S' is moving with a velocity 0.8 c along X-axis. So, the length of the vector will change only along X-axis while its length along Y-axis will remain the same.

From Lorentz length contraction relation

$$L = L_0\sqrt{(1-v^2/c^2)}$$

Along X-axis, $L_0 = 8$ and $v = 0.8\ c$

$$\therefore \quad L = 8\sqrt{\left[1-\left(\frac{0.8c}{c}\right)^2\right]} = 8\times\sqrt{0.36} = 4.8$$

So, in the system S the vector may be represented by

4.8*i* + 6*j* Ans.

Example 24. *The average life-time of a neutron as a free particle at rest is 15 minutes. It disintegrate spontaneously into an electron and proton. What is the average minimum velocity with which a neutron must leave the sun in order to reach the earth before breaking up? Distance of sun from earth is equal to 11 × 10^{10} metres.*

Solution. Let t_0 be the average life time of a neutron in its own frame of reference, t that in the earth's frame of reference and v the average minimum velocity of neutron with which it must leave the sun, from time dilation relation

$$t = \frac{t_0}{\sqrt{(1-v^2/c^2)}}$$

In this case, since S = vt, So $t = 11\times 10^{10}/v$, t_0 = 15 minutes = 15 × 10 sec and $c = 3\times 10^8$ m/sec

$$\therefore \quad \frac{11\times 10^{10}}{v} = \frac{15\times 60}{\sqrt{[1-v^2/(3\times 10^8)^2]}}$$

Squaring both sides, we have

$$\frac{(11\times10^{10})^2}{v^2}=\frac{(15\times60)^2}{[1-v^2/(3\times10^8)^2]}$$

$$\frac{\left(1-\dfrac{v^2}{9\times10^{16}}\right)}{v^2}=\left(\frac{15\times60}{11\times10^{10}}\right)^2=6.694\times10^{-17}$$

$$\frac{1}{v^2}-\frac{1}{9\times10^{16}}=6.694\times10^{-17}$$

or $$\frac{1}{v^2}=6.694\times10^{-17}+1.111\times10^{-17}$$

$\therefore$ $$v^2=\frac{1}{7.805\times10^{-17}}=1.281\times10^{16}$$

and $$v = \mathbf{1.132\times10^8\ ms^{-1}}\ \textbf{Ans.}$$

Example 25. *A beam of particles of half life 2.0 × 10^{-8} sec travels in the laboratory with speed 0.96 times the speed of light. How much distance does the beam travel before the flux of the beam falls to 1/2 times the initial flux?*

Solution. Now from the problem, the actual half-life of the particles, that is, time-interval in the particle's own frame of reference in which its flux reduces to half of its initial flux is 2.0 × 10^{-8} sec. The ordinary half life t in the laboratory frame and actual half life t_0 of the particles are related according to time dilation formula is

$$t=\frac{t_0}{\sqrt{(1-v^2/c^2)}},\ \text{where } v = 0.96c$$

$\therefore$ $$t=\frac{2.0\times10^{-8}}{\sqrt{1-\left(\dfrac{0.96c}{c}\right)^2}}=\frac{2.0\times10^{-8}}{\sqrt{1-0.92}}=\frac{2.0\times10^{-8}}{0.283}$$

$$= 7.1\times10^{-8}\ \text{sec}$$

So, the distance travelled by the beam in this time in the laboratory frame

$$= vt = 0.96\times3\times10^8\times7.1\times10^{-8}\ \text{meters.}$$

$$= \mathbf{20.45\ m}\ \textbf{Ans.}$$

Example 26. *The mean life of a meson is 2 × 10^{-8} sec. Calculate the mean life of a meson moving with velocity 0.8 c.*

Solution. Let mean life of meson in its own frame of reference be to and let t that measured by an observer stationary frame, then from time dilation relation

$$t=\frac{t_0}{\sqrt{1-v^2/c^2}},\ \text{In this case } t_0 = 2\times10^{-8}\ \text{sec and } v = 0.8\ c$$

$\therefore$ $$t=\frac{2\times10^{-8}}{\sqrt{1-(0.8c/c)^2}}=\frac{2\times10^{-8}}{0.6}=\mathbf{3.33\times10^{-8}\ sec.}\ \textbf{Ans.}$$

Example 27. *A clock keeps correct time. With what speed should it be moved relative to an observer so that it may appear to lose 4 minutes in 24 hours?*

Solution. Let t_0 be the proper time interval on a clock at rest relative to an observer in clock frame and t that on clock in motion w.r.t to the stationary observer or clock at rest, then from to the time dilation formula

$$t = \frac{t_0}{\sqrt{(1 - v^2 / c^2)}} \quad \text{... (1)}$$

where v is the speed of the clock moving relative to an observer.

In this case Here $t_0 = 24 \times 60 = 1440$ min. is the proper time-interval measured by an observer moving with the clock and $t = 1444$ min is the same time-interval measured by a stationary observer. So, the moving clock appears to lose 4 min in 24 hours to the stationary observer.

We have from eq. (1)

$$1444 = \frac{1440}{\sqrt{(1 - v^2 / c^2)}} \text{ or } 1444 = 1440 \left(1 - \frac{v^2}{c^2}\right)^{-1/2}$$

Expanding by binomial theorem

$$1444 = 1440\left(1 + \frac{v^2}{2c^2} + ...\right)$$

For $v/c << 1$, the higher order terms may be neglected, So we have

$$\therefore \quad 1444 = 1440\left(1 + \frac{v^2}{2c^2}\right)$$

$$1 + \frac{v^2}{2c^2} = \frac{1444}{1440} = 1.003$$

or $$\frac{v^2}{2c^2} = 0.003 \text{ or } v^2 = 0.003 \times 2c^2$$

or $$v^2 = 0.003 \times 2 \times (3 \times 10^8)^2$$

or $$v = (5.4 \times 10^{14})^{1/2} = \mathbf{2.32 \times 10^7 \text{ m/sec Ans.}}$$

Example 28. *A man leaves the earth in a rocket ship that makes a round trip to the nearest star which is 4 light years away at a speed of 0.8 c. How much younger will the be on his return that his twin brother who preferred to stay behind?*

Solution. From the time dilation formula, the proper time t_0 and apparent time t are related to each other as

$$t_0 = t\sqrt{1 - v^2 / c^2}$$

The distance to the star is 4 light years and the speed is 0.8 c. So, for a person remaining on earth

$$t = \frac{2 \times 4 \text{ ly}}{0.8c} = 10 \text{ years}$$

According to the astronaut's clock

$$t_0 = \frac{2 \times 4 \text{ ly} \sqrt{1 - \frac{v^2}{c^2}}}{0.8c} = 10\sqrt{1 - \left(\frac{0.8c}{c}\right)^2} = 10 \times 0.6 = 6 \text{ years}$$

So $$10 - 6 = \mathbf{4 \text{ years Ans.}}$$

Example 29. *A wrist watch, keeping correct time on earth, is worn by the pilot of a space-ship. How much will it appear to lose per day with respect to an observer on the earth when spaceship leaves the earth with a constant velocity of 10^7 m/sec.*

Solution. Let t_0 be the proper time and t the apparent time of a wrist watch, from time dilation formula,

$$t = \frac{t_0}{\sqrt{(1 - v^2/c^2)}}$$

In this case $t = 24$ hours, $v = 10^7$ m/sec and $c = 3.0 \times 10^8$ m/sec

$$\therefore \quad 24 = \frac{t_0}{\sqrt{\left[1 - \left(\frac{10^7}{3 \times 10^8}\right)^2\right]}} \text{ or } t_0 = 24\left(1 - \frac{1}{900}\right)^{1/2}$$

$$\therefore \quad t_0 = 24 \times 0.9994 = 23.986 \text{ hrs}$$

Hence loss of time per day = 24 – 23.986 = 0.014 hrs = **50.4 sec Ans.**

Example 30. *Two spacecrafts A and B, moving in opposite directions are approaching the moon with respecting speeds of 2.4×10^8 m/sec and 2.7×10^8 m/sec as observed by an observer at the moon. Calculate the speed of A with which it is approaching (i) the moon (ii) the spacecraft B, as observed by an observer in the spacecraft A.*

Solution. (*i*) When the observer is at the spacecraft *A*, the relative velocity of *A* w.r.t an observer in spacecraft *A* will be zero, *i.e.*, $u' = 0$. The speed of spacecraft *A*, *u* with which it is approaching the moon is given by the law of addition of relativistic velocities in the form

$$u = \frac{u' + v}{1 + (u'v/c^2)} = \frac{0 + v}{1} = v = 2.4 \times 10^8 \text{ m/sec}$$

(*ii*) Let u' be the velocity of spacecraft *A* w.r.t the spacecraft *B*. The speeds of space crafts *A* and *B* moving relative to the moon are 2.4×10^8 m/sec and -2.7×10^8 m/sec respectively. So according to the relativistic velocity addition formula, we have

$$2.4 \times 10^8 = \frac{u' - 2.7 \times 10^8}{1 + \frac{-2.7 \times 10^8 \times u'}{(3 \times 10^8)^2}} = \frac{u' - 2.7 \times 10^8}{1 - 0.3 \times 10^{-8} u'}$$

or

$$2.4 \times 10^8 (1 - 0.3 \times 10^{-8} u') = u' - 2.7 \times 10^8$$

$$(1 + 2.4 \times 10^8 \times 0.3 \times 10^{-8})u' = 2.4 \times 10^8 + 2.7 \times 10^8$$

$$1.72u' = 5.1 \times 10^8$$

$$\therefore \quad u' = \mathbf{2.965 \times 10^8 \text{ ms}^{-1}} \text{ **Ans.**}$$

Example 31. *An experimenter observes a radioactive atom moving with a velocity of 0.25 c. The atom then emits a beta particle which has a velocity of 0.9 c relative to the atom in the direction of its motion. What is the velocity of the beta particle as observed by the experimenter?*

Solution. The velocity of β – particle , *u* as observed by the experimenter is found by using the law of addition of relativistic velocities in the form

$$u = \frac{u' + v}{1 + \frac{u'v}{c^2}}$$

In this case, the velocity of emitted β – particle relative to the atom in the direction of motion, $u' = 0.9$ and the velocity of the radioactive atom relative to an experimenter, $v = 0.25\ c$

$$\therefore \quad u = \frac{0.9c + 0.25c}{1 + \dfrac{0.9c \times 0.25c}{c^2}} = \frac{1.15c}{1.225} = 0.94\ \text{c}$$

So, the velocity of the β-particle as observed by the experimenter is 0.94 *c* Ans.

Example 32. *Calculate the relativistic energy and momentum of a proton which is moving with a speed 2.4 × 10⁸ m/sec.*

Solution. From mass-energy relation, the relativistic energy *E* of a proton may be written as

$$E = mc^2 = \frac{m_0 c^2}{\sqrt{(1 - v^2/c^2)}}$$

In this case $v = 2.4 \times 10^8$ m/sec and mass of proton, $m_0 = 1.673 \times 10^{-27}$ kg

$$\text{So} \quad E = \frac{1.673 \times 10^{-27} \times (3 \times 10^8)^2}{\sqrt{\left[1 - \left(\dfrac{2.4 \times 10^8}{3 \times 10^8}\right)^2\right]}} = \frac{1.673 \times 10^{-27} \times 9 \times 10^{16}}{\sqrt{1 - 0.64}} = 2.51 \times 10^{-10}\ \text{J}$$

Now momentum of proton may be written as

$$m = \frac{m_0}{\sqrt{1 - \dfrac{v^2}{c^2}}}$$

Here $m_0 = 10^{-24}$ kg; $v = 1.8 \times 10^8$ m/s; $c = 3 \times 10^8$ m/s; $\dfrac{v}{c} = 0.6$

$$m = \frac{10^{-24}\ \text{kg}}{\sqrt{1 - (0.6)^2}}; \quad \mathbf{m = 1.25 \times 10^{-24}\ kg.\ Ans.}$$

Example 33. *Show that* $\left(\dfrac{E^2}{c^2}\right) - p^2$ *is invariant.*

Solution. Let us consider two frames of reference *A* and *B* and let frame *B* be moving with a velocity *v* along the *X*-axis with respect to frame *A*.

Let *p* be the momentum and *E* the energy of a particle in reference frame *A* and *p′* and *E′* in reference frame *B*.

$$\therefore \quad p^2 = p_x^2 + p_y^2 + p_z^2 \qquad \text{... (i)}$$

$$\text{and} \quad p'^2 = p_x'^2 + p_y'^2 + p_z'^2 \qquad \text{... (ii)}$$

$$\lambda = \frac{1}{\sqrt{1 - \dfrac{v^2}{c^2}}} \qquad \text{... (iii)}$$

$$p'_y = p_y, p'_z = p_z$$

and $$E' = (E - vp_x)$$

$$\therefore \quad \frac{E'^2}{c^2} - p'^2 = \frac{\lambda^2}{c^2}(E - vp_x)^2 - [p_{x'}{}^2 + p_{y'}{}^2 + p_{z'}{}^2]$$

$$= \frac{\lambda^2}{c^2}(E - vp_x)^2 - \lambda^2\left(p_x - \frac{vE}{c^2}\right)^2 - p_y^2 - p_z^2$$

$$= \frac{\lambda^2}{c^2}\left[(E - vp_x)^2 - c^2\left(p_x - \frac{vE}{c^2}\right)^2\right] - p_y^2 - p_z^2$$

$$= \frac{\lambda^2}{c^2}\left[E^2 + v^2p_x^2 - 2vEp_x - p_x^2c^2 - \frac{v^2E^2}{c^2} + 2vEp_x - p_y^2 - p_z^2\right]$$

$$= \frac{\lambda^2}{c^2}\left[\left(E^2 - \frac{v^2E^2}{c^2}\right) + (-p_x^2c^2 + v^2p_x^2)\right] - py^2 - pz^2$$

$$= \frac{\lambda^2}{c^2}\left[E^2\left(1 - \frac{v^2}{c^2}\right) - p_x^2c^2\left(1 - \frac{v^2}{c^2}\right)\right] - p_y^2 - p_z^2$$

But $$\left(1 - \frac{v^2}{c^2}\right) = \frac{1}{\lambda^2}$$

$$\therefore \quad \frac{E'^2}{c^2} - p'^2 = \frac{\lambda^2}{c^2}\left[\frac{E^2}{\lambda^2} - \frac{p_x^2c^2}{\lambda^2}\right] - py^2 - pz^2$$

$$= \frac{E^2}{c^2} - p_x^2 - p_y^2 - p_z^2 = \frac{E^2}{c^2} - (p_x^2 + p_y^2 + p_z^2)$$

$$\frac{E^2}{c^2} - p'^2 = \frac{E^2}{c^2} - p^2 \quad \text{... (iv)}$$

Which shows that

$\frac{E^2}{c^2} - p^2$ **is an invariant. Ans.**

Example 34. *A particle of mass 10^{-24} kg is moving with a speed of 1.8×10^8 m/s. Calculate its mass when it is motion.*

Solution. $$m = \frac{m_0}{\sqrt{1 - \frac{v^2}{c^2}}}$$

Here $m_0 = 10^{-24}$ kg V = 1.8×10^8 m/sic = 3×10^8 ms; $\frac{v}{e} = 0.6$

$$m = \frac{10^{-24} \text{ kg}}{\sqrt{1 - (0.6)^2}}$$; $m = \mathbf{1.25 \times 10^{-24}}$ **kg Ans.**

5.19. Massless Particles

A particle which has zero rest mass (m_0) is called a massless particle. In classical physics, the existence of massless particle is not possible, but in relativistic mechanics, a particle with zero rest mass can exist.

The relativistic total energy E of a particle of rest mass m_0 in terms of its momentum p may be expressed as

$$E = \sqrt{m_0^2 c^4 + p^2 c^2}$$

For massless particle, $m_0 = 0$

So $E = pc$ or $p = \frac{E}{c}$

Since p is also equal to ($mv = Ev/c^2$, therefore, $v = c$, that is, the velocity of massless particle is same as that of light in free space.

The energy of the particle, $E = pc = mc^2$,

where m is the mass equivalent to energy E, that is, $m = E/c^2$.

From above, it is clear that every massless particle has energy pc and momentum E/c and it moves with the velocity of light.

The mass of the particle is F/c^2, *i.e.*, is massless particle has mass only when it is in motion. If it is stopped, then it ceases to exist *i.e.*, either it is absorbed completely or changed to heat at the surface. *Neutrinos photons and gravitons* are examples of massless particles.

5.20 Relativistic Relation between Kinetic Energy and Momentum

Kinetic energy of a parallel of mass m_0 is relativity to its to energy as

$$K = E - m_0 c^2$$

also $E = \sqrt{(m_0^2 c^4 + p^2 c^2)}$

So $K = \sqrt{(m_0^2 c^4 + p^2 c^2)} - m_0 c^2$

or $$K = \left[\left(1 + \frac{p^2}{m_0^2 c^2}\right)^{1/2} - 1\right] m_0 c^2$$

This is the required relativistic relation between kinetic energy and momentum of a particle. when $v << c$; *i.e.*, $p << mc^2$, in that case

$$K \cong m_0 c^2 \left[\left(1 + \frac{1}{2}\frac{p^2}{m_0^2 c^2} + \ldots\right) - 1\right]$$

Neglecting higher order terms of binomial expansion, we may write the expression for kinetic energy as

$$K = \frac{p^2}{2m_0}$$

If $v << c$, in that case $m = m_0$

So kinetic energy $K = \frac{p^2}{2m}$

which shows that for low values of velocity, kinetic energy expression reduces to classical value.

Example 35. *A particle is moving with a speed of 0.5 c. Calculate the ratio of the rest mass and the mass while in motion.*

Solution. $m = \dfrac{m_0}{\sqrt{1-\dfrac{v^2}{c^2}}}$ or $\dfrac{m_0}{m} = \sqrt{1-\dfrac{v^2}{c^2}}$

Here $v = 0.5\ c$

or $\dfrac{v}{c} = 0.5;\ \dfrac{m_0}{m} = \mathbf{0.866\ Ans.}$

Example 36. *The mean life of meson is 2×10^{-8} s. Calculate the mean life of a meson moving with a velocity of 0.8 c.*

Solution. $t = \dfrac{t_0}{\sqrt{1-\dfrac{v^2}{c^2}}}$

Here $t_0 = 2 \times 10^{-8}$ s; $v = 0.8\ c$

$$t = 2\times10^{-8} / \sqrt{1-(0.8)^2} = \frac{2\times10^{-8}}{0.6}$$

$$t = \mathbf{3.33 \times 10^{-8}\ s\ Ans.}$$

Example 37. *Calculate the velocity that one atomic mass unit will have if it a kinetic energy equal to twice the mass energy.*

Solution. Here $2m_0c^2 = mc^2 - m_0c^2$

$$3m_0c^2 = mc^2$$

$$m = 3m_0$$

but $m = \dfrac{m_0}{\sqrt{1-\dfrac{v^2}{c^2}}}\ \therefore\ 3m_0 = \dfrac{m_0}{\sqrt{1-\dfrac{v^2}{c^2}}}$

or $1-\dfrac{v^2}{c^2} = \dfrac{1}{9} \Rightarrow v^2 = \left(\dfrac{8}{9}\right)c^2 \Rightarrow v = \mathbf{0.941\ c\ Ans.}$

Example 38. *The rest mass of a proton is 2000 times the rest mass of an electron. Calculate the speed at which the electron should move so that its mass will be equal to the rest mass of the proton.*

Solution. $c = 3\times10^8$ m/s

$$m = \frac{m_0}{\sqrt{1-\dfrac{v^2}{c^2}}}$$

$$\frac{m}{m_0} = \frac{1}{\sqrt{1-\dfrac{v^2}{c^2}}}$$

Here $\dfrac{m}{m_0} = 2000$

$$2000 = \frac{1}{\sqrt{1-\frac{v^2}{c^2}}}$$

$$1-\frac{v^2}{c^2} = \left(\frac{1}{2000}\right) = \frac{1}{4\times10^6}$$

$$\frac{v^2}{c^2} = \left(1-\frac{1}{4\times10^6}\right)$$

$$\frac{v}{c} = \left(1-\frac{1}{4\times10^6}\right)^{1/2} = \left[1-\frac{1}{2\times4\times10^6}\right]$$

$$v = [1 - 1.25 \times 10^{-7}]c$$

$$v = (1 - 1.25 \times 10^{-7}) \times 3 \times 10^8$$

$$v = \mathbf{3 \times 10^8\ m/s^{-1}}$$ **Ans.**

Example 39. *The density of gold is 19.3 × 10³ kg/m³ in a frame S which is at rest. Calculate its density that an observer in a frame S′ would determine if the frame S′ is moving along the x-axis with a speed of 0.9 c.*

Solution. (*i*)
$$L = L_0\sqrt{1-\frac{v^2}{c^2}}$$

Here
$$\frac{v}{c} = 0.9$$

$$L = L_0\sqrt{1-(0.9)^2}$$

$$L = 0.4359\ L_0$$

Volume
$$V_0 = L_0 \times y \times z$$

$$V = L \times y \times z$$

$$= 0.4359\ L_0 \times y \times z$$

or
$$V = 0.4359\ V_0$$

Mass
$$m = \frac{m_0}{\sqrt{1-\left(\frac{v^2}{c^2}\right)}}$$

$$m = \frac{m_0}{\sqrt{1-(0.9)^2}} = \frac{m_0}{0.4559}$$

Original density,
$$\rho_0 = \frac{m}{V_0}$$

$$\frac{\rho}{\rho_0} = \left(\frac{m}{m_0}\right)\left(\frac{V_0}{V}\right) = \left(\frac{1}{0.4359}\right)\times\left(\frac{1}{0.4359}\right)$$

$$\rho = 19.3\times10^3\left(\frac{1}{0.4359}\right)^2 = \mathbf{101.6 \times 10^3\ kg/m^3}$$ **Ans.**

Example 40. *Electrons are accelerated upto a kinetic energy of 10^9 eV. Find (i) the ratio of their mass to the rest mass (ii) the ratio of their velocity to the velocity of light and (iii) the ratio of their energy to the rest mass energy.*

Solution. $E = 10^9 \text{ eV} = 10^9 \times 1.6 \times 10^{-19}$ J

Also $E = mc^2$

$$m = \frac{E}{c^2} = \frac{10^9 \times 1.6 \times 10^{-19}}{(3\times10^4)^2} = 1.77 \times 10^{-27} \text{ kg}$$

Rest mass of electron,

$$t\, m_0 = 9 \times 10^{-31} \text{ kg}$$

(*i*) $$\frac{m}{m_0} = \frac{1.77\times10^{-27}}{9\times10^{-31}} = 1.95\times10^3$$

(*ii*) $$m = \frac{m_0}{\sqrt{1-\left(\frac{v^2}{c^2}\right)}}$$

$$\frac{m}{m_0} = \frac{1}{\sqrt{1-\left(\frac{v^2}{c^2}\right)}}$$

$$1.95\times10^3 = \frac{1}{\sqrt{1-\left(\frac{v^2}{c^2}\right)}}$$

$$1-\left(\frac{v^2}{c^2}\right) = \left(\frac{1}{1.95\times10^3}\right)^2$$

$$\frac{v}{c} = \left[1-\left(\frac{1}{1.95\times10^3}\right)^2\right]^{1/2}$$

$$= \left(1-\frac{1}{2\times(1.95)^2\times10^6}\right) = [1 - 1.315 \times 10^{-7}] \simeq 1$$

(*iii*) Rest mass energy

$$E_0 = m_0c^2$$

$$E_0 = 9 \times 10^{-31} \times (3 \times 10^8)^2 = \mathbf{8.1 \times 10^{-14}\ J}$$

$$\frac{E}{E_0} = \frac{10^3\times1.6\times10^{-19}}{8.1\times10^{-14}} = \mathbf{1.975 \times 10^{-3}}.\ \textbf{Ans.}$$

Example 41. *The mean life time of a* μ *meson when it is at rest is* 2.2×10^{-6} *sec. Calculate the average distance it will travel in vacuum before decay, if it has velocity 0.9 c.*

Solution. Let t be the mean life period of meson, then the distance travelled by it before decay

$$= \gamma . v . t$$

where $$\gamma = \frac{1}{\sqrt{1-v^2/c^2}} = \frac{1}{\sqrt{1-81/100}} = \frac{10}{\sqrt{19}} = \frac{\mathbf{10}}{\mathbf{4.35}}$$

$\therefore$ Average distance travelled before decay

$$= \frac{10}{4.35} \times 0.9 \times 3 \times 10^{10} \times 2.2 \times 10^{-6} \text{ cm}$$

$$= \mathbf{1.36 \text{ km. Ans.}}$$

Example 42. *A particle in a stationary frame S lies in the X-Y plane and has a velocity 0.8 c, inclined at an angle of 60° to the axis of X. What will be the velocity of the particle, as observed by a person in a frame S' moving relative to frame S with a velocity 0.4 cm/sec along the X-axis?*

Solution. The velocity of the particle in the stationary frame S is $u = 0.8\ c$, components of velocity along two axis are given by

$$u_x = 0.8\ c \cos 60° = 0.8\ c \times \frac{1}{2} = 0.4\ c$$

and $$u_y = 0.8\ c \sin 60° = 0.8\ c \times (\sqrt{3/2}) = 0.4\sqrt{(3)}c$$

Given the velocity of moving frame S' relative to the stationary frame of reference S, $v = 0.4\ c$ along the X-axis.

From the law of addition of velocities, the values of velocity components, u'_x and u'_y as observed by the person in frame S' may be written as

$$u'_x = \frac{u_x - v}{(1 - u_x v / c^2)} = \frac{0.4c - 0.4c}{1 - \frac{0.4c \times 0.4c}{c^2}} = 0 \text{ and } u'_y = \frac{dy'}{dt'}$$

From Lorentz transformation equations: $y = y'$ and $t' = \dfrac{t - vx/c^2}{\sqrt{(1 - v^2/c^2)}}$

Which on differentiation will give $dy = dy'$ and $dt' = \dfrac{dt - vdx/c^2}{\sqrt{(1 - v^2/c^2)}}$

So $$u'_y = \frac{dy'}{dt'} = \frac{dy\sqrt{(1 - v^2/c^2)}}{dt - (vdx/c^2)}$$

or $$u'_y = \frac{(dy/dt)\left[\sqrt{(1 - v^2/c^2)}\right]}{\left(1 - \frac{v}{c^2}\frac{dx}{dt}\right)}$$

$$= \frac{u_y\left[\sqrt{(1 - v^2/c^2)}\right]}{\left(1 - \frac{uv_x}{c^2}\right)}$$

$\therefore$ $$u'_y = \frac{0.4\sqrt{3}c\sqrt{[1 - (0.4c/c)^2]}}{1 - \frac{(0.4c)(0.4c)}{c^2}} = \frac{0.4\sqrt{3}c\sqrt{(1 - 0.16)}}{1 - 0.16} = \frac{0.4\sqrt{3} \times \sqrt{(0.84)}c}{0.84}$$

or $$u'_y = \frac{0.4\sqrt{3}c}{\sqrt{0.84}} = \frac{0.6928c}{0.916} = 0.756\ c$$

So, the resultant velocity of the particle $= \sqrt{(u_x'^2 + u_y'^2)} = u'_y$ $(\because\ u'_x = 0)$

Hence, the velocity of a particle as observed by the person in frame S' = 0.756 c along y-axis. Ans.

Example 43. *A airforce rocket is chasing enemy's spaceship. From each, it is found that the speed of airforce rocket is 2.55 × 10[10] cm/sec while that of enemy's ship is 2.25 × 10[10] cm/sec. What is the relative velocity of*

(i) Enemy's ship as seen by airforce rocket.
(ii) Airforce rocket as seen by enemy's ship.
(iii) Airforce rocket with respect to enemy's ship as seen from earth.
(iv) Enemy's ship with respect to airforce rocket as seen from earth.

Solution. From the addition of relativistic velocities relation

$$u = \frac{u' + v}{1 + \dfrac{u'v}{c^2}}$$

In this case $u' = 2.55 \times 10^{10}$ cm/sec, $v = 2.25 \times 10^{10}$ cm/sec and $c = 3.0 \times 10^8$ m/sec = 3.0 × 10^{10} cm/sec

$$\therefore \qquad u = \frac{2.55\times10^{10} - 2.25\times10^{10}}{\left[1 - \dfrac{2.55\times10^{10}\times2.25\times10^{10}}{(3.0\times10^{10})^2}\right]} = \frac{0.30\times10^{10}}{(1-0.6375)} = 0.83 \times 10^{10}\ \text{cms}^{-1}$$

or $$= 0.83 \times 10^8\ \text{ms}^{-1}$$

(*i*) The relative velocity of enemy's ship as seen by airforce rocket will be **0.83 × 10^8 ms^{-1}. Ans.**

(*ii*) Similarly, the relative velocity of airforce rocket as seen by enemy's ship moving towards him, is also 0.83 × 10^8 ms^{-1}. **Ans.**

(*iii*) The relative velocity of airforce rocket with respect to enemy's ship as seen from the earth is

$$= u' - v = 2.55\times10^8 - 2.25\times10^8 = \mathbf{0.3 \times 10^8\ ms^{-1}}.\ \textbf{Ans.}$$

So, the observer situated at earth assuming enemy's ship at rest will see the rocket is moving towards the enemy's ship with a velocity **0.3 × 10^8 ms^{-1}.**

(*iv*) Considering airforce rocket to be at rest, the observer will see that the enemy's ship is approaching the airforce rocket with a velocity **0.3 × 10^8 ms^{-1}. Ans.**

Example 44. *A particle has a velocity, $\vec{u}' = 3i + 4j + 12k$ metre/sec in a coordinate system moving with velocity 0.8 c relative to laboratory along + ve direction of x-axis. Find $\vec{u}$ in laboratory frame.*

Solution. Given $\vec{u}' = 3i + 4j + 12k$, which means $u'_x = 3$, $u'_y = 4$, $u'_z = 12$ and $v = 0.8\,c$

From law of addition of velocities, the X, Y and Z components of $\vec{u}$ in the laboratory frame have the form

$$u_x = \frac{u'_x + v}{1 + \dfrac{vu'_x}{c^2}} = \frac{3 + 0.8c}{1 + \dfrac{(0.8c)(3)}{c^2}} = \frac{(3+0.8c)c}{(c+2.4)} \approx 0.8\,c = \mathbf{2.4 \times 10^8\ m/sec}\ \textbf{Ans.}$$

$$u_y = \frac{u'_y\sqrt{(1-v^2/c^2)}}{1+\dfrac{vu'_x}{c^2}} = \frac{4\sqrt{[1-(0.8c/c)^2]}}{1+\dfrac{(0.8c)(3)}{c^2}} = \frac{4\times\sqrt{(0.36)}c}{(c+2.4)} = \frac{4\times 0.6c}{c+2.4} \approx \frac{2.4c}{c} = \textbf{2.4 m/sec}$$

and $$u_z = \frac{u'_z\sqrt{(1-v^2/c^2)}}{1+\dfrac{vu'_x}{c^2}} = \frac{12\sqrt{[1-(0.8c/c)^2]}}{1+\dfrac{(0.8c)(3)}{c^2}} = \frac{12\times 0.6c}{c+2.4} \approx \frac{7.2c}{c} = \textbf{7.2 m/sec Ans.}$$

Example 45. *At what speed will the mass of a body be 2.25 times its rest mass?*

Solution. Let at a speed v, the mass of an object is 2.25 times to its value at rest. We have the relation

$$m = \frac{m_0}{\sqrt{(1-v^2/c^2)}}$$

In this case $m = 2.25\, m_0$ $\therefore$ $2.25\, m_0 = \dfrac{m_0}{\sqrt{(1-v^2/c^2)}}$

or $$1-\frac{v^2}{c^2} = \frac{1}{5.06} \quad \text{or} \quad \frac{v^2}{c^2} = 1-\frac{1}{5.06} = \frac{4.06}{5.06}$$

$\therefore$ $$v = \sqrt{\frac{4.06}{5.06}}\times c = \frac{3\times 10^8\times\sqrt{4.06}}{\sqrt{5.06}} = \mathbf{2.68\times 10^8\ m/sec\ Ans.}$$

Example 46. *A charged particle shows an acceleration of 4.2 × 10^{12} cm/sec^2 under an electric field at low speed. Compute the acceleration of the particle under the same field when the speed has reached a value 2.88 × 10^{10} cm/sec. The speed of light is 3.0 × 10^{10} cm/sec.*

Solution. The force acting on a charged particle due to an electric field E is given by

$$F = qE \qquad \text{... (1)}$$

At low speed ($v << c$) the effective mass of the particle is equal to its rest mass m_0. So the actual at low speed may be written as

$$a_0 = \frac{F}{m_0} = \frac{qE}{m_0} = 4.2\times 10^{12}\ \text{cm/sec}^2 \qquad \text{... (2)}$$

When the speed of the particle has reached a value 2.88 × 10^{10} cm/sec, its mass increases with velocity to an effective value m, as v is comparable to c. So, we have

$$m = \frac{m_0}{\sqrt{(1-v^2/c^2)}} = \frac{m_0}{\sqrt{\left[1-\left(\dfrac{2.88\times 10^{10}}{3.00\times 10^{10}}\right)^2\right]}} = \frac{m_0}{\sqrt{1-0.92}} = \frac{m_0}{\sqrt{0.08}} = \frac{m_0}{0.28}$$

Hence acceleration at high speed ($v \approx c$), $a = \dfrac{F}{m} = \dfrac{F}{(m_0/0.28)} = \dfrac{0.28F}{m_0}$

Putting the value of F/m_0 from equation (2), we get

$$a = 0.28\times 4.2\times 10^{12} = \mathbf{1.176\times 10^{12}\ cm/sec^2\ Ans.}$$

Example 47. *A man weigh 50 kg on the earth. When he is in rocket ship in flight, his mass is 50.5 kg as measured by an observer on earth. What is the speed of the rocket?*

Solution. From the variation of relativistic mass with velocity relation

$$m = \frac{m_0}{\sqrt{(1 - v^2/c^2)}} \quad \therefore \quad \sqrt{\left(1 - \frac{v^2}{c^2}\right)} = \frac{m_0}{m}$$

or
$$V = c\sqrt{1 - \left(\frac{m_0^2}{m^2}\right)}$$

Putting $m_0 = 50$ kg, $m = 50.5$ kg and $c = 3.0 \times 10^8$ m/s in above relation

$$\therefore \quad v = 3\times10^8\sqrt{1-\left(\frac{50}{50.5}\right)^2} = 3 \times 10^8 \times 0.141$$

$$= \mathbf{4.23 \times 10^7 \ m/sec} \text{ **Ans.**}$$

Example 48. *A particle is moving with 90% of the velocity of light. Compare its relativistic mass with its rest mass.*

Solution. We have the relation

In this case $v = (90/100)\, c = 0.9\, c$, $\quad m = \dfrac{m_0}{\sqrt{1 - v^2/c^2}}$

where m_0 is the rest mass of the particle $m = \dfrac{m_0}{\sqrt{\left[1 - \dfrac{(0.9c)^2}{c^2}\right]}}$

or
$$\frac{m}{m_0} = \frac{1}{\sqrt{1-0.81}} = \frac{1}{\sqrt{0.19}} = \mathbf{2.294} \text{ **Ans.**}$$

Example 49. *A certain young lady decides on her twenty-fifth birthday that it is time to slenderize. She weights 100 kg. She has heard that if she moves fast enough, she will appear thinner to her stationary friends.*

(*i*) *How fast must she move to appear slenderized by a factor of 50%?*

(*ii*) *At this speed what will her mass appear to be to her stationary friends?*

(*iii*) *If she maintains her speed until the day she calls her twenty-ninth birthday, how old will her stationary friends claim she is according to their measurements?*

Solution. (*i*) From the dorentz length contraction formula

$$L = L_0\,(1 - v^2/c^2)^{1/2}$$

In this case the dimensions of young lady are reduced to 50%,

Which means 50% of L_0 or $L = \dfrac{L_0}{2}$

$$\therefore \quad \frac{L_0}{2} = L_0\sqrt{(1 - v^2/c^2)} \text{ or } 1 - \frac{v^2}{c^2} = \frac{1}{4}$$

or
$$\frac{v^2}{c^2} = 1 - \frac{1}{4} \text{ or } v^2 = \frac{3}{4}c^2 \text{ or } v = \frac{\sqrt{(3)}c}{2} = \mathbf{0.866\, c} \text{ **Ans.**}$$

(*ii*) From the variation of mass with velocity relation, the mass of a particle moving with velocity v is given by relation

$$m = \frac{m_0}{\sqrt{(1-v^2/c^2)}}$$

Putting $m_0 = 100$ kg nad $v = 0.866\ c$

$$\therefore \quad m = \frac{100}{\sqrt{\left[1-\left(\frac{0.866c}{c}\right)^2\right]}} = \frac{100}{\sqrt{1-(0.75)}} = \frac{100}{0.5} = 200 \text{ kg}$$

So, **mass appeared to her stationary friend = 200 kg Ans.**

(*iii*) Now from time dlation formula

$$t = \frac{t_0}{\sqrt{(1-v^2/c^2)}}, \text{ where } t_0 \text{ is the dilated time}$$

$$\therefore \quad = \frac{4}{\sqrt{1-\left(\frac{0.866c}{c}\right)^2}} = \frac{4}{0.5} = 8 \text{ years}$$

So **the lady will appear 25 + 8 = 33 years old Ans.**

Example 50. *How fast must an electron move in order that its mass equals the rest mass of the proton?*

Solution. We have the relation

$$m = \frac{m_0}{\sqrt{(1-v^2/c^2)}} \quad \text{or} \quad \sqrt{\left(1-\frac{v^2}{c^2}\right)} = \frac{m_0}{m}$$

or $$v = c\sqrt{1-\left(\frac{m_0}{m}\right)^2}$$

In this case $m_0 = m_e = 9.11 \times 10^{-31}$ kg, $m = m_p = 1.67 \times 10^{-26}$ kg and $c = 3 \times 10^8$ m/sec

$$\therefore \quad v = 3\times10^8\sqrt{1-\left(\frac{9.11\times10^{-31}}{1.67\times10^{-27}}\right)^2} = \mathbf{2.99 \times 10^8\ ms^{-1}\ Ans.}$$

$$= 2.99 \times 10^8 \text{ m/sec}$$

Example 51. *The rest mass of a proton is 2.67×10^{-27} kg. At what speed will its mass be double its rest mass?*

Solution. We have the relation

$$m = \frac{m_0}{\sqrt{(1-v^2/c^2)}} \quad \text{or} \quad \sqrt{\left(1-\frac{v^2}{c^2}\right)} = \frac{m_0}{m}$$

$$\therefore \quad v = c\sqrt{1-\left(\frac{m_0}{m}\right)^2}$$

In this case $m_0 = 2.67 \times 10^{-27}$ kg, $m = 2m_0$ and $c = 3 \times 10^8$ m/sec

$$\therefore \quad v = 3\times10^8\sqrt{1-\left(\frac{m_0}{2m_0}\right)^2} = \frac{3\times10^8\times\sqrt{3}}{2} = \mathbf{2.6 \times 10^8\ m/sec\ Ans.}$$

Example 52. *If the kinetic energy of a body is twice its rest mass energy, find its velocity.*

Solution. The relativistic Kinetic energy may be written as

$$K = (m - m_0)c^2 \text{ or } mc^2 = K + m_0c^2$$

In this case $K = 2m_0c^2 \therefore mc^2 = 3m_0c^2$ or $m = 3m_0$

From the variation of mass with velocity relation

$$m = \frac{m_0}{\sqrt{(1 - v^2/c^2)}} \text{ or } 1 - \frac{v^2}{c^2} = \left(\frac{m_0}{m}\right)^2$$

or
$$v = c\sqrt{\left[1 - \left(\frac{m_0}{m}\right)^2\right]} \quad \therefore v = 3\times10^8\sqrt{1 - \left(\frac{1}{3}\right)^2}$$

Hence $v = 3 \times 10^8 \times 0.943 =$ **2.829×10^8 m/sec Ans.**

Example 53. *The mass of a moving electron is 11 times its rest mass. Find its kinetic energy and momentum.*

Solution. The kinetic energy of the particle is given by the relation

$K = mc^2 - m_0c^2$, in this case $m = 11m_0$, $m_0 = 9.11 \times 10^{-31}$ kg and $c = 3 \times 10^8$ m/sec

So $K = 11m_0c^2 - m_0c^2 = 10m_0c^2 = 10 \times 9.11 \times 10^{-31} \times (3 \times 10^8)^2$

$$= 8.2 \times 10^{-13} = \frac{8.2\times10^{-13}}{1.6\times10^{-19}} \text{ eV} = 5.12 \times 10^6 \text{ eV}$$

The momentum p of the particle is given by, $p = mv$

From the variation of mass with velocity relation

$$m = \frac{m_0}{\sqrt{(1 - v^2/c^2)}} \text{ or } 1 - \frac{v^2}{c^2} = \left(\frac{m_0}{m}\right)^2 \text{ or } v = c\sqrt{1 - \left(\frac{m_0}{m}\right)^2}$$

So
$$v = 3.0\times10^8\sqrt{1 - \left(\frac{m_0}{11m_0}\right)^2} = 3\times10^8 \times 0.996 = 2.99 \times 10^8 \text{ m/sec}$$

Hence, $p = 11m_0 \times 2.99 \times 10^8 = 11 \times 9.11 \times 10^{-31} \times 2.99 \times 10^8 =$ **2.99×10^{-21} kg-m/sec Ans.**

Example 54. *A particle of rest mass m_0 moves with speed $c/\sqrt{2}$. Calculate its mass, momentum, total energy and kinetic energy.*

Solution. From variation of mass with velocity relation, the relativistic mass m of the particle may be written as

$$m = \frac{m_0}{\sqrt{(1 - v^2/c^2)}} = \frac{m_0}{\sqrt{\left(1 - \frac{c^2}{2c^2}\right)}} = \sqrt{2}m_0 = 1.41\, m_0$$

Also momentum p of the particle is

$$p = mv = m_0\sqrt{2} \times \frac{c}{\sqrt{2}} = m_0c$$

and total energy E of the particle is given by Einstein's relation

$$E = mc^2 = (1.41\, m_0)c^2 = 1.41\, m_0c^2$$

So kinetic energy K of the particle is given by

$$K = E - m_0c^2 = 1.41\, m_0c^2 - m_0c^2 = \mathbf{0.41\, m_0c^2} \textbf{ Ans.}$$

Example 55. *If the total energy of a particle is exactly thrice its rest energy, what is the velocity of the particle?*

Solution. From Einstein mass-energy relation, $E = mc^2$

In this case total energy = 3 × rest energy, which means

$$E = 3m_0c^2 \text{ or } 3m_0c^2 = mc^2 \text{ or } m = 3m_0$$

Using the variation of mass with velocity relation

$$m = \frac{m_0}{\sqrt{(1 - v^2/c^2)}} = 3m_0$$

or $$\sqrt{(1-v^2/c^2)} = \frac{1}{3} \text{ squaring, } \frac{1-v^2}{c^2} = \frac{1}{9}$$

or $$v^2 = \frac{8}{9} \times c^2 \quad \text{or} \quad v = \frac{2\sqrt{2}}{3}c$$

or $$v = \frac{2\sqrt{2} \times 3 \times 10^8}{3} = \mathbf{2.828 \times 10^8 \text{ m/sec}} \textbf{ Ans.}$$

Example 56. *Calculate the energy equivalent to amu in MeV. Given Avogadro number = 6.02 × 10^{23}/gm mole.*

Solution. In this case

$$\text{mass } m = 1 \text{ amu} = \frac{1}{6.02 \times 10^{23}} \text{ gm} = \frac{10^{-3}}{6.02 \times 10^{23}} \text{ kg} = \frac{10^{-26}}{6.02}$$

From mass-energy relation, we have

$$E = mc^2 = \frac{10^{-26}}{6.02} \times (3 \times 10^8)^2 \text{ J} = \frac{1 \times 10^{-26} \times (3 \times 10^8)^2}{6.02 \times 1.6 \times 10^{-19}} = 934.5 \times 10^6 \text{ eV}$$

or $\mathbf{E = 934.5 \text{ MeV}}$ **Ans.**

Example 57. *Find the mass and speed of 2 MeV electron.*

Solution. The energy of a particle to related is masses

$$E = mc^2 \text{ or } m = E/c^2$$

In this case E = 2 MeV = 2 × 10^6 × 1.6 × 10^{-19} joule and c = 3 × 10^8 m/sec

$$\therefore \quad m = \frac{2 \times 10^6 \times 1.6 \times 10^{-19}}{(3 \times 10^8)^2} = 3.55 \times 10^{-30} \text{ kg}$$

From the variation of mass with velocity relation

$$m = \frac{m_0}{\sqrt{(1-v^2/c^2)}} \text{ or } 1 - \frac{v^2}{c^2} = \left(\frac{m_0}{m}\right)^2$$

or $$v = c\sqrt{1 - \left(\frac{m_0}{m}\right)^2} = 3 \times 10^8 \times \sqrt{1 - \left(\frac{9.11 \times 10^{-31}}{3.55 \times 10^{30}}\right)^2} = 3 \times 10^8 \times 0.967$$

$$= \mathbf{2.90 \times 10^8 \text{ ms}^{-1}} \textbf{ Ans.}$$

Example 58. *Calculate the amount of work to be done to increase the speed of an electron from 0.6 c to 0.8 c. Given that rest energy of electron = 0.5MeV.*

Solution. The relativistic kinetic energy of an electron may be written as

$$K = (m - m_0)\,c^2 = c^2\left[\frac{m_0}{\sqrt{(1-v^2/c^2)}} - m_0\right] = c^2 m_0\left[\left(1-\frac{v^2}{c^2}\right)^{-1/2} - 1\right]$$

Initial K.E. of an electron, having velocity 0.6 c, is given by

$$K_1 = m_0c^2\left[\left(1-\frac{(0.6c)^2}{c^2}\right)^{-1/2} - 1\right]$$

$$= m_0c^2[(1-0.36)^{-1/2}-1] = m_0c^2[(0.64)^{-1/2}-1]$$

$\therefore$ $K_1 = m_0c^2[1.25-1] = 0.25\times m_0c^2 = 0.25\times 0.5\times 10^6$ eV $= 1.25 \times 10^5$ eV

Final K.E. of electron when velocity increases to 0.8 c will be

$$K_2 = m_0c^2\left[\left\{1-\left(\frac{0.8c}{c}\right)^2\right\}^{-1/2} - 1\right]$$

or $K_2 = m_0c^2[(1-0.64)^{-1/2}-1] = m_0c^2[(0.36)^{-1/2}-1]$

$= m_0c^2[1.67-1] = m_0c^2\times 0.67 = 0.67\times 0.5\times 10^6$ eV

$= 3.35 \times 10^5$ eV

So the amount of work to be done to increase the velocity of electron from 0.6 c to 0.8 c will be

$K_2 - K_1 = 3.35 \times 10^5 - 1.25 \times 10^5 = 2.1 \times 10^5$ eV

$= 2.1 \times 10^5 \times 1.6 \times 10^{-19} = \mathbf{3.36 \times 10^{-14}}$ **J** **Ans.**

Example 59. *Compute the mass m and speed v of an electron having kinetic energy 1.5 MeV. Given rest mass of electron,* $m_0 = 9.11 \times 10^{-31}$ *kg and velocity of light in vacuum,* $c = 3 \times 10^8$ *m/sec.*

Solution. The relativistic relation for kinetic energy is given by $K = (m - m_0)c^2$

In this case K = 1.5 MeV = $1.5\times 10^6 \times 1.6\times 10^{-19}$ Joule and $m_0 = 9.11 \times 10^{-31}$ kg

So $1.5\times 10^6 \times 1.6\times 10^{-19} = (m - 9.11\times 10^{-31})(3\times 10^8)^2$

$\therefore$ $m - 9.11 \times 10^{-31} = \dfrac{1.5\times 10^6\times 1.6\times 10^{-19}}{(3\times 10^8)^2}$

or $m = 9.11\times 10^{-31} + \dfrac{1.5\times 10^6\times 1.6\times 10^{-19}}{(3\times 10^8)^2} = 9.11\times 10^{-31} + 2.667\times 10^{-30}$

So $m = 3.58 \times 10^{-30}$ kg

From the variation of mass with velocity formula.

$$m = \frac{m_0}{\sqrt{(1-v^2/c^2)}} \text{ or } 1-\frac{v^2}{c^2} = \left(\frac{m_0}{m}\right)^2$$

Which gives $v = c\sqrt{\left[1-\left(\dfrac{m_0}{m}\right)^2\right]} = 3\times 10^8 \times \sqrt{1-\left(\dfrac{9.11\times 10^{-31}}{3.58\times 10^{-30}}\right)^2}$

or $$v = 3 \times 10^8 \times \sqrt{(0.935)} = 3 \times 10^8 \times 0.967 = \mathbf{2.9 \times 10^8\ ms^{-1}}$$ **Ans.**

Example 60. *Calculate the velocity of electrons accelerated through a potential of 10^6 volts. Rest mass of an electron = 9×10^{-31} kg.*

Solution. We have the relation for kinetic energy

Kinetic Energy K = Total Energy – Rest Energy

$$= mc^2 - m_0c^2 = \frac{m_0c^2}{\sqrt{(1-v^2/c^2)}} - m_0c^2$$

$$\therefore \quad K = m_0c^2[(1-v^2/c^2)^{-1/2} - 1]$$

Now kinetic energy = 1 MeV = 10^6 eV = $10^6 \times 1.6 \times 10^{-19}$ J = 1.6×10^{-13} J.

$$\therefore \quad 1.6 \times 10^{-13} = 9 \times 10^{-31} \times (3 \times 10^8)^2\ [(1 - v^2/c^2)^{-1/2} - 1]$$

$$(1 - v^2/c^2)^{-1/2} - 1 = \frac{1.6 \times 10^{-13}}{9 \times 10^{-31} \times 9 \times 10^{16}} = 1.975$$

or $$(1 - v^2/c^2)^{-1/2} = 2.975 \text{ or } \frac{1}{2.975} = \sqrt{(1-v^2/c^2)}$$

or $$1 - \frac{v^2}{c^2} = \frac{1}{(2.975)^2} \text{ or } \frac{v^2}{c^2} = 1 - 0.113 = 0.887$$

$$\frac{v}{c} = \sqrt{(0.887)} = 0.942 \ \therefore\ v = 0.942 \times c$$

$$= 0.942 \times 3 \times 10^8\ ms^{-1}$$

$$\therefore \quad v = \mathbf{2.836 \times 10^8\ ms^{-1}}$$ **Ans.**

Example 61. *How much does a proton gain in mass when accelerated to a kinetic energy of 500 MeV?*

Solution. Relation for K.E. from relativity is given by

$$K = (m - m_0)c^2 \text{ or } mc^2 = K + m_0c^2.$$

So gain in mass, $m - m_0 = K/c^2 = \Delta m$

In this case K = 500 MeV = $500 \times 10^6 \times 1.6 \times 10^{-19}$ joule, $m_0 = 2.67 \times 10^{-27}$ kg and $c = 3.0 \times 10^8$ m/sec

So $$\Delta m = \frac{500 \times 10^6 \times 1.6 \times 10^{-19}}{(3 \times 10^8)^2} = \mathbf{8.89 \times 10^{28}\ kg}$$ **Ans.**

Example 62. *Find the speed of 0.1 MeV electron according to relativity mechanics.*

Solution. From classical mechanics the kinetic energy of a particle is given by

$$K = \frac{1}{2}mv^2 \text{ or } v = \sqrt{(2K/m)}$$

In this case K = 0.1 MeV = 0.1×10^6 eV = $0.1 \times 10^6 \times 1.6 \times 10^{-19}$ J and $m = 9.11 \times 10^{-1}$

$$\therefore \quad v = \sqrt{\left(\frac{2 \times 0.1 \times 10^6 \times 1.6 \times 10^{-19}}{9.11 \times 10^{-31}}\right)} = \mathbf{1.87 \times 10^8\ ms^{-1}}$$

From relativstic mechanics, K.E. of an electron may be written as

$$K = mc^2 - m_0c^2 \text{ or } mc^2 = K + m_0c^2$$

$$\therefore \quad \frac{m_0c^2}{\sqrt{1-v^2/c^2}} = K + m_0c^2 \text{ or } (1 - v^2/c^2) = \frac{m_0c^2}{K+m_0c^2}$$

Hence $$v = c\sqrt{1-\left(\frac{m_0c^2}{K+m_0c^2}\right)^2}$$

Now rest mass of electrons $m_0c^2 = 0.512$ MeV

So $K + m_0c^2 = 0.1 \text{ MeV} + 0.512 \text{ MeV} = 0.612 \text{ MeV}$

$$\therefore \quad v = 3.0 \times 10^8 \sqrt{1-\left(\frac{0.512}{0.612}\right)^2} = 3 \times 10^8 \times 0.548 = \mathbf{1.644 \times 10^8\ ms^{-1}} \textbf{ Ans.}$$

Example 63. *Show that the rest mass of a particle is given by*

$$m_0 = \frac{p^2c^2 - T^2}{2Tc^2}$$

where P and T devote the momentum and kinetic energy of the particle

Solution:

Total energy of the particle is given by

$$E = m_0c^2 + T \qquad \text{... (1)}$$

Also $$E^2 = p^2c^2 + m_0^2c^4 \qquad \text{... (2)}$$

Squaring (1) and equating with (2) we get

$$m_0^2c^4 + T^2 + 2T\,m_0\,c^2 = p^2c^2 + m_0^2c^4$$

$$\therefore \quad m_0 = \frac{p^2c^2 - T^2}{2T\,c^2} \quad \textbf{Ans.}$$

Example 64. *Calculate the length and orientation of a metre rod in a frame of reference which is moving with a velocity 0.6 c in a direction making an angle 30° with the rod.*

Solution:

x - component of length of metre for $l_x = \cos 30° = 0.866\ m.$

y – component of meter rod

$Ly = 1. \sin 30° = 0.5$ m.

Component of length along x – axis in moving frame is

$$L_x = Lx\sqrt{1-v^2/c^2}$$

$$= 0.866\sqrt{1-(0.6)^2}$$

$$= 0.693\ m.$$

Component of length I to the motion in moving frame

$$L'y = Ly = 0.5\ m$$

Length of rod in moving frame is

$$L = \sqrt{L_x'^2 d + L_y'^2} = \sqrt{(0.5)^2 + (0.693)^2}$$

$$= 0.854 \text{ m.}$$

Let ν be the angle that the rod appears to make in moving frame

Then $$\tan\theta = \frac{L'_x}{L'_y} = \frac{0.5}{0.693}$$

$$= 0.72$$

∴ $$\nu = \tan\theta\ (0{,}72)$$

$$= 35° \ 50 \quad \textbf{Ans.}$$

5.21. Minkowski's Space (Four dimension space-time continum)

The basis of the theory of relativity is Lorentz transformation equations concerning length, mass and time. Minkowski put forward the concept of time to account *for four dimensional space.*

Minkowski took space and time together and according to him the universe is not made up of points with only x, y and z co-ordinates, but it also consists of *events involving x, y, z and time*. Any event is supposed to be denoted by its position in space and the time of its occurrence. So we live in a four-dimensional world *i.e.,* in a world of events rather than the world of points and this is called a space-time continuum *i.e.,* space and time are in separable.

Consider two co-ordinate systems x, y, z and x', y' and z' with O as common origin (Fig. 5.10). From three dimensional space, the position of P w.r.t. co-ordinate system is given by

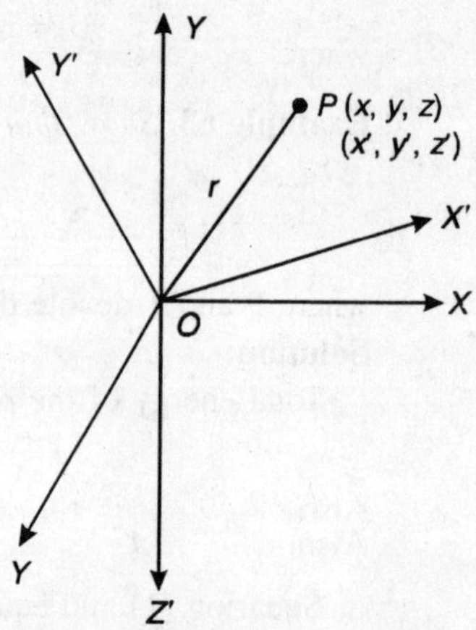

Fig. 5.11

$$r^2 = x^2 + y^2 + z^2$$

and $$r^2 = x'^2 + y'^2 + z'^2$$

i.e., the value of r^2 remains same, whatever may be the values of x, y, z for co-ordinate axes.

Minkowski introduced fourth dimension τ, where τ is variable involving time and is given by

$$\tau = ict$$

∴ $$r^2 = x^2 + y^2 + z^2 + \tau^2 \qquad \text{... (1)}$$

and $$r^2 = x'^2 + y'^2 + z'^2 + \tau'^2 \qquad \text{... (2)}$$

From eqs. (1) and (2), we have

$$r^2 = x^2 + y^2 + z^2 - c^2t^2 \qquad \text{... (3)}$$

and $$r^2 = x'^2 + y'^2 + z'^2 - c^2t'^2 \qquad \text{... (4)}$$

Equations (3) and (4) show that space time continuum differs from the Euclidean space in a particular way.

Also as may be seen τ^2 (or τ'^2) has a negative sign and r^2 may be zero since $x^2 + y^2 + z^2$ will have positive value. This is however not possible in three dimensional space.

It may be clearly noted that **Michelson-Morely** experiment gave a negative result. This negative result can be explained on the basis of four dimensional space-time continuuum. Minkowski termed the four dimensional-space time continum as world and velocity and acceleration were consequently **world velocity and world acceleration.**

5.22 Minkowski's World Velocity

Let u_1, u_2, u_3 and u_4 be the component of velocity in four dimensional space-time continuum.

$$u_1 = \frac{dx}{dt'} = \frac{dx}{dt} \times \frac{dt}{dt'} = u_x \times \alpha \quad \text{... (1)}$$

$$u_2 = \frac{dy}{dt'} = \frac{dy}{dt} \times \frac{dt}{dt'} = u_y \times \alpha \quad \text{... (2)}$$

$$u_3 = \frac{dz}{dt'} = \frac{dz}{dt} \times \frac{dt}{dt'} = u_z \times \alpha \quad \text{... (3)}$$

$$u_4 = \frac{d\tau}{dt'} = \frac{d\tau}{dt} \times \frac{dt}{dt'} = \frac{d(ict)}{dt} = ic \times \alpha \quad \text{... (4)}$$

where $\alpha = \dfrac{1}{\left(1 - \dfrac{v^2}{c^2}\right)^{1/2}}$

We can get world velocity u by comparing the four components

$$u = \sqrt{u_1^2 + u_2^2 + u_3^2 + u_4^2}$$

$$= \left[\alpha^2 (u_x^2 + u_y^2 + u_z^2 - c^2)\right]^{1/2}$$

Also $v^2 = u_x^2 + u_y^2 + u_z^2$

$\therefore \quad u = [\alpha^2 (v^2 - c^2)]^{1/2}$

$$= \left(\frac{v^2 - c^2}{1 - v^2/c^2}\right)^{1/2} = ic \quad \text{... (5)}$$

It may be noted that in the space time continuum, there is one world velocity $u = i.c$ known as *four velocity* and it is invariant. In other words; u_1, u_2, u_3 may have any magnitude but u_4 will be such that together, the resultant will always be equal to ic.

QUESTIONS

1. Derive the Galilean transformation formulae. State the fundamental postulates of special theory of relativity. Deduce the Lorentz transformation equations.
2. What was the objective of conducting the Michelson-Morley experiment? Describe the experiment. How is the negative result of the experiment interpreted?
3. What is Galilean transformation? Obtain the transformation equations for velocity and acceleration in frame S' moving with constant velocity v relative to an inertial frame S.
4. Describe Michelson-Morley experiment. How is the negative results of the experiment interpreted?
5. Discuss Galilean transformation for position, velocity and acceleration.
6. State the fundamental postulates of special theory of relativity and deduce from them the Lorentz transformation equations and discuss how these account for the phenomenon of length contraction. What is proper length?
7. Discuss the theoretical back ground of Michelson-Morley experiment. Briefly describe the experimental set up and what conclusions of importance have been obtained from this experiment.
8. Derive Lorentz transformation equations. Show how Lorentz transformation equations account for contraction of bodies in motion.
9. What do you understand by inertial and non-inertial frames of reference? Establish Galilean transformation equations for space and time.

10. Deduce the relativistic velocity addition theorem. Show that it is consistent with Einstein's second postulate.
11. What do you understand by time dilation? What is proper interval of time? Explain why a moving clock appears to go slow to a stationary observer.
12. Show that if the variation of mass with velocity is taken into account, the kinetic energy of a particle of rest mass m_0 and moving with a velocity v is given by
$$K = m_0c^2 [(1 - v^2/c^2)^{-1/2} - 1].$$
13. Derive the formula for the variation of the mass of a particle with its velocity.
14. Show that the mass of a body depends on its velocity. Deduce an expression for the variation of mass with velocity. Establish Einstein's mass-energy relation.
15. Obtain Lorentz transformation equations. Use them to obtain the law of addition of velocities.
16. Discuss the time dilation on the basis of Lorentz transformation equations. How the time dilation has been experimentally verified?
17. Obtain relativistic formula for the addition of velocities and show that the addition of velocity to the velocity of light merely reproduces the velocity of light.
18. Show how the relativistic invariance of the law of conservation of momentum leads to the concepts of variation of mass with velocity and the equivalence of mass and energy.
19. Derive the Inverse Lorentz transformation equations. Derive the Lorentz-Fitzerald contraction formula. Explain its significance.
20. Show that the apparent length of a rigid body in the direction of its motion with uniform velocity v is reduced by the factor $\sqrt{(1-v^2/c^2)}$. Discuss the result.
21. Deduce an expression for time dilation on the basis of Lorentz transformation equations. Give an example to show that time dilation is real effect.
22. What are massless particles? Show that the massless particles can exist only if they move with the speed of light and that their energy E and momentum p must have the relation, $E = pc$.
23. Deduce Einstein's mass-energy relation, $E = mc^2$ and discuss it. Give some evidence showing its validity.
24. Derive the formula, $E^2 = m_0^2c^4 + p^2c^2$, where symbols have their usual meanings. Hence obtain relativistic relation between kinetic energy and momentum.

NUMERICAL PROBLEMS

1. Calculate the fringe-shift in Michelson-Morley experiment. Given that $l = 11$ m, $v = 30$ km/sec and $\lambda = 6 \times 10^{-5}$ cm. **[Ans.** 0.37]
2. To an observer at rest on the ground, a body thrown vertically upwards in a uniformly moving frame S appears to describe a parabolic path. What is the path of the body as it would appear to an observer in another reference frame S' in a uniform motion, parallel to S, when S' has (*i*) an identical velocity with S, (*ii*) a velocity equal and opposite to that of S and (*iii*) a velocity twice that of S. (**Ans.** (*i*) Vertical motion (*ii*) parabola (*iii*) parabola).
3. In the Michelson-Morley experiment, the wavelength of the monochromatic light used is 5000 Å. What will be the expected fringe shift on the basis of the stationary ether hypothesis if the effective length of each path is 5 m? (Velocity of the earth = 3×10^4 m/sec and $c = 3 \times 10^8$ m/sec). (**Ans.** 0.2)
4. For $l = 50$ cm and $v = 1000$ m/sec, calculate the path difference Δ between longitudinal and lateral journeys in Michelson-Morley experiment, in terms of wavelength of visible light $\lambda = 5 \times 10^{-5}$ cm. (**Ans.** 1.1×10^{-5} wavelength)
5. A rocket ship is 100 m long on the ground. When it is in flight, its length is 99 m to an observer on the ground. What is its speed? (**Ans.** 4.23×10^9 cm s^{-1})
6. A spaceship 50 m long passes the earth at a speed of 2.8×10^8 m/sec. What will be its apparent length? ($c = 3 \times 10^8$ m/sec). (**Ans.** 18 m)
7. A certain process requires 10^{-6} sec to occur in an atom at rest in laboratory. How much time will this process require to an observer in the laboratory, when the atom is moving with a speed of 5×10^7 m/sec. (**Ans.** 1.013×10^{-6} sec)

8. What is the mean life time of π^+ mesons moving with speed 0.73 c, when the proper life time is 2.5 $\times 10^{-8}$ sec. (**Ans.** 3.7×10^{-8} sec)
9. Calculate the speed of a rocket ship so that one day on the ship corresponds to three days on the earth. (**Ans.** 2.83×10^8 ms^{-1})
10. An inertial frame S' is moving with velocity 1.8×10^8 m/sec with respect to the frame S along the common XX' axes in such a way that the origins of the two frames coincide at $t = t' = 0$. Two events are recorded in the frame S, one occurs at the origin at $t_1 = 0$ and the other occurs on the X-axis at $x_2 = 300$ m and $t_2 = 1 \times 10^{-6}$ sec. Find the time interval between the events as measured in the frame S'. (**Ans.** 0.5×10^{-6} sec)
11. Prove that $1 = bmv^2$, where $m = m_0 / \sqrt{(1 - v^2/c^2)}$, does not equal the kinetic energy of a particle moving at relativistic speed.
12. Two spacecrafts A and B, are moving in the same direction with respective speeds of 2.70×10^8 m/sec and 2.85×10^8 m/sec as observed by an observer at the earth. Calculate the relative speed of the spacecrafts as observed by the observers in the two spacecrafts. (**Ans.** 1.03×10^8 ms^{-1})
13. A burst of μ-mesons having a proper mean life of 2.0×10^6 sec is produced at some height in atmosphere. These μ-mesons travel towards earth at a speed of 0.99 c. If only 1% of μ-mesons produced reach earth, estimate the height at which the burst was produced. (**Ans.** 1.95×10^4 m)
14. Two particles move towards each other with velocity 0.5 c and 0.75 c relative to a laboratory frame of reference. Find the approach velocity. (**Ans.** 0.909 c)
15. Calculate the energy equivalent to 1 kg mass. (**Ans.** 5.6×10^{35} e.v)
16. Calculate the mass and momentum of a proton of restmass 1.67×10^{-24} gm, moving with a velocity of 0.8c. If it collides and sticks to a stationary nucleus of mass 5.0×10^{-13} gm, find the velocity of the resultant particle. (**Ans.** 2.78×10^{-24} gm, 6.67×10^{-14} gm cm s^{-1}, 1.26×10^9 cm s^{-1}).
17. Deduce the velocity at which the mass of a particle will be five times the mass at rest. (**Ans.** 2.94×10^8 ms^{-1}).
18. A man has a mass of 100 kg on the earth. Calculate his mass when he is moving with a speed of 4.0 $\times 10^7$ m/sec relative to the earth. (**Ans.** 100.9 kg).
19. Rockets A and B are observed from the earth to be travelling with velocities 0.8 c and 0.7 c in the same line in the same direction. What is the velocity of B as seen by an observer on A?(**Ans.** 0.23 c)
20. A spaceship moving away from earth with speed 0.9 c fires a missile in same direction as its motion, with a speed 0.9 c relative to spaceship. What is missile's speed relative to earth. (**Ans.** 0.994 c)
21. How much work must be done in order to increase the speed of an electron from 1.8×10^8 m/sec to 2.4×10^8 m/sec? (**Ans.** 3.4×10^{-14} J).
22. The velocity of a particle is $5i + 5j + 20k$ m/sec in a frame of reference moving with uniform velocity 0.5 c with respect to the laboratory along + ve X-axis. Find out the velocity of the particle in the laboratory.
23. In the laboratory two particles are observed to travel in opposite directions each with speed 2.8×10^{10} cm/sec. Deduce the relative speed of the particles ($c = 3.0 \times 10^{10}$ cm/sec). (**Ans.** 2.99×10^{10} cm s^{-1})
24. Find the energy produced in kilowatt hours when one kg of matter is converted into energy. (**Ans.** 2.5×10^9 kwh).
25. Two particles came towards each other with speed 0.8 c with respect to laboratory. What is their relative speed? (**Ans.** 0.972 c).

6

X-RAYS AND COMPTON EFFECT

6.1. Discovery of X-rays

In 1895, Roengten discovered X-rays, which are produced when a beam of cathode rays strikes a solid target. He found that the operation of a cathode-ray tube produced fluorescence in a screen covered with barium-platinocyanide and placed at some distance from the tube. In studying the properties of these new type of radiations, Roengten noticed that if opaque materials were placed between the tube and the screen the intensity of the fluorescence decreased but did not disappear, showing that X-rays can penetrate opaque substances. It was also found that the X-ray radiations can blacken a photographic plate and produce ionisation in any gas through which they pass. The X-rays were found to travel in straight lines from the source and could not be deflected by electric and magnetic field, from which it was concluded that they are not charged particles. Later work showed that X-rays can be reflected, refracted and diffracted, and, therefore, these rays are an electromagnetic radiations like light but of much shorter wavelength.

6.2. Production of X-rays (Roentgen Tube)

It consists of an evacuated glass bulb containing a concave cathode of Aluminium *C*. Concave cathode focusses a beam of cathode rays on the tungsten anti cathode. This anticathode is inclined a an angle of 45° to the path of the cathode rays. Anticathode is sometimes called target. The anode A is an aluminium rod connected externally to the anti cathode. The anode helps to make the emission of X-rays steady. Since the cathode rays produce a large heating effect, the anti-cathode is usually fitted with either a water cooling or air cooling device. This tube is worked by an induction coil producing a voltage ranging from 10^3 to 10^5 volts. When the voltage applied is small the X-rays produced are *soft* and when the voltage applied is high the X-rays produced are *hard*. An inductance coil does not give a strictly unidirectional current, therefore, a rectifying value arrangement is included in the secondary circuit of the induction coil. The cathode-rays fall on the anticathode and hence X-rays are produced.

The Anticathode should possess the following properties

(*i*) It should have high atomic number.

(*ii*) It should have high melting point.

(*iii*) It should have high conductivity towards heat and electricity.

(*iv*) It should have low vapour pressure at high temperature.

Tantalum, platinum and tungsten fulfil these requirements, and of these tungsten is the best. A heavy block of this metals is used as anti cathode as shown in Fig. 6.1.

The Function of Anode

The function of the anode is very complex, but when used, it has the following effects :

(*i*) The life of the tube is increased.

(*ii*) The discharge becomes more smooth and hence the intensity of the X-rays beam is more constant.

(*iii*) The efficiency of the tube is increased, *i.e.,* more of the energy of the cathode-rays is converted into X-rays.

(*iv*) It avoids the damage of the tube, when it is run with an induction coil, whose voltage is not unidirectional.

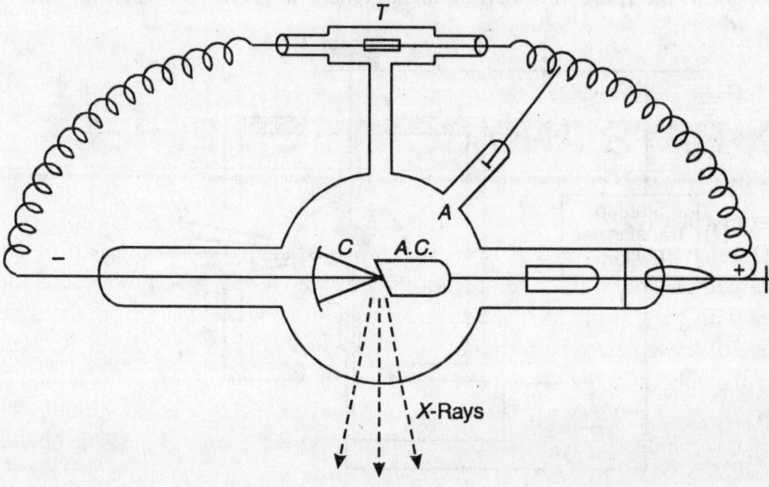

Fig. 6.1

The functioning of the tube depends upon the pressure of the gases inside the tube and the potential difference applied. At low pressure, say of the order of 10^{-4} cms., a comparatively low voltage is needed and, therefore, X-rays produced are called *'soft X-rays'*. If the vacuum is high, of the order of 10^{-5} cms., a comparatively high voltage is needed to run the tube and the X-rays produced have large penetrating power and are said to be *'hard X-rays'*.

During the working of the X-ray tube, the positive ions of the gas are produced. These positive ions should be removed, and hence the pressure inside the tube is lowered, therefore, a greater potential is needed to run the discharge, and the emitted X-rays are harder. If the pressure falls considerably, the tube may even refuse to allow the discharge. To avoid this, 'softening' devices are employed. At first, Palladium rod was used, one half of which was inside the tube. On heating the outer part of the rod with a Bunsen burner, the occluded hydrogen is liberated and enters the X-ray tube and thus increases its conductivity. A better softening device is a side tube *T* containing platinised asbestos in which gases have been occluded. When the tube becomes too hard, the discharge passes through the side tube *T*, liberating some occluded gases and restore the working conditions.

The main defects of the Roentgen Tube are:

(*i*) The intensity of X-rays cannot be controlled.

(*ii*) The softening devise is not perfect.

(*iii*) The quality of X-rays cannot be controlled.

(*iv*) The applied voltage is required to be too high.

(*v*) The cathode is likely to be damaged, when the tube is run by an induction coil by the reverse current which may produce great amount of heat on the cathode and cause excitation of cathode rays in the opposite direction.

The defects have been removed in the Coolidge X-ray tube.

6.3. Coolidge X-rays Tube

The Coolidge X-rays tube was designed by Cooling in 1912. The essential parts of this tube are shown in Fig. 6.2. It consists of hot cathode which produces a beam of electrons which can be used to control both the intensity and the quality of the X-rays. The cathode is heated by 6 volts battery or by a separate tapping from the trasformer *T*. The filament current is adjusted by a rheostat *R*, and is read by an ammeter *A*. The intensity of X-rays depend upon the value of current, *i.e.*, a larger filament current producing a more intense X-ray beam. The electron beam is focussed on the target by surrounding the filament by a cylindrical tube *S* called the shield which is charged negative with respect to the filament. Such an arrangement is known as electron gun.

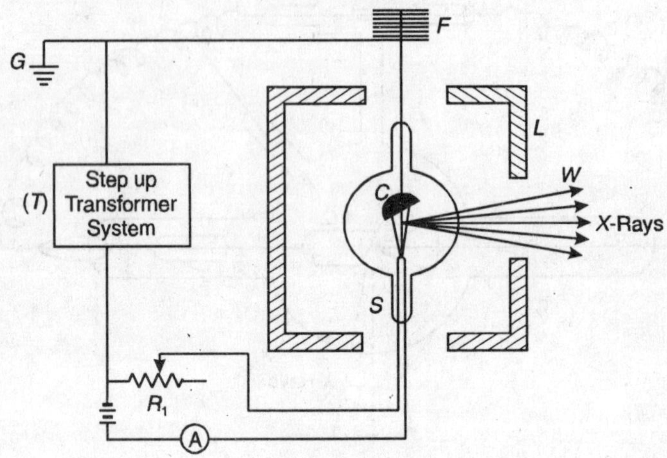

Fig. 6.2

The anticathode or the target is cooled by means of radiating fans *F*. In better types of tubes water cooling arrangement is provided. The high voltage is taken from a large step-up transformer *T.* The X-ray tube works as its own rectifier. The current will only pass when the anticathode is positive with respect to the filament, since the emitted electrons can move only from the filament (negative) to the anti-cathode (positive) and not *vice-versa*. The anode current through the tube is read by the milliammeter and is regulated by the rheostat (not shown in figure). This regulates the applied voltage across the tube and controls the quality of X-rays produced. A larger voltage or current in milliammeter produces harder and more penetrating X-rays.

The tube is enclosed in read box *L* about 0.5 cm thick and is provided with a small window *W* through which the *X*-rays emerge.

In an improved design of the tube cooling is done by rotating anode. The anode, in this case, consists of a tungsten disc which is rotated at high speed by a rotating magnetic field applied to an iron rotor on which the disc is mounted. Thus each part of the target is bombarded and heated for a very small fraction of a revolution and cools during the rest of its journey. By this arrangement, large tube currents are possible and the X-rays produced are harder. These are generally used for short intense exposure as *per radiographs of the heart.*

The highest voltage X-ray tube uses about a million volts, and produces X-rays of 10^{-10} cm wavelength.

6.4. Properties of X-rays

1. X-rays are not deflected by electric and magnetic fields. This property shows that X-rays do not consists of any charge.
2. They are highly penetrating and can pass through many solids which are opaque towards light *e.g.*, wood, flesh etc. Soft X-rays have small penetrating power and are easily observed, and hard X-rays have relatively high penetrating power. The transparency depends upon the density of the material. The denser the substance the less transparent it is to X-rays. A sheet of lead one millimeter thick absorbs these X-rays, whereas the rays can pass through the same thickness of aluminium.

The penetrating power of X-rays depends upon:

(*a*) The applied potential difference.

(*b*) The nature of the anti-cathode material. The greater the potential difference and higher the atomic weight of the anti-cathode, the more penetrating X-rays are produced.

3. They cause flourescence in many substances, *e.g.*, barium platinocyanide, cadmium tungsten, zinc sulphide etc.
4. They affect a photographic plate and are even more effective than light.
5. They ionise a gas and also knockout electrons from certain metals on which they fall. This effect is known as photoelectric effect.
6. They travel in straight line with the velocity of light, *i.e.*, 3×10^8 ms^{-1}
7. Like light, X-rays consist of electromagnetic waves of very short wavelength and show reflection, refraction, interference, diffraction and polarisation in a similar way.
8. When X-rays (primary) coming from the anticathode of an X-ray tube fall on a plate of some chosen material, they give rise to a complex phenomenon of secondary radiations which consist of three different types as shown in Fig. 6.3.

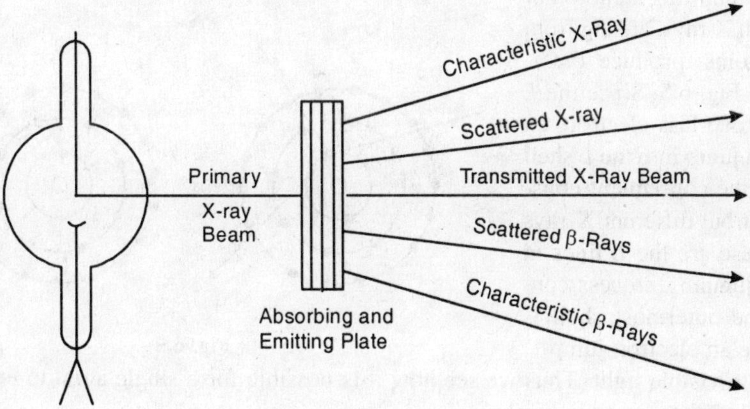

Fig. 6.3

(*i*) Scattered X-rays. They are practically of the same nature and wavelength as the primary X-rays. Their properties do not depend upon the nature of the scsattering substance.

(*ii*) **Characteristic X-rays.** The wavelength of the characteristic X-rays is equal to or less than the primary X-rays but it is characteristic of the substance upon which the primary X-rays are incident and do not depend upon the wavelength of the primary X-rays. *The characteristic X-rays from substances of high atomic weight are of shorter wavelength than those from substance of low atomic weight and hence are more penetrating.* Every substance gives rise to a particular set of characteristic X-rays consisting of one or more series known as *K, L, M* etc. The X-rays in *K* series of element have smaller wavelength than in *L, M* or *N* series, whereas the wavelength of the X-rays in the same series, say *K* or *L,* produced from an element of higher atomic weight is less thant that produced from an element of low atomic weight.

(*iii*) **β-rays.** They consist of fast moving electrons produced by the photoelecetric effect. *Their properties are independent of the nature of the scattering substance but depend upon the quality of primary X-rays.*

9. They have destructive effect on living tissues. Exposure of human body to X-rays causes reddening of the skin, like the sun burn, while prolonged exposure produces surface sores and injures the tissues of the body. The white corpuscles in the blood are killed by prolonged exposure to X-rays and may develop into gangrene. To shield the effect of X-rays, the operators use lead-lined aprons and lead glasses.

6.5. The Origin of X-rays

X-rays, like visible light, originate from the jumping of an electron from one orbit to another. When high-speed electrons from the cathode of an X-ray tube strike the target, they ionize many atoms composing the surface lyers of the metal.

Owing to their very high speeds (about 10% the velocity of light), the electrons penetrate the atoms and remove an electron from the inner shells by collision. This is illustrated in Fig. 6.4 where an electron is knocked out of the *K* shell.

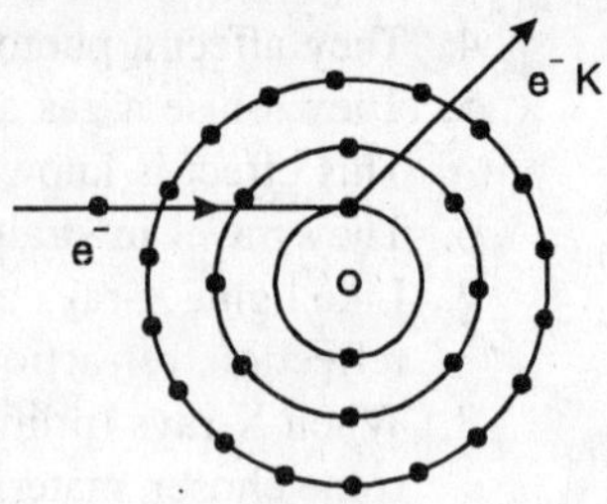

Fig.6.4

The designations *K*, *L*, *M*, *N*, *O*, *P* etc., for the various electron shells originated with the X-ray spectroscopist and are identical with the quantum numbers n = 1, 2, 3, 4, 5, 6 etc. When an electron is missing in the innermost *K* shell, a nearby electron from the next shell beyond jum into the vacant space, simultaneously emitting a photon of energy $h\nu$. Such X-rays, arising from millions of atoms, produce the *K* lines shown in Fig. 6.5. Since the *L* shell now has one less electron, an *M* electron can jump into the *L* shell vacancy, with the consequent emission of another but different X-rays frequency. These are the *L* lines in Fig. 6.5. The jumping process continues until the outermost shell is reached, where an electron jumping gives rise to visible light. Thus we see how it is possible for a single atom to emit X-rays of different wavelengths.

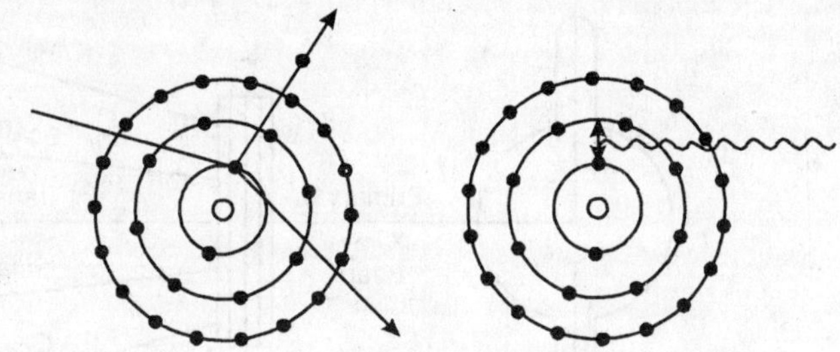
Fig. 6.5

The continuous X-ray spectrum is due to another phenomenon often referred to as "Breamsstrabhlung". These radiations are due to the *slowing down of high speed electrons* as they pass close to the nuclei of the atoms within the target of the X-ray tube. The process is illustrated in Fig. 6.6. As the electron passes passes through the atom, it is attracted by the positive charge of the nucleus and is deflected in its path.

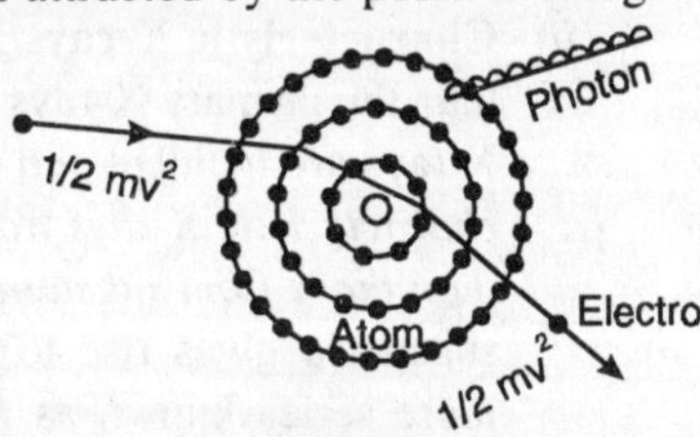

Fig. 6.6

During the deflection of the electron in the strong electric field of the nucleus, a light wave of energy $h\nu$ is emitted. Since the law of conservation of momentum must hold for such a collision, the electron is deflected off to one side of the atom, and the photon off to the other. Since the law of conservation of energy must hold, some of the energy of the incoming electron $1/2mv^2$ is given up to the newly created photon $h\nu$, and the remainder $1/2mv'^2$ is retained by the electron. Thus the electron is slowed down to a velocity v' by the encounter.

The closer the electron comes to the nucleus, the greater is its loss in velocity and energy, and the greater is the frequency and energy of the radiated photon. By the conservation of energy

$$\boxed{\frac{1}{2}mv^2 - \frac{1}{2}mv'^2 = h\nu} \qquad \text{...(1)}$$

The highest frequency that is possible is one in which the electron is completely stopped by the atom. In this special case,

$$\frac{1}{2}mv^2 = h\nu_{max} \qquad \text{...(2)}$$

Since the kinetic energy of the electrons in the beam striking the target is given by the voltage V applied to the tube, we may use equation,

$$Ve = \frac{1}{2}mv^2 \quad ...(3)$$

and obtain

$$Ve = hv_{max} \quad ...(4)$$

While equation (3) must be replaced by the relativistic equation, when voltages are greater than 20,000 volts, Eq (4) holds true for all voltages.

So the production of continuous spectrum can be considered to be inverse photoelectric effect because in this case electron K.E. $\frac{1}{2}mv^2$ *is converted into photon energy* $hv_{max.}$

6.6. Mosley's Law

In 1913-14, Mosley carried out a systematic study of characteristic X-ray spectra of various metallic elements by using them as targets in X-ray tube. A Bragg's spectrometer was used for this purpose and following facts were found:

(*i*) The characteristic X-ray spectra of different elements are similar to each other in the sense that each consists of K, L, and M series.

(*ii*) The frequency of lines (in each series) produced from an element of higher atomic number is greater than that produced by an element of lower atomic number. The reason for this is that binding energy of electron increases as we go from one element to another of higher atomic number *i.e.*, in case of an element of higher atomic number, larger amount of energy is required to liberate an electron from K, L, and M shells respectively of that element. Thus, in case of K_α line of characteristic X-ray spectrum, higher is atomic number of the target material, higher is the frequency of K_α line.

Mathematically, we can represent the relation between frequency and charge number Z as

$$v \propto (Z-a)^2$$

$$v = b(Z-a)^2 \quad ...(1)$$

Where v = frequency of characteristic radiation

b = constant, which is different for different series

a = constant, known as screening constant and is different for each series.

For K_α line, b was found to be equal to 3/4 R, where R is Rydberg constant and α was found to be 1. So for K_α line

$$v_{K_\alpha} = \frac{3}{4}R(Z-1)^2 \quad ...(2)$$

Relation (1) is known as *Mostley law* or *Mostly equation*. So, Mosley law may be stated as follows:

"*The frequency of a speactral line in characteristics X-ray speactrum varie directly as the square of atomic number Z of the element emitting it.*"

Figure 6.7 Shows Mosley diagram for K_α and K_β lines obtained by plotting, $\sqrt{v}$ *versus* atomic number of different elements of periodic table. The graph is linear one as expected.

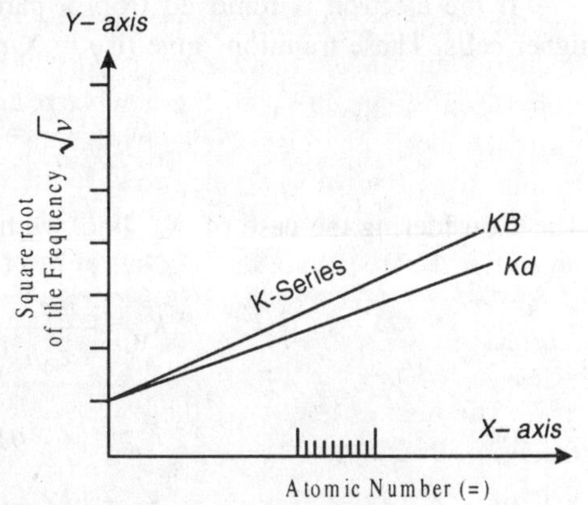

Fig. 6.7

The exact form of Mosley law is

$$\frac{1}{\lambda} = R(Z-\sigma)^2\left(\frac{1}{n^2}-\frac{1}{p^2}\right) \qquad ...(3)$$

where σ is a correction factor and n and p are the principal quantum numbers of the energy levels between which the transition takes place.

Derivation of Mosley's law on the basis of Bohr's theory of atom. Consider the atom of an element with atomic number Z. From Bohr's theory, the energy of an electron moving in an orbit of principal quantum number n_1 is given by the relation

$$E_n = \frac{mZ^2e^4}{8\varepsilon_0^2 n^2 h^2} \qquad ...(4)$$

Similarly, the energy of an electron moving in an orbit of principal quantum number n_2 is given by

$$E_p = \frac{mZ^2e^4}{8\varepsilon_0^2 p^2 h^2} \qquad ...(5)$$

Let us consider the effect of electrons which screen the positive charge of the nucleus reducing thereby the value of Z to $(Z - a)$, the above expressions (4) and (5) can be written in the form

$$E_n = \frac{m(Z-a_1)^2e^4}{8\varepsilon_0^2 p^2 h^2} \qquad ...(6)$$

$$E_p = \frac{m(Z-a_2)^2e^4}{8\varepsilon_0^2 p^2 h^2} \qquad ...(7)$$

Also energy difference $E_n - E_p$ is given by

$$E_n - E_p = \frac{me^4}{8\varepsilon_0^2 h^2}\left[\frac{(Z-a_2)^2}{p^2}-\frac{(Z-a_1)^2}{n^2}\right]$$

where Z is very high $a_1 \equiv a_2 \approx a_{02}$, so we have the energy difference as

$$\Delta E = E_n - E_p = \frac{me^4}{8\varepsilon_0^2 h^2}(Z-a)^2\left[\frac{1}{p^2}-\frac{1}{n^2}\right]$$

If the electron is removed from a particular shell, transition can occur to that shell from higher cells. These tranitions give rise to X-ray photons whose frequency ν is given by

$$\nu = \frac{\Delta E}{h} = \frac{me^4(Z-a)^2}{8\varepsilon_0^2 h^3}\left(\frac{1}{p^2}-\frac{1}{n^2}\right) \qquad ...(8)$$

Considering the case of K_α line, we have $p = 1$ and $n = 2$.

$$\therefore \qquad {}^{\nu}K_\alpha = \frac{me^4}{8\varepsilon_0^2 h^3}(Z-a)^2\left(\frac{1}{p^2}-\frac{1}{2^2}\right)$$

or $$ {}^{\nu}K_a \propto (Z-a)^2$$

or $$ {}^{\nu}K_a = b(Z-a)^2 \qquad ...(9)$$

where $$b = \frac{me^4}{8\varepsilon_0^2 h^3}\left(\frac{1}{1^2} - \frac{1}{2^2}\right) = \text{constant.}$$

Eq. (9) represents the Mosley law.

6.7. Importance of Mosley Law

According to Mosley's law, it is the atomic number of an element and not the atomic weight which accounts for the physical and chemical properties of an element. Thus this law provides the guideline that elements must be arranged in periodic table according to their atomic number and not according to their atomic weights.

Before Mosley's work, Mandeleev arranged the elements in increasing atomic weight to form the periodic table. It was found that if the ascending order of atomic weights is to be preserved, cobalt (${}_{27}CO^{58.94}$) with atomic weight 58.94 should be placed after nickel (${}_{28}Ni^{58.69}$) with atomic weight 58.69. Similarly, potassium (${}_{19}K^{39}$) should come before argon (${}_{18}A^{40}$). Mandeleev observed that to preserve the periodicity of chemical and physical properties, their orders should be reversed. This difficulty was removed by Mosley law because as per their atomic numbers, their order should be opposite to what it is according to Mandeleev.

By considering the X-ray spectra of rare earths, this law has been helpful in determining their atomic numbers and also fixing their positions in periodic table at appropriate positions.

This law also led to the discovery of some new elements like hafnium (72), prometheum (61) etc.

6.8. X-ray Spectrum

As explained above X-ray spectrum consists of

(*a*) continuous spectrum and (*b*) line spectrum. These two are shown in Fig. 6.8 (*b*)

(*a*) **Continuous Spectrum**

(*i*) It is produced due to the de-acceleration of high velocity electrons when they are deflected while passing near the positively charged nucleus of an atom of the target material.

(*i*) It has a sharply-defined short-wavelength limit given by

$$\lambda_{min} = \frac{12{,}400}{V} \times 10^{-10} \text{ m or } \frac{12{,}400}{V} \text{Å}$$

(*iii*) The cut-off wavelength λ_{min} is independent of the nature of the target material but is inversely proportional to the potential difference between the cathode and anode of an X-ray tube. The value of λ_{min} decreases as this potential difference is increased (Fig. 6.8).

(*iv*) The intensity of the continuous spectrum is found very nearly proportional to the *square of the applied voltage* for a given target and to the atomic number of the target material when a constant potential difference is applied.

(*v*) There is a shift of the maximum intensity position towards the short wavelength side as voltage is increased.

(*b*) **Line Spectrum**

(*i*) It is produced when electrons are dislodged from the innermost orbits of the atoms of the target material followed by electron jumps from outer orbits.

(*ii*) It consists of discrete spectral lines which constitute *K*-series, *L*-series and *M*-series etc. *K*-series consists of those lines for which electron jumps end at *K*-level
Similarly, *L*-series and *M*-series are produced when jumps end at *L* and *M*-levels (Fig. 6.8)

(*iii*) K-series being most energetic consititute the hard X-rays whereas *L* and *M*-series form the 'soft' X-rays.

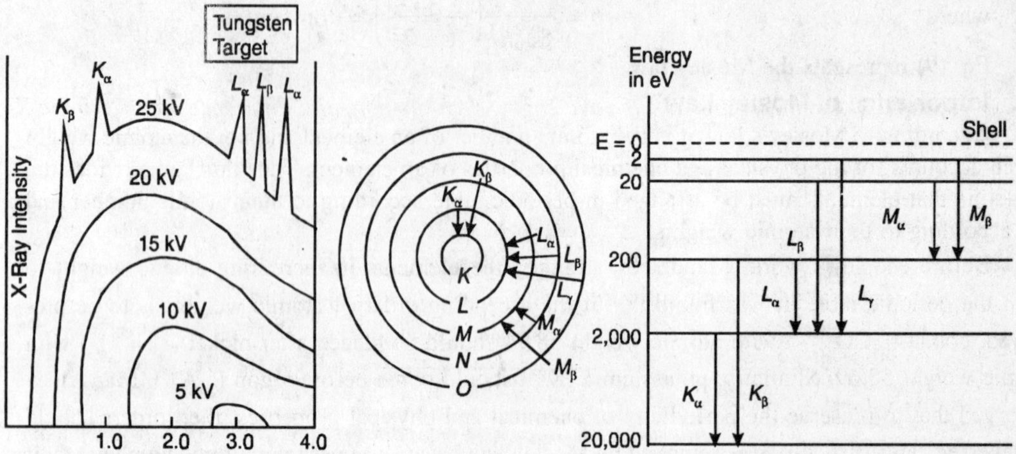

Fig. 6.8

(*iv*) Line spectrum is *characteristic of the target material used.* In fact, X-rays constituting the line spectrum are known *as characteristic* X-rays. The number of lines present in the spectrum depends *both on the nature of target material and the excitation voltage.*

(*v*) There is a regular shift towards shorter wavelength in the *K*-spectrum as the atomic number of the target is increased (Fig. 6.9). The exact relationship, as found by Mosley, is given as

$$\frac{\lambda_2}{\lambda_1} = \frac{\nu_1}{\nu_2} = \frac{(Z_1 - 1)^2}{(Z_2 - 1)^2}$$

where ν_1 is the frequency of the K_α line for a target material having an atomic number of Z_1 and ν_2 and Z_2 are similar quantities for some different target material.

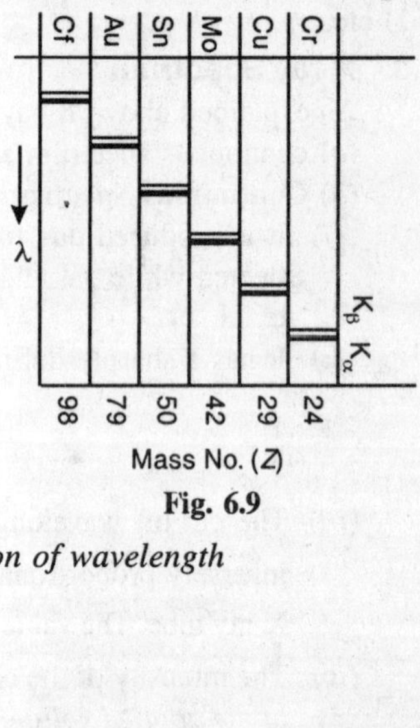

Fig. 6.9

Example 1. *Calculate the energy in electron volts of a photon of wavelength* 5500×10^{-10} *m.*

Given: $h = 6.62 \times 10^{-34}$ *J-s.*

Solution. Given : $\lambda = 5500 \times 10^{-10}$ m.

$h = 6.62 \times 10^{-34}$ J-s.

Velocity of light $c = 3 \times 10^8$ ms^{-1}

If ν be the frequency of radiation emitted, then we have

$$c = \nu\lambda$$

$$\therefore \quad \nu = \frac{c}{\nu} = \frac{3 \times 10^8}{5500 \times 10^{-10}} = \frac{3}{55} \times 10^{16}$$

$$= 5.5 \times 10^{14} \ Hz.$$

Energy of photon $= h\nu$

$$= 6.62 \times 10^{-34} \times 5.5 \times 10^{14} \text{ joules}$$

$$= 36.41 \times 10^{-20} \text{ joules}$$

$$= \frac{36.41 \times 10^{-20}}{1.6 \times 10^{-19}} = 2.275 \text{ eV}$$

$\therefore$ **Energy of photon = 2.275 eV. Ans.**

Example 2. *Calculate the cut-off wavelength when the potential difference applied to the X-ray tube is 25 kV.*

Solution. We have the relation

$$\lambda_{min} = \frac{12,400}{V} = \frac{12,400}{25,000} = \mathbf{0.496\ Å} \quad \textbf{Ans.}$$

Note : Corresponding maximum frequency of the emitted X-rays is

$$f_{max} = \frac{c}{\lambda_{min}} = \frac{3 \times 10^8}{0.496 \times 10^{-10}} = \mathbf{6.05 \times 10^{18}\ Hz\ \ Ans.}$$

Example 3. *Electron bombarding the anode of a Coolidge tube, produce X-rays of wavelength 1Å. Find the energy of each electron at the moment of impact given $h = 6.62 \times 10^{-34}$ J-s. $e = 1.6 \times 10^{-19}$ C.*

Solution Given that
$\lambda = 10^{-10}$ m
$h = 6.62 \times 10^{-34}$ J-s
$e = 1.6 \times 10^{-19}$ coulomb

Velocity of light $c = 3 \times 10^8$ ms^{-1}

If V be the potential difference applied between cathode and anti-cathode, then

$$Ve = \frac{hc}{\lambda} \qquad ...(i)$$

But,

$$Ve = K.E.$$

$$Ve = \frac{1}{2} mV^2 \qquad ...(ii)$$

From (*i*) and (*ii*), we get

$$\therefore \quad K.E. = \frac{hc}{\lambda}$$

$$= \frac{6.62 \times 10^{-34} \times 3 \times 10^8}{10^{-10}} \text{ joules}$$

$$= \frac{19.86 \times 10^{-16}}{1.6 \times 10^{-19}} \text{ eV}$$

$$= \frac{19.86 \times 10^3}{1.6} = \text{eV}$$

$\therefore$ **Energy = 12.41 keV. Ans.**

Example 4. *An X-ray tube operates at 50 kV. Calculate the shortest wavelength of X-ray produced.*

Given: $h = 6.62 \times 10^{-27}$ ergs-sec.

Charge on electron $e = 4.8 \times 10^{-10}$ e.s.u.

Solution. Given that $V = 50$ kV $= 50 \times 10^3$ volts

$$= \frac{50 \times 10^3}{300} \text{e.s.u} = \frac{500}{3} \text{e.s.u.}$$

$$e = 4.8 \times 10^{-10} \text{ e.s.u.}$$
$$h = 6.62 \times 10^{-27} \text{ ergs-sec.}$$

Velocity of light $c = 3 \times 10^{10}$ cm/sec.

Energy of X-ray with charge e and potential difference V is given by

$$E = Ve$$

But $$E = h\nu \qquad ...(1)$$

or $$E = \frac{hc}{\lambda} \qquad \left(\because\ c = \nu\lambda \text{ or } V = \frac{c}{\lambda}\right) \qquad ...(2)$$

From (1) and (2), we get $\frac{hc}{\lambda} = Ve$

or $$\lambda = \frac{hc}{Ve} = \frac{6.62 \times 10^{-27} \times 3 \times 10^{10}}{300 \times 4.8 \times 10^{-10}} \times \frac{3}{1}$$

$$= \frac{6.62 \times 9}{500 \times 4.8} \times 10^{-7} = 0.2482 \times 10^{-8} \text{ cm}$$

$$\lambda = 0.2482 \text{ Å}$$

Therefore, the wavelength of X-ray is **0.2482 Å. Ans.**

Example 5. *If the potential difference applied across an X-ray tube is 5 kV and the current through it is 2mA, calculate*

(i) the number of electrons striking the target per second and

(ii) the speed at which they strike it.

Take $e = 1.6 \times 10^{-19}$ *C and* $m = 9.1 \times 10^{-31}$ *kg.*

Solution. (*i*) If n is the number of electrons striking the anode per second, then

$$I = ne$$

$$n = \frac{I}{e} = \frac{2 \times 10^{-3}}{1.6 \times 10^{-19}} = \mathbf{1.25 \times 10^{16}\ s^{-1}}$$

(*ii*) We have the relation $\frac{1}{2}mv^2 = eV$

$$v = \sqrt{\frac{2eV}{m}} = 5.93 \times 10^5 \text{m/s}$$

$$= 5.93 \times 10^5 \times \sqrt{5000} = \mathbf{4.15 \times 10^7\ ms^{-1}} \textbf{ Ans.}$$

Example 6. *Deduce the wavelength of* K_α *line for an atom of Z = 92, using Mosley's law. Given rydberg constant* $R = 1.1 \times 10^7 m^{-1}$

Solution. The spectral line K_α is emitted when an electron jumps from an orbit of principal quantum number 2 to 1. Its wavelength λ is given by

$$\bar{\nu} = \frac{1}{\lambda} = \frac{\nu}{c} = R\,(z-a)^2 \left(\frac{1}{1^2} - \frac{1}{2^2}\right)$$

But for K series $a = 1$. Therefor,

$$\frac{1}{\lambda} = \frac{3}{4} R\,(z-1)^2$$

Putting the given values of R and Z,

$$\frac{1}{\lambda} = \frac{3}{4} R\,(1.1 \times 10^7)\,(92-1)^2$$

$$= 6.83 \times 10^{10}\ \text{m}^{-1}$$

$$\lambda = \frac{1}{6.83 \times 10^{10}} = 0.146 \times 10^{10}\ \text{m}$$

$$= \mathbf{0.146\ Å\ \ Ans.}$$

Example 7. *An X-ray tube works at 18 kV. Find the maximum speed of electrons striking the anticathode. given mass of electron 9 × 10^{-28} gm and charge of electron = 4.8 × 10^{-10} e.s.u. and*

$$1\ volt = \frac{1}{300}\ e.s.u.$$

Solution. Given that

potential difference $V = 18$ kv $= 18 \times 10^3$ Volts

$$= \frac{18 \times 10^3}{300} = 60\ \text{e.s.u.}$$

Mass of electron = 9 × 10^{-28} gm

Charge on electron e = 4.8 × 10^{-10} *e.s.u.*

If the velocity fo electron = v, then

$$K.E. = \frac{1}{2} mv^2 = Ve$$

$$mv^2 = 2\ Ve$$

$$v = \sqrt{\frac{2\ Ve}{m}} = \sqrt{\frac{2 \times 60 \times 4.8 \times 10^{-10}}{9 \times 10^{-28}}}$$

$$= \sqrt{64} \times 10^9$$

$$v = \mathbf{8 \times 109\ cm/sec. = 8 \times 10^7\ ms^{-1}\ \ Ans.}$$

Example 8. *Calculate the minimum applied potential required to produce X-rays of 1Å wavelength.*

Solution. We have $\lambda_{min} = 12{,}400/V$ Å

$$\therefore \qquad V = 12{,}400/\lambda_{min} = \frac{12400}{1} = \mathbf{12.4\ kV.\ \ Ans.}$$

Example 9. *An X-ray tube operated at 40 kV emits a continuous X-ray spectrum with a short wavelength limit λ_{min} = 0.310Å. Calculate the Planck's constant.*

Solution. we know

$$Ve = h\nu_{max} = \frac{h\,c}{\lambda_{min}}$$

or

$$h = \frac{Ve\,\lambda_{min}}{c}$$

Substituting $V = 40$ kV $= 40 \times 10^3$ volts, $e = 1.6 \times 10^{-19}$ Coulomb

$\lambda_{min} = 0.310$ Å $= 0.310 \times 10^{-10}$ m and $c = 3 \times 10^8$ m/s

$$h = \frac{40 \times 10^3 \times 1.6 \times 10^{-19} \times 0.310 \times 10^{-10}}{3 \times 10^8}$$

$$= \mathbf{6.61 \times 10^{-34}\ Joule - sec.\ \ Ans.}$$

Example 10. *Stopping potentials of 24, 100, 110 and 115 kV are measured for photo-electrons emitted from a certain element whin it is irradiated with monochromatic X-rays. If this element is used as a target in an X-ray tube, what will be the wavelength of the K_α line ?*

Solution. The stopping potential energy, eV_s, is equal to the difference between the energy of the incident photon and the binding energy of the electron in particular shell.

$$eV_s = E_P - E_B$$

The different stopping potentials arise from electrons being emitted from different shells, with the smallest value (24 kV) corresponding to direction of a K-shell electron. subtracting the expression for the two smallest topping potentials, we obtain

$$eV_{sL} - eV_{sK} = (E_P - E_{BL}) - (E_P - E_{BK}) = E_{BK} - E_{BL}$$

or $$100 \text{ keV} - 24 \text{ keV} = E_{BK} - E_{BL}$$

This difference, 76 keV, is the energy of the K_α line. The corresponding wavelength is

$$\lambda = \frac{hc}{E_{BK} - E_{BL}} = \frac{12.4 \text{ KeVÅ}}{76 \text{ keV}} = \mathbf{0.163\ Å} \quad \textbf{Ans.}$$

Example 11. *When 0.50 Å X-rays strike a material, the photo-electrons from the K shell are observed to move in a circle of radius 23 mm in a magnetic field of 2 × 10^{-2} T. what is the binding energy of K-shill electrons ?*

Solution. The velocity of the photo-electrons is found from $F = ma$

$$evB = m\frac{v^2}{R} \quad \text{or} \quad v = \frac{e}{m}$$

This kinetic energy of the photo-electrons is then

$$K = \frac{1}{2}mv^2 = \frac{1}{2}\frac{e^2 B^2 R^2}{m}$$

$$= \frac{1}{2}\frac{(1.6\times10^{-19}\text{C})^2 (2\times10^{-2}\text{T})^2 (23\times10^{-3}\text{m})^2}{(9.11\times10^{-31}\text{kg})}$$

$$= 2.97 \times 10^{-15} \text{ J}$$

or $$K = (2.97 \times 10^{-15} \text{ J})\frac{1 \text{ keV}}{1.6\times10^{-16}} = 18.6 \text{ keV}$$

The energy of the incident photon is

$$E_V = \frac{hc}{\lambda} = \frac{12.4 \text{ keV.Å}}{0.50 \text{ Å}} = 24.8 \text{ keV}$$

The binding energy is the difference between these two values

$$BE = Ev - K = 24.8 \text{ ke V} - 18.6 \text{ ke V} = \mathbf{6.2 \text{ ke V.}} \quad \textbf{Ans.}$$

Example 12. *In Zn (Z = 30) the ionization (binding) energies of the K and L shells are, respectively, 9.659 keV and 1.021 keV. Determine the kinetic energy of an Auger electron emitted from the L shell by a K_α X-ray.*

Solution. The energy of K_α x-ray is

$$E_{K_\alpha} = EL - E_K = -1.021 \text{ keV} - (-9.659 \text{ keV}) = 8.639 \text{ keV}$$

The kinetic energy of the Auger electron will be equal to the difference, between the K_α photon energy and binding energy of the electron in the L shell:

$$K = E_{K_\alpha} - E_{BL} = 8.638 \text{ keV} - 1.021 \text{ keV} = \mathbf{7.617 \text{ keV}} \ \textbf{Ans.}$$

6.9. Absorption of X-rays

We can study the absorption of X-rays with an apparatus shows in Fig. 6.10. A narrow beam of monochromatic X-rays when passes through matter, some part of it is absorbed while the rest is transmitted. X-rays are made to pass through slits S_1 and S_2 of lead plates. So that a narrow beam of it is allowed to fall on a crystal and due to Bragg reflection, we get a monochromatic beam and then this beam is made to enter ionization chamber in which the current of ionization produced will measure the intensity of X-rays.

After knowing the intensity of X-ray a sheet of absorbing material is put in the path of X-rays before it enters the ionization chamber resulting in the reduction of intensity of X-rays due to absorption of X-rays by the absorber.

The ratio $\frac{I}{I_0}$ *i.e.*, the currents before and after absorption is called absorption coefficient of the material.

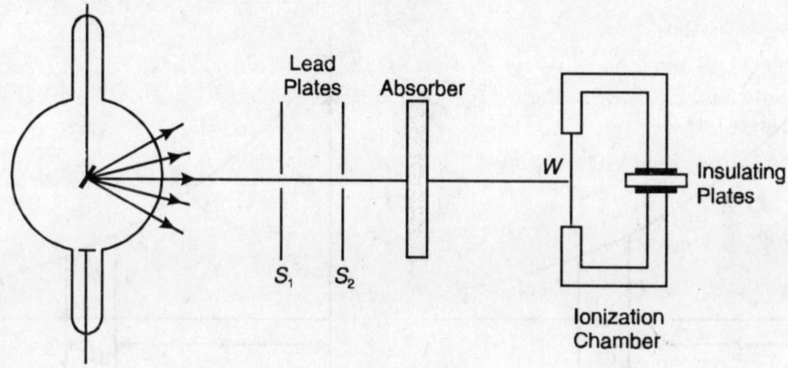

Fig. 6.10

We can get a relation for absorption coefficient by considering I and I_0 as the intensity of rays before and after it passes through an absorber of thickness x.

Let dI be the further decrease in intensity when X-rays travel through an additional thickness dx of the absorber, then we have $-\frac{dI}{dx}$ as the decrease of intensity with increase in distance dx of the absorber and this will be proportional to the original intensity of X-rays *i.e.*,

$$-\frac{dI}{dx} \propto I$$

or

$$\frac{dI}{dx} = -\mu I \qquad ...(1)$$

where μ is the constant of proportionality and is called linear absorption coefficient of the absorber or linear attenuation coefficient of the absorber. Sometimes it is also called macroscopic absorption coefficient.

From (1), we have

$$\frac{dI}{I} = -\mu \, dx$$

Integrating both sides, we have

$$\int_{I_0}^{I} \frac{dI}{I} = -\mu \int_0^x dx$$

$$\therefore \qquad \log_e \frac{I}{I_0} = -\mu x$$

or $$I = I_0 \, e^{-\mu x} \qquad \text{...(2)}$$

which shows that the intensity of X-rays decreases exponentially with thickness of the material.

We can define the absorption coefficient from eq. (2) as

$$\mu = -\frac{\frac{dI}{I}}{dx}$$

i.e., it is equal to the fractional decrease in the intensity of X-rays per unit thickness of the absorber. It is found that μ depends on the wavelength of X-rays nearly by the relation $\mu \propto \lambda^3$.

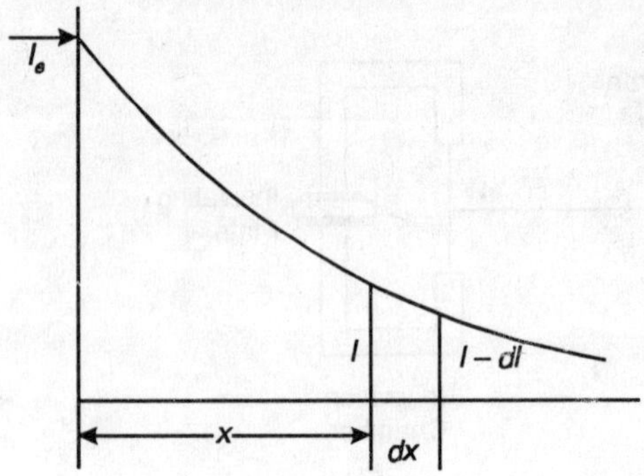

Fig. 6.11

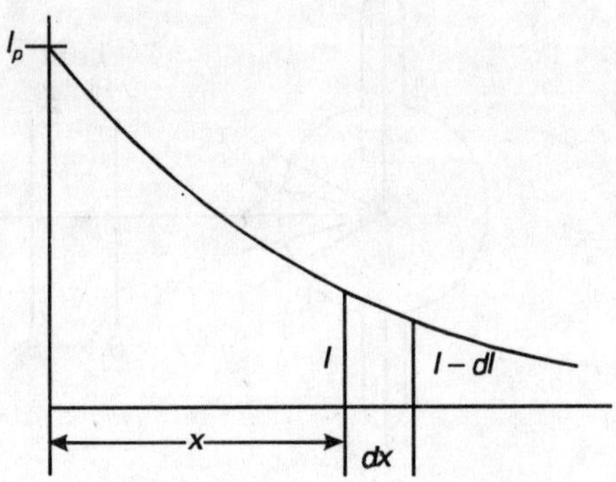

Fig. 6.12

The value of μ can be found by finding the thickness of the absorber sheet which reduces the intensity of X-ray beam to 1/2 of its initial value, such thickness is called half value layer (*hvl*) and is denoted by T.

If we put $\frac{I}{I_0} = \frac{1}{2}$ and $x = T$ in the eq. (2), we get

$$\therefore \quad e^{-\mu T} = \frac{1}{2}$$

$$\therefore \quad e^{\mu T} = 2$$

$$\mu T = \log_e 2 = 0.693$$

$$\therefore \quad \mu = \frac{0.693}{T}$$

which gives another definition of absorption coefficient in terms of *hvl*.

In practice, however, it is $\frac{\mu}{\sigma}$ called *mass absorption coefficient* μ_m *is more important than* μ *and may be defined as equal to the fractional decrease in intensity of a beam of unit cross-section in passing through a unit mass of the absorbing material.*

μ_m is found to depend on the wavelength λ of X-rays and also on atomic number Z of absorber. A relation of it is given by

$$\mu_m = D\lambda^3 Z^3 \qquad \text{...(4)}$$

6.10. X-ray Diffraction

Due to the short wavelength of X-ray, which is of the order of 10^{-8} cm ordinary grating which has 6,000 lines per cm cannot produce any appreciable diffraction with X-rays. Therefore, a crystal grating in which the atoms are arranged in a regular pattern is used for X-ray diffraction. The arranged atoms in the crystal grating correspond to the *grating lines* and the distance between two atoms, which is of the order of 10^{-8} cm, from the grating element. *The difference between crystal grating and optical grating is that the atomic centres of diffraction in the crystal grating are not all in one plane but distributed in space, while in case of optical grating they are limited to one plane.*

The crystal is thus a *three-dimensional space grating* rather than a two-dimensional plane grating. Wen a narrow pencil of X-rays is made to pass through a thin crystal, the diffracted rays emerged in various well-defined directions to form a symmetrical pattern of spots on a photographic plate placed perpendicular to the beam.

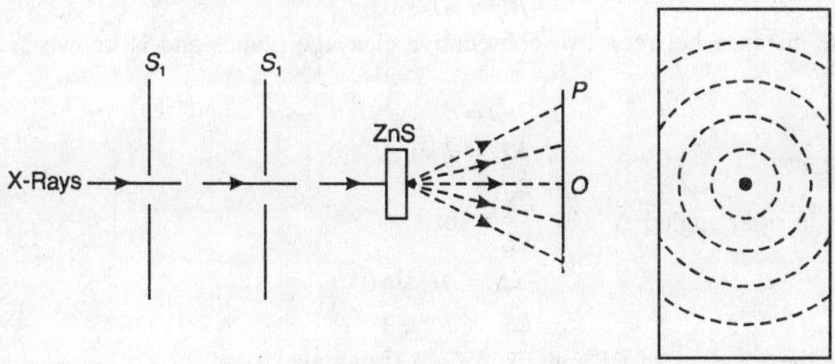

Fig. 6.13

A narrow beam of X-rays is allowed to pass through a thin crystal of zinc blende (ZnS) and transmitted beam is received on a photographic plate *P* as shown in Fig. 6.13. After an exposure of many hours, it can be found on developing the plate that in addition to a central black spot at *O* where the direct beam struck the plate, there are many other fainter spots regularly arranged, indicating that the X-rays, have been diffracted from the various planes of the crystal. These spots, known, as *Laue spots* form a geometrical pattern depending upon the particular crystalline system to which the crystal belongs since different Laue pattern are obtained with different crystalline configurations.

6.11. Bragg's Law

Since the crystal constitutes a three-dimensional grating, therefore, the theory of Laue spots is very complex. Prof. W.L. bragg has given a very simple interpretation of the diffraction pattern obtained by supposing that reflection takes place from the various planes in the crystal. The atoms in a crystal are arranged in regular fashion having a constant space between them. When a beam of X-rays falls on the crystal, each atom acts as a scattering centre and emits secondary wavelets whose envelope gives rise to the reflected wave-front according to Huygen's construction as in optics. Since, X-rays are much more penetrating than ordinary light, therefore, only partial reflection takes place at each layer of atoms.

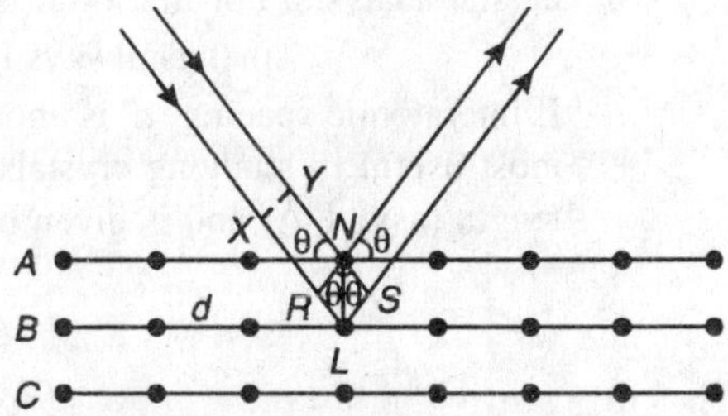

Fig. 6.14

The condition for measurement of wavelength of the X-ray from a set of parallel layers in given plane is obtained as follows:

Let a narrow beam of monochromatic X-rays of wavelength λ be incident on cleavage planes A, B, C ...etc., in a crystal and let XY be an advancing wave front which meets the cleavage plane of the crystal at a glancing angle θ at X. the ray XL is reflected at the cleavage plane B whereas the ray YN is reflected at the cleavage plane A. If the waves reflected from these planes of atoms are in phase with each other at N they will reinforce each other, but if they are in opposite phases they will cancel other's effect and there will be no reflected beam.

For reinforcement the path difference between the two coming waves from any two crystal planes should be an integral multiple of λ, *i.e.*, the wavelength of X-rays.

The path difference between them is given by ($RL + LS$). Therefore, for reinforce condition.

$$RL + LS = n\lambda \qquad \text{... (3)}$$

In right angled $\Delta\, RLN$ $\dfrac{RL}{NL} = \sin\theta$

$$\therefore \qquad RL = NL \sin\theta$$

If d is the distance between two consecutive cleavage planes and is known as lattice constant.

$$\therefore \qquad NL = d$$

$$\text{and} \qquad RL = d \sin\theta$$

Similarly, in right angled $\Delta\, NLS$ $\dfrac{LS}{NL} = \sin\theta$

$$\therefore \qquad LS = NL \sin\theta$$

$$\text{or} \qquad LS = d \sin\theta$$

Substitute these values of RL and LS in Eq. (3), we get

$$d \sin\theta + d \sin\theta = n\lambda$$

$$2d \sin\theta = n\lambda \qquad \text{...(4)}$$

where n in an integral number 0, 1, 2, 3 etc.

This relation is known as Bragg's law which is similar to the relation obtain for the ordinary plane diffraction grating. It indicates that for given values of n, λ and d there is reflection only in a particular direction defined by θ. In other directions there will be destructive interference. For $n = 1$ we get first order spectrum, for $n = 2$ we get the second order spectrum and so on.

6.12. Bragg's Condition for X-ray Diffraction

1. The relation (4) shows why radiations in X-ray region of spectrum are most useful for crystal analysis. For first order spectrum *i.e.*, $n = 1$.

   $\sin\theta$ is always less than 1.

   If interatomic spacing 'd' is about 2 Å, then X-rays with wavelengths less than 2 Å are most useful in studying crystals. The energy of X-ray photon is related to this wavelength (asy) 1 Å) and is given by

$$E = h\nu = \frac{hc}{\lambda} = \frac{6.62\times10^{-34}\times3\times10^{8}}{1\times10^{-10}}$$

$$= 19.86 \times 10^{-16} \text{ joule} = \frac{19.86\times10^{-16}}{1.6\times10^{-19}}$$

$$= 12.4 \times 10^{3}\ eV = 12.4 \text{ keV}$$

   Therefore, X-ray from 10 to 50 KeV range should be used.

2. Bragg reflection can occur only for wavelength $\lambda \le 2d$. This is the reason why visible wavelength cannot be used in diffraction.

3. For the same order and spacing the angle θ decreases as the wavelength decreases. Consequently for gamma rays glancing angles must be used.

6.13. Bragg's X-ray Spectrometer

Bragg devised an apparatus known as the crystal X-ray spectrometer, where the crystal is not used as a transmission grating but as a reflection grating. It is used to measure glancing angle θ. It is also to investigate the structure of a single crystal by using X-rays.

Construction: X-ray spectrometer is similar in construction to optical spectrometer. The essential parts of Bragg's X-ray spectrometer are shown in Fig. 6.15. X-rays from the anti-cathode A, limited to a narrow pencil by two adjustable slits S_1 and S_2 are made to fall upon the crystal C. The crystal C is mounted in wax on the table of the spectrometer. A vernier scale V is attached to table on which the crystal is placed. This vernier scale V is capable tɔ move on the circular scale S and determines the position of the crystal.

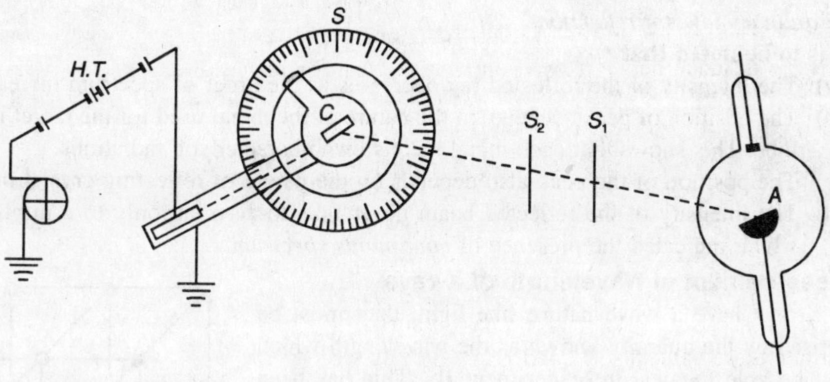

Fig. 6.15

The reflected rays after passing through the slit S_3 enter the ionisation chamber I through a narrow aluminium window W. The chamber is mounted on an arm R and its position is determined by a second vernier scale V_1. One of the plates of ionisation chamber is connected to the positive of af $H.T.$ battery. The negative of which is connected to a quadrant electrometer through a condenser. The other plate of the ionsiation chamber is cathode.

Intensity of X-rays *vs* glancing angle: Curve is obtained as shown in fig. 6.16. It is called *X-ray spectrum.* It is seen that only for certain values of glancing angle (θ) the intensity of X-rays is maximum as indicated by sharp maxima at elecrtometer is a measure of ionisation and consequently gives the intensity of reflected X-rays. To increase ionisation, the chamber is some times filled with sulphur di-oxide or methyl iodide.

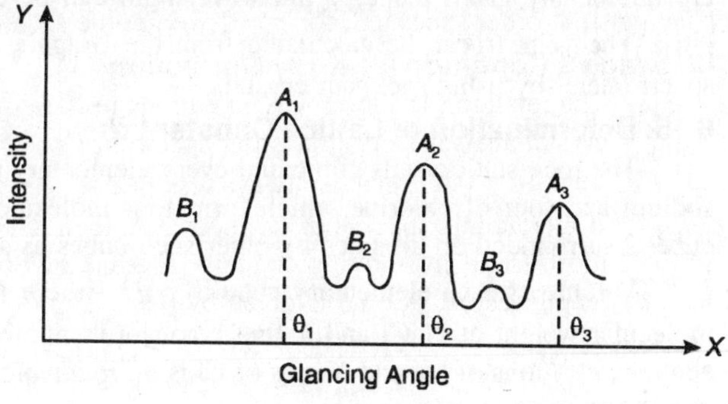

Fig. 6.16

Working: In using the apparatus the crystal is placed in such a position that glancing angle $\theta = 0$ and the ionisation chamber I is adjusted to receive the X-rays.The crystal and the ionisation chamber are now moved in small steps so that the angle through which the chamber is mved is twice the angle through which the crystal is rotated. The rate of deflection of the electrometer measures the intensity of the reflected beam. The observations are repeated for other glancing angles, always taking care to rotate the ionisation chamber through twice the angle through which the crystal is rotated. If we plot a graph taking glancing angle along X-axis and intensity of X-rays along Y-axis. A_1, A_2 and A_3 corresponding to glancing angle θ_1, θ_2 and θ_3. It is found that

$$\sin\theta_1 : \sin\theta_2 : \sin\theta_3 :: 1 : 2 : 3$$

which indicates that the peaks refer to the first, second and third order reflections respectively. In a similar manner, measuring the angles corresponding to the smaller peaks B_1, B_2 and B_3 the same proportionality between the sines of these angles is found. Using Bragg's relation *i.e.*, $2d \sin\theta = n\lambda$ it can be shown that A_1, A_2 and A_3 refer to X-rays of the same wavelength, similarly B_1, B_2 and B_3 refer to the same wavelength but different from that of A_1, A_2 and A_3. A_1B_1 belongs to the first order spectrum, A_2B_2 to the second order and so on. The Bragg with his crystal X-ray spectrometer was able to prove experimentally the relation $2d \sin\theta = n\lambda$ and thereby establish that *X-ray also obey the ordinary laws of reflection.*

It is to be noted that :

(*a*) The intensity of the reflected ray decreases as the order of spectrum increases.

(*b*) The position of peaks depend on the nature of the metal used for the target in the X-ray tube. This shows that each metal has its own characteristic radiations.

(*c*) The position of the peak also depends on the nature of reflecting crystal plane.

(*d*) The intensity of the reflected beam never falls to zero but only to a minimum value, which indicated the presence of *continuous spectrum.*

6.14. Measurement of Wavelength of X-rays

If X-rays have a wave nature like light, they must be characterised by the quantity known as the wavelength which one must be able to measure experimentally. This has been done to a high degree of accuracy with the crystal X-ray spectrometer.

The wavelength of X-rays is determined by using the Bragg's relation $2d \sin\theta = n\lambda$. If we know the value of glancing angle θ for any given order, 'd' the lattice constant of the crystal for any given plane, λ the wavelength can be calculated. The value θ can be calculated from the Bragg's X-ray spectrometer by using rock-salt crystal.

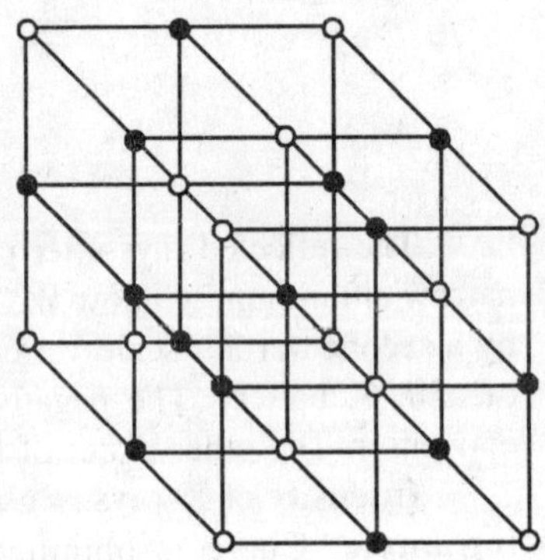

NaCl Crystal Structure

Fig. 6.17

6.15. Determination of Lattice Constant *d*

The rock salt crystals cubic and every elementary cube has eight ions at its corners, four of sodium and four of chlorine which form four molecules of sodium chloride. Every elementary cube is surrounded on all sides by other such cubes as shown in Fig. 6.17.

Volume of each elementary cube is $\rho\, d^3$, where ρ is the density of the crystal. If M is the molecular weight of NaCl and N the Avaogadro's number, since there are $2N$ ions of diffracting centres *i.e*, N ions of Na and N ions of Cl is *a gram molecule,* the mass associated with each ion is $\frac{M}{2N}$. As there is only one ion for each cube in the system, either Na or Cl.

$$\therefore \qquad \frac{M}{2N} = \rho d^3 \qquad \text{or} \qquad d = \left(\frac{M}{2N\rho}\right)^{\frac{1}{3}}$$

Knowing M, N and ρ ,d can be found. Thus for NaCl,

$M = 23 + 35.5 = 58.5 \qquad N = 6.06 \times 10^{23}$ and $\rho = 2.17$

$$\therefore \qquad d = \left(\frac{58.5}{2 \times 6.06 \times 10^{23} \times 2.17}\right)^{\frac{1}{3}} = 2.81 \times 10^{-8} \text{ cm} = 2.81 \text{ Å}$$

If the glancing angle θ for the first order (*i.e*,. $n = 1$) is 11.8° as experimentally found with NaCl, the wavelength of the X-rays used is given by

$$1 \times \lambda = 2d \sin \theta$$

$$\lambda = 2 \times 2.81 \times 10^{-8} \times \sin 11.8°$$

$$= 2.62 \times 10^{-8} \times 0.204$$

$$= 1.51 \times 10^{-8} \text{ cm} = 1.51\text{Å}$$

In the accurate deternination of λ , the following important corrections have to be made:

(*a*). (D)ue to the thermal expansion of crystals, the value of d vary with the temperature. Therefore, the experiment must be performed only at 18°C which is the temperature at which the value of d are usually referred. At other temperature a correction must be applied.

(*b*) refraction of X-rays takes place along with the reflection. Therefore, a correction for reflection of X-rays has to be made.

6.16. Determination of Crystal Structure Using X-rays

The crystal structure can be determined by using Bragg's X-ray spectrometer. In this, the crystal face used for reflecting the X-rays can be so cut that it is parallel first to one set of planes, then to another and so on, in the crystal. A monochromatic beam of X-rays of known wavelength is used and observations are made to obtain the glancing angle θ for a given order of spectrum (say $n = 1$) for the each set of the chosen planes.

Then, by using Bragg's relation,

$$n\lambda = 2d \sin \theta$$

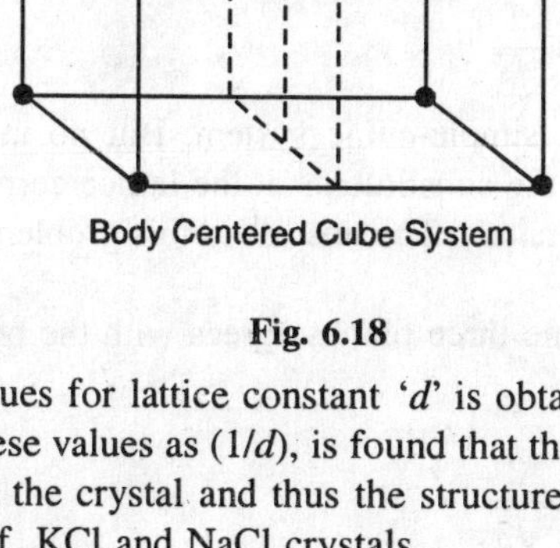

Body Centered Cube System

Fig. 6.18

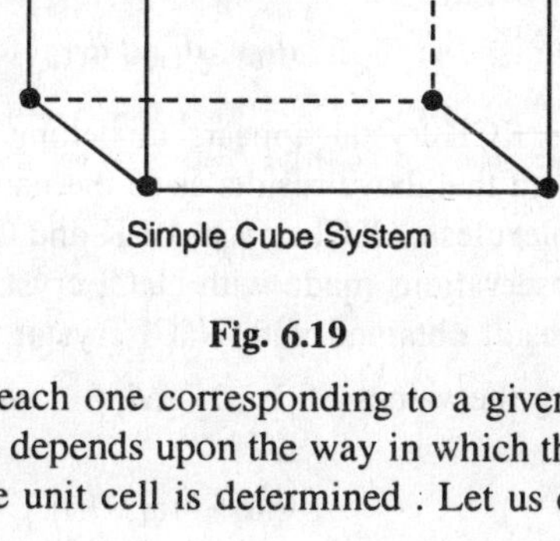

Simple Cube System

Fig. 6.19

a set of values for lattice constant 'd' is obtained, each one corresponding to a given plane. Expressing these values as ($1/d$), is found that the ratio depends upon the way in which the atoms are arranged in the crystal and thus the structure of the unit cell is determined . Let us consider the examples of KCl and NaCl crystals.

6.17. Structure of KCl and NaCl Crystal

These two crystals belong to crystals in which there are three sub-types with different concentrations of atom and with different lattice constant. They are

(*a*) Simple cube system.
(*b*) Body-centred cube system.
(*c*) Face centred cube system.

(*a*) **Simple cube system.** In this system an atom lies at each corner of the cube as shown in fig. 6.19.

(*b*) **Body-centred cube system.** In this system an additional atom at the centre of the cube as shown in Fig. 6.18.

(*c*) **Face-centred cube.** In this system an additional atom lies at the centre of each of the six faces of the cube as shown in Fig. 6.20.

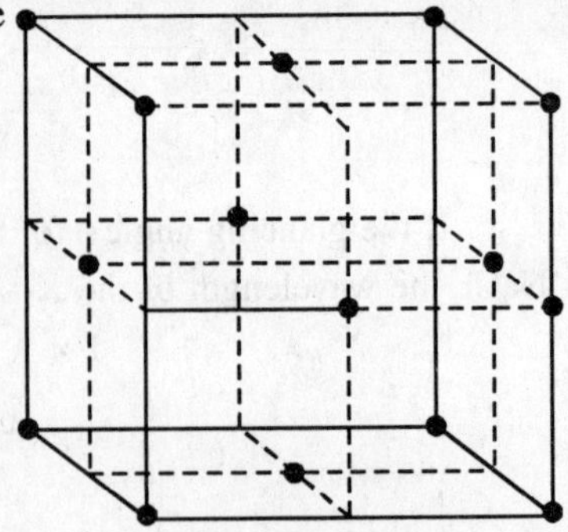

Face-centered Cube

Fig. 6.20

In these three types, the actual spacing d of the planes rich in atoms is given as (1, 0, 0), (1, 1, 0) and (1, 1, 1). The ratio between them is given as follows :

	d_{100}	:	d_{100}	:	d_{111}
Simple cube...	1	:	$\frac{1}{\sqrt{2}}$	:	$\frac{1}{\sqrt{3}}$
Body-centred...	1	:	$\frac{2}{\sqrt{2}}$	:	$\frac{1}{\sqrt{3}}$
Face-centred ...	1	:	$\frac{1}{\sqrt{2}}$	:	$\frac{2}{\sqrt{3}}$

With the KCI crystal, using the three planes (1, 0, 0) (1, 1, 0) and (1, 1, 1) one after the other for reflecting a given monochromatic ray in the Bragg's X-ray spectrometer, the glancing angles θ_1, θ_2 and θ_3 for the first order are measured.

∴ From bragg's law

$$\lambda = 2d_{100} \sin\theta_1 = 2d_{110} \sin\theta_2 = 2d_{111} \sin\theta_3$$

$$\therefore \qquad d_{100} : d_{110} : d_{111} :: \frac{1}{\sin\theta_1} : \frac{1}{\sin\theta_2} : \frac{1}{\sin\theta_3}$$

The values of θ_1, θ_2 and θ_3 obtained from the experiment are 5.22°, 7.3° and 9.05° respectively, which gives

$$d_{100} : d_{110} : d_{111} :: 1 : \frac{1}{\sqrt{2}} : \frac{2}{\sqrt{3}}$$

Hence KCI crystal appears to belong to the simple-cubic system. But no information is available from the above results as to the nature of the constituents at the lattice corners, whether they are molecules of KCI or ions of K and CI alternately. The answer to this problem is obtained from the observations made with NaCI crystal.

The result obtained with NaCI crystal for same three planes agrees with the previous case. By obtaining the values of θ_1, θ_2 and θ_3, we get

$$d_{100} : d_{110} : d_{111} :: 1 : \frac{1}{\sqrt{2}} : \frac{2}{\sqrt{3}}$$

which shows that NaCI crystal belongs to the face-centred cube system.

Example 13. *Monochromatic X-rays of wavelength 1.4Å are incident on a crystal having 1.5Å as interatomic spacing. Find the various orders in which the diffraction takes place.*

Solution. $\lambda = 1.4\text{Å}$, $d = 1.5$ Å

By using Bragg's relation

$$2\,d\sin\theta = n\lambda$$

$$n = \frac{d\sin\theta}{\lambda} = \frac{2\times 1.5\sin\theta}{1.4}$$

$$n = \frac{3}{1.4}\sin\theta \quad \text{or} \quad n = 2.14\sin\theta \quad \textbf{Ans.}$$

Since, $\sin\theta$ is always less than 1.Therefore, maximum order of fthe spectrum can be two *i.e.*, not more than 2nd order.

$$n \le 2$$

Example 14. *The glancing angles of reflection for K_α-X-rays from palladium are 5.4° from (100) planes; 7.6° from (110) planes and 9.4° from (111) planes. From the above data determine the cubic lattice structure of the crystal.*

Solution. From Bragg's formula

$$2d\sin\theta = 1.\lambda \qquad \text{(for first order } n = 1)$$

$$\therefore \quad d = \frac{\lambda}{2\sin\theta}$$

$$\text{or} \quad \frac{1}{d} = \frac{2\sin\theta}{\lambda}$$

For 100 plane

$$\frac{1}{d_1} = \frac{2\times\sin 5.4°}{\lambda} = \frac{2\times 0.0941}{\lambda}$$

For 110 plane

$$\frac{1}{d_2} = \frac{2\times\sin 7.6°}{\lambda} = \frac{2\times 0.1323}{\lambda}$$

For 111 plane

$$\frac{1}{d_3} = \frac{2\times\sin 9.4°}{\lambda} = \frac{2\times 0.1633}{\lambda}$$

$$\therefore \quad \frac{1}{d_1} : \frac{1}{d_2} : \frac{1}{d_3} = 0.0914\,;\, 0.1323 : 0.1633$$

$$= 1 : 1.412 : 1.732 : = \mathbf{1 : \sqrt{2} : \sqrt{3}} \quad \textbf{Ans.}$$

Which shows that the given crystal has simple cubic lattice.

Example 15. *Calculate the wavelength of X-ray beam incident at 12° for the first order reflection from a calcite crystal if the inneratomic spacing d for the crystal is 3.035. Å*

Solution. Given that $\theta = 12°$

Order of spectrum $n = 1$

Interatomic spacing $d = 3.035\text{Å} = 3.035 \times 10^{-8}\text{cm}$

By using Bragg's equation we have

$$2d\sin\theta = n\lambda$$

$$\lambda = \frac{2d\sin\theta}{n}$$

$$= \frac{2\times 3.035\times 10^{-8}\times \sin 12°}{1}$$

$$= 2 \times 3.035 \times 10^{-8} \times 0.2079$$

$$= \mathbf{1.262\ Å}\ \ \textbf{Ans.}\ (\because \sin 12° = 0.2079)$$

Example 16. *The Bragg angle corresponding to the first order reflection from (1, 1, 1) planes in a crystal is 30° when X-rays of wavelength 1.75Å are used, Calculate the interatomic spacing.*

Solution. The interplanar distance d for a plane having Miller indices (h,k,l) is given by

$$d_{hkl} = \frac{a}{\sqrt{h^2 + k^2 + l^2}}$$

where a is the interatomic spacing.

In case of plane having Miller indices (1, 1, 1)

$$h = k = l = 1$$

$$d_{111} = \frac{a}{\sqrt{3}}, \quad \theta = 30°, \quad \therefore \quad \sin\theta = \frac{1}{2}$$

$$n = 1 \qquad \lambda = 1.75Å$$

By Bragg's law

$$2d_{111}\sin\theta = n\lambda$$

$$2\times\frac{a}{\sqrt{3}}\times\frac{1}{2} = 175Å$$

$$a = 1.75 \times \sqrt{3}Å = 3.031\ Å$$

$$= \mathbf{3.031 \times 10^{-8} cm} \quad \textbf{Ans.}$$

6.18. Experimental Methods to Find Crystal Structures in X-ray Diffraction

The Bragg's law, $n\lambda = 2d\sin\theta$ for X-ray diffraction depends upon the values of θ and λ. To satisfy the Bragg's law it is necessary to provide experimentally for a continuous range of values of either λ or θ. The standard methods used for this are:

(1) *Laue Method:* In this method a crystal is held stationary in the path of continuous X-ray beam. The crystal selects out and diffracts the discrete values of λ for which planes of spacing d exist and incidence angle θ satisfy the Bragg's law.

(2) *Rotating Crystal Method:* In this method a single crystal is rotated about and fixed axis in a beam of monochromatic X-rays. The variation in the value of θ brings different atomic planes into position of reflection.

Fig. 6.21

(3) *Powder Method (Debye-Scherrer-Hull Method):* In this method a powdered sample of the crystalline material is places in fixed position in a monochromatic beam. Only for certain orientations the angle of incidence satisfied the Bragg's law.

In these days, the above methods are used but with some modifications (*i*) The Laue method is suitable for the rapid determination of crystal orientation and symmetry and also for studying crystalline imperfection under mechanical and thermal treatment. (*ii*) The rotating crystal method is the most suited method for the structure determination when single crystal specimen is available. (*iii*) The powder method is employed to known the structure of metals.

1. **Laue Method.** The principle of this method is shown in Fig.6.21. The beam falls upon a small crystal of thin crystal section and the diffracted beams are recorded on a photographic plate placed at a distance from the crystal. The rays are limited to a narrow pencil by the slit system, the crystal is set on a holder which permits adjustment of its orientation and photographic plate is placed in the other holder.

A continuous beam of X-ray is obtained by using a bulb with tungsten anticathode and a Potential of about 65,000 volts. Laue photograph obtained is shown in Fig. 6.21. The various crystal planes reflect the X-rays as if they constituted a number of mirrors set at different angles.

The relation : $$n\lambda = 2d\sin\theta$$

can be satisfied, although the angle θ is fixed for each set of planes ; because radiation of appropriate wavelength can be selected from the continuous range. If the crystal is rotated, the spots change their position upon the screen, as if they were the reflections of a narrow beam of light by the faces of a cut-glass stopper. All the spots which lie on an ellipse are due to reflections by planes parallel to common zone axis. Incident rays fall upon a crystal having zone axis parallel to ZZ′, and we consider rays which are reflected by a plane of corresponding zone. Suppose a reflecting plane a be rotated about ZZ-axis, the reflected ray will trace a curve *BB′* on the plate. If the incident rays make an angle θ with ZZ′-axis, the reflected ray will make the same angle with it, so that they lie on a cone of semi-vertical angle θ about the zone axis. The ellipses in the Laue photograph are the intersection of such cones. Each spot is at an intersection of *ellipce*, since a phane is parallel to two zone axes in a crystal plane and can reflect the X-ray beam.

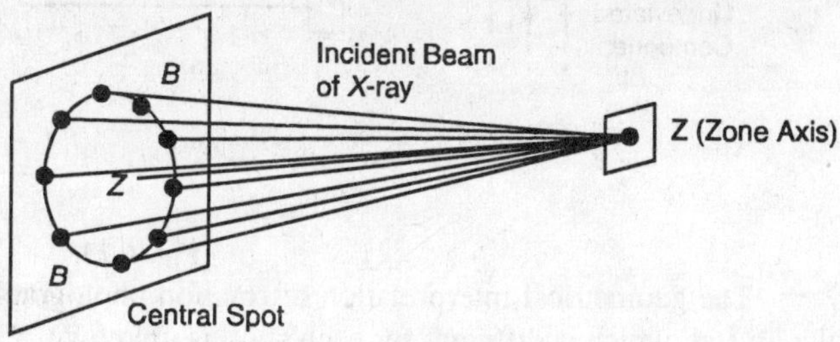

Fig. 6.22

The deduction of the crystal structure from the appearance of Laue photograph is a complicated process, as the intensities of the spots do not depend on the structure alone. Because of the extended range of wavelengths it is possible for several wevelengths to reflect in different orders from a single plane so that different orders to reflection may superimpose to give a single spot. They are also influenced by the different blackening effect of radiation of different wavelengths, and complications arise here owing to the absorption of the X-rays by silver and bromine in the photographic plate. *In spite of these difficulties, the Laue photograph is used oftenly in the determination of crystal orientation and symmetry.*

2. **Rotating Crystal Method :** In the rotating crystal method a single crystal is rotated about a fixed axis in a beam of monoenergetic X-rays. The variation in the angle θ brings different atomic planes into position for reflection. A modified rotating crystal method is used for structure determination when a single crystal specimen is available.

The diffracted beams are recorded photographically upon a plate or upon a film bent into a cylinder around the axis of rotation. The dimensions of the crystal should not be greater than 1 mm. The incident X-ray beam is made nearly monochromatic by a filter or by reflection from a monochromatic crystal. The beam is diffracted from a given crystal plane whenever in the courses of rotation the value of θ satisfies the Bragg's Equation. Beams from all planes parallel to the vertical rotation axes will lie in the horizontal plane. Planes with other orientations will reflect in layers above and below the horizontal plane. As the crystal rotates, one plane after another is brought in such position that it reflects the radiation. The film when developed show a number of spots due to various planes which have passed through requisite positions. The apparatus shown in Fig. 6.23 can be used for both Laue and rotation photographs. In this the crystal is rotated by a clockwork mechanism contained in a metal box.

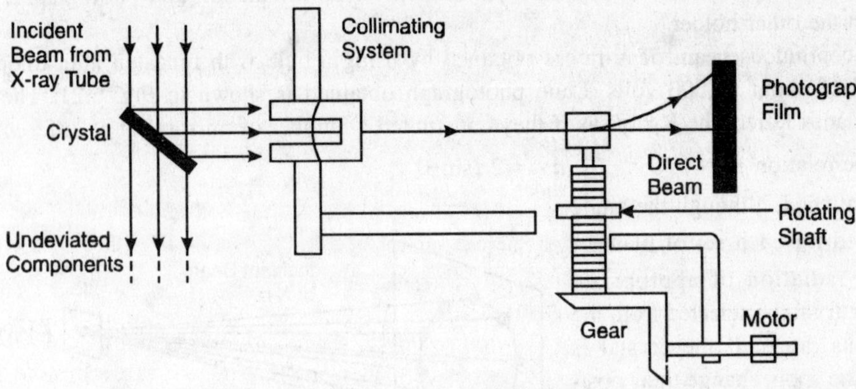

Fig. 6.23

The geometrical interpretation of rotation photographs is more complex. The orientation of the crystal, which is different for each spot, is unknown. We only know that it lies within a certain range when the crystal has been turned through limited arc. The distance between two successive lattice points along the zone axis. *PO* is represented in Fig. 6.24. The incidental rays are at right angles to this axis. The necessary condition for diffraction is that the path difference for rays scattered by *P* and *O* should differ by a whole number of wavelength so that the path difference *OQ* is $n\lambda$, this condition determining the angle which the diffracted rays make with the zone axis. As the crystal is rotated it will pass through a number of positions suitable for diffraction which have in common a path difference of $n\lambda$ between *O* and *P*. All the corresponding spots lie on a layer line which will be the *n*th above or below the equational line in photograph.

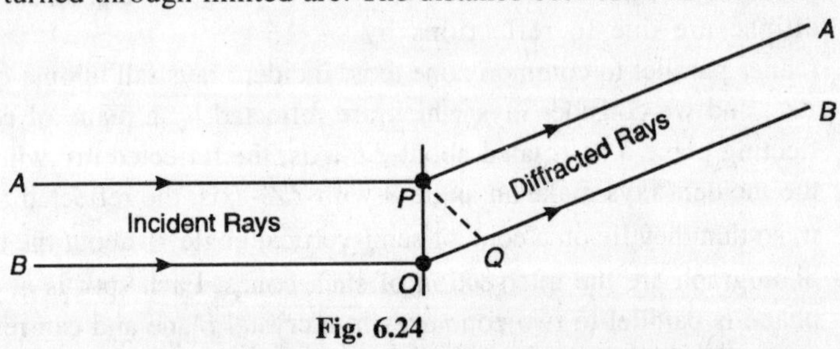

Fig. 6.24

If *R* is axis about which the crystal is rotated, when an incident beam of X-ray falls on the crystal, reflection will take place with indices (*h* ,*k* ,*l*) which move upwards making an ϕ_1 angle with the horizontal given by r sin $\phi_1 = \lambda$. Such reflections constitute first layer line. In general, reflections with indices (*h, k, l*) lie on the layer line *l*. Since the vertical spacing of the layer lines depends only upon the distance between lattice points on the axis about which crystal is rotated; the distance (*r*) can be immediately deduced from the photograph.

If rotation photograph is taken separately about three translation axes, *a, b, c* and hence the dimensions of the unit cell of crystal structure is determined. The rotating crystal method is thus very powerful method of giving the size of the unit cell.

We will obtain so many spots with complete rotation which will cause confusion, therefore, it is the useful practice to rotate the crystal. backwards and forwards through an angular range of only 15° in order to have limited number of spots on the photographic plate.

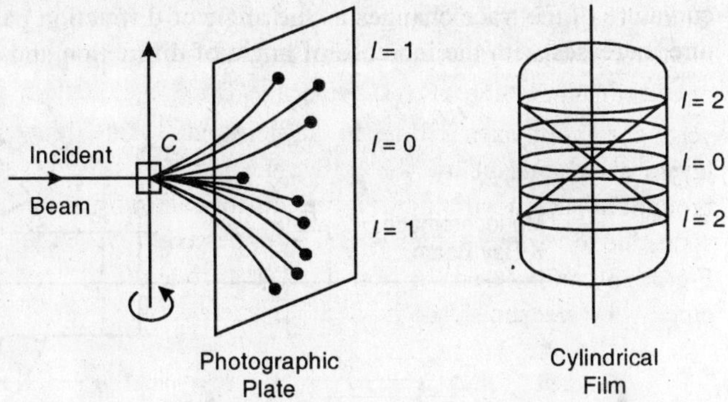

Fig. 6.25

A better technique was given by Weissenberg. According to him crystal is rotated through 180° and back again continuously and at the same time the cylindrical camera containing the film moves at a constant speed backward and forward parallel to the axis of rotation. The rotation of camera and film is adjusted so that a given position of the camera corresponds accurately to definite angular position of the crystal as its rotation. The co-ordinates of a spot on the film thus give both the angle of reflection as well as the position of the reflecting plane and thus allows the spot to be indexed without doubt. A metal cylinder provided with equiatorial slit a few millimeters wide is placed between the crystal and film is placed in such a position that the spot of only layer line can pass through it.

In modern methods we use *scintillation counters or proportional counters* to detect the diffracted radiation. These methods allow automatic collection of data and are more useful in the study of *dislocation and chemical bonding in crystal.*

3. Powder-Photograph Method. In the powder-photograph method the incident monochromatic radiation strikes a finely powdered specimen or a fine grained polycrystalline specimen contained in a thin-walled capillary tube. The distribution of crystalline orientations will be nearly continuous. The powder method is convenient because single crystal is not required. Diffracted rays go out from individual crystals width happens to be oriented which planes making an incident angle θ with the beam satisfying the Bragg's equation. Diffracted rays leave the specimen along the generator of cones concentric with the original beam. The generator makes an angle 2θ with the direction of the original beam, where θ is the Bragg's angle. The cones intercept the film in a series of concentric rings as shown in Fig. 6.26.

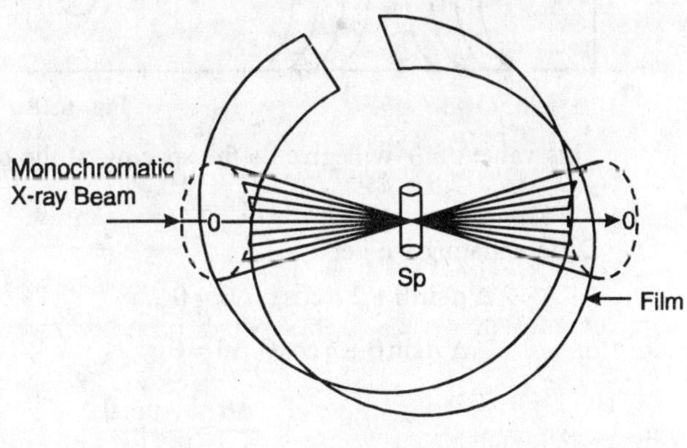

Fig. 6.26

The arrangement of experiment is shown in Fig. 6.27. The X-rays are made monochromatic by passing them through zirconium oxide filter *F*. A narrow beam of monochromatic X-rays limited by slits S_1 and S_2 falls on the powdered specimen *C*. The photographic film fits round the inner surface of the camera covering nearly the whole in circumference in order to record beams diffracted upto nearly 180°. The strip of the film is usually long in its horizontal dimensions, so that only arcs of circles appear on the film on developing. The X-rays after falling on the powder passes out of the camera through a hole cut in the film, in order to minimise the fogging produce due to direct scattering. The traces obtained on the photographic plate are shown in Fig.6.27. The curvature of the trace changes as the angle of diffraction passes through 90°. The radius of curvature increases with the increase of angle of diffraction and at 90° straight lines are obtained.

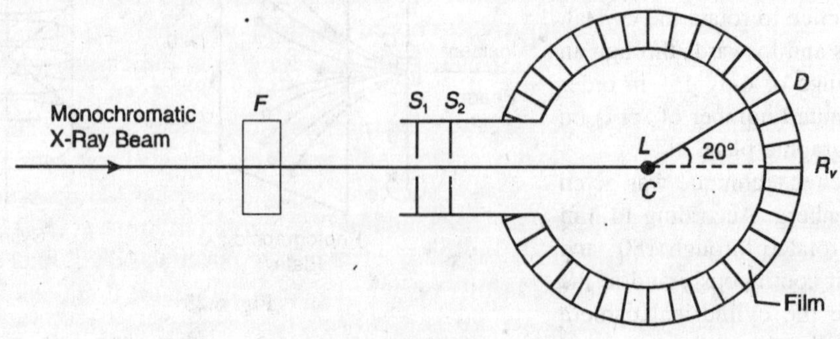

Fig. 6.27

The crystal structure can be obtained from the arrangement of the traces and their relative intensities. Consider a spectrum with glancing angle θ at a point D which is at a distance R from O where the direct incident beam strikes the film and if L be the radius of the cylindrical camera, then

$$\theta = \frac{R}{2L}$$

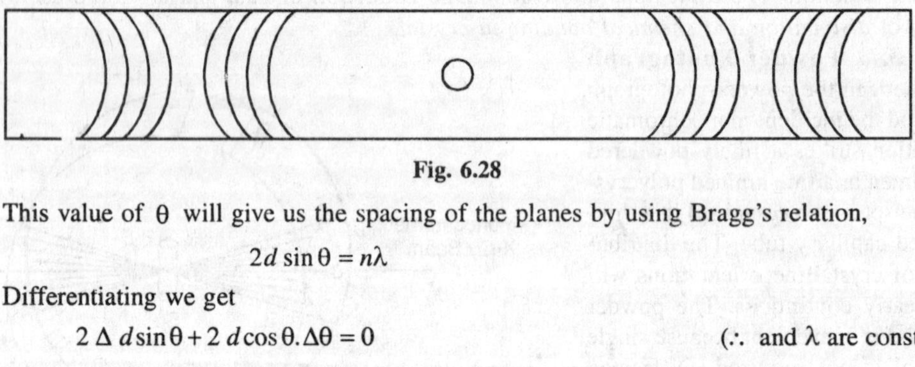

Fig. 6.28

This value of θ will give us the spacing of the planes by using Bragg's relation,

$$2d \sin\theta = n\lambda$$

Differentiating we get

$$2\,\Delta\, d\sin\theta + 2\, d\cos\theta.\Delta\theta = 0 \qquad (\therefore \text{ and } \lambda \text{ are constants})$$

or $$\Delta\, d\sin\theta + d\cos\theta.\Delta\theta = 0$$

$$\therefore \qquad \frac{\Delta\theta}{\Delta d} = -\frac{\tan\theta}{d}$$

when θ tends to 90°, $\frac{\Delta\theta}{\Delta d}$ becomes very great, so that small variation in d produces large variations in θ. Therefore, high resolution when the rays are reflected back through nearly 180°. This is the reason why spectra for which the semi-vertical angle is greater than a right angle are obtained by the use of the long cylindrical film nearly surrounding the crystal powder.

This method is applicable for all kinds of crystalline matter, however, small the individual crystal may be. This method is specially used to determine the structure of simple cubic crystals for which the spacing of the unit cell are all equal. It is most useful for the investigation of metals and alloys structure, while single crystal are not possible.

In spite of X-ray diffractions we can also use Electron Diffraction and Neutron diffraction to determine crystal structure. In case of electron diffraction, the amount of matter required to produce electron diffraction, however, is extremely small compared with that required for X-ray diffraction, and the time of exposure required for a photographic record are measured in fraction of a second. Due to these facts. Electron diffraction is very useful to apply where it is difficult or impossible to use X-rays, *e.g.*, in the structure of surface films or to the analysis of the molecules of gas. In case of neutron diffraction intensities depend largely upon scattering from nuclei, whereas in case of X-ray diffraction intensities depend upon scattering from the orbital electrons of the atom.

Neutron diffraction tells us whether two scattering amplitudes have the same size or not but does not tell us for any given nuclei whether amplitude is positive or negative. Neutron diffraction is also used in a long graphite rod to provide source of "*cold neutron*". Cold neutrons are those whose probable speed is much below the speed required for thermal neutrons at (300 K). Another important use of neutron diffraction is the study of magnetic materials. Shull and Smart used neutron diffraction for the detection of antiferromagnetism in MnO sample first time. In the study of magnetic properties of the material we use the fact that a neutron can be scattered by interaction of its magnetic moment with the atomic or ionic magnetic moment of the sample material. The diffraction of particles not only confined to electrons and neutrons but also possible for atoms and molecules, the corresponding wavelength being given by the de Broglie's relation.

Results obtained by X-ray Crystallography. The following important results have been obtained by X-ray crystallography:

1. *The crystal structure of metals.* X-ray analysis has clearly pròved that most metals and alloys are crystalline in structure. Number of metals and alloys have been found to be cubic, tetragonal and hexagonal types of crystals.

2. *The internal structure of the atom.* With X-rays we can see the interior structure of atom. By the analysis of the photoelectric effect of X-ray, it is possible to determine the distribution of the electrons around the nucleus.

3. *The nature of all matter.* With X-ray crystallography we can know the architecture of matter in the solid state, which is of fundamental scientific importance.

4. *Type of chemical bond in solids.* X-ray crystallography gives us a fair knowledge of chemical bonds in solids *i.e.*, which types of forces are acting between the ions and the atoms constituting molecules *e.g.*, in case of NaCl there are electrostatic forces between Na^+ and Cl^-. This is known as *heteropolar binding* whereas in case of Hydrogen (H_2) there is *homopolar binding.*

6.19 Industrial Applications of X-rays

X-rays are widely used in industry such as:

(*i*) To locate defects in internal structure of a body. It is also used to lacate the imperfection in mould forgings and castings which could not otherwise be examined internally.

(*ii*) To study the structure of rubber, plastics in rubber and plastic industries. The diffraction of X-rays by these substances leads to valuable information about their molecular grouping.

(*iii*) In megtallurgy X-rays are used to analyse the structure of various alloys by determining the structure of in-gots with the help help of X-ray diffraction.

(*iv*) In heavy industries X-rays are used to test the gross structure of the material and welding etc. For this purpose the X-ray photographs are taken which reveal defects in the manufacture of heavy steel plates and also locate the cracks and blow holes in metal plates. X-rays are also used to detect flaws in the wings of aeroplanes and parts of other fast moving machines.

From this it is clear that X-rays have become the important part of all the heavy industries.

6.20. Other Uses of X-rays

X-rays have many applications in many other branches such as for scientific research and for medical purposes.

Applications of X-rays for scientific Research:

(*i*) X-rays are used to study the crystal structure in crystallography.

(*ii*) X-rays are used in atomic physics to study the atomic structure.

(*iii*) X-rays are used to find the atomic number of the elements and also for the identification of the elements.

(*iv*) X-rays are used to study the structure of complex organic compounds.

Applications of X-rays for Medical Purposes : In medicine, X-rays are used for two main purposes.

(*a*) **Radiography.** X-rays are widely used in *radiography*. In radiography we use X-ray as the source and a photo film upon which we take the X-ray photos. Since bones and other solids are more dense and hence more opaque to X-rays than flesh and blood, therefore, we can detect fractures, tumours, the presence of foreign matter like bullet etc., in human body by interposing it between the X-ray tube and photo film. This process of taking X-ray photos on photo films is known as radiography.

(*b*) **X-Ray Theraphy.** Many types of skin diseases, malignant sores, cancers, tumours etc. have been cured by controlled exposure to X-rays of suitable quality. In this, we destroy the diseased portion of the body by exposing to X-ray for a suitable time. Long exposure of human body to X-rays may cause unnecessary incurable injuries and destroy white cells in the blood.

X-rays have been used for the identification of different types of cells and tissues and for bringing about genetic mutations.

6.21. Auger Effect

In the basic explanation of X-rays it was assumed that the vacancy of an inner shell, say *'K'* shell is filled by an electron from some outer shell, along withe the emission of a photon. However, experimentally it is found that flourescence yield $\omega = n_p / n_e$ where $n_p \rightarrow$ no, of K photons and $n_e \rightarrow$ no. of K electrons knocked out (*i.e.*, no of K-shell vacancies) is < 1 which may range between 0.1 for light elements to 0.95 for uranium.

Auger in 1925 gave explanation for the above observation in the following way:

The ejection of electron from K-shell is often accompanied by the *ejection* of another electron. It is due to the fact that an electron that may undergo transition to the vacant K-shell may knock out another electron on its passage to the same shell. This is called *Auger transition* and the emitted electron is known as *Auger electron.*

It should be clearly noted that *Auger electron* is not because of the absorption of a photon given out by an electron when it comes from a higher energy level to a lower energy level but it comes directly as a part of readjustment of the atom. For example, the vacancy in the K-shell may be filled by an electron of the L_{I}-shell and another electron from L_{II}-shell may be knocked out thereby creating two vacancies in the L-shell. So the de-excitation of the atom may be accompanied by either the emission of *photon* (characteristics radiation) of an *electron* (Auger electron). These two processes together account for the number of vacancies that exist in K-shell. In *nuclear transition*, the *Auger effect* is also known as *internal conversion* or Auto-ionization. So under certain conditions of energy, an atom in a state of single ionization may undergo spontaneously an *Auger transition* which leaves it in a state of double ionization. The electron released in such a transition is expelled from the atom and left with an amount of kinetic energy E given by

$$E = E_i - E_f \qquad ...(1)$$

where E_i is the initial energy of the singly ionized atom and E_f is the final energy of the doubly ionized atom. This type of atomic transition does not involve the emission of a photon and, therefore, is often called a "nonradiative" transitions. Nonradiative transitions are possible, in general, only when an electron can be ejected from the atom; and for this to be possible it is necessary that E as defined by Eq. (1) be zero or more than zero. The electrons ejected by such radiationless transitions were detected by *Auger* as fat-short tracks made by them on the gas of a cloud chamber.

Example 17. *The short wavelength limit of the continuous X-ray spectrum emitted by an X-ray tube operates at 30 kV is 0.414 Å. Calculate the Plancks's constant.*

Solution. We have the relation

$$\lambda_{min} = \frac{hc}{eV} \quad \text{or} \quad h = \frac{eV}{c}\lambda_{min}$$

In this case $e = 1.6 \times 10^{-19}$ coulomb, $V = 30$ kV $= 30 \times 10^3$ volts,

$$\lambda_{min} = 0.414 \text{ Å} = 0.414 \times 10^{-10} \text{ m}$$

and $$c = 3 \times 10^8 \text{ m/sec}$$

$$\therefore \quad h = \frac{1.6\times10^{-19}\times30\times10^3\times0.414\times10^{-10}}{3\times10^8}$$

$$= \mathbf{6.624 \times 10^{-34}\ J\text{-}sec} \quad \textbf{Ans.}$$

Example 18. *An X-ray tube operates at 20 kV. Find*

(*a*) *the maximum speed of the electrons striking the anticathode.*

(*b*) *the shortest wavelength of X-ray produced.*

Solution. We have the relation for max. speed of electrons as

$$\frac{1}{2}mv^2_{max} = eV \quad \text{or} \quad v^2_{max} = \sqrt{\frac{2eV}{m}}$$

In this case $$V = 20\ kV = 20 \times 10^3 \text{ volts}$$

$$\therefore \quad v_{max} = \sqrt{\frac{2\times1.6\times10^{-19}\times20\times10^3}{9.1\times10^{-31}}}$$

$$= \mathbf{0.838 \times 10^8\ m/sec.} \quad \textbf{(Ans).}$$

(*b*) Relation for shortest wavelength of X-ray is given by

$$\lambda_{min} = \frac{hc}{eV} = \frac{6.63\times10^{-34}\times3\times10^8}{1.6\times10^{-19}\times20\times10^3} = \mathbf{6.216 \times 10^{-11}\ m}$$

$$= 0.6216 \times 10^{-10} \text{ m} = \mathbf{0.6216\ Å}$$

Example 19. *If the K_α radiation of Mo (Z = 42) has a wavelength of 0.71 Å. Calculate the wavelength of the corresponding radiation of Cu (Z = 29).*

Solution. From Mosley's law

$$\nu = a\,(Z - b)^2$$

For K_α radiation, the screening constant $b = 1$ and $\upsilon = c/\lambda$, So we have

$$1/\lambda \propto (Z-1)^2$$

Let λ_{Cu} and λ_{Mo} be the wavelengths corresponding to the radiation of Cu and Mo respectively,

$$\therefore \quad \frac{\lambda_{Cu}}{\lambda_{Mo}} = \frac{(Z_{Mo}-1)^2}{(Z_{Cu}-1)^2} = \frac{(41)^2}{(28)^2}$$

Now $\lambda_{Mo} = 0.71$ Å,

$$\lambda_{Cu} = 0.71 \times \left(\frac{41}{28}\right)^2 = \mathbf{1.52\ Å} \quad \textbf{Ans.}$$

Example 20. *An X-rays tube operates on 5 kV. Calculate the shortest wavelength of the X-ray produced.*

Solution. We have the relation for shortest wavelength with applied potential as

$$\lambda_{min} = \frac{hc}{eV}$$

Now $h = 6.63 \times 10^{-34}$ J-sec, $c = 3.0 \times 10^8$ m/sec, $e = 1.6 \times 10^{-19}$ coulomb
and $V = 5\ kV = 5 \times 10^3$ volts.

$$\lambda_{min} = \frac{6.63 \times 10^{-34} \times 3.0 \times 10^8}{1.6 \times 10^{-19} \times 5 \times 10^3} = 2.486 \times 10^{-10} \text{ m} = \mathbf{2.486\ Å} \quad \textbf{Ans.}$$

Example 21. *If the series limit of the Balmer series for hydrogen is 364.6 nm, calculate the atomic number of the element which given X-ray wavelength down to 0.1 nm.*

Solution. In hydrogen spectrum the Balmer series is given by

$$\frac{1}{\lambda} = R\left(\frac{1}{2^2} - \frac{1}{n^2}\right), \text{ where n = 3, 4, 5...}\infty$$

For the series limit ($n = \infty$), we have

$$\frac{1}{\lambda_\infty} = \frac{R}{4} \qquad \therefore R = \frac{4}{\lambda_\infty} = \frac{4}{364.6 \text{ nm}}$$

The X-ray wavelengths of the *K*-series are given by the relation

$$\frac{1}{\lambda} = R\ (Z-1)^2\left(\frac{1}{1^2} - \frac{1}{n^2}\right)$$

when $n = \infty$, the minimum wavelength of this *K*-series occurs

$$\frac{1}{\lambda} = R(A-1)^2 \qquad \text{or} \qquad (Z-1)^2 = \frac{1}{\lambda R}$$

Now $\lambda = 0.1$ nm and $R = \dfrac{4}{364.6 \text{ nm}}$

$$\therefore \qquad (Z-1)^2 = \frac{1 \times 364.6}{0.1 \times 4} = 911.5$$

or Z – 1 = 30.19 $\therefore$ **Z = 31** which corresponds to element Gallium. **Ans.**

Example 22. *Find the nuclear screening constant for the L-series of X-rays if it is known that the X-rays with a wavelength of $\lambda = 1.43$ Å are emitted when an electron in a tungsten atom (Z= 74) is transferred from the M-level to L-level. Take Rydberg constant, R = 10.97 × 10⁶ m⁻¹.*

Solution. From mosley's law, the spectral line L_a emitted when an electron jumps from the *M*-level to *L*-level is given by

$$\frac{1}{\lambda} = R(Z-b)^2\left(\frac{1}{2^2} - \frac{1}{3^2}\right)$$

[$\because$ For *M*-level, $n_2 = 3$ and *L*-level, $n_1 = 2$]

$$\therefore \qquad \frac{1}{\lambda} = \frac{5R}{36}(Z-b)^2 \qquad \text{or} \qquad (Z-b)^2 = \frac{36}{5R\ \lambda}$$

In this case $R = 10.97 \times 10^6 \text{m}^{-1}$ and $\lambda = 1.43 \text{ Å} = 1.43 \times 10^{-10} \text{m}$

$$\therefore \quad (74 - b)^2 = \frac{36}{5 \times 10.97 \times 10^6 \times 1.43 \times 10^{-10}} = 4589.76$$

or $74 - b = 67.748$ or. $\mathbf{b = 6.25}$ **Ans.**

Example. 23. *In Bragg's reflection of X-ray, a reflection was found at 30° glancing angle with lattice planes of spacing 1.87 Å. If this is a second order reflection. Calculate the wavelength of X-rays.*

Solution. From Bragg's law

$$2d \sin\theta = n\lambda$$

Now $\theta = 30°$, $n = 2$ and $d = 1.87$ Å

$$\therefore \quad \lambda = \frac{2d \sin\theta}{n} = \frac{2 \times 1.87 \times \sin 30°}{2} = 1.87 \times 0.5 = \mathbf{0.935 \text{ Å}} \text{ **Ans.**}$$

Example 24. *An X-ray beam of wavelength 0.97 Å is obtained in the third order after reflection at 60° from the crystal plane. Another beam is obtained in the first order after reflection at 30° from the same crystal plane. Find the wavelength of the second X-ray beam.*

Solution. From Bragg's law of diffraction, we have

$$2d \sin\theta = n\lambda$$

For the first X-ray beam, $\theta = 60°$, $n = 3$ and $\lambda = 0.97$ Å

$$\therefore \quad 2d \sin 60° = 3 \times 0.97 \qquad ...(2)$$

Let λ' be the wavelength of the second X-ray beam, As the crystal is same, so therefore

$$2d \sin\theta' = n'\lambda'$$

Now $\theta' = 30°$ and $n = 1$

$$2d \sin 30° = 1 \times \lambda' \qquad ...(3)$$

Dividing equation (3) by (2), we obtain

$$\frac{\lambda'}{3 \times 0.97} = \frac{\sin 30°}{\sin 60°} \quad \text{or} \quad \lambda' = 3 \times 0.97 \times \frac{(1/2)}{(\sqrt{3}/2)}$$

Which gives $\lambda' = \dfrac{3 \times 0.97 \times 1}{\sqrt{3}} = \mathbf{1.68 \text{ Å}}$ **Ans.**

Example 25. *The first order reflection from the plane of NaCl is obtained at an angle* $2\theta = 20°$ *with the incident beam. If d = 2.82 Å, calculate the wavelength of X-ray used.*

Solution. From Bragg's equation for the diffraction of X-rays, $2d \sin\theta = n\lambda$

For first order $n = 1$, $\therefore$ $2d \sin\theta = \lambda$

Here $2\theta = 2°$ or $\theta = 10°$ and $d = 2.82 \times 10^{-10}$ m

$$\therefore \quad \lambda = 2 \times 2.82 \times 10^{-10} \sin 10° = 2 \times 2.82 \times 10^{-10} \times 0.1736$$
$$= 9.791 \times 10^{-11} \text{ m} = \mathbf{0.9791 \text{ Å}} \text{ **Ans.**}$$

Example 26. *A set of lattice planes reflects X-rays of wavelength 1.32 Å at a glancing angle of 9°30′. Deduce the possible spacing of this set of planes.*

Solution. We have the equation of Bragg's law

$$2d \sin\theta = n\lambda$$

For first order reflection, $n = 1$

$$\therefore \quad 2d\sin\theta_1 = \lambda \qquad ...(1)$$

For second order reflection. $n = 2$

$$\therefore \quad 2d\sin\theta_2 = 2\lambda \qquad ...(2)$$

For third order reflection, $n = 3$

$$\therefore \quad 2d\sin\theta_3 = 3\lambda \qquad ...(3)$$

Similarly, for fourth order reflection, $n = 4$

$$2d\sin\theta_4 = 4\lambda \qquad ...(4)$$

Dividing equation (2) by equation (1), we get

$$\frac{\sin\theta_2}{\sin\theta_1} = 2 \quad \text{or} \quad \sin\theta_2 = 2\sin\theta_1 \qquad ...(5)$$

Similarly we can get from eqs (3) and (1) and (4) and (1)

$$\sin\theta_3 = 3\sin\theta_1 . \quad \text{and} \quad \sin\theta_4 = 4\sin\theta_1$$

So for the first order reflection the spacing d is given by

$$\frac{d}{n} = \frac{\lambda}{2\sin\theta_1} = \frac{1.32\times10^{-10}}{2\times\sin 9°30'}$$

or

$$d = \frac{1.32\times10^{-10}}{2\times0.1650} = \mathbf{4.0 \times 10^{-10}\ m.}$$

Again If $n = 2$,

$$\frac{d}{n} = \frac{\lambda}{2\sin\theta_2} = \frac{\lambda}{2\times2\sin\theta_1} = \frac{1}{2}\left(\frac{\lambda}{2\sin\theta_1}\right)$$

$$\therefore \quad \frac{d}{n} = \frac{4.0\times10^{-10}}{2}\text{m} = \mathbf{2.0 \times 10^{-10}\ m}$$

and, if , = 3

$$\frac{d}{n} = \frac{\lambda}{2\sin\theta_3} = \frac{\lambda}{2\times3\sin\theta_1} = \frac{1}{3}\times4.0\times10^{-10}\text{m}$$

or

$$\frac{d}{n} = 1.33\times10^{-10}\text{ m}$$

and, if $n = 4$

$$\frac{d}{n} = \frac{\lambda}{2\sin\theta_4} = \frac{\lambda}{2\times4\sin\theta_1}$$

$$= \frac{1}{4}\times4.0\times10^{-10} = \mathbf{1.0 \times 10^{-10}\ m} \quad \textbf{Ans.}$$

Similarly we can get the other values going in the same way.

6.22 Non-destructive Tests

The purpose of destructive test is to know the quality of the material so that it may be considered suitable for a particular job or may be rejected if it is not upto the specifications:

Term 'non-destructive testing' is used in general from as this method uses testing or inspections without changing the structural properties of the material so that it may still be used in future. These tests have the following servicibility.

1. it has no effect on future servicibility.
2. Profits or productivity may be increased after the tests.
3. Safety norms nay be applied as per the requirements after the tests.

Non-Destructive tests involve the following essential parts.

(*i*) application of the testing or inspection material.

(*ii*) modifications of the materials after identifying the various defects.

(*iii*) using suitable device to identify the defects and their positions.

(*iv*) Conversion of the change into a suitable from and then getting the correct interpretation of the results.

In addition; one has to be sure that the interpretation of the result is sure for better marketing avenues.

Non destructive tests are generally used to *find and identify the defects in the* materials while destructive tests pertain to finding the *mechanical properties like hardness, toughness, compressive strength, yield strength, tensile strength etc.*

We will briefly describe some of the important methods used in *non-destructive tests.*

6.23 Magnetic Method

In this method the specimen is magnitized and the distortion of magnetic field due to defects is studied. The magnetic field distortion may be studied in many ways; but the most widely used method employs *magnetic particles* such as iron oxide magnetic flux distortion due to a defect will change the *induced voltage* generated in the search coil. Flaws in magnetized pipe causes a change in magnetic flux which is turn produces a change in induced voltage. By suitable parameters it is possible to distinguished between *inside and outside* fatigue cracks inside the pipe. Using this technique defects of thickness is 0.7 cm to 1.25 cm can be identified. Further inspection may let us know about the presence or absence of defects. An experienced operator can easily identify the position and size of the defects. Some factors like sharpness, depth, direction can indicate the location of defects. However by this method *hairline cracks* are difficult to locate unless they are below the surface.

6.24 Electrical Method

These methods are based on the princo;e of different materials having different resistivities. In electrostatic method, variation of electric fields can be deleted by motion of charged particles.

When a coil carrying high frequency alternating current is moved near a conductor, *eddy currents* are induced in the *conductor.* Also a magnetic field is associated with induced currents. Some defects, flows, physical and chemical constituents and other structural variations causes change of resistivity thereby the eddy currents also very resulting in the variations of *induced magnetic fields.* These variations in magnetic fields can either be detected by search coil or by their effect on coils inducing the currents.

Various techniques used for non-destructive electrical tests are:

(*i*) Change of electrical resistance (*ii*) Thermoelectric effect

(*iii*) Triboelectric effect (*iv*) Variation of static field

These techniques are useful to find lamination in rolled sheets, casting defects, joint defects, defective welding spots, cracks in electrical bushing as well as electrical insulation defects. .

(*i*) A special type of *Kelvin double bridge* know as electrical crack detector has been used to measure *resistance* heavier parts and to know the propagation of cracks, fatigues and fatigues. Potentials drop technique is used to study the depth of the cracks. Electrical resistance techniques is used to study corrosion ratios using the method of loss of metal from a probe installed in *corrosive system under study.*

(*ii*) *Thermo Electric effects* is the setting up of e.m.f. in a circuit of two dissimilar metals forming two junctions, one being kept at higher temperature and the other at a lower temperature. This arrangement is called *thermocouple*. If no. of thermocouples are connected in series, more e.m.f. is produced and the arrangement is called hermopiple. In a thermoelectric metal comparator one probe is heated to a fixed temperature and the other is at the specimen, so that it attains the temperature of specimen. Signal from the probe is amplified and fed to a phase sensitive detector. When probe is placed on specimen, the temperature drops and the amount of fall in temperature depends on the thickness and thermal conductivity of the specimen. This techniques is used in measuring thickness for non-conductor films.

(*iii*) *Tribo Electric effects* corresponds to the phenomenon when *two metallurgically different metals are brought in contact and agitated mechanically thus generating a voltage.* This voltage depends upon the nature of metals or alloys in contact. Heat treatment produces chemical changes in most alloys and so this techniques can be used to distinguish between *heat treated and hot worked alloys.* This technique has been successfully used to sport our *copper base alloys* according to their lead content and *lead base alloys* according to antimony content.

(*iv*) Static field technique is used to check the defects in insulator or dielectrics. If there is a defect in a dielectric, its dielectric constant will change and so the *capacitance* will change. Capacitance change can be detected either by statiflux method, similar to magnetic particle technique or by using or oscillator which measures the frequency changes due to change in dielectric properties.

6.25 Radiographic Methods

Radiographic method used the technique of *X-rays* or *γ-radiations*. In these days even *electron and neutron beams* are being used.

On passing through matter, these high radiation beams are modified by the various defects in the material. *Phenomenon of differential absorption forms the basis of use of radiography.* The contrast i.e. the density difference on the developed film between the image of an area containing *a defect and image of defects free* part of specimen allows us to know the flaw.

Decrease in intensity ΔI is found to be proportional to the original intensity of the beam and thickness Δx of the absorber.

$$\therefore \qquad \Delta I = -\mu I \, \Delta x$$

Where μ is constant of proportionality and negative sign shows that with increase of thickness, intensity decrease. Many types of film are available for radiography, each having different values of speed and contrast densities. Normal X-ray sources can find flaws upto a thickness of 20 cm while for higher thicknesses, more input voltage ~ 2000 V may be used to study inner structures. X-ray scattering, X-rays differaction techniques are having used to study the various flaws as per the requirement of the job. We can summarized some of the techniques being used in non-destructive testing using Radiography as under-

1. A *penetrameter* is used to check the *radiographic definition.* Radiographic definition refers to sharpness with which radiograph shows any *discontinuities* in the specimen. A *penetrameter* consists of a strip of the same material as the specimen having thickness equal to some definite percentage of the specimen thickness. *Flat, bead, wire and step* are four types of penetrameters. Penetrameter having sensitivity of 1% to 2% can be got with the known techniques for a wide range thickness of steel.

2. *Weld radiography* can find flaws, slag inclusion, incomplete fusion, cracks, pin holes in welds etc. The term 'Slag' on metallic inclusion is used to describe the oxide and other solids which are found along with the weld.

3. Multiple film techniques are being used to scan total specimen in one exposure. This avoid retakes or camera.

4. In pipe radiography, three techniques *viz* single wall, double wall and variation of double wall used.

5. X- rays can produce floursence fluo on materials. Using this *fluoroscopy* is developed in which X-rays are converted into visible light. This is a fast and economical technique. However contrast and sensitivity in that, case are *lower* than the conventional radiography.

6. Casting radiography is having used to find defects like gas holes, cold shuts, hot cracks, cold crack, cavitiescavities, misruns etc. *Misruns* are due low powering temperatures which result in the *molten metal* to solidify before filling the mold cavity. X-ray films will show the defects in castings.

7. High speed radiography, xero-radiography and stress biography are modern methods are given better resolution than ordinary X-ray films.

8. Micro radiography is being used to study very small defects and to identify thin sections having minute voids.

9. Electron radiography uses fast electrons and is based on the principles of *surface electron emission* due to Secondary photoelectrons. This technique is more useful for metals having high atomic number such as *lead.*

10. In neutron radiography, neutrons are used instead of X-rays. Neutron radiography is very useful for *light plastic, rubber* etc where X-rays an γ-rays can not be used.

11. Gamma Radiography being more economical than X-rays is being used in many fields. However, it has lower contrast as compared to X-ray radiography. Radium, Cs Radon, Cobalt-60 are some of they γ-ray sources. Cs is suitable to study the radiography of steel.

It is important to note that all the high energy beam used in above techniques are highly radioactive. So only trained persons having sufficient safeguards against the hazards of radiations should perform the above tests.

6.26 Comparison of γ–ray Radiography and X– radiography

1. For small defects in castings, less than about 50 mm, X-rays given better results than γ-rays.

2. X-rays give quick result in seconds while γ-rays may take many hours.

3. For thick defects in castings, γ-rays give better results than the X-rays.

4. Due to minimum scattering, γ-rays scan better than X-rays, Moreover, γ-rays give food results even for varying thicknesses unlike the X-rays which are more suitable for uniform thickness of defects. Also γ-rays can detect many defects at the same time.

5. γ-apparatus in small and portable os it is more useful for field testing and inspections.

6.27 Ultrasonic Methods

Ultrasonic waves find us in the detection of cracks or cavities in metal castings due to the property of reflection of sound waves when a change of medium takes place.

Methods consists of sending a strong ultrasonic beam through an object as shown in fig.6.29 Due to a crack or cavity there is a change of medium and so the beam gets reflected. The reflected beam is defected on a CRO (Cathode Ray Oscilloscope) and time of transit of the beam is noted. Knowing the *velocity of sound waves in the medium of solid,* position of crack can be detected.

By inspecting the metal specimen from all directions, exact position of the crack which may be as small as 0.005 mm can be detected.

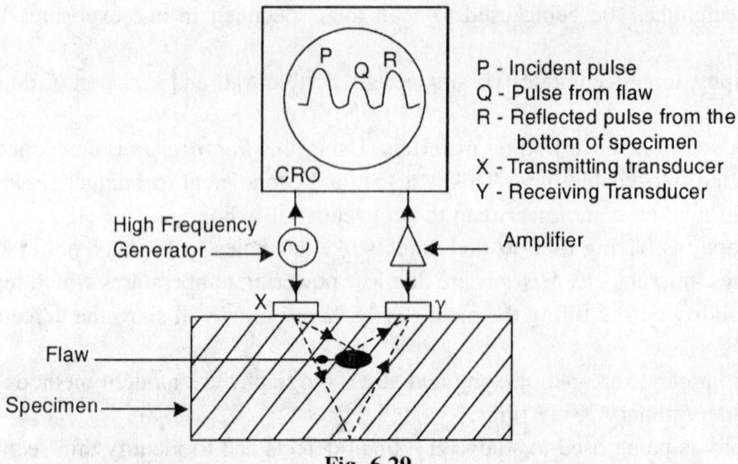

Fig. 6.29

An *important advantage* of ultrasonic method is that it can test very large size castings without destroying the material. Also it involves very low cost and speed of the operation in rapid. This method of locating flaws in more sensitive and less costly than the radiography method. In sections of large castings, rails, heavy axles as an locomotives, forgiving etc. are successfully by ultrasonics.

6.28 Thermal Methods

Basic principle in thermal tests in to apply heat to the specimen and measure the *heat distribution*. In case, there are flows, heat distribution and consequently temperature measurement will differ as flaws.change the heat distribution. Heat may be supplied directly or by electrical heating or by infra red sources. Temperatures may be measured by thermocouples, Bolometers or photoconductive devices.

Forst test is one of the tests that tests the nuclear fuels. In this case Diphenyl is *Frost material* and this test can locate defects of 0.7 cm^2 having minimum length of 0.5 cm. **Thermal** resistance of the material for this test should be high.

6.29 Thermography

This corresponds to optical thermal testing using non-destructive method. This technique involves scanning of thousands of points per second in the form of picture, known as *Thermogram*. In case there is any flaw, the intensity of radiation will change from that surface.

Thermography is a good and effective tool for diagnostic evaluation of electronic circuits. It helps us to identify the defective modulus and to isolate the defect without affecting any change in the circuit, or signal tracing, voltage variations, soldering and unsoldering etc.

Optical material like Germanium and Silicon lenses require infra red testing as they are opaque to visible light. Contact thermography shows the discontinuity in the bonding of metal to metal contacts. Thermocolor paints are used to locate flaws in metal. As temperature of the surface reaches a particular value, colour is changed. *Evapography* provides a two dimensional heat picture from intra-red radiations given by the sample. Tempilstile, Tempilag and thermisters are very sensitive of temperature variations and so find applications in may fields on thermography.

6.30 Liquid penetration method:

Liquid penetrant test is used to delect flows inside the surface e.g. cracks seams porosity etc. This tests is useful for metals as well as ceramic, plastics glass etc.

Principles of liquid penetrant test is that liquids used for the test enter the material through pores or cracks by capillary action. Normally liquid dyes or fluorescent liquid penetrants are used to check the specimen. The liquid is allowed to remain in the specimen for sufficient long time so that it enters the cracks are defective part of the specimen. The specimen is than cleansed and a developer is put so that the dye becomes visible in some particular colour in the defective part. This method allows to locate and know the extent of developed part of the specimen.

6.31. γ-ray Absorption by Matter.

γ-rays may be absorbed by an atom in one of the following ways:

(*a*) **Photo-electric absorption.** In this case, γ-rays striking an atom may be absorbed and the energy imparted to the atom may be sufficient to liberate an electron from an orbit surrounding the nucleus of the atom. In this case, the γ-ray energy is converted to electronic energy (*i.e.*, used for taking out an electron from the atom). So it is called photoelectric absorption of the γ-rays.

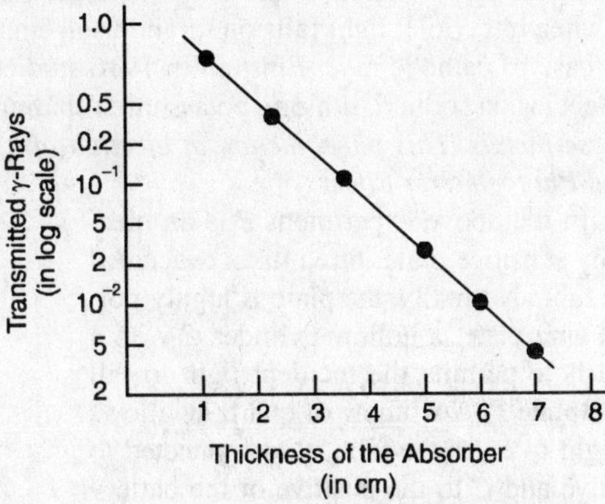

Fig. 6.30

(*b*) **Compton effect.** In this case the energy given by the γ-rays is more than the energy required to liberate an electron from the atom. So in addition to the electron we get photon (or γ-rays) with different energy along with the original-ray.

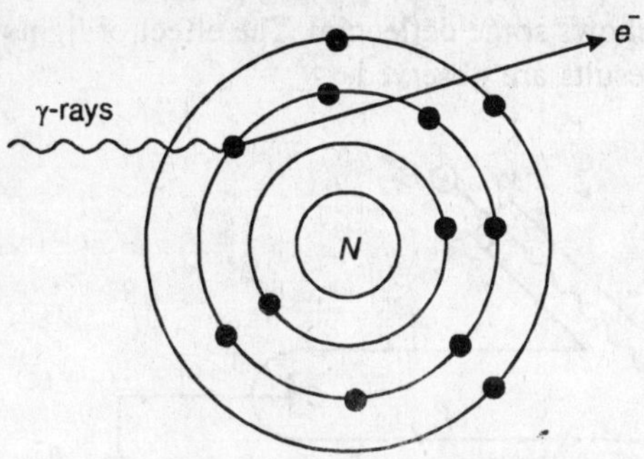

Fig. 6.31 (a)

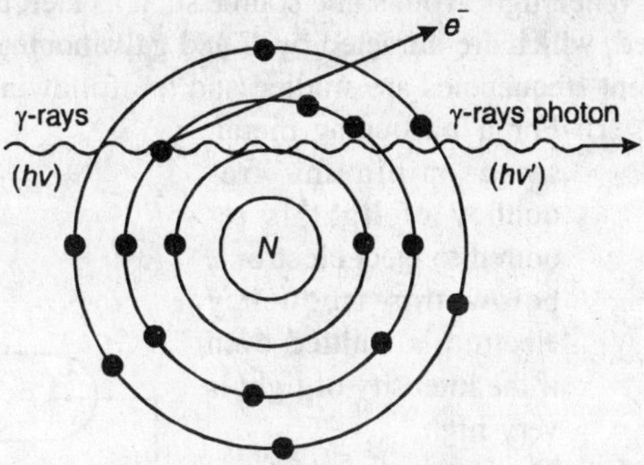

Fig. 6.31 (b)

(*c*) **Electron-positron pair production.** In this case, the γ-ray is absorbed by the atom on which it falls and the energy is liberated in the form of electron-positron pair. It should be kept in mind that the pair production cannot be had if the energy of the γ-rays is less than 1.02 MeV (0.51 MeV for electron liberation and 0.51 MeV for positron liberation).

6.32. Photoelectric Effect

Hertz was the first scientist who discovered in 1887 that a metallic surface is capable of emitting electrons, when a light of very small wavelength falls on it. He also noted that the air in the spark gap became a good conductor when it was illuminated by ultraviolet light. It was further found later on that

(*i*) When ultraviolet light was incident on a neutral zinc plate, the plate became positively charged.

(*ii*) When ultraviolet light was incident on a negatively charged zinc plate, it rapidly lost its charge.

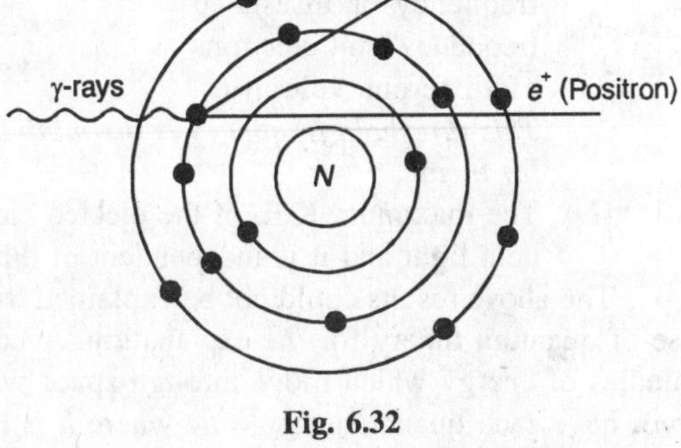

Fig. 6.32

(*iii*) When ultraviolet light was incident on a positively charged zinc plate, it became more positively charged.

It was also concluded that only the negatively charged particles can be emitted by the súrface when ultraviolet light falls on it. Thomson enunciated that *e/m* of these particles was the same as in case of cathode rays. Einstein in 1916 studied the above effect for visible light of different frequencies on sodium, lithium, potassium, rubidium etc. and found that *electrons were ejected out of these metals. This phenomenon of ejection of electrons when light falls on a metal surface is called Photoelectric Effect.*

In the above experiment *P* is an electrically sensitive plate, fitted in an evacuated glass tube. Normally, the plate is highly polished zinc plate, a hollow cylinder *C* with a small hole permits the incident light to fall on the plate *P*. A window of quartz *W* allows the light to enter the tube. *P* is connected to negative and *C* to the positive of the battery and the deflection can be noted by a sensitive galvanometer *G*.

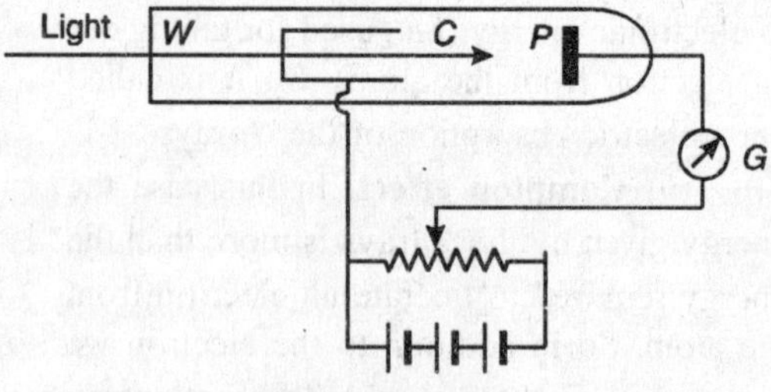

Fig. 6.33

When light from some source such as mercury is made to fall on plate *P*, photo electrons are emitted, which are attracted by *C* and galvanometer shows some deflection. The effect of lights of different frequencies are studied and the following results are observed.

(*i*) For a particular metal, some minimum frequency of light is required to eject electrons, below this frequency electron is emitted even if the intensity of light is very high.

(*ii*) The number of electrons ejected depends on the intensity of light and consequently photoelectric current depends on the intensity of incident light.

(*iii*) Light of frequencies greater than the critical frequency or threshoed frequency emit electrons of different velocities.

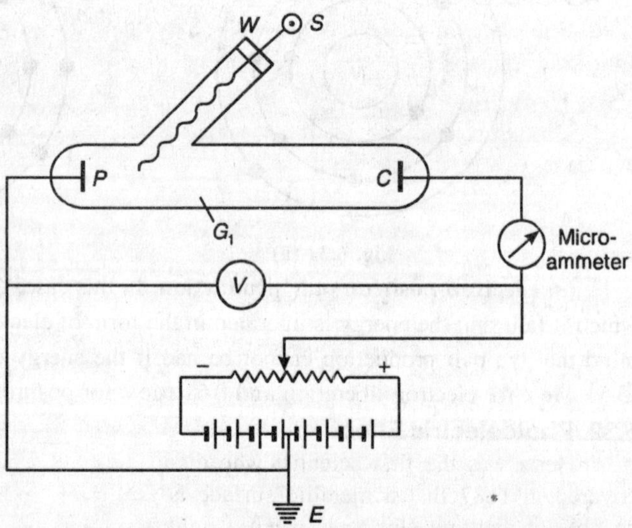

Fig. 6.34

This threshold frequency is the minimum frequency that eject electrons from a particular metal.

(*iv*) The maximum K.E. of the ejected electrons increases linearly with frequency of incident light and it is independent of the intensity of light.

The above results could not be explained from the wave theory of light and so Einstein made use of quantum theory for the explanation. According to this theory, light consists of quantas or bundles of energy which move through space with velocity of light. These particles, called *photons*, have each quanta energy $= h\nu$ where h is Planck's constant and ν is the frequency of light. When a photon strikes the metal surface, there is transfer of energy and this transfer is only in quantas *i.e.*, either a photon gives total energy to the electron or there is no transfer of energy.

6.33. Einstein's Photoelectric Equation

According to Einstein, when light frequency v falls on a metallic surface, the photons of energy hv strike the surface. A photon gives energy to the electron which is ejected out of the metal. However, some minimum frequency called *threshold frequency* is required for ejection of an electron. If a photon carries more energy than the threshold energy (of threshold frequency), then a part of the energy is used for ejection of electron and rest is transferred as K.E. to the electron. Let the mass of the electron be m and velocity is V. If v is the frequency of incident light and v_0 the threshold frequency, then

$$hv - hv_0 = \frac{1}{2}mv^2 \qquad \text{...(1)}$$

In equation (1) $w_0 = hv_0$ is called photoelectric work function i.e., the energy required to liberate an electron from metal surface. So eqn. (1) may be written as

$$hv - w_0 = \frac{1}{2}mv^2$$

or

$$hv = w_0 + \frac{1}{2}mv^2 \qquad \text{...(2)}$$

This is called Einstein's Photoelectric equation.

From (1) and (2), we get

$$hv - hv_0 = eV_0 \qquad [\because eV_0 = \frac{1}{2}mv^2] \qquad \text{...(3)}$$

The value of the stopping potential V_0 depends on the frequency of incident light and on threshold frequency for that particular metal.

6.34. Experimental Verification

Einstein's photoelectric equation can be verified with the help of the millitean's apparatus as shown in Fig. 6.34

It consists of an evacuated tube G_1 fitted with a photo-sensitive plate P and a quartz window W which allows the ultraviolet light to fall on to plate P from the source S. Another plate C is also sealed at the other end of the tube. The middle point of the H.T. battery and the plate P are also seale4d at the other end of the tube. The middle point of the H.T. battery and the plate P are earth connected, so that the potential of plate P is zero. The plate C can be given a positive or a negative potential by the potential dividing arrangement as shown in the figure. A voltmeter V is connected across the plates P and C to measure the potential difference, whereas a micro-ammeter measures the current flowing through circuit. Light of different frequencies is allowed to fall on the plate P and the negative potential applied to C is adjusted till the micro-ammeter just shows zero current. The value of this stopping potential is noted from the voltmeter. A graph is then plotted between the stopping potential and the frequency of light. The graph is a straight line which verifies the equation, because

$$V = \frac{h}{e}v - \frac{h}{e}v_0$$

has the same form as the equation of a straight line.

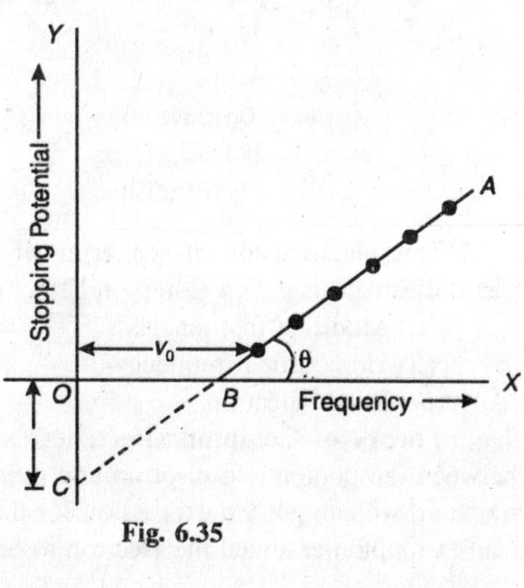

Fig. 6.35

$$y = mx + c$$

(*i*) The slope of the straight line is given by

$$\tan\theta = \frac{h}{e}$$

$$\therefore \qquad h = e\tan\theta$$

The value of Planck's constant derived from this relation agrees with the value obtained from measurements of black body radiations.

(*ii*) For $V = 0$, $\nu = \nu_0$ which is represented by the intercept OB on the X-axis.
In this case the threshold frequency $= \nu_0 = OB$ on the X-axis.

(*iii*) For $\nu = 0$, we have

$$V = -\frac{h\nu_0}{e} = -\frac{w_0}{e}$$

This is represented by the intercept OC on the Y-axis. So the photoelectric work function may be given as

$$w_0 = eV = e \times OC$$

6.35. Compton Effect

Prof, Compton, in 1921, discovered that when a beam of monochromatic radiation of high frequency is allowed to fall on a substance capable of scattering the beam, then scattered radiation contains two components.

(*i*) One component having lower frequency and larger wavelength

(*ii*) The other component having the same frequency of wavelength.

The high frequency radiations used were radiations like X-rays, γ-rays etc. The radiations having the lower frequencies are called modified radiations, while those having the same frequencies are called unmodified radiations.

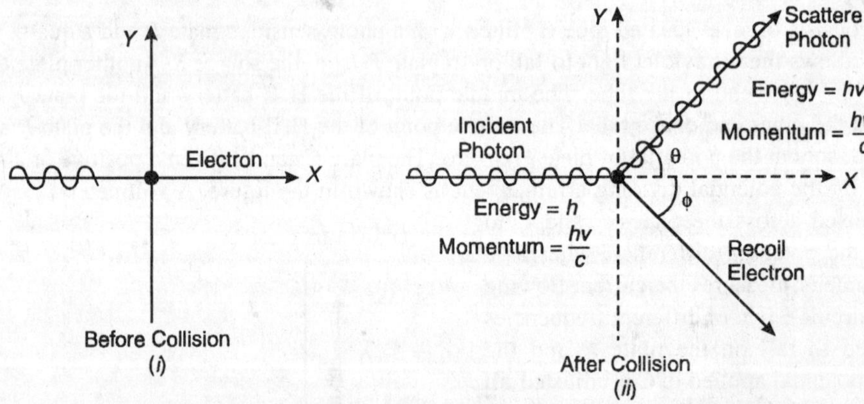

Fig. 6.36

This phenomenon of scattering of incident radiations into scattered radiations and some recoil electrons is called Compton Effect and we have the following results observed:

(*i*) Modified frequencies

(*ii*) Unmodified frequency

(*iii*) Recoil electron.

Theory of Compton effect: let us consider the phenomenon of scattering as a collision between the photon and electron and then applying the laws of conservation of energy and momentum, we can get the expression for the change in wavelength.

Compton assumed the electron to be free and at rest before collision with the incident high energy photon. After the collision has taken place, the relativistic mass of the electron is taken. This is a justified assumption because only a small part of the energy of the incident radiations (or high energy photons) gives sufficient energy to the recoil electron.

Let us consider a photon of energy $h\nu$ colliding with an electron at rest. We assume that even when the electron is bound to the nucleus around which it is revolving, it receives sufficient energy from the incident photon and so it can be treated as practically free. During the collision process, some part of the energy is imparted to the electron which in turn gains kinetic energy and so recoils. In Fig. 6.36 the process of scattering and recoil electron is shown where angle of recoil is given as ϕ and angle of scattering is given by θ.

Let us consider that before the collision

(*i*) Energy of incident photon = $h\nu$

(*ii*) Momentum of incident photon = $h\nu/c$

(*iii*) Rest mass of electron = m_0, so that rest mass energy = m_0c^2.

(*iv*) Rest mass momentum of electron = 0

And after the Collision

(*i*) Energy of scattered photon = $h\nu'$

(*ii*) Momentum of scattered photon = $\dfrac{h\nu'}{c}$

(*iii*) Mass of the electron moving with velocity v = m, so the relativistic energy of electron = mc^2

(*iv*) Momentum of the recoil electron = mv.

Now applying the principle of conservation of energy, before and after collision, we have

$$h\nu + m_0c^2 = h\nu' + mc^2 \qquad ...(1)$$

Again applying the principle of conservation of momentum along the direction of incident photon, before and after the collision, we have

$$\frac{h\nu}{c} + 0 = \frac{h\nu'}{c}\cos\theta + mv\cos\theta \qquad ...(2)$$

and conservation of momentum in a direction ⊥ to the direction of incident photon, we have

$$0 + = \frac{h\nu'}{c}\sin\theta + mv\sin\phi \qquad ...(3)$$

Now from eqn. (2), we have

$$h\nu = h\nu'\cos\phi + mvc\cos\theta \qquad ...(4)$$

or

$$mvc\cos\phi = h\nu - h\nu'\cos\theta$$

And from eqn. (3), we have

$$mvc\sin\phi = h\nu'\sin\theta \qquad ...(5)$$

Squaring eqns. (4) and (5) and adding, we get

$$m^2vdc^2 = (h\nu - h\nu\cos\theta)^2 + (h\nu'\sin\theta)^2 \qquad ...(6)$$

or

$$m^2v^2c^2 = h^2\nu^2\cos^2\theta + 2h^2\nu\nu'\cos\theta + h^2\nu^2\sin^2\theta$$

$$= h^2\nu^2 + h\nu\nu' - 2 + h^2\nu\nu'\cos\theta$$

$$= h^2(\nu^2 + \nu'^2\nu^2 - 2\nu\nu'\cos\theta)$$

Also from eqn. (1), we have

$$mc^2 = h(\nu - \nu') + m_0c^2$$

Squaring, we get

$$m^2c^2 = h^2(v^2 - 2vv' + m_0^2c^2) + 2hm_0c^2(v - v) \quad \text{... (7)}$$

Subtracting (6) from (7), we have

$$m^2c^2(c^2 - v^2) = 2h^2vv'(\cos\theta - 1) + 2h\,(v - v')\,m_0c^2 + m_0^2\,c^4$$

$$= 2h^2vv'(1 - \cos\theta) + 2h(v - v')m_0c^2 + m_0^2c^4$$

or
$$= \frac{m_0^2}{(1 - v^2/c^2)} \times c^2(c^2 - v^2)$$

$$= -2h^2vv'(1 - \cos\theta) + 2h(v - v')mc^2 + m_0^2c^4$$

$$\left[\because \; m = \frac{m_0}{(1 - v^2/c^2)^{1/2}}\right]$$

or
$$m_0^2c^4 = 2h^2vv'(1 - \cos\theta) + 2h(v - v')m_0c^2 + m_0^2c^4$$

or
$$2h^2vv'(1 - \cos\theta) = 2h(v - v')m_0c^2$$

$$\frac{v - v'}{vv'} = \frac{h}{m_0c^2}(1 - \cos\theta)$$

or
$$\frac{1}{v'} - \frac{1}{v} = \frac{h}{m_0c^2}(1 - \cos\theta) \quad \text{...(8)}$$

which may also be written as

$$\lambda' - \lambda = \frac{h}{m_0c}(1 - \cos\theta)$$

or
$$\frac{c}{v'} - \frac{c}{v} = \frac{h}{m_0c}(1 - \cos\theta)$$

$$= \frac{h}{m_0c} 2\sin^2\theta/2 \quad \text{...(9)}$$

from equation (9) it is clear that

(*i*) Change in wavelength is independent of the wavelength of the incident photon but depends on only the angle of scattered photon.

(*ii*) When $\theta = 0$, then

$\lambda' - \lambda = \Delta\lambda = 0$ *i.e.*, no scattering takes place along the direction of incident photon.

(*iii*) When $\theta = \pi/2$, then

$$\lambda' - \lambda = \Delta\lambda = \frac{h}{m_0c}$$

$$= \frac{6.6 \times 10^{-34}}{9.1 \times 10^{-32} \times 3 \times 10^8} = 0.024 \text{ Å}$$

This change in wavelength is called *Compton's shift* and as is clear from above it is constant.

(*iv*) When $\theta = \pi$, then

$$\lambda' - \lambda = \Delta\lambda = \frac{2h}{m_0c} = 0.048 \text{ Å}$$

so when θ varies between 0° and π, the wavelength of scattered photon varies between λ to $\lambda + \dfrac{2h}{m_0c}$ provided the incident photon is of *very small wavelength*.

6.36. Direction of the Recoil Electron

Dividing eqn. (4) by eqn. (5), we have

$$\tan\phi = \frac{h\nu'\sin\theta}{h\nu - h\nu'\cos\theta} = \frac{\nu'\sin\theta}{\nu - \nu\cos\theta} \quad ...(10)$$

Also from eqn. (8), we have

$$\frac{1}{\nu'} = \frac{1}{\nu} + \frac{h}{m_0c^2}(1-\cos\theta)$$

$$= \frac{1}{\nu} + \frac{h}{m_0c^2} 2\sin^2\frac{\theta}{2}$$

or

$$\frac{\nu}{\nu'} = 1 + \frac{h\nu}{m_0c^2} 2\sin^2\frac{\theta}{2}$$

or

$$\nu' = \frac{\nu}{1 + \left(\frac{h\nu}{m_0c^2}\right) \times 2\sin^2\frac{\theta}{2}}$$

Putting this value of ν' in equation (10), we get

$$\tan\phi = \frac{\nu\sin\theta \Big/ \left(1 + \alpha.\ 2\sin^2\frac{\theta}{2}\right)}{\nu - \left[\frac{\nu}{\left(1+\alpha.\ 2\sin^2\frac{\theta}{2}\right)}\right].\cos\theta}$$

where

$$\alpha = \frac{h\nu}{m_0c^2}$$

$\therefore$

$$\tan\phi = \frac{\sin\theta}{1 + 2\ \alpha\sin^2\frac{\theta}{2} - \cos\theta} = \frac{2\sin\frac{\theta}{2}.\cos\frac{\theta}{2}}{1 + 2\ \alpha\sin^2\frac{\theta}{2}\cos\theta}$$

$$= \frac{2\sin\frac{\theta}{2}.\ \cos\frac{\theta}{2}}{(1-\cos\theta) + 2\ \alpha\sin^2\frac{\theta}{2}} = \frac{2\sin\frac{\theta}{2}.\ \cos\frac{\theta}{2}}{2\sin^2\frac{\theta}{2} + 2\ \alpha\sin^2\frac{\theta}{2}}$$

$$= \frac{2\sin\frac{\theta}{2}.\ \cos\frac{\theta}{2}}{2\sin^2\frac{\theta}{2}(1+\alpha)} = \frac{\cot\frac{\theta}{2}}{1+\alpha}$$

or

$$\tan\phi = \frac{\frac{\cot\theta}{2}}{\left(1 + \frac{h\nu}{m_0c^2}\right)} \quad ...(11)$$

which gives the direction of the recoil electron in terms of the frequency of the incident photon and the direction of the scattered photon.

6.37. Energy of Recoil Electron

In the Compton effect, K.E. of recoil electron is equal to the decrease in energy of the incident photon. It is because some energy is imparted to the electron, so that it leaves electronic orbit.

Let the energy of incident proton = $h\nu$

and energy of scattered proton = $h\nu'$

where $\nu > \nu'$

So the decrease in energy of the photon = $h\nu - h\nu' = h(\nu - \nu')$ ...(1)

Let m_0c^2 and mc^2 be the rest mass energy and recoil energy of the electron respectively.

The K.E. of recoil electron $= mc^2 - m_0c^2$

$= (m - m_0)\, c^2$ ...(2)

This recoil energy is equal to the decrease in energy of the incident photon.

So K.E. of recoil electron = $h(\nu - \nu')$

$$= h\nu\left(1 - \frac{\nu'}{\nu}\right) = h\nu\left(1 - \frac{\lambda}{\lambda'}\right) \qquad [\because c = \nu\lambda]$$

$$h\nu\frac{\lambda' - \lambda}{\lambda'} = h\nu\frac{\Delta\lambda}{\lambda + \Delta\lambda}$$

where $\Delta\lambda$ is change in wave length.

Now $$\Delta\lambda = \frac{h}{m_0c}(1 - \cos\theta)$$

$\therefore$ K.E. of recoil electron is given by

$$E = \frac{h\nu.\ \dfrac{h}{m_0c}(1-\cos\theta)}{\lambda + \dfrac{h}{m_0c}(1-\cos\theta)}$$

$$= \frac{h\nu \times \dfrac{h}{m_0c\lambda}(1-\cos\theta)}{1 + \dfrac{h}{m_0c\lambda}(1-\cos\theta)}$$

$$= \frac{\dfrac{h^2\nu^2}{m_0c^2}(1-\cos\theta)}{1 + \dfrac{h\nu}{m_0c^2}(1-\cos\theta)} \qquad (\because c = \nu\lambda)$$

Let us study the different cases

(*i*) If $\theta = 0$; $E = 0$

which means that no electron will recoil for incident photon at angle $\phi = 0$

(*ii*) if $\theta = \pi/2$,

$$E = \frac{\dfrac{h^2\nu^2}{m_0c^2}}{1 + \dfrac{h\nu}{m_0c^2}}$$

Let us put $$\alpha = \frac{h\nu}{m_0c^2}$$

$$\therefore \qquad E = \frac{h\nu}{(1+a)}.\alpha \qquad ...(3)$$

(*iii*) If $\phi = \pi$,

$$E = \frac{\frac{2h^2\nu^2}{m_0c^2}}{1+\frac{2h\nu}{m_0c^2}} = \frac{2h\nu\alpha}{1+2\alpha}$$

$$= h\nu\frac{2\alpha}{1+2\alpha} \qquad ...(4)$$

From the above cases we have the conclusion that the minimum energy transferred to electron by the photon is zero (when $\phi = 0$) and maximum energy transferred to recoil electron is

$$h\nu.\frac{2\alpha}{1+2\alpha}$$

Also because $\frac{2\alpha}{1+2\alpha} < 1.$

So the energy transferred by photon to electron is less than the energy of incident proton ($h\nu$). *So photon is unable to transfer its whole energy to the electron in photo emission.*

6.38. Compton Effect with Visible Light

From compton equation, the % change in wave length of a scattered photon due to compton scattering may be written as

$$\frac{\lambda'-\lambda}{\lambda}\times 100 = \frac{h}{m_0\, c\lambda}(1-\cos\theta)\times 100$$

% change will be maximum when θ = 180°,
In that case

Max. % change in wavelength $\lambda = \frac{200h}{m_0c\lambda}$

$$= \frac{200\times 0.0246}{\lambda}$$

$$\left(\text{where } \frac{h}{m_0c} = 0.0246 \text{ (comption wavelength)}\right)$$

In case X-ray photons are used for compton scattering $(\lambda \sim 1\text{Å})$, max. % change in wavelength will be

$$= \frac{200\times 0.02426\text{Å}}{1\text{Å}} \simeq 4.82\%$$

This change can easily be recorded. However in case white light $\lambda - 600\text{Å}$ is used, then Max. % change in wavelength will be

$$= \frac{200\times 0.02426\text{Å}}{600\text{Å}} = 8.08\times 10^{-4}\%$$

which is very small end cannot be detected. So compton effect is not possible with visible light.

6.39. Experimental Set Up of Compton Effect

Figure 6.37 shows the experimental setup for study of Compton effect. A monochromatic beam of X-rays is allowed to fall on a carbon block *B* which acts as scatterer. Normally, Compton effect can be best observed in case of elements of low atomic number because the binding energies of the electrons in these elements are negligible as compared to the quantum energy of the incident photon. The scattering of X-ray photons can take place in different directions and their intensities and wavelengths can be measured by Bragg's spectrometer.

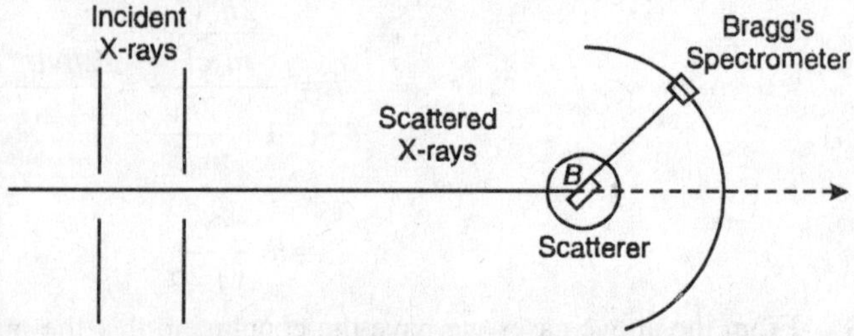

Fig. 6.37

In a typical experiment, K_α line of molybdenum was scattered by graphite and the distribution of intensities was studied at different angles. If a graph is plotted between relative intensities and wavelengths, then we observe that for each value of θ there are distinct intensity peaks for two wavelengths one of which corresponds to the incident radiation λ and the other has a higher value λ'. The peak corresponding to λ' is called modified peak.

6.40. Pair Production

Anderson in 1932, observed the phenomenon by which γ – rays are absorbed and **electron-positron** pair is emitted. This phenomenon confirms the conversion of radiant energy into mass.

For pair production phenomenon to take place the incident photon must carry a minimum energy of 1.02 MeV (since both electron and positron have an energy of 0.51 MeV each). It may be noted that this phenomenon occurs mainly in the nuclear field and to some extent in the field to electrons. However, the presence of nucleus is necessary for the conservation of momentum.

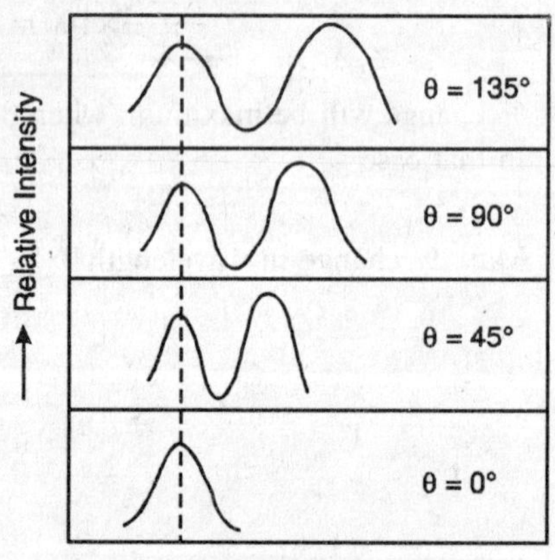

Fig. 9.38

The **pair production** phenomenon can be explained theoretically from Dirac's electron theory, which considers positrons as **holes** in otherwise completely filled sea of negative state of **electrons.** This feature is shown in Fig. 6.40.

Here we observe that to take a negative electron out of **negative energy state**, an energy of 2 m_0c^2 is required to cross the **barrier**. By this process, a negative electron is created in positive energy state and so it becomes observable. The vacancy in the negative energy region is also observable, having the same properties as the negative electron (also referred as **negatron**) except that it has opposite charge. The total energy between the electron and positron is seen to be multiple of $h\nu$, confirming the quantum theory and mass-energy relationship.

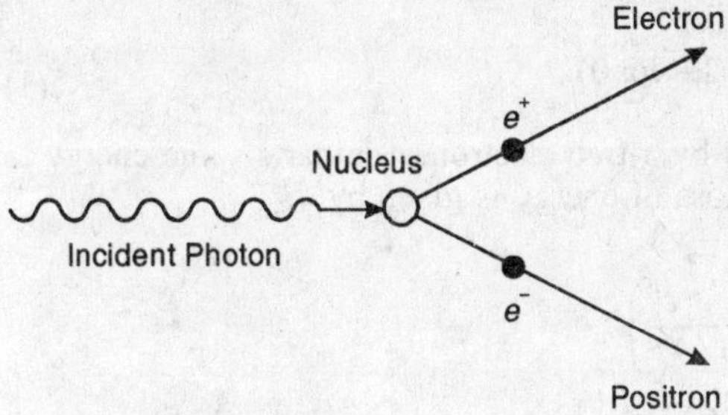

Fig. 6.39

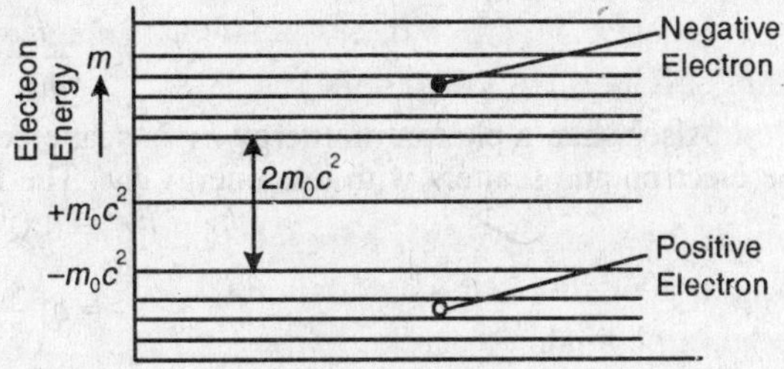

Fig. 6.40

6.41. Distinction between γ-rays and X-rays

γ-rays and X-rays are both electromagnetic radiations which differ from each other in following manner:

(*i*) γ-rays are produced spontaneously by radioactive emission, whereas X-rays are given out by the transition of electrons from higher to lower energy level.

(*ii*) γ-rays are produced due to spontuneous emission of radiations from the nucleus of the atom, whereas X-rays are produced by making fast moving electrons strike solid targets.

(*iii*) The wavelength of γ-rays is much shorter than that of X-ray wavelength.

(*iv*) The penetrating power of γ-rays is much higher than that of X-rays. X-rays can pass through flesh only, whereas γ-rays can penetrate through 30 cm of thick iron sheet.

(*v*) γ-rays are monochromatic electromagnetic radiations, while X-rays consist of three components hard, medium and soft X-rays.

(*vi*) X-rays produce photoelectric effect and scattering effect, whereas γ − rays produce photoelectric effect, scattering effect as well as positron-electron pairs.

Example 27. *In Compton experiment, the wavelength of X-ray radiation scattered at an angle of 45° is 0.022 Å. Calculate the wavelength of the incident X-rays.*

Solution. We have the relation

$$\lambda' - \lambda = \frac{h}{m_0 c}(1-\cos\theta)$$

In this case, $\theta = 45°, \lambda' = 0.02 \text{ Å} = 0.022 \times 10^{-10}\text{ m}$

$h = 6.6 \times 10^{-34}$ joule-sec, $m_0 = 9.1 \times 10^{-31}$ kg

$c = 3 \times 10^8$ metre/sec

$$\therefore \quad \lambda = \lambda' - \frac{h}{m_0 c}(1-\cos\theta)$$

$$= 0.022 \times 10h^{-10} - \frac{6.6 \times 10^{-34}}{9.1 \times 10^{-31} \times 10^8}(1-0.707)$$

$$= (0.022 - 0.007) \times 10^{-10} = \mathbf{0.015\text{ Å}} \ \textbf{Ans.}$$

Example 28. *Show that the energy lost by a photon in Compton interacting which a free stationary electron can be written as*

$$h\nu\left[\frac{\alpha(1-\cos\theta)}{1+\alpha(1-\cos\theta)}\right]$$

where $\alpha = \frac{h\nu}{m_0 c}$

Solution. We have the change in wavelength given by relation

$$\lambda' - \lambda = \frac{h}{m_0 c}(1 - \cos\theta) \quad ...(1)$$

Also when a photon of energy hv is scattered by a free electron, it imparts some energy to the electron and scatters with less energy hv'. The loss of energy is given by

$$hv - hv' = h(v - v')$$

$$= h\left(\frac{c}{\lambda} - \frac{c}{\lambda'}\right)$$

$$= hc\frac{(\lambda' - \lambda)}{\lambda\lambda'} \quad ...(2)$$

Putting the values of λ' from eqn. (1) in eqn. (2), we get

$$= \frac{hc\left[\dfrac{h}{m_0 c}(1 - \cos\theta)\right]}{\lambda\left[\lambda + \dfrac{h}{m_0 c}(1 - \cos\theta)\right]}$$

$$= \frac{h^2}{m_0}\left[\frac{(1 - \cos\theta)}{\lambda^2\left[1 + \dfrac{h}{m_0 c\lambda}(1 - \cos\theta)\right]}\right]$$

$$= \frac{h^2 v^2}{m_0 c^2}\left[\frac{(1 - \cos\theta)}{\left[1 + \dfrac{h}{m_0 c\lambda}(1 - \cos\theta)\right]}\right]$$

$$= hv \times \frac{hv}{m_0 c^2}\left[\frac{1 - \cos\theta}{1 + \dfrac{hv}{m_0 c^2}(1 - \cos\theta)}\right]$$

$$= \mathbf{hv}\left[\frac{\alpha(1 - \cos\theta)}{1 + \alpha(1 - \cos\theta)}\right] \quad \textbf{Ans.}$$

Example 29. *Find the wavelength of X-ray photons which produce recoil electrons of energy 5 keV in Compton effect. Assume the direction of recoil electron to be in the direction of incident photon while photon is scattered through an angle of 180°.*

Solution. In this case, the direction of scattered photon $\theta = 180°$

Also $$\frac{hc}{\lambda} - \frac{hc}{\lambda'} = 5 \text{ keV} = 5.000 \times 1.6 \times 10^{-19} \text{ joule}$$

$\therefore$ $$\frac{hc}{\lambda} - \frac{hc}{\lambda'} = 8 \times 10^{-16} \text{ joule} \quad ...(1)$$

From conservation law of momentum in the direction of incident photon,

$$\frac{h}{\lambda} + 0 = \frac{h}{\lambda'}\cos 180° + mv\cos 0°$$

$$= \frac{h}{\lambda'} \cos 180° + mv$$

or
$$\frac{h}{\lambda} + \frac{h}{\lambda'} = \sqrt{\frac{1}{2} mv^2 \times 2m}$$

$$= \sqrt{8 \times 10^{-10} \times 9.1 \times 10^{-31}}$$
$$= 38.1 \times 10^{-24}$$

Multiplying both sides by c

$$\frac{hc}{\lambda} + \frac{hc}{\lambda'} = 3 \times 10^8 + 38.1 \times 10^{-24} \quad ...(2)$$
$$= 11.448 \times 10^{-17} \text{ joule}$$

Adding eqns. (1) and (2), we get

$$\frac{2hc}{\lambda} = 8 \times 10^{-16} + 11.448 \times 10^{-17}$$
$$= 122.48 \times 10^{-16} \text{ joule}$$

or
$$\lambda = \frac{2hc}{122.48 \times 10^{-16} \text{ J}}$$

$$= \frac{(2 \times 6.6 \times 10^{-34} \text{ J})(3 \times 10^8 \text{ m/s})}{122.48 \times 10^{-16} \text{ J}}$$
$$= 0.324 \times 10^{-10} \text{ m} = \mathbf{0.324 \text{ Å}} \quad \textbf{Ans.}$$

Example 30. *A photon of energy 1.02 MeV is scattered through 90° by a free electron. Calculate the energy of photon and electron after interaction.*

Solution. We have the relation

$$\Delta\lambda = \lambda' - \lambda = \frac{h}{m_0 c}(1 - \cos\theta)$$

$$= \frac{6.6 \times 10^{-34}}{9.1 \times 10^{-31} \times 3 \times 10^8}(1 - 0)$$
$$= 2.42 \times 10^{-12} \text{ m}$$

Change in frequency of photon $\quad \Delta v = \frac{c}{\Delta\lambda} = \frac{3 \times 10^8 \text{ m}}{2.42 \times 10^{-12} \text{ m}}$

Change in energy of photon $\quad \Delta E = h\,\Delta v = \frac{6.6 \times 10^{-34} \times 3 \times 10^8}{2.42 \times 10^{-12}} = \mathbf{0.51 \text{ MeV}}.$

This energy is transferred to the free electron which in turn will be the K.E. of the recoil electron.

$\therefore$ Energy of the photon after interation = 1.02 – 0.51 = **0.51 MeV**. **Ans.**

Example 31. *In a diode valve, the anode has a potential of 100 volts relative to the cathode. Find the wavelength associated with the electron as it reaches the anode. Mass of electron m = 9.11 × 10^{-28} gm, charge e = 4.803^{-10} e.s.u. and h = 6.62 × 10^{-27} ergs-sec.*

Solution. Potential difference $\quad V = \frac{100}{300} = \frac{1}{3}$ e.s.u.

Let v be the velocity acquired by the electron on reaching the anode, then energy of the electron is given by

$$\frac{1}{2} mv^2 = Ve$$

$$v = \sqrt{\frac{2Ve}{m}}$$

Now, wavelength 'λ' of the electron in given by

$$\lambda = \frac{h}{mv} = \frac{h}{\sqrt{\frac{2Ve}{m}}} = \frac{h}{\sqrt{2Vem}}$$

$$\lambda = \frac{6.62 \times 10^{-27}}{\sqrt{2 \times 1/3 \times 4.803 \times 10^{-10} \times 9.11 \times 10^{-28}}}$$

$$\lambda = \frac{6.62 \times 10^{-27}}{\sqrt{2 \times 1.601 \times 10^{-10} \times 9.11 \times 10^{-28}}}$$

$$= \mathbf{1.226 \times 10^{8}\ cm} \quad \textbf{Ans.}$$

Example 32. *A photoelectric surface has a work function of 4 eV. What is the maximum velocity of the photoelectrons emitted by light of frequency 10^{15} Hertz incident on the surface ? ($h = 6.6 \times 10^{-34}$ joule sec, $e = 1.6 \times 10^{-19}$ Coul. $m = 9 \times 10^{-31}$ km).*

Solution. We have the relation

$$hv - w_0 = \frac{1}{2}mv^2$$

In this case,

$$h = 6.6 \times 10^{-34} \text{ Joule sec.}$$

$$v = 10^{15} \text{ Hertz}$$

$$w_0 = 4 \text{ eV} = 4 \times 1.6 \times 10^{-19} \text{ Joule}$$

$$= 6.4 \times 10^{-79} \text{ joule}$$

$$m = 9 \times 10^{-31} \text{ kg.}$$

$$\therefore\ 6.6 \times 10^{-34} \times 10^{15} - 6.4 \times 10^{-19} = \frac{1}{2} \times 9 \times 10^{-31} v^2$$

or

$$v = \mathbf{2.108 \times 10^{5}\ m/sec} \quad \textbf{Ans.}$$

Example. 33. *X-ray with $\lambda = 1\ \AA$ are scattered from a carbon block. The scattered radiation is viewed at 90° to the incident beam. (a) What is the Compton shift $\Delta\lambda$? (b) What K.E. is imparted to the recoiling electron?*

Solution. (*a*) We have the Compton shift given by

$$\Delta\lambda = \frac{h}{m_0 c}(1 - \cos\theta)$$

In this case,

$$\theta = 90°, \cos 90° = 0$$

$$\Delta\lambda = \frac{h}{m_0 c}(1 - \cos\theta)$$

In this case,

$$\theta = 90° = 0$$

$$\therefore\ \Delta\lambda = \frac{h}{m_0 c} = \frac{6.62 \times 10^{-34} \text{ Joule} - \text{sec}}{9.1 \times 10^{-31} \text{kg} \times 3 \times 10^{8} \text{m/sec}}$$

$$= 2.43 \times 10^{-12} \text{ m}$$

$$= \mathbf{0.0243\ \AA} \quad \textbf{Ans.}$$

(*b*) From the conservation law of energy, we have

$$\frac{hc}{\lambda} = \frac{hc}{\lambda'} + K$$

where K is the K.E. of recoil electron

Also $$\lambda' = \lambda + \Delta\lambda$$

$\therefore$ $$\frac{hc}{\lambda} = \frac{hc}{\lambda + \Delta\lambda} + K$$

or $$K = \frac{hc\ \Delta\lambda}{\lambda(\lambda + \Delta\lambda)}$$

$$= \frac{6.62 \times 10^{-34} \text{ joule sec} \times 3 \times 10^{8} \text{ m/sec} \times 2.43 \times 10^{-12} \text{ m}}{1 \times 10^{-10} \text{ m} \times (1 + 0.0243) \times 10^{-10} \text{ m}}$$

$$= 4.72 \times 10^{-17} \text{ joule}$$

$$= \frac{4.72 \times 10^{-17}}{1.6 \times 10^{-19}} \text{ eV}$$

$$= \mathbf{0.295\ eV.}$$ **Ans.**

Example 34. *The photoelectric work function of potassium is 2.0 eV. If light having a wavelength 3600 Å falls on potassium, find (i) the kinetic energy in eV of the most energetic electrons ejected (ii) the stopping potential (iii) the velocity of these electrons.*

Solution. Work function $w_0 = 2$ eV.

$$= 2 \times 1.6 \times 10^{-12} \text{ ergs}$$

$$h = 6.624 \times 10^{-27} \text{ ergs-sec}$$

$$c = 3 \times 10^{10} \text{ cm/sec}$$

Now, $$w_0 = \frac{hc}{\lambda_0} \text{ where } \lambda_0 \text{ is the threshold wavelength}$$

$$\lambda_0 = \frac{hc}{w_0} = \frac{6.62 \times 10^{-27} \times 3 \times 10^{10}}{2 \times 1.6 \times 10^{-12}} \text{cm}$$

$$= \frac{19.872}{3.2} \times 10^{-5} \text{cm}$$

$$= 6.2103 \times 10^{-5} \text{ cm}$$

$$= 6210.3 \times 10^{-8} \text{ cm} = 6210.3 \text{ Å}$$

$$\nu_0 = \frac{c}{\lambda_0} = \frac{3 \times 10^{10}}{6210.3 \times 10^{-8}} \times \frac{3}{6210.3} \times 10^{18}$$

$$\nu = \frac{c}{\lambda} = \frac{3 \times 10^{10}}{3600 \times 10^{8}} \frac{1}{12} \times 10^{16}$$

(*i*) Now $$\text{K.E.} = \frac{1}{2} m v^2 = h\nu - h\nu_0$$

$$E = h\left[\frac{1}{12} \times 10^{16} - \frac{1}{2070} \times 10^{18}\right]$$

$$= h \times 10^{16}\left[\frac{2070 - 1200}{24840}\right]$$

$$= h\times10^{16}\times\frac{870}{24840}\text{ergs}$$

$$= 6.624\times10^{-27}\times10^{17}\frac{870}{24840}$$

$$= \frac{6.624\times870}{24840\times1.6\times10^{12}}\times10^{-11}\text{eV}$$

$$= \frac{6.624\times870}{2484\times1.6}\text{eV} = \mathbf{1.45\ eV.\ Ans.}$$

(*ii*) Stopping potential $E = 1.45$ eV.

So stopping potential = **1.45 volts**

(*iii*) Velocity of the electron

$$\frac{1}{2}mv^2 = 1.45\text{ eV} = 1.45\times1.6\times10^{-12}\text{ ergs.}$$

$$v^2 = \frac{2\times1.45\times1.6\times10^{-12}}{9.1\times10^{-28}}$$

$$v = \mathbf{7.1\times10^7\ cm/sec.\ Ans.}$$

Example 35. *The photoelectric threshold of tungsten is 2300 Å. Determine the energy of the electrons ejected from the surface by ultraviolet light of wavelength 1800 Å.*

Solution. Here $\lambda_0 = 2300\text{ Å} = 2300\times10^{-8}$ cm.

$$\lambda = 1800\text{Å} = 1800\times10^{-8}\text{cm.}$$

$$\nu_0 = \frac{c}{\lambda_0} = \frac{3\times10^{10}}{2300\times10^{-8}} = \frac{3}{23}\times10^{16}$$

$$\nu = \frac{c}{\lambda} = \frac{3\times10^{10}}{1800\times10^{-8}} = \frac{1}{6}\times10^{16}$$

$$\text{Now kinetic energy} = \frac{1}{2}mv^2 = h\nu - h\nu_0$$

$$E = h\nu - h\nu_0 = h\left[\frac{10^{16}}{6} - \frac{3\times10^{16}}{23}\right]$$

$$= h\times10^{16}\left[\frac{23-18}{138}\right]$$

$$= 6.62\times10^{-27}\times10^{-16}\times\frac{5}{138}$$

$$= 6.62\times\frac{5}{138}\times10^{-11}\text{ergs}$$

$$= \frac{33.10}{138}\times10^{-11}\times\frac{1}{1.6\times10^{-13}}\text{eV}$$

$$\therefore \quad E = \mathbf{1.52\ eV\ \ Ans.}$$

Example 36. *The energy of photoelectrons emitted from a sensitive plate is 1.56. If the threshold wavelength is 2500 Å, calculate the wavelength of the incident light; given that*

$$1 eV = 1.6 \times 10^{-12} \text{ ergs}$$
$$h = 6.62 \times 10^{-27} \text{ ergs-sec}$$
$$\text{Velocity of light } c = 3 \times 10^{10} \text{ cms./sec.}$$

Solution. $\lambda_0 = 2500\text{Å} = 2500 \times 10^{-8}\text{cm.}$

$$c = 3 \times 10^{10}\text{cm/sec.}$$

The threshold frequency $\nu_0 = \frac{c}{\lambda_0} = \frac{3\times10^{10}}{25\times10^{-8}} = \frac{3}{25}\times10^{16}$

$$h\nu_0 = \frac{3}{25}\times10^{16}\times6.62\times10^{-27}\text{ ergs.}$$

$$= \frac{3\times6.62\times10^{-11}}{25\times1.6\times10^{-12}}eV = 4.965\text{ eV.}$$

Let ν be the incident frequency

E, the energy of the electron ejected = 1.56 eV

Thus $= h\nu_0 + E \qquad h\nu = h\nu_0 + E$

$$= (4.965 + 1.56)\text{ eV} = 5.525\text{ eV}$$

But $\nu = \frac{c}{\lambda}$

$$\frac{hc}{\lambda} = 5.525\times1.6\times10^{-12}\text{ ergs}$$

$$\lambda = \frac{hc}{5.525\times1.6\times10^{-12}}\text{cm}$$

$$= \frac{6.62\times10^{-27}\times3\times10^{10}}{5.525\times1.6\times10^{-12}}\text{cm}$$

$= 2246 \times 10^{-8}$ cm. = **2246Å. Ans.**

Example 37. *Calculate the longest wavelength of incident radiation which will eject electrons from Platinum. The work function of Platinum = 6.3 eV., $h = 6.624 \times 10^{-27}$ ergs-sec., $c = 3 \times 10^{10}$ cm/sec.*

Solution. Work function $w_0 = 6.3\ eV$

$$w_0 = 6.3\ 1.6 \times 10^{-12}\text{ ergs}$$
$$= 10.08 \times 10^{-12}\text{ ergs}$$
$$h = 6.624 \times 10^{-27}\text{ ergs-sec}$$
$$c = 3 \times 10^{10}\text{ cm/sec}$$

Now $w_0 = h\nu_0$

If λ_0 is the threshold wavelength of incident light

$$w_0 = \frac{hc}{\lambda_0}$$

$$= \frac{6.624\times10^{27}\times3\times10^{10}}{10.08\times10^{-12}}$$

$$= \frac{19.872}{10.08}\times10^{-5}$$

$= 1.972 \times 10^{-5}$cm

$= 1972 \times 10^{-8}$cm. = **1972Å Ans.**

Example 38. *Calculate the photoelectric work function for sodium metal in electron volts; given that the threshold wavelength is 6800 Å, h = 6.6 × 10^{-7} ergs.sec.*

Solution. Here threshold wavelength $\lambda_0 = 6800$Å

$= 6800 \times 10^{-8}$cm.

∴ Threshold frequency $\quad v_0 = \dfrac{c}{\lambda_0} = \dfrac{3\times10^{10}}{6800\times10^{-8}} = \dfrac{3}{68}\times10^{16}$

Let w_0 be the work function for metal, then

$$w_0 = hv_0$$

$$= 6.62\times10^{-27}\times\frac{3}{68}\times10^{16}\text{ ergs.}$$

$$= 2.92 \times 10^{-12}\text{ ergs.}$$

$$= \frac{2.92\times10^{12}}{1.6\times10^{-12}}\text{ eV}$$

w_0 = **1.825 eV. Ans.**

Example 39. *Calculate the energy of a quantum of light of wavelength* λ *= 5.4 × 10^{-5} cm and express it in eV. If such a light falls on calcium whose work function is 1.9 eV, find retarding potential to stop the most energetic electrons.*

Solution. $\quad \lambda = 5.4\times10^{5}$cm.

$c = 3 \times 10^{10}$ cm/sec

Now energy of one quantum of light

$$hv = \frac{hc}{\lambda} = \frac{6.62\times10^{-27}\times3\times10^{10}}{5.4\times10^{-5}}$$

$$= \frac{6.62\times3\times10^{-12}}{5.0\times1.6\times10^{-12}}\text{ eV} = 2.21\text{ eV.}$$

Let V be the stopping potential

Work function $\quad w_0 = 1.2\ eV = 1.9 \times 1.6 \times 10^{-12}$ ergs.

Now $\quad$ energy of electron $E = hv - w_0$

$$Ve = hv - w_0$$

$$V = \frac{hv - w_0}{e}$$

$$= \frac{2.21 - 1.6\times10^{-12}\times1.9\times10^{-12}}{4.8\times10^{-10}}$$

$$= \frac{1.6\times10^{-12}}{4.8\times10^{-10}}(2.21-1.9)$$

$= .13 \times 10^{-2}$ e.s.u.

$= 0.13 \times 10^{-2} \times 300$ volts = **0.39 Volts Ans.**

6.42. Raman Effect

It corresponds to scattering with change in wavelength somewhat similar to fluorescence. However, it is different from fluorescence in two ways:

(*i*) The light which is incident of the scattering materials must have a wavelength the does not correspond to one of the absorption lines to bands of the material (as otherwise fluorescence phenomenon occurs).

(*ii*) Intensity of scattered light is much less than the intensity obtained in fluorescence phenomenon.

It is because of these reasons that *Raman effect* is difficult to observe and the observations have to be done by photography.

It may be mentioned here that the scattering that occurs without change in wavelength or frequency is *Rayleigh scattering.*

Raman, in 1928 observed that light scattered from molecules at right angles to the incident direction has component of radiation in addition to the incident frequency radiation v some frequencies $\nu \pm \nu'$. So the *scattering of light from molecules with a change in frequency or wavelength is called Raman effect.*

It may be mentioned that in the Raman spectrum, we observe in the scattered light a line having frequency equal to the frequency of the incident light and in addition *few weak lines* on either side of incident line are observed. If the frequency ν of incident light is changed, the weak lines are once again observed on either side of the Rayleigh line with the same difference in frequency.

It is evident that the frequency difference ν' between the incident and the scattered light in Raman effect is determined by the *nature of the scattering molecules* and is independent of the frequency of the incident light.

6.43. Theory of Raman Effect

Quantum theory of light states that light consists of packets of energy in the form of photons each of which carries energy in quantum $h\nu$. When this light falls on a molecule, the photon will undergo a collision with molecule. This collision may be:

(*a*) **Elastic Collision :** When there is no transfer of energy from photon to the molecule. So in this case scattering of photon will be without any change in energy. It means that the frequency of the scattered photon will be identical with the frequency of the incident photon. This will be the case of the presence of *unmodified* line in scattered light spectra.

(*b*) **Inelastic collosion :** In this case there is transfer of energy which may take place in either of the following two ways:

(*i*) The molecule loses enegry ΔE during the collosion with photon. It will correspond to that case when the molecule is in an excited state. Let the frequency of the scattered photon be ν_2. Then from the law of conservation of energy, we have

$$h\nu_1 = h\nu + \Delta E$$

or

$$\nu_1 = \nu + \nu' \qquad \dots (i)$$

$$\nu_1 = \nu + \nu'$$

where

$$\nu' = \frac{\Delta E}{h}$$

which shows that the frequency of scattered photon is more than the frequency of the incident photon.

(*ii*) The molecule in the ground state gets engery from the photon and goes to an excited state. Let ΔE be the energy absorbed by the molecule during collision and let ν_2 be frequency of the scattered photon. Then from the law of conservation of energy, we have

$$h\nu = h\nu_2 + \Delta E$$

or
$$\nu = \nu_2 + \frac{\Delta E}{h} = \nu_2 + \nu'$$

$$\nu_2 = \nu - \nu'$$

which shows that the frequency of scattered photon is less than the frequency of the incident photon.

The line with a frequency lower than that of the incident frequency is called *Stock's line* while that with a frequency higher than the incident frequency is called *Anti Stocke's line.*

Selection Rules. Let us consider a scattering molecule in the rotational state α and having the rotational quantum number J. Ordinary in the rotational spectrum we have the selection rule

$$\Delta J = \pm 1$$

If we study Raman spectrum quantum mechanically, we observe that a *Raman jump* is possible between two states α and β only if there exists a third state γ, such that in ordinary case jumps are allowed between a and γ and β and γ states. Which in other words may be expressed as a molecular transition can take place from level α and β by going thoroughly γ.

Let a molecular transition take place from α state (having rotational quantum number J'') via a γ state (having rotational quantum number J').

Now from the selection rule of rotational spectrum a translation from α to γ state can take place with quantum number J changing by ± 1. So the rotational quantum number J' or γ state will be

$$J' = J \pm 1 \qquad \text{...(1)}$$

Again a transition from γ to β state is allowed if a change in rotational quantum number is ± 1. So we have the rotational quantum number of β state given by

$$\Delta'' = J'' \pm J \qquad \text{... (2)}$$

So the total change in quantum number J is going from α to β state is

$$\Delta J = J'' - J \qquad \text{... (3)}$$

Putting the value of J'' from eq. (2) in eq. (3), we get

$$\Delta J = J' \pm 1 - J \qquad \text{... (4)}$$

Putting the value of J' from eq. (1) in eq. (4), we get

$$\Delta J = J \pm 1 = 1 - J$$

or
$$\Delta J = \pm 2, 0$$

Now $\Delta J = 0$ corresponds to change in frequency and so will represent Rayleigh scattering. $\Delta l = \pm 2$, represents Raman scattering. So in the Raman spectrum, we observe transitions which observe the selection rule $\Delta J = \pm 2$. It may be mentioned that the spectral lines in rotational spectra which are not allowed by selection rules of rotational spectra are allowed in Raman spectra (selection rule $\Delta J = \pm 2$).

Experimental study of Raman Scattering. Raman effect has been studied by many scientists. The basic principle involved in their researches in to illuminate the experimental substance with an intense monochromatic source of light and then to photograph the scattered light by means of spectrograph arranged in transversed direction. They used strong light sources as otherwise the Raman lines will be of very low intensity. Recently Lasers are put to use for this purpose. The experimental arrangement is shown in fig. 6.41.

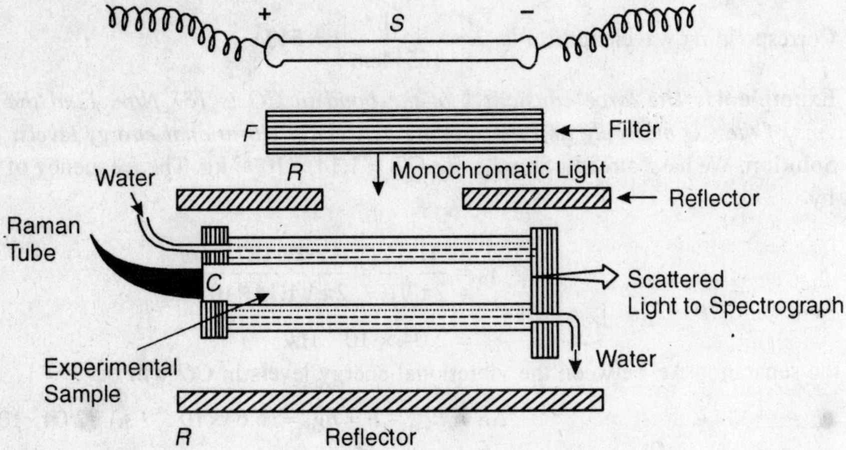

Fig. 6.41

In the figure, S is mercury are lamp *i.e.,* an intense source of light. The light is passed through a filter F to get a monochromatic beam. The light is then passed through an opening a metallic reflector and then on Raman tube. The Raman tube consists of a glass tube of about 1 or 2 cm in diameter and 10 to 15 cm in length. One end of the tube has flat glass surface through which the scattered light comes out. The other end is drawn out into horn-shaped and blackened outside to provide a black background. The tube is surrounded by a cold water jacket *J* to prevent overheatting the sample in Raman tube due to presence of the exciting are. The experimental sample is placed inside Raman tube. The scattered beam emerges from the flat end of the Raman tube and is photographed.

It is enclosed in tube at a pressure of a few cm of mercury. In the case of solids, it is made into powder form and put in some solvent like paraffin. In case of gas sample, it is put in a tube at few cm of Hg pressure.

Example 40. *The exciting line in an experiments is 5460 A and the stokes's line it at 55220 A. Calculate the wavelength of antistokes line.*

Solution. Raman shift is given by

$$\Delta\nu - \nu_i - \nu_s$$

where $\nu \rightarrow$ wave number of incident radiation and $\nu_5 \rightarrow$ wave number of scattered radiation.

For stokes lines $\nu_s < \nu_i$

$\therefore$ In this case $$\nu_i = \frac{1}{5460\text{Å}} = \frac{1}{5460\times10^{-8}}$$

$$= 18315 \text{ cm}^{-1}$$

and $$\nu_s = \frac{1}{5520\text{ Å}} = \frac{1}{5520\times10^{-8}}$$

$$= 18116 \text{ cm}^{-1}$$

$\therefore$ Shift $$\Delta\nu = \nu_i - \nu_s$$

$$= 18315 - 18116 = 199 \text{ cm}^{-1}$$

As the stokes and antistockes lines have same wave number displacement w.r.t. exciting line ν_i and for antiscokes line $\nu_s > \nu_i$ we have

$$(\nu_5) \text{ antistokes} = \nu_i + \Delta\nu$$

$$= 18315 + 199 = 18514 \text{ cm}^{-1}$$

Corresponding wavelength $\lambda = \dfrac{1}{18514 cm^{-1}} = \mathbf{540Å}$

Example 41. *The force constant k of the bond in CO is 187 N/m. Find the frequency 0 vibrations of the CO molecule and the spacing between it vibrational energy levels.*

Solution. We have already found μ for CO = 1.14 × 10^{-28} kg. The frequency of vibrations is given by

$$\nu_0 = \frac{1}{2\pi}\sqrt{\frac{k}{\mu}} = \frac{1}{2\pi}\sqrt{\frac{187 N/m}{1.14\times10^{-26}\,\text{kg}}}$$

$$= 2.04 \times 10^{13} \text{ Hz.}$$

the separation ΔE between the vibrational energy levels in CO will be

$$\Delta E = E_{v+1} - E = h\nu_0 = (6.63\times10^{-34} J.s.)\ (2.04\times10^{13} S^{-1})$$

$$= 1.35\times10^{-20} J = \frac{1.35\times10^{-20}}{1.6\times10^{-19}} = \mathbf{8.44\times10^{-12}\ eV}$$

Example 42. *A substance shows a Raman line at 4567Å when exciting line 4538Å is used. Deduce the position of stokes and antistokes lines for that substance when the exciting 4047 Å is used.*

Solution. Raman shift is given by

$$\Delta\nu = \nu_i - \nu_s = \frac{1}{4358Å} - \frac{1}{4567Å}$$

$$= \frac{1}{4358\times10^{-8}} - \frac{1}{4567\times10^{-8}} = 1050 \text{ cm}^{-1}$$

The wave number of other exciting line

$$\nu_i = \frac{1}{4047Å} = 24710 \text{ cm}^{-1}$$

∴ Wave number of

(*i*) Stoke's line $= \nu_i^1 - \Delta\nu$

$= 24710 - 1050 = 23660 \text{ cm}^{-1}$

(*ii*) Anti Stokes line $= \nu_i + \Delta\nu$

$= 24710 + 1050 = 25760 \text{ cm}^{-1}$

Corresponding wavelengths are

$$\lambda_{\text{Stokes}} = \frac{1}{23660 \text{ cm}^{-1}} = 4226.5 \text{ A}$$

and $$\lambda_{\text{Antistokes}} = \frac{1}{25760 \text{ cm}^{-1}}$$

Corresponding wavelengths are

$$\lambda_{\text{Stokes}} = \frac{1}{23660 \text{ cm}^{-1}} = 4226.5 \text{ A}$$

and $$\lambda_{\text{Antistokes}} = \frac{1}{25760 \text{ cm}^{-1}} = \mathbf{3882\ Å}$$ **Ans.**

6.44. Importance of Raman Effect

Raman effect finds many applications in physics and chemistry and confirms the *quantum theory of radiation.* This effect is useful to study the moleculae structure and chemical analysis of solids, liquids, gases and their molecular symmetry. Raman spectroscopy is finding good number of applications in *Molecular Biology* and *Lattice dynamics* of solids Raman effect has also been applied to study some nuclear characteristics like spin, statistics and isotopic constituents of molecules and has been successful in experimental verification of the characteristics possessed by homo nuclear molecules like H_2, D_2, N_2, O_2 etc.

6.45. Raman Effect and Compton Effect

Both Raman effect and compton effect are interaction of radiation with matter and are based on *scattering phenomenan*. However, in Compton effect, a loosely bound electron of an atom acquires energy from incident photon and the *electron leaves the atom* while incident shoton gets scattered with a reduced energy (or frequency). *Nucleus of atom [practically does not participate in Compton effect.*

Raman effect in special case of interaction of radiation with matter where *entire molecule takes* part. Molecule simply takes energy from incident photon and jumps to a higher state. In this case no electron leaves the atom and the energy transfer in due to *rotation or vibration of the interacting molecules.*

6.46. Raman Effect and Fluorescence

Raman effect and fluorescence resemble each other because in each case light suffers change of wavelength after falling on a scatterer. However, Raman effect differs from fluorescence in the following ways.

1. In fluorescent spectrum, the frequencies are always less than the incident frequency while in Raman spectrum lines with higher frequencies (and-Stokes lines) and lower frequencies (Stoke's lines) are observed.

2. In case of fluorescence, frequencies of scattered radiation independent of frequencies of exciting radiation while in Raman Spectroscopy the scattered radiation is directly related to the incident radiation frequency.

3. Fluorescent materials fixes the frequency of scattered radiation while in case of Raman spectra frequency *shift depends* on the scatter and not the actual frequencies.

4. Fluorescent lines are unlarized while *Raman lines are strongly polarized.*

5. In general, Raman frequencies are either actual infrared frequencies of the scatterer or different in such frequencies which do not hold true in fluorescence phenomenan.

6.47. Raman Effect and Rayleigh Scattering

Rayleigh scattering *i.e.,* scattering $\propto \frac{1}{\lambda^4}$ is *coherent scattering* with means that scattered light has same frequency as that of incident light. This in other words, may be explained as : no exchange of energy takes place between incident and scattered photon *i.e.,* in this case the *collision in elastic.*

In the case of Raman effect, the frequency change occurs on scattering (Stokes and ant Stokes lines). Which means that the scattered photon may have more or less energy as compared to the incident photon. So exchange of energy takes place in this case and hence the *collision is inelastic* in Raman *scattering effect.*

QUESTIONS

1. Explain how X-rays are produced. State the properties of X-rays.
2. How are continuous and characteristic X-rays produced ? How can you control intensity and penetrating power of X-rays ? Discuss some industrial and engineering applications.
3. How do the characteristic X-ray spectral lines arise ? Give Moseley's law and how can it be explained on the Bohr's theory ?
4. Find out the relation between minimum wavelength and accelerating potential with respect to continuous X-ray emitted in a X-ray tube.
5. What are characteristic X-rays ? Explain their production. Describe an X-ray energy level diagram for an atom.
6. Draw an energy level diagram to show the emission of K-series of X-rays. Discuss the importance of excitation and absorption limit in X-ray spectra.
7. What is meant by absorption coefficient of X-rays ? Describe the experimental arrangement of its measurement.
8. How does the frequency of characteristic X-rays depend upon atomic number of element ?
9. What is Moselay's law ? How can it be explained on the basis of Bohr's theory ? What is its importance ?
10. What are characteristic X-rays ? How is their production accounted for ? What important conclusions are drawn by Moseley from the study of characteristic X-rays ?
11. Derive Bragg's equation for reflection of X-rays by crystal planes.
12. Describe Bragg's X-ray spectrometer and deduce the necessary formula, Explain how it is used to determine the wavelength of X-rays ?
13. What information does the sharp absorption edges in X-ray spectra give regarding the atomic structure.
14. Discuss Laue's principle of X-ray diffraction and obtain the diffraction condition for a simple cubic lattice.
15. Describe Bragg's spectrometer and explain how it is used to study the structure of crystals.
16. In what respect does the X-ray spectra of elements differ from their optical spectra ?
17. Discuss Compton effect and derive an expression for the Compton shift. Why are both the modified and the unmodified lines simultaneously present for no-zero scattering angles ?
18. Show that the Compton shift

$$\Delta\lambda = \frac{h}{m_0 c^2}(1 - \cos\theta)$$

where symbols have their usual meaning.

19. What is the importance of non-destructive tests? Give suitable methods used by Radiography.
20. Give a method of non-destructive tests using magnetic particles.
21. How non-destructive tests can detect flaws using electrical methods? Give the various methods in supports to your answer.
22. How will you compare the various NDT (non-destructive tests) techniques? Illustrate your answer by describing due of the techniques.
23. What is radiography? Describe its importance in NDT techniques.
24. How X-ray radiography differs from γ-rays radiography? Illustrate your answer with importance of each technique.
25. How many forms of beam can be used to study NDT by radiography? Give the importance of using electrons and neutrons in said of conventional X-rays or γ-rays.
26. What is fluoroscopy? How is it important in NDT technique?
27. What is thermography? Explain is principle and method of use for NDT technique.
28. Draw a neat diagram of an ultrasonic flaw detector. In what way this technique is preferred over radiography.
29. Give significance of liquid penetration method for NDT techniques.

NUMERICAL PROBLEMS

1. The potential difference between the cathode and the target in a Coolidge tube is 120 kV. What can be the minimum wavelength (in Å) of the X-rays emitted by this tube ? What will be the momentum of this photon ? **[Ans.** 0.1031 Å, 6.4 × 10^{-23} kg – m–51]
2. An X-ray beam has a wavelength of 1 Å. Calculate (*i*) the energy, (*ii*) momentum, (*iii*) mass, and (*iv*) the rest mass of the X-ray photon. **[Ans.** (*i*) 1.98 × 10^{-15} J (*ii*) 6.6 × 10^{-24} NS (*iii*) 2.2 × 10^{-32} kg. (*iv*) zero]
3. An X-ray tube is operated at 15 kV. Calculate the upper limit of the speed of the electrons striking the target (anticathode) and the lower limit to the wavelength of the X-rays produced. **[Ans.** 0.1 Å, 6.4 × 10^{-23} kg m/sec.]
4. An X-ray tube operated at 40 kV emits a continuous X-ray spectrum with a short wavelength limit λ_{min} = 0.310 Å. Calculate the Planck's constant. **[Ans.** 6.61 × 10^{-34} J-S]
5. Calculate the longest wavelength that can be analysed by a rock salt crystal of spacing d = 2.82 Å in the first order. **[Ans.** 5.64 Å]
6. When a nickel target is used in an X-ray tube, the two shortest wavelengths emitted are found with a Bragg crystal spectrograph to be diffracted at angle of 15.1° and 17.1° respectively. Find their wavelengths, assuming crystal spacing of 2.81 Å. **[Ans.** 1.464 A°, 1.652 Å]
7. 10 kV electrons are passed through a thin film of a metal for which the atomic spacing is 5.5 × 10^{-10} m.
What is the angle of deviation of the first order maximum ? **[Ans.** 6°31']
8. The largest distance between the interatomic planes of a crystal is 10 Å. What is the upper limit for the wavelength of X-rays which can be usefully studied with this crystal ? **[Ans.** 20 Å]
9. In Compton effect, if the incident photon has wavelength 2.0 × 10^{-8} cm and $\theta = 90°$, deduce
(*i*) the wavelength of the scattered photon,
(*ii*) the energy of the recoil electron,
(*iii*) the angle at which are recoil electron appears. **[Ans.** (*i*) 2.0243 Å (*ii*) 74.5 eV (*iii*) $\tan^{-1}$ (0.99)]
10. In Compton scattering the incident photons have wavelength 3 × 10^{-10} m. Calculate the wavelength of scattered radiation if they are viewed at an angle of 60° to the direction of incidence. **[Ans.** 3.012 Å]
11. In a Compton scattering experiment an electron attains a kinetic energy of 0.100 MeV when an X-ray of energy 0.500 MeV strikes it. Determine the wavelength of the scattered photon if the electron is initially at rest. **[Ans.** 31 × 10^{-3} A°]
12. A carbon block scatters a monochromatic beam of x-rays of wavelength 0.124 Å. Find the wavelength of X-rays scattered through (*i*) 45°, (*ii*) 90°, and (*iii*) 180°. Also find the wavelength of scattered radiation so that the recoil electron has the maximum kinetic energy.(*i*) 0.127 Å (*ii*) 0.148 Å, (*iii*) 0.172 Å
13. In an experiment on Compton scattering, X-rays of wavelength 10^{-10} m are incident. Calculate the wavelength of the X-rays scattered in the backward direction. **[Ans.** 1.048 Å]
14. In Compton scattering the incident radiation has a wavelength of 2 Å, and that scattered through 180° has 2.048 Å. Calculate the energy of the recoil electron which scatters radiation through 60°. **[Ans.** 37 eV]
15. Calculate the change in wavelength of a photon for 90° scattering from a neutron. (h = 6.6 × 10^{-34} joule-sec, c = 3 = 3 × 10^{8} m/sec, neutron mass = 1.0087 amu and 1 amu = 1.66 × 10^{-17} kg).

7

MAXWELL'S EQUATIONS

7.1. Introduction

Maxwell's equations bear the same relation to electromagnetism that Newton's laws of motion do to *mechanics*. There is, however, an important difference. Einstein presented his special theory of relativity in 1905, about 200 years after Newton's laws were predicted and about 40 years after Maxwell's equations. However Newton's laws had to be drastically modified for cases in which relative speeds approached the speed of light while no changes were required for Maxwell's equations. Maxwells's equations were totally consistent with special theory of relativity. In fact, Einstein's theory grew out of the deep and careful study of Maxwell's *e.m.* Equations.

Maxwell's equations predict the existence of *e.m.* waves (travelling electric and magnetic fields) propagating in free space with the speed of light. These equations show how the *e.m.* wave are gives as under :

1. Electromagnetic waves are propagated in the form of electric and magnetic fields such that both the fields are at right angles to each other and also to the direction of propagation of waves. It means that the *e.m.* waves are transverse in nature.
2. In vacuum, the *e.m.* waves move with a definite velocity given by
$$c = \frac{1}{(\mu_0 \varepsilon_0)^{1/2}}$$
which is the velocity of light in free space.
3. As the waves spread out in free space, the amplitude of the electric and magnetic fields viz. $\vec{E}$ and $\vec{B}$ decreases steadily. $\vec{E}$ and $\vec{B}$ in steady state are related by $c = \frac{\vec{E}}{\vec{B}}$.
4. When two disturbances are superimposed, they observe the laws of superposition *i.e.*, at any instant, the resultant electric field or magnetic field is the vector sum of the *individual field.*
5. *E.M.* waves of different frequencies can exist. However, all such waves would be propagated with the speed of light.
6. Unlike other types of waves *i.e.*, water waves, sound waves, stretched string waves etc,. no material particle moves with the e.m. waves. The only changes are in the *field intensities.*
7. Wiener's experiment on stationary waves have shown that it is the electric vector which *plays the role of light vector.*

As is clear from the discussions on wave propagation in medium (*e.g.* string, rod or a transmission line), a wave carries energy and the physical medium is visualized as a continuum of coupled oscillator. We know that a simple oscillator is characterized by three parameters, two of them being responsible for string and exchanging energy (inertia and elasticity in case of mechanical medium while L and C in an electrical circuit. The third parameter governs the mechanism of energy dissipation (friction in a mechanical oscillator and resistor in the case of an

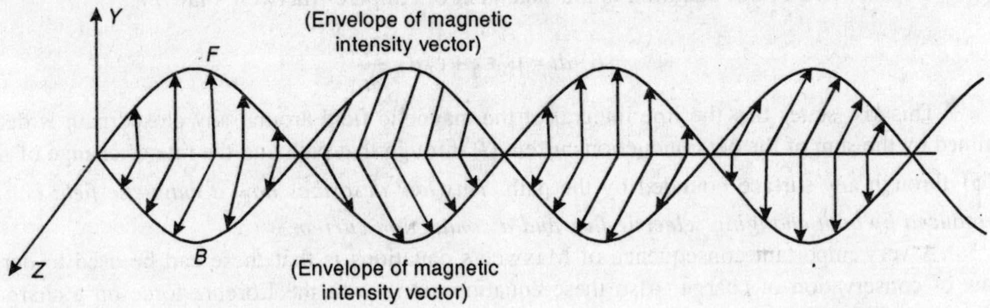

Fig. 7.1

electrical oscillator), propagation of *e.m.*, waves, however, is constituted by the periodic variation of *electric and magnetic fields*. Though, these are separate, yet they are closely interrelated. These vector fields can exist in empty space and no medium is *compressed or rarefied* during their propagation. The *inertia* or the *inductive* property of the medium is provided by the *permeability* μ, the *elasticity or capacity parameters* by the *permittivity* ε. The medium stores the magnetic energy through permeability and electric energy through permittivity. A dielectric medium has finite magnitude of μ and ε and its conductivity $\sigma = 0$. However. In the case of a conductor, all the three parameters have to be accounted for to describe the *e.m.* wave.

7.2. Importance of Maxwell's Equations

1. Maxwell's first equation is the statement of Gauss's law in electrostatics *i.e.*

$$\oint \vec{E}.\vec{ds} = \frac{q}{\varepsilon_0} \quad ...(1)$$

which states that the total electric flux through any closed surface equals the net charge inside that surface divided by ε_0. This law gives the relation between *electric field and charge distribution*. Electric field lines start from positive charges and end at negative charges. In fact, Gauss's law is equivalent to coulomb's inverse square law in electrostatics which has been found to be experimentally true thereby confirming the validity of the equation.

2. Maxwell's second equation describes Gauss's law of electrostatics *i.e.* $\oint \vec{B}.\vec{ds} = 0$ .It states that the net magnetic flux through a closed surface is zero. It means that the number of magnetic lines entering a closed surface is equal to the number of field lines leaving the surface. It also implies that isolated monopoles do not exit. Since isolated magnetic monopoles are not observed in nature so this equation also stands confirmed.

3. The Maxwell's third equation describes Faraday's law of electromagnetic induction *i.e.*,

$$\oint \vec{E}.\vec{dl} = -\frac{d\phi_B}{dt} \quad ...(3)$$

This equation relates on electric field with changing magnetic flux (ϕ_B). According to this law, the line integral of electric field around any closed path (*i.e. e.m.f.*) is equal to the rate of change of magnetic flux through any surface bounded by that. So equation (3) may be written as

$$e = -\frac{d\phi_B}{dt}$$

Which is Faraday's law of electromagnetic induction. This law has been verified experimentally thereby confirming the validity of the Maxwell's third equation.

4. Maxwell's fourth equation is the statement of Ampere-Maxwell's law *i.e.*,

$$\vec{B} \cdot \vec{dl} = \mu_0 \varepsilon_0 + \varepsilon_0 \mu_0 \frac{d\phi}{dt} \qquad ...(4)$$

This law states that the line integral of the magnetic field around any closed path is determined by the sum of the net conduction current (I) through that path and the rate of change of flux (ϕ) through any surface bounded by the path. *This law describes how a magnetic field can he produced by both changing electric flux and a conduction current.*

A very important consequence of Maxwell's equations is that these can be used to derive law of conservation of charge. Also these equation along with the Lorentz force on a charge q moving in electric and magnetic fields given by the equation $\vec{F} = q(\vec{E} + v \times \vec{B})$ provide a complete description of all the *electromagnetic interactions.*

7.3. Speed of e.m. Wave

From Faraday's law of *electromagnetic induction*

$$\oint \vec{E} \cdot \vec{dl} = -\frac{d\phi_B}{dt} \qquad ...(1)$$

and from Ampere-Maxwell's law

$$\oint \vec{B} \cdot \vec{dl} = \mu_0 \varepsilon_0 + \mu_0 \varepsilon_0 \frac{d\phi}{dt} \qquad ...(2)$$

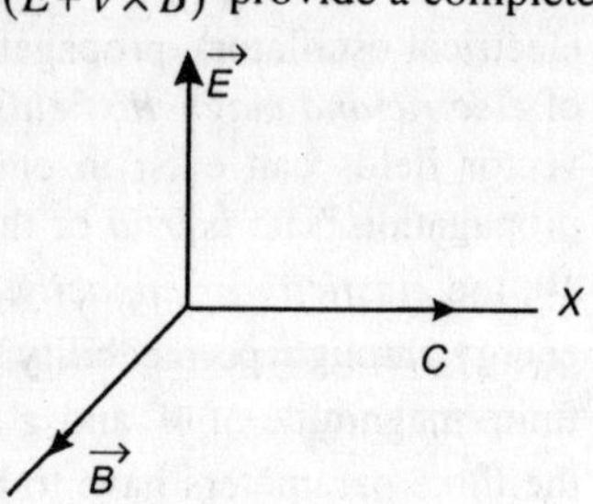

Fig. 7.2

Where ϕ is electric flux and ϕ_B is the magnetic flux.

For a plane polarexed light travelling along x- direction eqn (1) and (2) give

$$\frac{\partial \vec{Ey}}{\partial x} = -\frac{\partial \vec{B_z}}{\partial t}$$

or

$$\frac{\partial \vec{E}}{\partial x} = -\frac{\partial \vec{B}}{\partial t} \qquad ...(3)$$

and

$$\frac{\partial \vec{B_z}}{\partial x} = -\mu_0 \varepsilon_0 \frac{\partial \vec{E_y}}{dt}$$

or

$$\frac{\partial \vec{B}}{\partial x} = -\mu_0 \varepsilon_0 \frac{\partial \vec{E}}{\partial t}$$

Where we have dropped the subscript of components Ey and Bz for simplification.

Taking partial derivative of eq. (3), we have

$$\frac{\partial}{\partial x}\left(\frac{\partial \vec{E}}{\partial x}\right) = -\frac{\partial}{\partial x}\left(\frac{\partial \vec{B}}{\partial t}\right) = \frac{\partial}{\partial t}\left(\frac{\partial \vec{B}}{\partial x}\right)$$

$$\therefore \quad \frac{\partial^2 \vec{E}}{\partial x^2} = -\frac{\partial}{\partial t}\left(-\mu_0 \varepsilon_0 \frac{\partial \vec{E}}{\partial t}\right) = \mu_0 \varepsilon_0 \frac{\partial^2 \vec{E}}{\partial t^2} \qquad ...(5)$$

Similarly

$$\frac{\partial^2 \vec{B}}{\partial x^2} = \mu_0 \varepsilon_0 \frac{\partial^2 \vec{B}}{\partial t^2} \qquad ...(6)$$

Also we have the general wave equation

$$\frac{\partial^2 f}{\partial x^2} = \frac{1}{v^2}\frac{\partial f}{\partial t^2}$$

Where v is the speed of wave and f is the wave amplitude, we get,

$$\frac{1}{v^2} = \mu_0 \varepsilon_0$$

or $$v = \frac{1}{\sqrt{\mu_0 \varepsilon_0}}$$

Now $$\mu_0 = 4\pi \times 10^7 \, Wb\, A^{-1} m^{-1}$$

and $$\varepsilon_0 = 8.85 \times 10^{-12} C^2 N^{-1} m^{-2}$$

$\therefore$ $$v = 3 \times 10^8 ms^{-1}$$

which is the speed of light. So we can replace v by c and write the speed of *e.m.* wave as

$$\mathbf{c} = \frac{\mathbf{1}}{\sqrt{-\mu_0 \varepsilon_0}}$$

7.4. Ratio of Electric Field and Magnetic Field

The sinusoidally varying electric and magnetic fields of a plane *e.m.*wave are given as

$$\vec{E} = \vec{E}_0 \cos(kx - \omega t) \quad ...(i)$$

$$\vec{B} = \vec{B}_0 \cos(kx - \omega t) \quad ...(ii)$$

Where E_0 and B_0 are the maximum values of the electric and magnetic fields associated with the e.m. waves. Also k, the propagation constant = $\frac{2\pi}{\lambda}$, where λ is the wavelength of *e.m.* wave and $\omega(2\pi v)$ is angular frequency of the wave.

$\therefore$ $$\frac{\omega}{k} = \frac{2\pi v}{2\pi/\lambda} = v\lambda = c \quad ...(iii)$$

Taking partial differentiation of eq. (*i*) and (*ii*) we have

$$\frac{\partial E}{\partial x} = -k\,\vec{E}_0 \sin(kx - \omega t)$$

and $$\frac{\partial \vec{B}}{\partial x} = -\omega\,\vec{B}_0 \sin(kx - \omega t)$$

Now since $$\frac{\partial \vec{E}}{\partial x} = -\frac{\partial \vec{B}}{\partial t}$$ [From Faraday's law]

$\therefore$ $$-k\,\vec{E}_0 \sin(kx - \omega t) = -\omega\,\vec{B}_0 \sin(kx - \omega t)$$

or $$k\,\vec{E}_0 = \omega\,\vec{B}_0$$

or $$\frac{E_0}{B_0} = c$$ [From eq. (*iii*)

From eq. (*i*) and (*ii*)

$$\frac{E}{B} = \frac{E_0}{B_0} = c$$

So at every instant the *ratio of the electric field to the magnetic field of an* **e.m. wave is** *equal to the speed of light.*

7.5. Poynting Vector

E.M. waves carry energy with them, when these waves pass through space, they transfer energy to the objection there path. The rate of flow of energy in an *e.m.* wave is given by a vector $\vec{S}$, called Poynting vector, (named after **J.H.** Poynting). It is given by

$$\vec{S} = \frac{\vec{E} \times \vec{B}}{\mu_0} \quad \text{...(1)}$$

For a plane *e.m.* wave, $\vec{E} \perp \vec{B}$

$$\therefore \quad S = \frac{EB}{\mu_0} \quad \text{...(2)}$$

Units of *S*

$$[S] = |E| \quad |H|$$
$$= [Vm^{-1}]\,[Am^{-1}] = VAm^{-2} = wm^{-2} = Js^{-1}m^{-2}$$

So units of *S* are J/sec/m^2

Also $\quad c = \dfrac{E}{B}$

$$\therefore \quad S = \frac{E^2}{\mu_0 c} = \frac{cB^2}{\mu_0} \quad \text{...(3)}$$

We have the equation for electric and magnetic fields as

$$\vec{E} = \vec{E}_0 \cos(kx - \omega t)$$

and $\quad \vec{B} = \vec{B}_0 \cos(kx - \omega t)$

$$\therefore \quad S = \left[\frac{E_0 B_0}{\mu_0}\right] \cos^2(kx - \omega t)$$

If S_{av} is the average value of *S* taken over a complete cycle,

$$S_{av} = \frac{1}{T}\int_0^T S.dt = \frac{1}{T}\int_0^T \frac{E_0 B_0}{\mu_0} \cos^2(kx - \omega t)dt$$

$$= \frac{1}{T} \times \frac{E_0 B_0}{\mu_0} \times \frac{T}{2} \qquad \left[\therefore \int_0^T \cos^2(kx - \omega t)dt = \frac{T}{2}\right]$$

$$= \frac{E_0 B_0}{2\mu_0} = \frac{E_0^2}{2\mu_0 c} = \frac{cB_0^2}{2\mu_0} \qquad \left[\therefore = \frac{E_0}{\mathbf{B}_0} = c\right] \quad \text{...(4)}$$

where S_{av} is called the *wave intensity. The instantaneous energy density associated with an electric field is given by*

$$U_e = \frac{1}{2}\varepsilon_0 E^2 \quad \text{...(5)}$$

and instantaneous energy density associated with a magnetic field is given by

$$U_m = \frac{B^2}{2\mu_0}$$

$$\therefore \quad U_m = \frac{(E/c)^2}{2\mu_0} = \frac{E^2/c^2}{2\mu_0} \times (\mu_0 E_0)$$

$$= \frac{1}{2}\varepsilon_0 E^2 = U_e \qquad \left[\because \frac{E}{B} = c \text{ and } c = \frac{1}{\sqrt{\varepsilon_0 \mu_0}}\right] \qquad ...(6)$$

which means that for an e.m. wave, the instantaneous energy density associated with the magnetic field is equal to the instantaneous energy density associated with the electric field .

Which is other words mean that in a given volume, *the energy is shared equally by two fields* relation between total energy density and average energy density.

Relation between wave intensity and total energy density

If U be the total energy density,

then $$U = Ue + U_m = \varepsilon E^2 = \frac{B^2}{\mu_0} \qquad ...(7)$$

If U_{av} is the total energy density (*i.e.*, U_{av} is the average value of U taken over a complete cycle), then

$$U_{av} = \frac{1}{T}\int_0^T U.dt = \frac{1}{T}\int_0^T \varepsilon_0 E^2 dt$$

$$= \frac{\varepsilon_0}{T}\int_0^T E_0^2 \cos^2(kx - \omega t)dt$$

$$= \frac{\varepsilon_0}{T}\cdot E_0^2 \cdot \frac{T}{2} \qquad \left[\int_0^T \cos^2(kx - \omega t)dt = \frac{T}{2}\right]$$

$$\therefore \quad U_{av} = \frac{1}{2}\varepsilon_0 E^2$$

Also $$\frac{E_0}{B_0} = c$$

$$\therefore \quad U_{av} = \frac{B_0^2}{2\mu_0}$$

$$\therefore \quad U_{av} = \frac{1}{2}\varepsilon_0 E^2 = \frac{1}{2}\frac{B_0^2}{\mu_0}$$

From eq. (4)

$$S_{av} = c.\frac{1}{2}\frac{B_0^2}{\mu_0}$$

$$\therefore \quad S_{av} = c.U_{av}$$

which means that the intensity of an e.m. wave is equal to the product of average energy density with the speed of light.

7.6. Interpretation of Divergence and Curl of a Vector Point

Divergence. *Divergence of a continuously differentiable vector point function F is denoted by div F and is defined by the equation.*

Div $\vec{F} = \vec{\nabla}.\vec{F} = \hat{i}\frac{\partial F}{\partial x} + \hat{j}\frac{\partial F}{\partial y} + \hat{k}\frac{\partial F}{\partial z}$

If we consider $\vec{F} = f\hat{i} + \phi\hat{j} + \Psi\hat{k}$

then div $\vec{F} = \vec{\nabla}.\vec{F} = \left(\hat{i}\frac{\partial}{\partial x} + \hat{j}\frac{\partial}{\partial y} + \hat{k}\frac{\partial}{\partial z}\right)(f\hat{i}\phi j + \Psi\hat{\kappa})$

$$= \frac{\partial f}{\partial x} + \frac{\partial \phi}{\partial y} + \frac{\partial \Psi}{\partial z}$$

Curl. The curl of a continuously differentiable vector point function F is defined by the equation

$$\text{Curl F} = \vec{\nabla}\times\vec{F} = \hat{i}\frac{\partial F}{\partial x} + \hat{j}\frac{\partial F}{\partial y} + \hat{k}\frac{\partial F}{\partial z}$$

if $\vec{F} = f\hat{i} + \phi\hat{j} + \Psi\hat{k}$

then curl $\vec{F} = \vec{\nabla}\times\vec{F}$

$$= \left(\hat{i}\frac{\partial}{\partial i} + \hat{j}\frac{\partial}{\partial y} + \hat{k}\frac{\partial}{\partial z}\right)(f\hat{i} + \phi\hat{j} + \Psi\hat{k})$$

$$= \begin{vmatrix} \hat{i} & \hat{j} & \hat{k} \\ \frac{\partial}{\partial x} & \frac{\partial}{\partial y} & \frac{\partial}{\partial z} \\ f & \phi & \Psi \end{vmatrix}$$

$$= \hat{i}\left(\frac{\partial \psi}{\partial y} - \frac{\partial \phi}{\partial z}\right) + \hat{j}\left(\frac{\partial f}{\partial z} - \frac{\partial \phi}{\partial x}\right) + \hat{k}\left(\frac{\partial \phi}{\partial z} - \frac{\partial f}{\partial y}\right)$$

7.7. Physical Interpretation of Divergence

Consider the motion of a fluid having velocity $\vec{U} = U_x\hat{i} + U_y\hat{j} + U_z\hat{k}$ at a point $P(x, y, z)$. Consider a small parallelopiped with edges S_x, S_y, S_z parallel to the axes in the mass of the fluid, with one of it corners at P,

So the amount of fluid entering the face PB' per unit time

$$= V_{y+dy}\delta_y, \delta_x$$

and the amount of fluid leaving the foce P, B in unit time $V_{y+}\ \delta_y, \delta_z, \delta_x$

$$= \left(U_y + \frac{\partial U_y}{\partial y}.\delta y\right).\delta z.\delta x$$

So the net decrease in the amount of fluid due to the flow across these two faces

$$= \frac{\partial U_y}{\partial y}.\delta x.\delta y.\delta z$$

Going an the same lines, we have the contribution due to the other two pairs of faces. So the total decrease of amount of fluid inside the parallelopiped per unit time

$$= \left(\frac{\partial U_x}{\partial x} + \frac{\partial U_y}{\partial y} + \frac{\partial U_z}{\partial z}\right).\delta x.\delta y.\delta z$$

Hence the rate of loss of fluid/unit volume

$$= \frac{\partial U_x}{\partial x} + \frac{\partial U_y}{\partial y} + \frac{\partial U_z}{\partial z} = \text{div}\,\vec{V}$$

i.e., div V gives the rate at which the fluid is originating per unit Volume.

In the same way if V represents an electric flux, *div V* as the amount of flux which diverges per unit volume per unit time. Again if V represents *heat flux*, div V will represent the amount of heat issuing from a point per unit volume per unit time.

In general, divergence of a vector point function representing any physical quantity gives at each point the rate/units volume at which the physical quantity is issuing from the point.

It is due to this fact that it is called divergence of a vector point function.

If a fluid is incompressible, there will not be any gain or loss in volume of the element. In that case div. $V = 0$. This in hydrodynamics is called *equation of continuity for an incompressible field.*

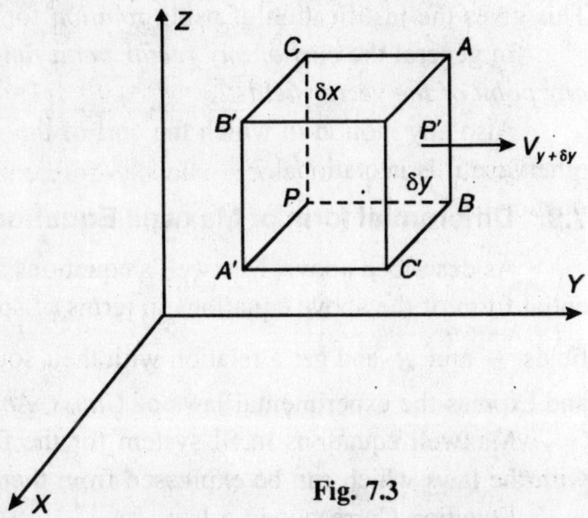

Fig. 7.3

If the flux entering any element of space is the same as leaving it *i.e.*, div $V = 0$ every where, then such a function is called *a solenoidal vector function.*

7.8 Physical Interpretation of Curl

Consider the motion of a rigid body rotating about a fixed axis through O. If Ω be the angular velocity then the velocity $\vec{V}$ of any particle P (R) of the body is given by

$$\vec{V} = \vec{\Omega} \times \vec{R}$$

if $$\vec{\Omega} = \hat{i}\omega_1 + \hat{j}\omega_2 + \hat{k}\omega_3$$

and $$\vec{R} = \hat{i}x + \hat{j}y + \hat{k}z$$

then $$\vec{V} = \vec{\Omega} \times \vec{R} = \begin{vmatrix} \hat{i} & \hat{j} & \hat{k} \\ \omega_1 & \omega_2 & \omega_3 \\ x & y & z \end{vmatrix}$$

$$= \hat{i}(\omega_2 z - \omega_3 y) + \hat{j}(\omega_3 x - \omega_1 z) + \hat{k}(\omega_1 y - \omega_2 x)$$

$$\text{Curl } \vec{V} = \begin{vmatrix} \hat{i} & \hat{j} & \hat{k} \\ \frac{\partial}{\partial x} & \frac{\partial}{\partial y} & \frac{\partial}{\partial z} \\ \omega_2 z - \omega_3 y & \omega_3 x - \omega_1 z & \omega_1 y - \omega_2 x \end{vmatrix}$$

$$= \hat{i}(\omega_1 + \omega_1) + \hat{j}(\omega_2 + \omega_2) + \hat{k}(\omega_3 + \omega_3)$$

$$= 2\omega_1\hat{i} + 2\omega_2\hat{j} + 2\omega_3\hat{k}$$

$$= 2(\omega_1\hat{i} + \omega_2\hat{j} + 2\omega_3\hat{k})$$

$$= 2\vec{\Omega}$$

$$\vec{\Omega} = \frac{1}{2}\text{curl}\,\vec{V}$$

So the angular velocity of rotation at any point is equal to half the curl of velocity vector. This gives the justification of using *rotation* for the curl of a vector.

In general the *curl of any vector point function gives the measure of the angular velocity at any point of the vector field.*

Also any motion in which the curl of the velocity vector is not zero is said to be rotational, otherwise it is irrotational.

7.9. Differential form of Maxwell Equations

As described above, Maxwell's equations are a set of four equations. We can have the differential form of the above equations in terms of spatial derivatives (curl or divergence) of the vector fields $\vec{E}$ and $\vec{B}$ and get a relation with their sources viz charge density ρ and current density $\vec{J}$ and express the experimental laws of *Gauss, Ampere and Faraday*.

Maxwell equations in SI system for the fields in vacuum are summarized as below, along with the laws which can be expressed from them.

Equation	Corresponding law
(*i*) $\nabla.\vec{D} = \rho$	(1) Gauss's law
(*ii*) $\nabla.\vec{B} = 0$	(2) Absence of isolated magnetic fields
(*iii*) $\nabla.\vec{E} = -\frac{\partial \vec{B}}{\partial t}$	(3) Faraday's law
(*iv*) $\vec{\nabla}\times\vec{H} = J + \frac{\partial \vec{D}}{\partial t}$	(4) Ampere's circuital law.

where ρ and $\vec{J}$ are charge and current densities and constitute the sources for the field vector $\vec{E}$ and $\vec{B}$. The *microscopic vector* $\vec{E}$ and $\vec{B}$, on the introduction of a material medium become *macroscopic* vectors viz the electric displacement vector $\vec{D}$ and magnetic field $\vec{H}$ given by the relation;

$$\vec{D} = \varepsilon_r.\varepsilon_0\,\vec{E} = \varepsilon\,\vec{E} \qquad \text{... (5)}$$

and

$$\vec{B} = \mu_r.\mu_0\,\vec{H} = \mu\,\vec{H} \qquad \text{... (6)}$$

where ε_0 and ε_r are the permittivity of vacuum and relative permittivity of the dielectric medium, while ε corresponds to the absolute permittivity of the medium.

Similar comments can be applied for μ.

The conduction current density $\vec{J}$ is related to vector $\vec{E}$ though *Ohm's law* having the form

$$\vec{J} = \sigma\vec{E} \qquad \text{... (7)}$$

where σ is the conductivity of the medium.

The importance of the above equations lie from *Helmholtz's theorem of vector analysis*. The field vectors $\vec{E}$ and $\vec{B}$ are cross coupled with each other, which in other words mean that *each is generated by the time variation of the other*. From helmholtz's theorem, vector field is known completely if its curl and *divergence are given.*

As may be observed, the equations (1) and (2) are steady state equations while eqns (3) and (4) are time varying and together, they explain *all the known properties of* $\vec{E}$ *and* $\vec{B}$.

Maxwell showed that with some manipulation, these equations lead to wave equation. He again showed that these waves could propagate in free space with wave velocity which could be deduced from electrical measurements which has no connection with the wave phenomenon. The value thus calculated is the same as the value of velocity of light in vacuum. *This establishes the electromagnetic nature of light.* The velocity of light c in vacuum is the *maximum limit* of velocity of propagation of interactions and is called *signal velocity.*

7.10. Physical Significance of Maxwell's Equations:

I steady state equation

$$\text{(I)}\quad \vec{\nabla}.\vec{D} = \rho \qquad \text{... (1)}$$

This equation is a restatement of Gauss's law in electrostatics. Putting eq. (1) in the form $\vec{\nabla}.\vec{E} = \frac{\rho}{\varepsilon}$ (as $\vec{D}$ is the displacement vector given by $\vec{D} = \varepsilon\vec{E}$ and ρ is the charge density and integrating over volume V, we get

$$\iiint \vec{\nabla}.\vec{E}\, dV = \frac{1}{\varepsilon}\iiint \rho dV \qquad \text{... (2)}$$

From the divergence theorem of vector analysis.

$$\iiint \vec{\nabla}.\vec{E}\, dV = \oint\!\!\oint_s \vec{E}.d\vec{s} \qquad \text{... (3)}$$

where $\oint\!\!\oint_s$ represents integration over a closed surface

Comparing eq. (2) and (3), we have

$$\oint\!\!\oint \vec{E}.d\vec{s} = \frac{1}{\varepsilon}\iiint \rho dV.$$

Now $\oint\!\!\oint \vec{E}.d\vec{s}$ represents the electric flux through a closed surface and $\iiint \rho dv = q$ represent the charge in the closed surface.

$$\therefore \qquad \oint\!\!\oint \vec{E}.d\vec{s} = q/\varepsilon$$

which means that the total flux through a closed surface is equal to the total charge inside the surface divided by ε.

This is the Gauss's law of electrostatics.

II. Second steady state equation: In vectors form the second steady state equation is given by

$$\vec{\nabla}.\vec{B} = 0$$

where $\vec{B}$ is the magnetic induction vector. This equation is the restatement of Gauss's law in magnetism and proves that *magnetic poles exist only in pairs. i.e.,* isolated magnetic poles can not be found.

For a closed surface, we have

$$\iiint \nabla . \vec{B} . dv = 0$$

Applying the divergence theorem in vector analysis, we may have

$$\iiint \vec{\nabla} . \vec{B} = \oiint \vec{B} . d\vec{s}$$

$$\therefore \qquad \oiint \vec{B} . d\vec{s} = 0 \qquad \text{... (5)}$$

Now $\oiint \vec{B}$ represents the magnetic flux through a closed surface, so the magnetic flux through a closed surface is zero. In other words it means that a magnetic flux entering into the volume is equal to the magnetic flux leaving the volume. So there is no source of sink in the volume, which in other words mean that *isolated magnetic poles (monopoles cannot exist).*

III. First time varying equation: This equation is the restatement of Faraday's law of electromagnetic induction.

The equation is given by

$$\vec{\nabla} \times \vec{E} = \frac{-\partial \vec{B}}{\partial t} = -\mu \frac{\partial \vec{H}}{\partial t} \qquad \text{... (6)}$$

Integrating eq. (5) over a bounded surface bounding a path S, we get

$$\int_s (\vec{\nabla} \times \vec{E}) . d\vec{s} = -\oint_s \frac{\partial \vec{B}}{\partial t} . d\vec{s} \qquad \text{... (7)}$$

From Stokes's theorem of vector analysis

$$\int_s (\vec{\nabla} \times \vec{E}) . d\vec{s} = -\oint_s \vec{E} . d\vec{l}$$

So eq. (7) may be written as

$$\oint \vec{E} . d\vec{l} = \oint_s \frac{\partial \vec{B}}{\partial t} . d\vec{s}$$

where $\oint \vec{E} . d\vec{l}$ represent as *induced e.m.f.* in a closed circuit and $-\oint_s \frac{\partial \vec{B}}{\partial t} . d\vec{s}$ represents *negative rate of change of magnetic flux through the circuit.*

This is Faraday's law of electromagnetic induction. It shows that time variation of magnetic induction $\vec{B}$ *generates the electric field* $\vec{E}$

(IV) Second time varying equation: In vector form second time varying equation is given by

$$\vec{\nabla} \times \vec{H} = \vec{J} + \frac{\partial \vec{D}}{\partial t}$$

where J is the conduction current density given by $\vec{J} = \sigma + \vec{E}$, σ being the conductivity of the medium and $\frac{\partial \vec{D}}{\partial t}$ is the displacement current density.

Physical Significance

This equation is the restatement of Amperes law combined together for *displacement current and conduction current, and states that a changing electric field produces a changing magnetic field.*

Also we know that changing magnetic field produces a changing electric field. So we may conclude that:–

A changing electric field produces a changing magnetic field and a changing magnetic field produces a changing electric field. *Alternate production of electric and magnetic fields give rise to the propagation of electromagnetic waves.*

7.11. Distinction between Conduction and Displacement Current

Conduction current is the actual current that flows through the conducting medium and so obeys ohm's law *i.e.*, I = V/R as well as its vector form $\vec{J} = \sigma \vec{E}$ where J is the *conduction current density,* σ the *conductivity* of the medium and $\vec{E}$ the *electric field intensity.*

Displacement current is the current which is set up in the dielectric medium (i.e. $\sigma = 0$) due to the variation of induced displacement charge produced by the varying electric field applied across the dielectric. Displacement charge density is given by $\frac{\partial \vec{D}}{\partial t} = \varepsilon \frac{\partial \vec{E}}{\partial t}$ and it has the limits of current/unit area as shown under :-

Maxwells second time varying equation for e.m. field is given by

$$\nabla \times \vec{H} = \vec{J} + \frac{\partial \vec{D}}{\partial t}$$

where $\vec{J} = \sigma \vec{E}, \sigma$ being the conductivity of the medium.

$\vec{D} = \varepsilon \vec{E}, \varepsilon$ Being the permittivity of the medium having units. Farad/meter $\vec{E}$ being the electric field vector having units volt/meter

$\frac{\partial \vec{D}}{\partial t}$ is called displacement term and $\vec{D}$ is the electric displacement vector.

$\therefore$ Dimensions of $D = \frac{\text{Farad}}{\text{meter}} \times \frac{\text{Volt}}{\text{meter}} = \frac{\text{Coulomb}}{(\text{meter})^2}$ and dimensions of $\frac{\partial D}{\partial t} = \frac{\text{Coulomb}}{\text{meter}^2} \times \frac{1}{\text{sec}}$

$= \text{Ampere/meter}^2$

So the dimensions of displacement term are that of current/unit area or current density.

7.12. Major Contribution of Maxwell Equations to e.m. Wave Theory

Maxwell introduction of *displacement vector* in a polarizable dielectric medium corresponding to the electric vector in vacuum or free space explained many relevant phenomenon as shown under:

Displacement vector $\vec{D}$ when relates to electric field vector $\vec{E}$ is given by

$$\vec{D} = \varepsilon \vec{E}$$

where ε is the permittivity of the medium.

Now from the Maxwell's idea of *displacement current,* we have

$$\frac{\partial \vec{D}}{\partial t} = \frac{\partial}{\partial t}(\varepsilon \vec{E}) = \varepsilon \frac{\partial \vec{E}}{\partial t}$$

So Maxwell's second time varying equation for e.m. wave in modified form will be

$$\nabla \times \vec{H} = \vec{J} + \frac{\partial \vec{D}}{\partial t} = \sigma \vec{E} + \varepsilon \frac{\partial \vec{E}}{\partial t}$$

The term $\frac{\partial \vec{D}}{\partial t} = \varepsilon \frac{\partial \vec{E}}{\partial t}$ accounts for all radiative for all radiative processes from a charge (positive or negative) undergoing an acceleration. Various ways for them to happen may be when an electron:

(*i*) jumps from high energy level to low energy level due to the de-excitation of an atom.
(*ii*) suffers collision and its path changes and so gets accelerated.
(*iii*) undergoes oscillatory motion in an alternating field as in the case of a wireless antina.

7.13. Representation of Light Vector

Light is a form of e.m. wave and the force exerted by its vector $\vec{E}$ on a particle of charge q is given by $q\vec{E}$ (magnitude is qE). If the particle is moving with velocity v, then the force exerted by magnetic vector $\vec{B}$ is given by $q(\vec{v} \times \vec{B})$ and its maximum value is given by the magnitude qvB.

For an e.m. wave

$$\frac{|\vec{E}|}{|\vec{H}|} = \sqrt{\frac{\mu}{\varepsilon}} = \mu\sqrt{\frac{1}{\mu\varepsilon}} = \mu.c.$$

or
$$\frac{|\vec{E}|}{|\vec{B}|} = c \text{ or } B = E/c.$$

$$\therefore \qquad qvB = q.v.\frac{E}{c} = q\,E.v/c.$$

which means that *magnetic force* is equal to $\frac{v}{c}$ times *electric force*. Since c is very large and compared to v, so *electric force* is much greater than the *magnetic force*. It is due to this reason that *light wave is characterised by its electric field vector.*

It may be noted that although a wave cannot be exist unless both $\vec{E}$ and $\vec{B}$ are present, yet $\vec{E}$ is more important than $\vec{B}$ from the point of view of effects due to $\vec{E}$ and $\vec{B}$. We can ignore the effects of $\vec{B}$ as compared to those of $\vec{E}$ as the wave passes through matter.

As an example, let us take the intensity of light from bright sun on earth which is 4/kw/m^2 which means electric vector has an amplitude of 1 KV/m; which is quite measurable quantity.

On the other hand the amplitude of corresponding magnetic field is about 3 micro tesla $(3\mu T)$. Which is ~ $\frac{1}{10}$th of the Earth's magnetic field and so is very difficult to measure.

The above example clearly shows that importance of electric field vector $\vec{E}$ over magnetic field vector $\vec{H}$. It is due to this reason that in *polarization of light*. (*i.e.*, when light wave passes through some polarizing materials like calcite tourmaline etc., one of its components is selectively absorbed). **As the electric field vector $\vec{E}$ has much greater magnitude than the magnetic field vector $\vec{H}$, *polarization is determined by the electric field vector.***

7.14. e.m. Waves in a Medium of Finite Permeability μ and Permittivity ε = 0

As $\sigma = 0$, let us consider the case of a ion-conducting medium (insulutor or dielectric having finitu permeability μ and panittivity ε.). In such a case as there is no charge in the medium, so charge density $\rho = 0$. In that case Maxwell equations for vectors $\vec{E}$ and $\vec{H}$ may be written as

$$\varepsilon \vec{\nabla}.\vec{E} = \text{or } \vec{\nabla}.\vec{E} = 0 \quad \text{... (1)}$$

$$\vec{\nabla}.\vec{B} = 0; \quad \mu \vec{\nabla}.\vec{H} = 0 \quad \text{or} \quad \vec{\nabla}.\vec{H} = 0 \quad \text{... (2)}$$

$$\vec{\nabla}\times\vec{E} = -\mu\frac{\partial \vec{H}}{\partial t} \quad \text{... (3)}$$

$$\vec{\nabla}\times\vec{H} = \vec{J} + \varepsilon\frac{\partial \vec{E}}{\partial t} \quad \text{or} \; \vec{\nabla}\times\vec{H} = \sigma\vec{E} + \varepsilon\frac{\partial \vec{E}}{\partial t} \quad \text{... (4)}$$

As $\sigma = 0$, the above equation for a non-conducting medium may be written as

$$\vec{\nabla}\times\vec{E} = \varepsilon\frac{\partial \vec{E}}{\partial t} \quad \text{... (5)}$$

For electric field $\vec{E}$. Take the curl of equation (3), and putting the value of $\vec{\nabla}\times\vec{H}$ from eq. (5), we get

$$\vec{\nabla}\times(\vec{\nabla}\times\vec{H}) = \vec{\nabla}\times\left(-\mu\frac{\partial \vec{H}}{\partial t}\right) = -\mu\left(\vec{\nabla}\times\frac{\partial \vec{H}}{\partial t}\right) = -\mu\frac{\partial}{\partial t}(\vec{\nabla}\times\vec{H})$$

$$= -\mu\frac{\partial}{\partial t}\left(\varepsilon\frac{\partial \vec{E}}{\partial t}\right) = -\mu\varepsilon\frac{\partial^2 \vec{E}}{\partial t^2} \quad \text{... (6)}$$

Let us use the vector triple identity

$$\vec{A}\times(\vec{B}\times\vec{C}) = \vec{B}(\vec{A}.\vec{C}) - \vec{C}(\vec{A}.\vec{B}) \text{ to the case of } \vec{\nabla}\times(\vec{\nabla}\times\vec{E})$$

we get $\vec{\nabla}\times(\vec{\nabla}\times\vec{E}) = \vec{\nabla}(\vec{\nabla}.\vec{E}) - \vec{E}(\vec{\nabla}.\vec{\nabla}) = -\nabla^2\vec{E}$ ... (7)

As $\vec{\nabla}\cdot\vec{E} = 0$ according to equation (1) and $-\vec{E}(\vec{\nabla}.\vec{\nabla}) = -\nabla^2\vec{E}$

So we may have from eq. (6) and (7)

$$-\mu\varepsilon\frac{\partial^2 \vec{E}}{\partial t^2} = -\nabla^2\vec{E}$$

or

$$\nabla^2\vec{E} = \mu\varepsilon\frac{\partial^2 \vec{E}}{\partial t^2} \quad \text{... (8)}$$

Eq. (8) is the general eq. for the vector field E in a non-conducting medium.

For magnetic field $\vec{H}$. Let us take curl of eq. (5) and putting the value of $\vec{\nabla}\times\vec{E}$ in eq. (3) we get

$$\vec{\nabla}\times(\vec{\nabla}\times\vec{H}) = \vec{\nabla}\times\left(\varepsilon\frac{\partial \vec{E}}{\partial t}\right) = \varepsilon\left(\vec{\nabla}\times\frac{\partial \vec{E}}{\partial t}\right) = \varepsilon\frac{\partial}{\partial t}(\vec{\nabla}\times\vec{E})$$

$$= \varepsilon \frac{\partial}{\partial t}\left(-\mu \frac{\partial \vec{H}}{\partial t}\right) = -\mu\varepsilon \frac{\partial^2 \vec{H}}{\partial t^2} \quad \text{... (9)}$$

Again from vector identity

$$\vec{A}\times(\vec{B}\times\vec{C}) = \vec{B}\times(\vec{A}.\vec{C}) - \vec{C}(\vec{A}.\vec{B})$$

Applying to $\vec{\nabla}\times(\vec{\nabla}\times\vec{H})$, we get

$$\vec{\nabla}\times(\vec{\nabla}\times\vec{H}) = \vec{\nabla}(\vec{\nabla}.\vec{H}) - \vec{H}(\vec{\nabla}.\vec{\nabla}) = -\nabla^2\vec{H} \quad \text{... (10)}$$

As $\vec{\nabla}.\vec{H} = 0$ [From eq. (2)]

$$\therefore \quad -\vec{H}(\vec{\nabla}.\vec{\nabla}) = -\nabla^2\vec{H}$$

From eq. (9) and (10), we get

$$-\mu\varepsilon \frac{\partial^2 \vec{H}}{\partial t^2} = -\nabla^2\vec{H}$$

or
$$\nabla^2\vec{H} = \mu\varepsilon \frac{\partial^2 H}{\partial t^2} \quad \text{... (11)}$$

Eq. (11) is the general wave equation for the vector field $\vec{H}$ in case of a *non-conducting* medium.

7.15. *e.m.* Waves in a Conducting Medium.

Consider a conducting medium having finite permeability μ, permittivity ε and conductivity σ.

If the medium is free from a charge source, in that case $\rho = 0$ and Maxwell's equations for vectors $\vec{E}$ and $\vec{H}$ may be written as

$$\varepsilon\vec{\nabla}.\vec{E} = 0 \quad \text{or} \quad \vec{\nabla}.\vec{E} = 0 \quad \text{... (1)}$$

$$\mu\vec{\nabla}.\vec{H} = 0 \text{ or } \vec{\nabla}.\vec{H} = 0 \quad \text{... (2)}$$

$$\vec{\nabla}\times\vec{E} = -\mu\frac{\partial \vec{H}}{\partial t} \quad \text{... (3)}$$

$$\vec{\nabla}\times\vec{H} = \vec{J} + \varepsilon\frac{\partial \vec{E}}{\partial t}$$

$$= \sigma\vec{E} + \varepsilon\frac{\partial \vec{E}}{\partial t} \qquad [\because \vec{J} = \sigma\vec{E}] \quad \text{... (4)}$$

For electric field $\vec{E}$. Take curl of eq. (3) and put the value of $(\vec{\nabla}\times\vec{H})$ from eq. (4), we get

$$\vec{\nabla}\times(\vec{\nabla}\times\vec{E}) = \vec{\nabla}\times\left(-\mu\frac{\partial \vec{H}}{\partial t}\right) = -\mu\left(\vec{\nabla}\times\frac{\partial \vec{H}}{\partial t}\right)$$

$$= -\mu\frac{\partial}{\partial t}(\vec{\nabla}\times\vec{\nabla}) = -\mu\frac{\partial}{\partial t}\left(\sigma\vec{E} + \sigma\frac{\partial \vec{E}}{\partial t}\right)$$

$$= -\mu\sigma\frac{\partial \vec{E}}{\partial t} - \mu\varepsilon\frac{\partial^2 \vec{E}}{\partial t^2} \quad \text{... (5)}$$

Again from triple vector identity

$$\vec{A}\times(\vec{B}\times\vec{C}) = \vec{B}(\vec{A}.\vec{C}) - \vec{C}(\vec{A}.\vec{B}) \quad \text{applied to } \vec{\nabla}\times(\vec{\nabla}\times\vec{E}) \text{, we get}$$

$$\vec{\nabla}\times(\vec{\nabla}\times\vec{E}) = \vec{\nabla}\times(\vec{\nabla}.\vec{E}) - \vec{E}(\vec{\nabla}.\vec{\nabla}) = -\nabla^2\vec{E} \quad \text{... (6)}$$

as $\vec{\nabla}.\vec{E} = 0$ from eq. (1) and $-\vec{E}(\vec{\nabla}.\vec{\nabla}) = -\nabla^2\vec{E}$

So from eq. (5) and (6) we have

$$-\nabla^2\vec{E} = -\mu\sigma\frac{\partial \vec{E}}{\partial t} - \mu\varepsilon\frac{\partial^2 \vec{E}}{\partial t^2}$$

or

$$\nabla^2\vec{E} = \mu\sigma\frac{\partial \vec{E}}{\partial t} + \mu\varepsilon\frac{\partial^2 \vec{E}}{\partial t^2} \quad \text{... (7)}$$

In case of a non-conducting medium

$$\sigma = 0 \text{, so we have } \nabla^2\vec{E} = \mu\varepsilon\frac{\partial^2 E}{dt^2} \quad \text{... (8)}$$

Comparing eq. (7) and (8), we see that $\mu\sigma\frac{\partial \vec{E}}{\partial t}$ is the *dissipative term which allows the current to flow through the medium.*

For magnetic field H. Taking curl of equation (4) and putting the value of $\vec{\nabla}\times\vec{E}$ from (3), we get

$$\vec{\nabla}\times(\vec{\nabla}\times\vec{H}) = \vec{\nabla}\times\left(\sigma\vec{E} + \varepsilon\frac{\partial \vec{E}}{\partial t}\right) = \sigma(\vec{\nabla}\times\vec{E}) + \varepsilon\frac{\partial}{\partial t}(\vec{\nabla}\times\vec{E})$$

$$= -\mu\sigma\frac{\partial \vec{H}}{\partial t} - \varepsilon\mu\frac{\partial^2 H}{\partial t^2} \quad \text{... (9)}$$

Applying the triple vector identity

$$\vec{A}\times(\vec{B}\times\vec{C}) = \vec{B}(\vec{A}.\vec{C}) - \vec{C}(\vec{A}.\vec{B}) \quad \text{to the case } \vec{\nabla}\times(\vec{\nabla}\times\vec{H}) \text{, we get}$$

$$\vec{\nabla}\times(\vec{\nabla}\times\vec{H}) = \vec{\nabla}(\vec{\nabla}.\vec{H}) - \vec{H}(\vec{\nabla}.\vec{\nabla}) = -\nabla^2\vec{H} \quad \text{... (10)}$$

since $\qquad \vec{\nabla}.\vec{H} = 0$ [from eq. (2)]

$\therefore \qquad -\vec{H}(\vec{\nabla}.\vec{\nabla}) = -\nabla^2\vec{H}$

From eq. (9) and (10), we have

$$-\nabla^2\vec{H} = -\mu\sigma\frac{\partial \vec{H}}{\partial t} - \varepsilon\mu\frac{\partial^2 \vec{H}}{\partial t^2}$$

or

$$\nabla^2 H = \mu\sigma\frac{\partial \vec{H}}{\partial t} + \mu\in\frac{\partial^2 \vec{H}}{\partial t^2} \quad \text{... (11)}$$

In case of non-conducting medium $\sigma = 0$ we have

$$\nabla^2 = \mu \in \frac{\partial^2 \vec{H}}{\partial t} \quad \text{... (12)}$$

Comparison of eq. (11) and (12) shows that $\mu\sigma \frac{\partial \vec{H}}{\partial t}$ is the *dissipative term* which allows the current to flow through the medium.

7.16. Velocity of e.m. Waves from Maxwell's Equations.

Wave equation for an e.m. wave in case of a non-conducting medium ($\sigma = 0$) having finite magnetic permeability μ and electric susceptibility ε in terms of vector field $\vec{E}$ is given by

$$\nabla^2 \vec{E} = \mu\varepsilon \frac{\partial^2 \vec{E}}{\partial t^2} \quad \text{... (1)}$$

and in terms of magnetic field vector $\vec{H}$, *it is given by*

$$\nabla^2 \vec{H} = \mu\varepsilon \frac{\partial^2 \vec{H}}{\partial t^2} \quad \text{... (2)}$$

The general wave equation in three dimensions has the form

$$\nabla^2 \Psi = \frac{1}{c^2} \frac{\partial^2 \Psi}{\partial t^2} \quad \text{... (3)}$$

where Ψ is the '*disturbance*' or '*wave function*' and is a function of x, y, z and t, c is the wave velocity.

Comparasion of eq. (1), (2) and (3) gives

Velocity of the electromagnetic wave $c = \frac{1}{\sqrt{\mu\varepsilon}}$

Equation (1) gives the wave equation for vector $\vec{E}$ and equation (2) for the vector $\vec{H}$. So the velocity of the vector $\vec{E}$ or $\vec{H}$ in electro magnetic waves propagating through free space is given by $\frac{1}{\sqrt{\mu_0 \varepsilon_0}}$

It can be seen that $\frac{1}{\sqrt{\mu\varepsilon}}$ has dimensions of velocity.

Taking the dimensions of μ and ε we have

$$\mu\varepsilon = \frac{\text{kg metre}}{\text{coulomb}^2} \times \frac{\text{coulomb}^2 \text{sec}^2}{\text{kg.metre}^2} = \frac{\text{sec}^2}{\text{m}^2}$$

$$\frac{1}{\sqrt{\mu\varepsilon}} = \frac{\text{metre}}{\text{sec}} = \text{m / sec}$$

which corresponds to the dimensions of velocity

7.17. Frardays laws of electromagnetic Induction

Faraday observed from the various experiments that

(*i*) In case a magnet is moved near a wire in the form of a closed circuit without a battery in the circuits, then a current is produced in the circuits so long as the movement of the magnet lasts and as the movement of the magnet is stopped, the current also disappears.

(*ii*) Same effect is also observed if the magnet is kept still and current is moved.

From above it is clear that for the production of current in the wire, relative movement of either the wire of the magnetic is required.

(*iii*) Faraday also observed that a current in induced in a loop of wire if the stationary current in an adjacent circuit is turned on or turned off.

In other words a transient current flows when the *flux through a circuit is altered.* The changing flux induced an electric field and consequently an e.m.f. in the circuit which helps the flow of current.

Faraday called this phenomenon as *electromagnetic induction* which may be stated as :

(*i*) When a magnetic flux through a circuit changes an e.m.f. in the circuit is induced which makes the induced current flow in the circuit till the change in flux persists.

(*ii*) magnitude of the induced e.m.f. is proportion to the rate of change of magnetic flux in the circuit.

Let e be the e.m.f. and φ, the magnetic flux, then

$$e \, \alpha \, \frac{d\varphi}{dt}$$

The direction of e.m.f. may be got from Lenz's law, which states that the *direction of e.m.f. produced is always opposite to the e.m.f. produced due to changing flux.* So we may write

$$e = -k \frac{d\varphi}{dt}$$

Where k is proportional constant and is unity in S.I. units

So we may write

$$\mathrm{e} = -\frac{d\varphi}{dt} \quad \text{... (1)}$$

This equation is called the *Faraday's law of induction.*

Induced e.m.f. is also called the back e.m.f. It is independent of the way of change of flux.

If φ be the instantaneous rules of magnetic flux, then

$$\varphi = \oiint \vec{B} . \vec{ds} \quad \text{... (2)}$$

Also the induced e.m.f. around a closed path is given by the line integral of the induced electric field $\vec{E}$ around the coil *i.e.*

$$e = \int_l \vec{E} . \vec{dl} \quad \text{... (3)}$$

If the circuit path l is bounded by an open surface S, in the case

$$e = \int_l \vec{E} . \vec{dl} = -\frac{d}{d} \int_l \oint_S \vec{B} . \vec{ds} \quad \text{... (4)}$$

So the Faraday's law states that the *line integral of the electric intensity around any fixed closed path is equal to the time rate of decrease flux of magnetic induction through the surface enclosed by the curve.*

We can write the differential form of Faraday's law by using the Stoke's law to replace the line integral of surface integral.

$$\int_l \vec{E} . \vec{dl} = \oiint_S (\vec{\nabla} \times \vec{E}) . \vec{ds} = -\frac{\partial}{\partial t} \oiint_S \vec{B} . \vec{ds}$$

$$\oiint \left(\vec{\nabla} \times \vec{E} + \frac{\partial \vec{B}}{\partial t} \right) \vec{ds} = 0$$

Where we have chosen S to be an arbitrary surface, so the integrand and must be zero. Thus

$$\vec{\nabla} \times \vec{E} + \frac{\partial \vec{B}}{dt} = 0$$

or $$\vec{\nabla} \times \vec{E} = -\frac{\partial \vec{B}}{dt} \quad \text{... (5)}$$

which is the differential form of Faraday's law.

7. 18. Derivation of Maxwell's field equations

(*i*) **Derivation** of $\vec{\nabla}.\vec{D} = \rho$.

From the Gauss's law, the integral of normal component of electric field *i.e.,* outward electric field over any closed surface is equal to the charge within the surface *i.e.,*

$$\int \vec{E}.d\vec{s} = q / \epsilon_0$$

where ϵ_0 is the permittivity of the free space. As ρ is the charge density so ρdV will be charge of volume element dv which contributes a charge $\rho \times \frac{dV}{\epsilon_0}$ to the surface integral.

If surface S encloses a volume V, in that case

$$\int_S \vec{E}.d\vec{S} = \frac{1}{\epsilon_0} \int_V \rho dV \quad \text{... (1)}$$

if S correspond to the surface sounding volume V of the *dielectric,* in that case total charge must be constituted by both *free and polarizing charges.* In that case the total charge density at a point in a small volume element. dV must be $(\rho + \rho')$, where ρ' corresponds to the *polarization charge density* and ρ is the free charge density.

So the above equation may be written as

$$\int_S \vec{E}.d\vec{S} = \frac{1}{\epsilon_0} \int_V (\rho + \rho')dV \quad \text{... (2)}$$

Now polarization charge density is related to *Polarization vector* . $\vec{p}$ by the relation

$$\rho' = div\ \vec{p}$$
$$= -\vec{\nabla}.\vec{P}$$

∴ e.g. (2) may be written as

$$\int_S \vec{E}.d\vec{S} = \frac{1}{\epsilon_0} \int_V (\rho - \vec{\nabla}.\vec{P})dV$$

or $$\int_S \epsilon_0\ \vec{E}.d\vec{S} = \int_V \rho dV - \int \vec{\nabla}.\vec{P}\ dV$$

Using Gauss's divergence theorem, we can change surface integral to volume integral and get

$$\int_S \in_0 \vec{E}.d\vec{S} = \int \text{div}(\in_0 \vec{E})dV = \int_V \nabla.(\in_0 \vec{E})dV$$

$$= \int_V \rho\, d\,V - \int (\vec{\nabla}.\vec{\rho})\, dV$$

or $$\int_V \rho\, d\,V = \int_V \nabla(\in_0 \vec{E} + \vec{\rho})dV$$

$\in_0 \vec{E} + \vec{P} = \vec{D}$ is the electric displacement vector.

so, $$\int_V \rho\, dV = \int_V \vec{\nabla}.\vec{D}dV$$

or $$\int_V (\vec{\nabla}.\vec{D} - \rho)dV = 0$$

This equation being true for any volume since V is chosen arbitrarily, so the integral must be zero *i.e.*

$$\vec{\nabla}.\vec{D} - \rho = 0$$

or $$\vec{\nabla}.\vec{D} = \rho$$

In the case of an isotropic medium $\vec{D}$. $\vec{E}$ and $\vec{P}$ are in the same direction. Also for small applied field, $\vec{D}$ is proportional to $\vec{E}$ *i.e.*

$$\vec{D} = \in \vec{E}$$

Where $\in$ is permittivity of the dielectric medium.
In the case of free space

$$\vec{D} = \in_0 \vec{E}$$

Where $\in$ is the permittivity of free space.

(*ii*) **Derivation** of $\vec{D}, \vec{B} = 0$

In case of magnetic lines of force, the number of lines of force entering any close arbitrary surface is equal to the number of lines of force leaving the surface. This is other words mean that total outward flux of magnetic induction $\vec{B}$ through any closed surface S is equal to zero *i.e.*

$$\int_0 \vec{B}.d\vec{S} = 0 \quad \text{... (1)}$$

Using Gauss's divergence theorem, we can convert surface integral to volume integral and get

$$\int_V (\vec{\nabla}, \vec{B})\, dV = 0$$

Since volume chosen is arbitrary so the integrand must be zero

$$\therefore \qquad \vec{\nabla}.\vec{B} = 0 \qquad \text{... (2)}$$

This is the Gauss's law of magnetism.

(*iii*) **Derivation or** $\vec{D} \times \vec{E} = \frac{\partial \vec{D}}{dt}$

From Faraday's law of electromagnetic induction when the magnetic flux through a circuit changes an e.m.f. is induced whose direction is always Opposite to the charge producing it (Lenz's law)

Let e be the induced e.m.f. and φ the magnetic flux then

$$e = -\frac{d\varphi}{dt} \qquad \text{... (1)}$$

Now induced e.m.f. is equal to the induced electric $\vec{E}$ field around the circuit, mathematically

$$e = \int_{l} \vec{E}.d\vec{l} \qquad \text{... (2)}$$

and magnetic flux through the circuit is

$$\varphi = \int_{S} \vec{B}.d\vec{S} \qquad \text{... (3)}$$

Where integral is over a bound surface of area S.

From the above equation, we have

$$\int \vec{E}.d\vec{l} = -\frac{d}{dt}\int_{S} \vec{B}.d\vec{S}$$

Since surface integral does not change with time, we may write the above equation.

$$\int_{l} \vec{E}.d\vec{l} = -\int_{S} \frac{d\vec{B}}{dt}.d\vec{S}$$

$$= -\int_{S} \frac{\partial \vec{B}}{\partial t}.d\vec{S} \qquad \text{... (4)}$$

where we have changed the total derivative into partial derivative as we are interested only in the change of $\vec{B}$ over a small area elements dS Eq. (4) shows that *e.m.f. around a closed path is equal to the time derivative of the magnetic displacement through any surface bounded by the path.*

Now using Stoke's theorem, we can transform the line integral into surface integral *i.e.*

$$\int_{l} \vec{E}.d\vec{l} = \int_{S} (\vec{\nabla} \times \vec{E}).d\vec{S} = -\int \frac{\partial \vec{B}}{\partial t}.d\vec{S}$$

$$= \int_{l} \left(\vec{\nabla}.\vec{E} - \frac{\partial B}{dt} \right).d\vec{S} = 0$$

The equation is valid for any surface field, so we may write

or
$$\left(\vec{\nabla}\times\vec{E}+\frac{\partial\vec{B}}{dt}\right)=0$$

$$\vec{\nabla}\times\vec{E}=-\frac{\partial\vec{B}}{dt}$$

Which is the differential form of Faraday's law of e.m. induction.

(*iv*) **Derivation of** $\vec{\nabla}\times\vec{H}=\vec{J}+\frac{\partial\vec{D}}{dt}$

This equation correspond to the Ampere's law in circuital form for a magnetic field accompanying an electric current.

This law states that the line integral of magnetic field around a closed path is equal to the total current crossing any surface bounded by the path of line integral.

Mathematically

$$\int_l \vec{H}\,d\vec{l}=\int_S \vec{J}.d\vec{S} \quad \text{... (1)}$$

Where we taken the surface integral corresponding to the charge flow that constitute the current I.

Changing line integral into surface integral using Stoke's theorem, we get

$$\int_l \vec{H}\,d\vec{l}=\int_S(\vec{\nabla}.\vec{H}).d\vec{S}=\int \vec{J}.d\vec{S}$$

$$\int_S\left[(\vec{\nabla}\times\vec{H})-\vec{J}.d\vec{S}\right]=0$$

For any arbitrary surface, we have

$$(\vec{\nabla}\times\vec{H})-\vec{J}=0$$

$$\therefore \quad \vec{\nabla}\times\vec{H}=\vec{J} \quad \text{... (2)}$$

The above equation is valid for steady state current, however, it can be modified for the time varying fields in which current density has to be modified.

Taking divergence on both sides, we get

$$\vec{\nabla}.(\vec{\nabla}\times\vec{H})=\nabla.J=0$$

As the divergence of a curl of a vector is zero. This is valid for steady state phenomenon. However, when the currents are changing with time, we have from the continuity equation.

$$\vec{\nabla}.\vec{J}+\frac{\partial\rho}{\partial t}=0$$

or
$$\vec{\nabla}.\vec{J}=-\frac{\partial\rho}{\partial t}$$

Where ρ is the volume charge density. It shows that $\nabla.J\neq 0$ *for time varying currents.*

Maxwell improved upon the above equation by introducing the concepts that the current density in incomplete and suggested that some factor must be added into $\vec{J}$ so that eq. (2) becomes

$$\vec{\nabla}\times\vec{H} = \vec{J}+\vec{J} \quad \text{... (3)}$$

Taking divergence of this equation, we get

$$\nabla.(\nabla\times\vec{H}) = [\nabla(\vec{J}+\vec{J})]$$

Since $\nabla.(\vec{\nabla}\times\vec{H}) = 0$

$$\nabla.\vec{J} = -\vec{\nabla}\vec{J}$$

But from eq. of continuity

$$\nabla.\vec{J} = -\frac{\partial\rho}{dt}.$$

Also we have $\vec{\nabla}.\vec{D} = \rho$, So we have

$$\vec{\nabla}.J' = \frac{\partial}{\partial t}(\vec{\nabla}.\vec{D}) = \vec{\nabla}.\left(\frac{\partial\vec{D}}{\partial t}\right)$$

$$\therefore \quad \vec{J'} = \frac{\partial\vec{D}}{\partial t}$$

Putting the value of J' in eg. (3) we get the modified form of Ampere's law as

$$(\vec{\nabla}\times\vec{H}) = \vec{J}+\frac{\partial\vec{D}}{\partial t} \quad \text{... (4)}$$

where term $\frac{\partial\vec{D}}{\delta t}$ is due to the change of electric displacement $\vec{D}$ with time and is called *displacement current density.* $\vec{J}$ is *called conduction current density.*

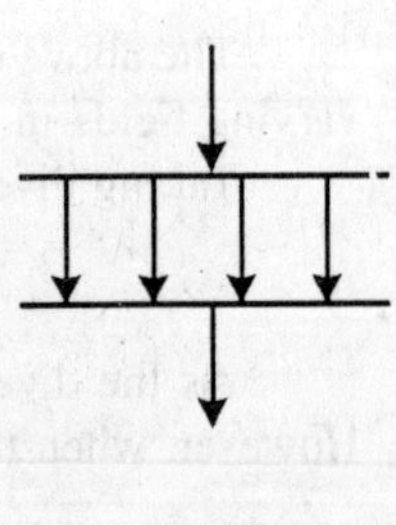

Magnetic flux $\vec{B}$ is related to magnetic field intensity $\vec{H}$ by the relation $\vec{B} = \mu\vec{H}$ where μ is the permeability. $\vec{B}$ in case of free space $\mu_0\vec{H}$ where μ_0 is the permeability of free space and has the value $4\pi \times 10^{-7}$ henry/meter

7.19. Displacement current

Displacement current has not the same significance as the actual current because in this case the charges actually do not move. To understand the displacement current concepts, let us consider a simple circuit in the case of a capacitor joined by a conducting wire.

Current is the circuit is equal to the rate of flow of charge i.e.

$$I = \frac{dq}{dt} \quad \text{... (1)}$$

Where q corresponds to the positive charge on a plate of capacitor and is related to electric field as

$$\vec{E} = \frac{\sigma}{\epsilon_0} = \frac{q}{\epsilon_0 A} \qquad \text{... (2)}$$

Where A is the area of the plate, σ is the surface charge density on the plate and ϵ_0 is the permittivity of the medium between the plate. In case of air $\epsilon_0 = 1$

Now $$1 = \frac{dq}{dt} = A\epsilon_0 \frac{d\vec{E}}{dt} = A\frac{\partial \vec{D}}{\partial t}$$

$\therefore$ $$\frac{1}{A} = \frac{\partial \vec{D}}{\partial t} \qquad \text{... (3)}$$

Now l/A corresponds to current density. So the quality $\frac{\partial \vec{D}}{\partial t}$ may be interpreted as the density of some *current that flows in the Space between the plates of the capacitor.* When the circuit is completed by the connecting wires connected to the two capacitor plates.

7.20. Wave Guides

Transmission of messages can be carried by means of (*i*) Transmission lines (*ii*) Radio waves. Transmission lines communication is generally termed as *Line communication* while communication by radio waves is called *radio communication.* We will briefly study these systems of communication as under:

Line communication: Most common form of line communication is *telephone and telex networks.* Communication is carried out with the help of *lines on poles or through burried cables.* Burried cables with capacities 4000 pairs are available. The pairs are twisted to avoid any *cross talk* among the subscribers. For long distance transmission *repeaters* may be used after some intervals to amplify the signals. Such cables are workable upto 500 *kHz.* For higher frequencies *coaxial cables* may be used.

A Coaxial cable is constituted by a pair of concentric conductors with some insulation filling the in between space. A number of such cables are bunched together inside a protective shield. The outer conductor of every co-axial cable provides an electrical shielding. However, shielding is effective only at frequencies above 60 kHz and below this frequency *cross-talk may start interfering.*

Co-axial cables have been used to provide *10800 telephone* lines at a frequency of *61.6 MHz.* For long distance communication repeaters may be used in this case as well.

Wave guides. Line communication in UHF region or above are carried with the use of *wave guides.* In wave guides, the signal propagation is done using radiowaves with the restriction that *cross-sectional dimensions of the waveguide rectangular hollow pipe maintain a proper relation with the wavelength of the wave.*

Wave guides are employed as transmission lines for *microwave communication* in frequency range of 2 GHz to 11 GHz. Wave guides do not transmit signals below *cut off frequency* and can have bandwidth over and above 20% of their operating frequencies. For example a wave guide of 10 GHz can have bandwidth of 2 GHz which can accomodate 500,000 channels if each chanel occupies of a frequency ~ 4 kHz.

Optical waveguides have been developed in the form of glass fibres to carry modulated optical signals at frequencies ~ 100 × 1012 Hz (100 Tera Hertz) with the help of LASERS. As is evident, the carrying capacity of the signals by optical fibres is increased enormously.

Salient features of wave guides. As illustrated earlier, wave guides are quite useful to minimize losses for *high power transmission at microwave frequencies.*

Wave guides consist of a single hollow metallic conductor in the shape of a *rectangular box or cylinder through which e.m. waves are propagated.* Wave guides are usually made of *copper, brass or aluminium* in various standard sizes corresponding to the signal frequencies to be propagated. **Inner surface of a wave guide is plated with thin layers of gold or silver to reduce transmission losses.**

Propagation through a wave guide is different than that taking place through open wire co-axial lines and takes place in the form of transverse electric (*TE*) or Transverse magnetic (*TM*) waves. The first one is called *H-mode*. While the second one is called *E-mode.*

Wave guides are usually coupled together by flanges bolted together and supported on metal stands at different intervals.

Important parameters of wave guides are *characteristic impedence* Z_0, *cut off wavelength* λc *and guide wave length* λ_g.

Wave Guide modes. Waves can travel in a wave guide in number of configurations which are termed as modes and are labelled as TE_{mn} or TM_{mn} depending whether the electric field or magnetic field is transverse to the direction of propagation. Letters m and n denote number of half wavelengths of intensity of electric fields in case of *TE modes* and magnetic fields for *TM modes* between each pair of wave guide walls. *m* is measured along *X-axis* while *n* is measured along *Y-axis* (or height) of waveguide cross-section.

Most commonly used transverse electric modes in *rectangular wave guides* are TF_{10}, TE_{20} and TE_{11}. While TE_{10} and TE_{20} wave guides use side walls for successive reflections and are not affected by top and bottom walls, TE_{11} *uses top and bottom* walls as well in addition to the side walls.

As an example, for TE10 mode of propagation,

$$Z_0 = 377 \text{ ohms}$$

where $$\lambda_g = \frac{1}{\sqrt{\frac{1}{\lambda^2} - \frac{1}{\lambda_c^2}}}$$

and $$\lambda_c = 2a$$

where $a \rightarrow$ width of wave guide mouth

and $\lambda \rightarrow$ free space wavelength of the signal.

Similar relation exist for all other modes of propagation and the nature of wave guide.

7.21. Cavity Resonators

Closed section of coaxial line or waveguide may be considered as forms of a more general type of resonator, called the *cavity of possible modes of resonance.* A cavity resonator for a specific application may usually be designed and excited in such a manner, however, that only one mode of resonance is obtained over a limited frequency range. Commonly used methods of tuning cavity resonators include changing the cavity shape, changing the values of lumped capacitance or inductance, and introducing conducting masses into the resonators in regions of high *electric or magnetic field intensity.*

Example 1. *Calculate the value poynting vector for a 60 w bulb at a distance of 0.5 m from it.*

Solution. Powers of the bulb = 60 *w*

Since the light from the bulb is spread in a sphere at radius r = 0.5 m, so the poynting vector SAv (average value of poynting vector over the surface of sphere) is given by

$$S_{Av} = \frac{P}{4\pi r^2} = \frac{60}{4\pi(0.5)^2} = \mathbf{19.1\ w/m^2}\ \textbf{Ans.}$$

Example 2. *The electric field in an electromagnetic wave is given by $E = E_0 \sin \omega(t - x/c)$, where E_0 = 100 Newton/coul. Find the energy contained in a cylinder of cross-section 10^{-3} m^2 and length 100 cm along the X-axis.*

Solution. The average electric energy density is given by the relation

$$u_{av} = (1/2)\varepsilon_0 E_0^2$$

Let V be the volume of a cylinder, then the energy contained in this volume will be

$$U = Vu_{av}$$

also V = cross-sectional area × length = $10^{-3} \times 100 \times 10^{-2} = 10^{-3}$ m³

∴ $U = (1/2) \times (8.85 \times 10^{-12}) \times (100)^2 \times 10^{-3} =$ **4.425×10^{-11} joule Ans.**

Example 3. *If the earth receives 2 cal min^{-1} cm^{-2} solar energy, what are the amplitudes of electric and magnetic fields of radiation?*

Solution. Poynting vector $\vec{S}$ gives the energy transported by e.m. wave/unit area/sec, so from the relation

$$\vec{S} = (\vec{E} \times \vec{H})$$

The energy flux per unit area per second is given by

$$|\vec{S}| = |\vec{E} \times \vec{H}| = \text{EH} \sin 90° = \text{EH (as } E \perp H)$$

The energy flux per unit area per second at the earth surface is given in the problem as

$$|\vec{S}| = 2 \text{ cal min}^{-1} \text{ cm}^{-2} = \frac{2 \times 4.2 \times 10^4}{60} \text{ joule m}^{-2} \text{ sec}^{-1}$$

∴ $$|S| = EH = \frac{2 \times 4.2 \times 10^4}{60} \text{ J/m}^2\text{/sec.} \quad \text{... (1)}$$

also $$\frac{E}{H} = \sqrt{\frac{\mu_0}{\varepsilon_0}} = \sqrt{\frac{4\pi \times 10^{-7}}{8.85 \times 10^{-12}}} = 376.72 \text{ ohms} \quad \text{... (2)}$$

Product of eq. (1) and (2) gives

$$EH \times (E/H) = 1400 \times 376.72 \quad \text{or} \quad E^2 = 5.274 \times 105$$

or $$E = 726.2 \text{ volt/m}$$

Put this value of E in equation (1), we get

$$(726.2)\ H = 1400 \quad \text{or} \quad \text{H} = \frac{1400}{726.2} = 1.928 \text{ amp-turn/m}$$

So, the amplitudes of *electric and magnetic fields of radiation* are

$E_0 = E\sqrt{2} = 726.2\sqrt{2}$ = **1026.8 volt/m** and $H_0 = H\sqrt{2} = 1.928\sqrt{2}$ = **2.726 amp-turn/m**

Ans.

Example 4. *From the sun energy received on the surface of earth is found to be 1.33 kw/m². Find the electric field associated with sunlight as received on the earth's surface. Assume the sun's light to be monochromatic $(\lambda = 6000\text{Å})$. Take the value of t E = 9 × 10^{-12} S.I. units.*

Solution. Energy received/sec/m² on the surface of Earth = Average value of poynting vector on the Earth's surface.

$$SAv = \frac{1}{2} c \in E_0^2)$$

$$\therefore \quad E_0 = \left[\frac{2S_{Av}}{c\in}\right]^{1/2} = \left(\frac{2\times 1.33\times 10^3}{3\times 10^8 \times 9\times 10^{-12}}\right)^{1/2} = \textbf{992.5 volt/meter Ans.}$$

Example 5. *The maximum electric field in a plane electromagnetic wave is 10^2 Newton/coul. The wave is going in the X-direction and the electric field is in the Y-direction. Find the maximum magnetic field in the wave and its direction.*

Solution. We have the relation for maximum electric field and maximum magnetic field as

$E_0 = cB_0$, where c is the velocity of light

$$\therefore \quad B_0 = \frac{E_0}{c} = \frac{10^2}{3\times 10^8} = 3.33\times 10^{-7} \text{ Weber/m2 or } 3.33 \times 10^{-7} \text{ tesla } \textbf{Ans.}$$

Example 6. *The relative permittivity of distilled water is 81. Calculate refractive index and velocity of light in it.*

Solution. We have the relation for refractive index $n = \sqrt{(\mu\varepsilon/\mu_0\varepsilon_0)}$ and $v = c/n$

In this case $\varepsilon = 81\varepsilon_0$ and for distilled water $\mu = \mu_0$

So $n = \sqrt{(81)} = 9$ and $v = \dfrac{3\times 10^8}{9} = \mathbf{3.33\times 10^7}$ **m/sec Ans.**

Example 7. *Calculate the poynting vector and the amplitude of the electric field if a laser beam of 500 w is concentrated using a bust on a cross-sectional area of 10^{-10} m^2. Given $\varepsilon_0 = 9 \times 10^{-12}$ F/m*

Solution. Poynting vector is given by

$$\vec{S} = \frac{\text{Power}}{\text{Area}} = \frac{500}{10^{-10}} = 5\times 10^{12}\,\text{w/m}^2$$

Average value of Poynting vector is given by the relation

$$S_{Av} = \frac{1}{2} c \in E_0^2$$

$\therefore$ Electric field $\quad E_0 = \left(\dfrac{2S_{Av}}{c\in}\right)^{1/2}$

Now $\in = 9 \times 10^{-12}$ F/m

$c = 3 \times 10^8$ m/sec

$$\therefore \quad E_0 = \left(\frac{2\times 5\times 10^{12}}{3\times 10^8 \times 9\times 10^{-12}}\right)^{1/2} = \mathbf{60.9\times 10^6} \textbf{ V/m Ans.}$$

[**Note:** Units of electric vector E may be got from eq. (1) as

$$E_0 = \left(\frac{W}{m^2}\times\frac{S}{m}\times\frac{m}{\text{farad}}\right)^{1/2} = \left(\frac{\text{w}\times\text{sec}}{m^2}\times\frac{\text{volt}}{\text{coulomb}}\right)^{1/2}$$

$$= \left[\frac{\text{Coulomb}\times\text{volt}\times\text{volt}}{\text{m}^2\times\text{coulomb}}\right]^{1/2} = \text{volt/meter}$$

Example 8. *Calculate the magnitude of the poynting vectors and the amplitude of the electric and magnetic fields at a distance of 10 km from a radio station which is radiating power of 10^5 watt uniformly over a hemisphere with radio station as centre.*

Given $\mu_0 = 4\pi \times 10^{-7}$ henry/meter and $\varepsilon_0 = 9\times 10^{-12}$ farad/meter

Solution. Total average power radiated by radio station $P = 10^5$ w

Radius of hemisphere = 10 kms = 10^4 m

Let SAv be the average poynting vector over the surface of hemisphere, then

$$S_{Av} = \frac{p}{2\pi r^2} = \frac{10^5}{2\times 3.142\times (10^4)^2} = 1.59\times 10^{-4} \text{ w/m}^2$$

For free space, average poynting vector is $S_{Av} = \frac{1}{2} c \in_0 E_0^2$

$$\therefore \quad E_0 = \left[\frac{2S_{Av}}{c\in_0}\right]^{1/2} = \left[\frac{2\times 1.59\times 10^{-4}}{3\times 10^8 \times 9\times 10^{-12}}\right]^{1/2} = 0.343 \text{ V/m}$$

Now $H_0 = \frac{B_0}{\mu_0} = \frac{E_0}{\mu_0 c} = \frac{0.343}{4.7\times 10^{-7}\times 3\times 10^8}$ = **9.1 × 10^{-4} Ampere turns/m Ans.**

Example 9. *Show that equation of continuity div* $\vec{J} + \frac{\partial \rho}{\partial t} = 0$ *is contained in Maxwell's equations.*

Solution. Maxwell's fourth equation has the differential form.

$$\vec{\nabla}\times \vec{H} = \vec{J} + \frac{\partial \vec{D}}{\partial t}$$

Taking divergence of both sides we have

$$\vec{\nabla}.(\vec{\nabla}\times \vec{H}) = \vec{\nabla}.\left(\vec{J} + \frac{\partial \vec{D}}{\partial t}\right)$$

But $\vec{\nabla}\times\vec{H} = 0$

$$\therefore \quad \vec{\nabla}.\left(\vec{J} + \frac{\partial \vec{D}}{\partial t}\right) = 0$$

or
$$\vec{\nabla}.\vec{J} + \vec{\nabla}.\left(\frac{\partial \vec{D}}{\partial t}\right) = 0$$

$$\vec{\nabla}.\vec{J} + \vec{\nabla}.\left(\frac{\partial \vec{D}}{\partial t}\right) = 0$$

or
$$\vec{\nabla}.\vec{J} + \frac{\partial}{\partial t}(\vec{\nabla}.\vec{D}) = 0$$

Now from maxwell's first equation $\vec{\nabla}.\vec{D} = \rho$ (where ρ is surface charge density)

So $\vec{\nabla}.\vec{J} + \frac{\partial \rho}{\partial t} = 0$

Hence the result Ans.

Example 10. *Assuming that all the energy from a 1000 watt lamp is radiated uniformly, calculate the average the intensities of electric and magnetic fields of radiation at a distance of 2m from the lamp.*

Solution. Let ρ_0 be the power radial uniformly in all direction, then the power or energy flux per unit, per second at a distance r from the point source (in this case lamp) will be

$$S_{av} = \frac{P_0}{4\pi r^2} = \frac{1000}{4\pi(2)^2} \text{ Watt/m}^2$$

Now from Poynting theorem, $|\vec{S}| = |\vec{E} \times \vec{H}| = \text{EH} \sin 90°$ $(\because \text{ as } F \perp H)$

$$\therefore \quad EH = \frac{1000}{16\pi} \quad \text{... (1)}$$

Now $$\frac{E}{H} = \sqrt{\frac{\mu_0}{\varepsilon_0}} = 376.72 \text{ ohm} \quad \text{... (2)}$$

Product of (1) and (2) equation will give

$$EH \times \frac{E}{H} = \frac{376.72 \times 1000}{16\pi}$$

or $$E = \left(\frac{376.72 \times 1000}{16 \times 3.14}\right)^{1/2}$$ or $E =$ **86.59 volt/m Ans.**

Put in eq. (1), we get

$$H = \frac{1000}{16 \times 3.14 E} = \frac{1000}{16 \times 3.14 \times 86.59} = \mathbf{0.23} \textbf{ amp-turn/m Ans.}$$

Example 11. *If the magnitude of $\vec{H}$ in a plane wave is 1 amp/meter, find the magnitude of $\vec{E}$ for plane wave in free space.*

Solution. We have the relation, $\frac{H_0}{E_0} = \sqrt{\frac{\varepsilon_0}{\mu_0}} \quad \therefore \quad E_0 = \sqrt{\frac{\mu_0}{\varepsilon_0}} H_0$

In this case $H_0 = 1$ amp/m, $\mu_0 = 4\pi \times 10^{-7}$ w/A-m and $\varepsilon_0 = 8.85 \times 10^{-12}$ C/N-m²

$$\therefore \quad E_0 = 1 \times \sqrt{\frac{4\pi \times 10^{-7}}{8.85 \times 10^{-12}}} = \mathbf{376.72 \text{ V/m Ans.}}$$

Summary

Forms and importance of maxwell's equations

INTEGRAL FORM

Name	*Equation*	*Importance*
1. Gauss's law of electricity	$\oint \vec{E}.d\vec{S} = \frac{q}{\varepsilon_0}$	It describes charge and electric field.
2. Gauss's law magnetism	$\oint \vec{B}.d\vec{S} = 0$	It describes magnetic of field.
3. Faraday's law	$\oint \vec{E}.\vec{dl} = -\frac{d\phi_{\vec{B}}}{dt}$	Describes an electric field produced by a magnetic field.
4. Maxwell's Ampere	$\oint \vec{B}.\vec{dl} = \mu_0\varepsilon_0 \frac{d\phi_{\vec{E}}}{dt} + \mu_0 \in_0$	A magnetic field produced by a changing electric field or by a current or both.

DIFFERENTIAL FORM

1. Gauss's law of electricity	$\nabla.\vec{D} = \rho$
2. Gauss's law of magnetism	$\nabla.\vec{B} = 0$
3. Faraday's law	$\nabla \times \vec{E} = -\frac{\partial \vec{B}}{\partial t}$
4. Ampere Maxwell's law	$\vec{\nabla} \times \vec{H} = \vec{J} + \frac{\partial \vec{D}}{\partial t}$

QUESTIONS

1. Deduce four Maxwell's equations in free space.
2. Show that if $\vec{B}$ is a magnetic field and its divergence $\vec{B}$ = 0 everywhere, it means that an isolated magnetic pole is not possible.
3. Write the Maxwell's equations. Show that the velocity of plane electromagnetic wave in the free space is given by $c = 1/\sqrt{(\mu_0\varepsilon_0)}$
4. Explain the concept of displacement current. Write four Maxwell's equations and explain their physical significance.
5. Write Maxwell's equations and discuss the propagation of electromagnetic waves in free space.
6. Show that the Faraday's law of electromagnetic induction can be expressed as, $\vec{\nabla} \times \vec{E} = -\frac{\partial \vec{B}}{\partial t}$ Write down its integral form.
7. Derive expressions for wave equation for electromagnetic waves with the help of Maxwell's equations.
8. The magnetic field in a plane electromagnetic wave is given by
   $B = (200\ \mu T) \sin [(4.0 \times 10^{15} \times 10^{15}\ \text{sec}^{-1})(t - x/c)]$

Find the maximum electric field and the average energy density corresponding to the electric field. **(Ans** B × 10^4 N/C, 0.016 Jm^{-3})

9. Show that the wave equation for the electric field in free space is given by $\nabla^2 \times \vec{E} = \mu_0 \varepsilon_0 \dfrac{\partial^2 \vec{E}}{\partial t^2}$
10. State and explain Poynting theorem for the flow of energy in electromagnetic waves.
11. Deduce the equation for the propagation of plane electromagnetic wave in the free space. Show that the electric, magnetic vectors are normal to each other and to the direction of the propagation of the wave.
12. Prove that the velocity of plane electromagnetic wave in the vacuum is given by $c = 1/\sqrt{(\mu_0 \varepsilon_0)}$.
13. Deduce the Maxwell's equation for free space and prove that the electromagnetic waves are transverse.
14. Show that the energy stored in a magnetic field per unit volume in free space is $B^2/2\mu_0$.
15. Deduce poynting theorem for the flow of energy in an electromagnetic field.
16. Discuss the propagation of electromagnetic waves in free space.
17. Show that the plane electromagnetic waves follow the laws of reflection just like light radiations.
18. Show that the equation for electric field E is given by $\nabla^2 \times \vec{E} = \mu_0 \varepsilon_0 \left(\dfrac{\partial^2 \vec{E}}{\partial^2 t} \right)$
19. Write down the equations of an electromagnetic field. Solve these equations for isotropic medium. Establish relations between $\vec{E}$ and $\vec{H}$ vectors.
20. Using $B = \mu_0 H$, find the ratio E_0/H_0 For a plane electromagnetic wave propagating through vacuum. Show that it has the dimensions of electric resistance. This ratio is a universal constant called the impedance of free space.

NUMERICAL PROBLEMS

1. If the electric amplitude of the wave is 5 V/m, what is he magnetic amplitude of this wave ? **(Ans.** 1.67 × 10^{-8} Wb/m^2)
2. A mercury lamp is radiating monochromatic light of 10 watts radiated power. Calculate the electric field strength at a distance 5 m from the lamp. **(Ans.** 4.9 v/m)
3. The magnetic field in a plane electromagnetic wave is given by B = (200 μT) sin [(4.0 × 10^{15} sec^{-1}) ($t - x/c$). Find the maximum electric field and the average energy density corresponding to the electric field. **(Ans.** 6 × 10^4 N/c, 0.016 Jm^{-3})
4. A light beam travelling in the X-direction is described by the electric field, Ey = (300 V/m) sin ω ($t - x$ c). An electron is constrained to move along the y-direction with a speed of 20. × 107 m/sec. Find maximum electric force and the maximum magnetic force on the electron. **(Ans.** (*i*) 4.8 × 10^{-17} N) (*ii*) 3.2 × 10^{-18} N
5. A parallel plate capacitor with plate area A and separation between the plate *d*, is charged by a constant current *i*. Consider a plane surface of area A/2 parallel to the plates and drawn symmetrically between the plates. Find the displacement current through this area. **(Ans.** *i*/2)
6. The maximum electric field in a plane electromagnetic wave is 600 N/coul. The wave is going in the *X*-direction and the electric field is in the *Y*-direction. Find the maximum magnetic field in the wave and its direction. **(Ans.** 2 × 10^{-6} Wb/m^{-6} along Z-direction).

8

MATTER WAVES AND UNCERTAINTY PRINCIPLE

8.1. Wave Properties of Particles

The successful explanation of the Compton and photoelectric effects and the experimental verifications of the predictions made on the basis of these explanations, established it beyond doubt that electromagnetic radiation travels not in the form of a continuous stream of energy but in the form of tiny packets or bundles of energy, following each other in quick succession. These packets of energy were called *photons* and behave almost exactly like a material particle.

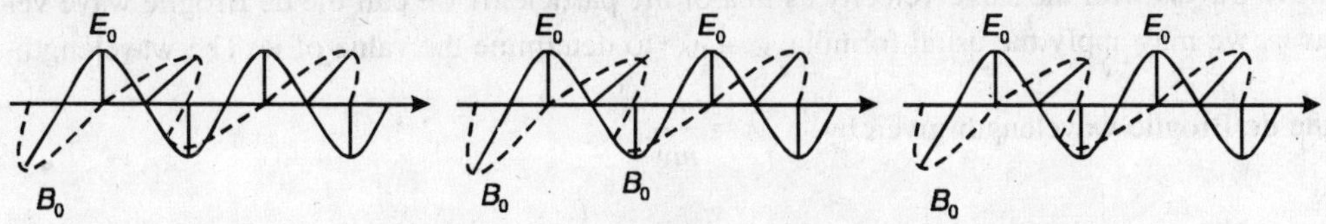

Fig. 8.1

On the other hand, the phenomenon like interference or diffraction of light or X-rays could not be explained unless electromagnetic radiation were assumed to possess a wave character. This resulted in the acceptance of the dual character of radiation, which is maintained till the present day. When radiation interacts with matter it exhibits its particle character whereas when radiation interacts with radiation, it exhibits wave character. Radiation never exhibit both the characters simultaneously.

An interesting outcome of the acceptance of duality in the character of *e.m.* radiation was the hypothesis put forward by L. de Broglie in 1924. He was guided by the following reasonings:

1. Nature is symmetrical.
2. Our observable universe is totally composed of light and matter and light has dual character.
3. There is a close parallelism between principle of least action in mechanics and principle of least time (Fermat's principle) in optics.

According to de Broglie "the wave-particle dualism should not be confined to radiation alone, but should also be extended to all the fundamental entities of physics". The electrons; protons; atoms and molecules, when in motion, should have some type of wave motion associated with them. de Broglie was led to this hypothesis from considerations based upon the special theory of relativity and quantum theory.

Energy of photon can be written as

$$E = h\nu = \frac{hc}{\lambda} \quad \text{...(1)}$$

and if a photon possesses some mass (which is by virtue of its motion *i.e.*, its rest mass would be zero) its energy according to the theory of relativity should be = (mass of photon) c^2

Hence $$E = \frac{hc}{\lambda} = \text{(mass of photon) } c^2$$

$\therefore$ $$\text{mass of photon} = \frac{1}{c^2}\left(\frac{hc}{\lambda}\right) = \frac{h}{c\lambda}$$

and momentum p of the photon $= p = \left(\dfrac{h}{c\lambda}\right) \cdot c$

i.e. $$p = \frac{h}{\lambda} \quad \text{or} \quad \lambda = \frac{h}{p} \qquad ...(2)$$

de Broglie carried these considerations over to the dynamics of a particle and said that the wave-length λ of the wave associated with a moving material particle having a momentum mv is given by

$$\lambda = \frac{h}{mv} \qquad ...(3)$$

8.2. Velocity of de Broglie Waves

We associate de Broglie wave with a moving particle, so it is reasonable to expect that this wave travels with the same velocity as that of the particle. If we call the de Broglie wave velocity as w, we may apply the usual formula $w = \nu\lambda$ to determine the value of w. The wavelength λ is the de Broglie wavelength given by $\lambda = \dfrac{h}{mv}$.

Now from the Planck's Law $E = h\nu$ or $\nu = \dfrac{E}{h}$

Also from mass energy relation $E = mc^2$, substituting above, we have

$$\nu = \frac{mc^2}{h}$$

de Broglie wave velocity is, therefore, given by

$$w = \nu\lambda = \frac{mc^2}{h} \times \frac{h}{mv} = \frac{c^2}{v} \qquad ...(4)$$

Since the particle velocity v, must be less than velocity of light c, de Broglie waves travel faster than the velocity of light. Such a situation can be visualised is case of electromagnetic waves *e.g., velocity of X-rays in different crystals and that of sodium light in sodium crystal is more than c.*

However, the problem in case of de Broglie waves was solved by showing that de Broglie waves always travel in the form of group of waves with a velocity called "group velocity", which is not only less than c, but is equal to the velocity of the particle.

Let a material particle, such as an electron be situated at some point $(x_0\ y_0\ z_0)$ in space. Imagine some quantity ψ whose periodic variation in space gives rise to matter waves such that at any instant t_0 the value of ψ in the immediate vicinity of the particle be given by

$$\psi = \psi_0 \sin(2\pi\, \nu_0\, t_0)$$

where ψ_0 is the amplitude at the point chosen and ν_0 is the frequency as observed by an observer at rest with respect to the particle.

Let now the particle be given a velocity v along the + ve direction of X-axis and the periodic variations of ψ as these now appear to the observer, who is still at rest at $(x_0\ y_0\ z_0)$, we have to apply the transformation equation of the theory of relativity to the time interval t_0.

$$t_0 = \frac{\left(t - \frac{vx}{c}\right)}{\sqrt{1 - \frac{v^2}{c^2}}} \quad \text{...(5)}$$

where t is the time interval as noted by an observer moving with the particle with velocity v.

Hence periodic variations of ψ as these appear to the observer who is still at rest at $(x_0\ y_0\ z_0)$ are given by

$$\psi = \psi_0 \sin \frac{2\pi V_0 \left(t - \frac{vx}{c^2}\right)}{\sqrt{1 - \frac{v^2}{c^2}}} \quad \text{...(6)}$$

Comparing this equation with the standard equation of Simple Harmonic wave

$$y = A \sin\ 2\pi\nu \left(t - \frac{x}{w}\right) \quad \text{...(7)}$$

where A is the amplitude; ν is the frequency and v is the velocity of the wave along + ve X-direction.

We have $\quad w = \dfrac{c^2}{v}$

and $\quad \nu = \dfrac{\nu_0}{\sqrt{1 - \frac{v^2}{c^2}}}$

Now from Einstein's mass energy relation

$$E = h\nu_0 = m_0 c^2 \qquad \therefore \nu_0 = \frac{m_0 c^2}{h}$$

or $\quad$ Also $\quad h\,\nu = mc^2$

or $\quad \therefore \quad \nu = \dfrac{mc^2}{h}$

Now the wavelength of matter waves is given by

$$\lambda = \frac{\text{velocity}}{\text{frequency}} = \frac{w}{\nu} = \frac{\frac{c^2}{v}}{\frac{mc^2}{h}} = \frac{h}{mv}$$

i.e. $\quad \lambda = \dfrac{h}{mv}$

This aspect of matter waves shows that with a material particle in motion two different velocities are involved *viz*, v, referring to the mechanical motion of the particle and w, the velocity with which the wave is propagated and they are related by

$$w = \frac{c^2}{v} \qquad ...(8)$$

8.3. Wave (Phase) Velocity and Group Velocity

The amplitude of the de Broglie waves associated with a moving particle, measures the probability of finding the particle at a particular place at a particular time. Evidently de Broglie waves cannot be described by a simple formula used to describe Simple Harmonic wave *viz*., $y = A \sin 2\pi\left(vt - \frac{x}{\lambda}\right)$ because Simple Harmonic waves are a series of waves all with same amplitude.

Instead we expect that wave associated with a moving particle should be in the form of a *"wave packet"* or *"group of waves"* shown in Fig. 8.2.

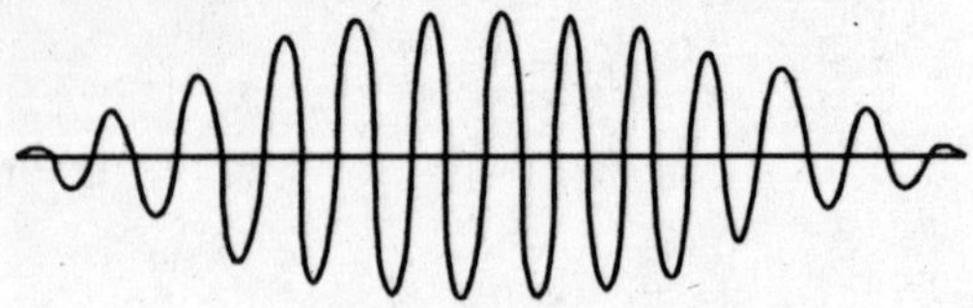

Fig. 8.2 Group of Waves

A familiar example of how groups of waves are formed can be got in the case of *beats* when two sounds of same amplitude, but of slightly different frequencies, are produced together, the resultant sound will have an average frequency, but its amplitude rises and falls number of times per second periodically, with two loudness peaks called beats as shown as in Fig 8.3.

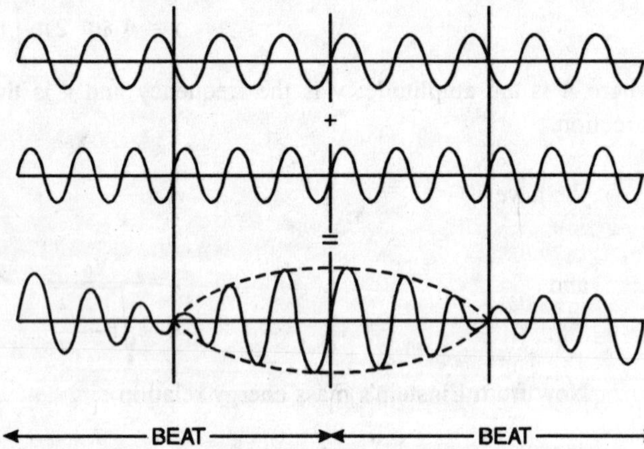

Fig. 8.3 Beat formation

Mathematically, a *"wave group"* is the result of superposition of individual waves each of different wavelength whose interference with one another results in the variation of amplitude that defines the group shape. If the speed of the individual waves are the same, the speed with which a *"wave group"* travels is identical, with the common speed. However, if the wave speed varies with wavelength. The different individual waves do not proceed together, and *"wave group"* are different from that of the waves which compose it.

Let us consider that a *"wave group"* arises from the combination of two waves, with the same amplitude A, but differing in angular velocity by $d\omega$ and an amount dk in wave number $\left(\frac{2\pi}{\lambda}\right)$.

So the two interferring waves may be represented as

$$y_1 = A \cos\ (\omega t - kx)$$

$$y_2 = A \cos\ [(\omega + d\omega)t - (k + dk)x]$$

The resultant displacement y at any time t and at any position x is the sum of y_1 and y_2; using identity

$$\cos\alpha + \cos\beta = 2\cos\frac{(\alpha+\beta)}{2}\cos\frac{(\alpha-\beta)}{2}$$

and since $(-\theta) = \cos\theta$, we have

$$y = y_1 + y_2 = 2A\cos\left[\frac{(2\omega+d\omega)t-(2k+dk)x}{2}\right]\times\cos\left(\frac{d\omega t - dkx}{2}\right)$$

Now since $d\omega$ and dk small as compared to ω and k respectively

$$\therefore \quad (2\omega + d\omega) \approx 2\omega$$

$$(2k + dk) \approx 2k$$

$$\therefore \quad y = 2A\cos(\omega - kx)\cos\left[\left(\frac{d\omega}{2}\right)i - \left(\frac{dk}{2}\right)x\right] \qquad ...(9)$$

Equation (9) represents a wave of angular velocity ω and wave number k which has superimposed upon it a wave (the process called modulation) of angular frequency $\left(\frac{d\omega}{2}\right)$ and wave number $\left(\frac{dk}{2}\right)$.

The effect of this modulation is to produce successive *"wave groups"* as shown in Fig 10.3. The *"phase velocity"* or *"wave velocity"* W is given by

$$W = v\lambda = \frac{2\pi(\nu)}{\left(\frac{2\pi}{\lambda}\right)} = \frac{\omega}{k} \qquad ...(10)$$

While the *"group velocity"* u of the *"group of waves"* is given by

$$u = \frac{d\omega}{dk} \qquad ...(11)$$

In general, depending upon the manner in which *"phase velocity"* or *"wave velocity"* W varies with the wave number in a particular medium, the *"group velocity"* u may be greater or less than the *"phase velocity"* or *"wave velocity"* If *"phase velocity"* W is the same for all wavelengths, the group and phase velocities are the same *e.g.*, in case of *e.m.* waves in vacuum.

The angular frequency ω and the wave number k of de Broglie waves associated with a particle of rest mass m_0 and moving with a velocity equal sign v are given by

$$\omega = 2\pi\nu = 2\pi\left(\frac{mc^2}{h}\right) = \frac{2\pi m_0 c^2}{h.\sqrt{1-\frac{v^2}{c^2}}} \qquad ...(12)$$

and

$$k = \frac{2\pi}{\lambda} = 2\pi\left(\frac{mv}{h}\right) = \frac{2\pi m_0 c^2}{h.\sqrt{1-\frac{v^2}{c^2}}} \qquad ...(13)$$

We find that ω and k are functions of the velocity v of the particle. The *"phase velocity"* or *"wave velocity"* W as we found earlier is $W = \frac{\omega}{k}$. Hence in case of de Broglie waves we get

$$W = \frac{2\pi m_0 c^2}{h.\sqrt{\left(1-\frac{v^2}{c^2}\right)}} \times \frac{h.\sqrt{1-\frac{v^2}{c^2}}}{2\pi m_0 (v)} \quad \text{or} \quad W = \frac{c^2}{v}$$

which exceeds both the velocity v of the particle and the velocity c of light. Since $v < c$.

The group velocity u of de Broglie wave associated with the particle is

$$u = \frac{d\omega}{dk} = \frac{d\omega / dv}{dk / dv}$$

From eq. (12),
$$\frac{d\omega}{dv} = \frac{2\pi m_0 c^2}{h}\left(-\frac{1}{2}\right)\left(1-\frac{v^2}{c^2}\right)^{-3/2}\left(-\frac{2v}{c^2}\right)$$

$$= \frac{2\pi m_0 (v)}{h\left(1-\frac{v^2}{c^2}\right)^{3/2}}$$

From eq. (13),
$$k = \frac{2\pi m_0 (v)}{h}\left(1-\frac{v^2}{c^2}\right)^{-1/2}$$

$\therefore$
$$\frac{dk}{dv} = \frac{2\pi m_0}{h}\left(1-\frac{v^2}{c^2}\right) + \frac{2\pi m_0 (v)}{h}\left[\left(-\frac{1}{2}\right)\left(1-\frac{v^2}{c^2}\right)^{-3/2}\left(-\frac{2v}{c^2}\right)\right]$$

$$= \frac{2\pi m_0}{h\left(1-\frac{v^2}{c^2}\right)^{\frac{1}{2}}} + \frac{2\pi m_0\left(\frac{v^2}{c^2}\right)}{h\left(1-\frac{v^2}{c^2}\right)^{\frac{3}{2}}}$$

$$= \frac{2\pi m_0\left[\left(1-\frac{v^2}{c^2}\right)+\left(\frac{v^2}{c^2}\right)\right]}{h\left(1-\frac{c^2}{c^2}\right)^{\frac{3}{2}}} = \frac{2\pi m_0}{h\left(1-\frac{v^2}{c^2}\right)^{\frac{3}{2}}}$$

$\therefore$ group velocity
$$u = \frac{dw / dv}{dk / dv} = \frac{2\pi m_0 v}{h\left(1-\frac{v^2}{c^2}\right)^{\frac{3}{2}}} \times \frac{h\left(1-\frac{v^2}{c^2}\right)^{\frac{3}{2}}}{2\pi m_0} = v$$

i.e.,
$$u = v \qquad ...(14)$$

The de Broglie *"wave group"* associated with a moving particle travels with the same velocity as that of the particle. The *"phase velocity"* or *"wave velocity"* W of de Broglie waves evidently has no simple physical significance in itself.

The *"wave velocity"* of the de Broglie waves differs from the velocity of light in one important aspect, that is even in free space the *"wave velocity"* is a function of the wavelength which may be shown as under

$$p = mv \qquad ...(i)$$

and total Energy
$$E = mc^2 \qquad ...(ii)$$

So that
$$p = \frac{E}{c^2}.v \qquad ...(iii)$$

Using the relativistic expression for the mass of a particle, the equation for its momentum becomes

$$p = \frac{m_0}{\left(1-\frac{v^2}{c^2}\right)^{\frac{1}{2}}} \cdot v \qquad ...(iv)$$

where m_0 is the rest mass of the particle. Eliminating v between the last two equations, we get

$$m_0^2 = \frac{E^2}{c^4} - \frac{p^2}{c^2} \qquad ...(15)$$

Now $p = \frac{h}{\lambda}$ and $E = h\nu$

So eq. (15) may be written as $m_0 = \frac{h}{c}\sqrt{\frac{\nu^2}{c^2} - \frac{1}{\lambda^2}}$

and for de Broglie waves $\text{v} = \nu\lambda$

Hence $$m_0 = \frac{h}{c}\sqrt{\frac{W^2}{c^2\lambda^2} - \frac{1}{\lambda^2}} \qquad ...(16)$$

From which $$W = c\sqrt{1 + \frac{m_0^2\, c^2}{\lambda^2}} \qquad ...(17)$$

The equation shows that for a particle of rest mass $m_0 > 0$, the wave velocity W is always greater than c; also the wave velocity of de Broglie waves is a functions of λ even in the free space.

As a special case of de Broglie waves, consider those waves which are propagated with a wave velocity $W = c$. This corresponds to the propagation of electromagnetic waves. The velocity of the associated particle *i.e.*, photon, is also equal to c. If $W = c$ is substituted in equation (16), we find that rest mass of the photon $m_0 = 0$, *i.e., there is no such thing as a photon at rest, photon always moves with the velocity c.*

8.4. Experimental Study of Matter Waves

(*i*) **Experiments of Davisson and Germer :** First of all in 1927 the two American physicists, davisson and Germer predicted experimentally the electron waves predicted by de Broglie, Davisson and Germer were studying the reflection of electrons from nickel target which was subjected to such a heat treatment that the crystal was transformed into a group of crystals. In this case the reflection became anomalous and the reflected intensity showed striking maxima and minima. They then suspected that the beam of electrons might be diffracted from the crystals like X-rays. This shows thai electrons behave like waves under certain cirumstances.

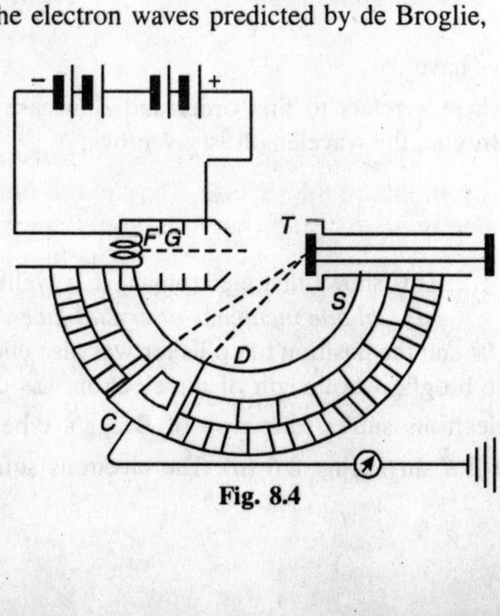

Fig. 8.4

The experimental arrangement is shown in Fig. 8.4. The electron beam is produced from what is known as electron gun. F is the filament which is heated to dull red. Electrons are emitted from the filament due to thermionic action. G is a system of electrodes with central holes maintained at increasing potentials. By this arrangement, electrons emerge as a well collimated beam. These monoenergetic electrons fall on the target T, a single crystal of nickel. Some of the scattered electrons entered the Faraday cylinder C knows as collector. The collector current was amplified and measured with a sensitive galvanometer G. The collector can be moved on a graduated circular scale S to receive electrons scattered in various directions. The collector has two walls insulated from each other. A retarding potential is applied between the inner and outer wall of the collector such that only fast moving electrons coming from electron gun may enter into the collector and not the secondary slow electrons coming from target. The slow electrons will be reflected from the retarding potential. The whole apparatus was enclosed and highly evacuated.

The experiment was conducted in two different ways, *i.e.*,

(*i*) for normal incidence.

(*ii*) for oblique incidence.

(*i*) *For normal incidence*. In this case the beam of electrons falls normally on the surface of the crystal. The collector was moved on the circular scale to various positions and galvanometer current is a measure of the intensity of diffracted beam. A graph was then plotted between the co-latitude (angle between the incident beam and the beam entering the collector) and galvanometer current. Several curves were obtained for different voltage electrons which are shown in Fig. 10.5.

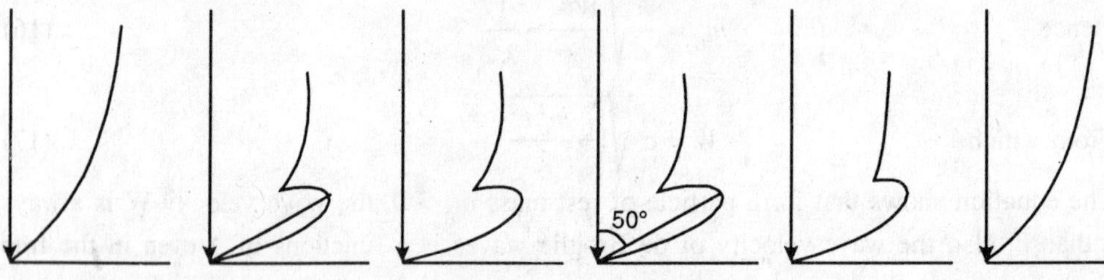

Fig. 8.5

It is observed that a bump begins to appear in the curve at 44-volt electrons. This bump moves upward as the voltage increases and attains the greatest development for 54 volts at co-latitude of 50°. Above 54 volts the bump again diminishes. The bump at 54 volts offers the evidence for the existence of electron waves.

Applying the relation for plane reflection grating:

$$n\,\lambda = D\sin\theta$$

we have

$$\lambda = 2.15\sin 50° = 1.67\ \text{Å}$$

where n refers to first order and d in case of nickel crystal is 2.15 Å. Again, according to de Broglie, the wavelength is given by

$$\lambda = \sqrt{\left(\frac{150}{v}\right)}\ \text{Å} = \sqrt{\left(\frac{150}{54}\right)}\ \text{Å} = 1.66\ \text{Å}$$

This shows that electrons have wavelike characteristics.

(*ii*) *Oblique incidence at crystal face* : Crystal is oriented so that the angle of incidence was 10° and the position of collector was also chosen corresponding to angle of reflection of 10°. The de broglie wavelength of the electron was changed by altering the acceleration potential V. The electrons suffer reflection of Bragg's type and law governing this type of diffraction is $n\lambda = 2\,d\,\sin\theta$. Fig. 8.6 (*a*). The electrons suffer diffraction from the successive parallel planes of atoms in the crystal. Now, as before, the current is measured as function of square root of acceleration potential V. Since $\lambda \propto \frac{1}{\sqrt{V}}$, the graph between current and $V^{1/2}$ also describes the relation between current and $\frac{1}{\lambda}$. See Fig. 8.6 (*b*).

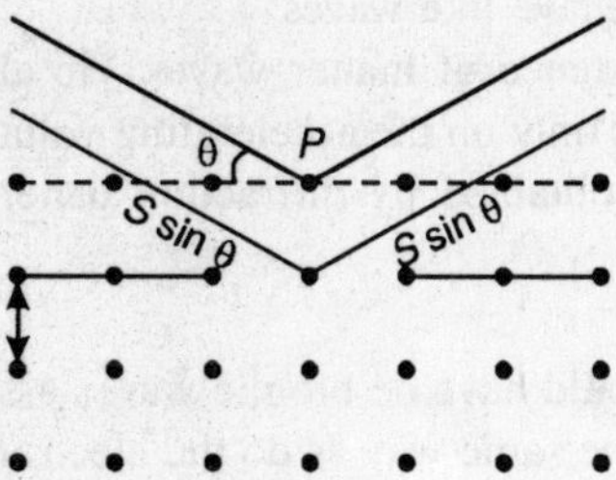

Fig. 8.6 (*a*) Bragg's reflection

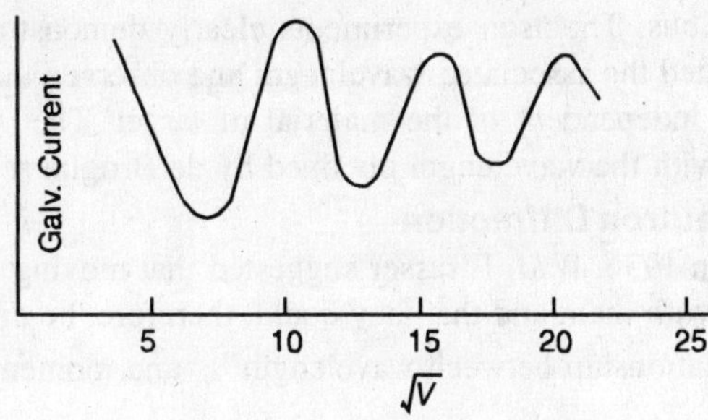

Fig. 8.6 (*b*)

Now, from de Broglie relation, $\lambda = \frac{h}{\sqrt{(2meV)}} \propto \frac{1}{V^{1/2}}$

and Bragg relation gives, $n\lambda = 2d \sin\theta = \text{constant}$

or $$n \propto \frac{1}{\lambda}$$

$$n \propto V^{1/2}$$

i.e., order of spectrum is proportional to $V^{1/2}$ ...(1)

The curve shows distinct maxima at regular intervals of $V^{1/2}$, confirming de Broglie theory.

8.5. *G.P.* Thomson Experiments

G.P. Thomson performed experiments with electrons accelerated from 10,000 to 50,000 volts. In his experiments, Thomson observed diffraction patterns exactly analogous to X-rays patterns, moreover, he was able to determine the wavelengths associated with Electrons. Since diffraction patterns are only possible provided the waves are associated with material particle hence experiments supported the concept of matter waves. His experimental arrangement is shown in Fig. 8.7.

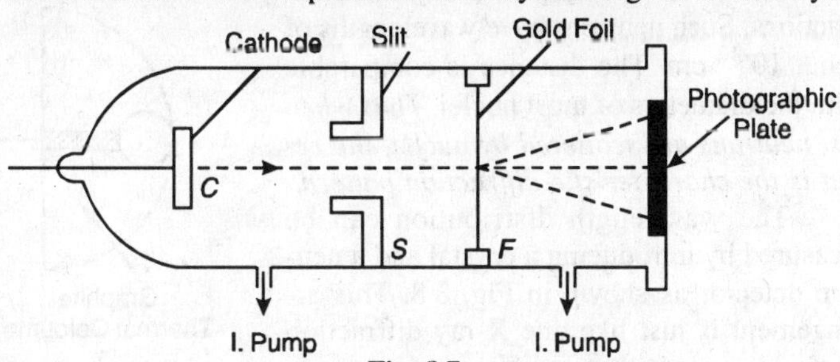

Fig. 8.7

The high energy electron beam is produced by the cathode C. The beam is excited with potentials upto 50,000 volts. A fine pencil of this beam is obtained by passing it through the slit or diaphragm S. The accelerated fine beam of electrons now falls on a thin film F (of the order of 10^{-6} cm) of gold or aluminium. The photograph of the beam from the foil is recorded on a photographic plate P. Whole of the apparatus is exhausted to high vacuum so that the electrons may not lose their energy in collisions with the molecules of the gas.

After developing the plate, a symmetrical pattern consisting of concentric rings about a central spot is obtained. This pattern is similar to that produced by X-rays. To be sure that this pattern is due to the electrons and not due to X-rays generated by the electrons in their passage through the foil, the cathode rays in the discharge tube are deflected by magnetic field. It was observed that the beam shifts correspondingly showing thereby that the pattern is produced by electrons and not by X-rays (X-rays pattern is not affected by electric and magnetic fields). The observed rings can only be interpreted by considering that the diffraction pattern of the incoming beam is due to the diffraction of electrons by the foil. As the diffraction pattern can only be produced by waves and not by the particles, so Thomson concluded that electrons behave like waves.

Thus, Thomson experiments clearly demonstrated the existence of matter waves. He also calculated the associated wavelength and observed that it depends only on the accelerating voltage and is independent of the material of target. The wavelengths obtained by diffraction patterns agree with the wavelength obtained by de Broglie relation.

8.6. Neutron Diffraction

In 1936, *W.M.* Elsasser suggested that moving neutrons should have de broglie waves associated with them and that they could, therefore, be diffracted in the same way as do the electrons. The relationship between wavelength λ and momentum p of the neutron is still given by

$$\lambda = \frac{h}{p} = \frac{h}{m_n v} \quad ...(1)$$

where m_n is the mass of the neutrons. We know that the mass of the neutron is about 2000 times as large as that of electron and hence the wavelength associated with neutron is about 1/2000 that of an electron of the same velocity. The relationship between energy E and wavelength λ is expressed as

$$\lambda = \frac{h}{\sqrt{(2m_n E)}} \quad \text{or} \quad \lambda = \frac{0.28}{\sqrt{E}}$$

where λ is expressed in Å and E in eV. When $E = 0.1\text{ eV}, \lambda = 1\text{ Å}$ which is required for diffraction work. It is also possible to confirm their wave properties for very much smaller wavelengths. For example, neutrons with energies of about 10 MeV are rather readily produced in nuclear reactions. Such neutrons have wavelengths of about 10^{-12} cm. The distance is comparable with the diameters of most nuclei. *Thus when fast neutrons are scattered by nuclei, the result is the characteristic diffraction pattern.*

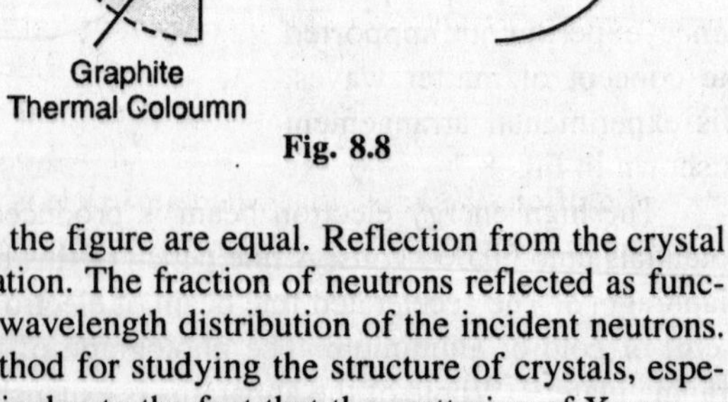

Fig. 8.8

The wavelength distribution can be measured by introducing a crystal and a neutron detector as shown in Fig. 8.8. This arrangement is just like one X-ray diffraction apparatus except that a different detector is used. The crystal and detector are turned in such a way so that the two angles shown as θ in the figure are equal. Reflection from the crystal occurs only when λ and θ satisfy the Bragg relation. The fraction of neutrons reflected as function of θ gives an experimental determination of wavelength distribution of the incident neutrons.

Neutron diffraction has proved a useful method for studying the structure of crystals, especially organic crystals containing hydrogen. This is due to the fact that the scattering of X-rays or electrons by hydrogen is very weak as atoms in a crystal cannot be detected. Neutrons, on the other hand, interact with nuclei through nuclear forces and are scattered by hydrogen atoms. Neutron diffraction is also frequently applied to the study of magnetic properties of solids and various other allied phenomenon.

Example 1. *Find the energy of the neutron in units of electron volt whose de Broglie wavelength is 1 Å.*

Given mass of the neutron = 1.674 × 10^{-27} kg
Planck's constant h = 6.60 × 10^{-34} Joule-sec.

Solution. We have the relation $\lambda = \frac{h}{mc} = \frac{h}{\sqrt{2mE}}$

or $$E = \frac{h^2}{2m\lambda^2}$$

In this case
$$m = 1.674 \times 10^{-27} \text{ kg}$$
$$\lambda = 1\text{Å} = 10^{-10}$$
$$h = 6.60 \times 10^{-34} \text{ Joule-sec.}$$

$$\therefore \quad E = \frac{(6.60 \times 10^{-34})^2}{1.674 \times 10^{-27} \times (10^{-10})^2} = 13.01 \times 10^{-21} \text{ jules}$$

$$= \frac{13.01 \times 10^{-24}}{1.6 \times 10^{-19}} = \mathbf{8.12 \times 10^{-2}\ eV.\ Ans.}$$

Example 2. *Compute the de Broglie wavelength of 10^{11} keV neutrons. Mass of neutron may be taken as 1.675 × 10^{-27} kg.*

Solution. *In this case kinetic energy of neutron*
$$= 10^{11}\ keV = 10^{14} \times 1.6 \times 10^{-19} \text{ joule.}$$
$$= 1.6 \times 10^{-5} \text{ joule.}$$

Now $$\frac{1}{2}mv^2 = 1.6 \times 10^{-5}$$

or $$v = \left(\frac{2 \times 1.6 \times 10\times^{-5}}{1.675 \times 10^{-27}}\right)^{\frac{1}{2}} \text{ m/s}$$

Again $$\lambda = \frac{h}{mv} = \frac{6.625 \times 10^{-34}}{1.675 \times 10^{-27}} \times \left(\frac{1.675 \times 10^{27}}{2 \times 1.6 \times 10^{-5}}\right)^{\frac{1}{2}}$$
$$= \mathbf{2.86 \times 10^{-15}\ m \quad Ans.}$$

Example 3. *An electron microscope uses 1.25 keV electrons. Find its ultimate resolving power cn the assumption that this is equal to the wavelength of the electron, given that*
$$e = \mathbf{4.8 \times 10^{-10}}\ \textit{e.s.u.}$$
$$m = \mathbf{9.0 \times 10^{-28}}$$
$$h = \mathbf{6.65 \times 10^{-27}}\ \textit{ergs sec.}$$

Solution. We know that when a particle is accelerated by *V* volts then it acquires V/300 ergs erergy and if its velocity is *v*, then we have
$$\frac{1}{2}mv^2 = \frac{eV}{300}$$

or $$mv = \sqrt{\left(\frac{meV}{150}\right)}$$

$$\therefore \qquad \lambda = \frac{h}{mv} = h\sqrt{\left(\frac{150}{meV}\right)}$$

In this case $E = 12.5$ Kev

$\therefore$ $V = 1250$ volts

and $$\lambda = 6.65\times10^{-27}\sqrt{\left(\frac{150}{9\times10^{-28}\times4.8\times10^{-10}\times1250}\right)} \text{ cm.}$$

$= \mathbf{0.351Å}$ **Ans.**

So resolving power of the microscope is 0.351Å.

Example 4. *What is de Broglie wavelength of an electron which has been accelerated from rest through a potential difference of 100 V.*

Solution. We have the relation

$$\lambda = \frac{12.25}{\sqrt{V}}\text{Å}$$

In this case $V = 100$ volts

$$\therefore \qquad \lambda = \frac{12.25}{\sqrt{100}} = \mathbf{1.225Å} \text{ **Ans.**}$$

Example 5. *Estimate the amount of acceleration voltage to which electrons are to be subjected in order to associate them with de Broglie wavelength of .150 Å. Given that*

$$e = 4.8\times10^{-10} e.s.u.$$
$$m = 9.0\times10^{-28} gm$$
$$h = 6.67\times10^{-27} \text{ ergs. sec}$$

Solution. Now if V is the accelerating potential measured in volts, then an electron will gain the energy E given by $E = \dfrac{eV}{300}$

and $$\lambda = \frac{h}{mv}$$

$$\lambda = \frac{h}{\sqrt{(2\,m\,E)}} \qquad \because \text{ also } E = \frac{eV}{300} \qquad \therefore \lambda = \frac{h}{\sqrt{\dfrac{meV}{150}}}$$

Therefore, substituting the values, we get

$$0.5\times10^{-3} = \frac{6.67\times10^{-27}}{\sqrt{\dfrac{meV}{150}}}$$

$$V = \frac{(6.67\times10^{-27})\times150}{(0.5\times10^{-8})^2\times9.0\times10^{-28}\times4.8\times10^{-10}}$$

$V = \mathbf{614.4\ Volts.}$ **Ans.**

Example 6. *Calculate the energy in electron volt of an electron wave of $\lambda = 3\times10^{-2}$ m. Given $h = 6.62\times10^{-34}$ Js.*

Solution. We know that energy is given by the relation

$$E = \frac{h^2}{2m\lambda^2}$$

In this case $h = 6.62 \times 10^{-34}$ Js, $m = 9.1 \times 10^{-31}$ kg

and $\lambda = 3 \times 10^{-2}$ m

$$\therefore \quad E = \frac{(6.62\times10^{-34})^2}{2\times9.1\times10^{-31}\times(3\times10^{-2})^2} \cong 0.27\times10^{-23} \text{ joules}$$

$$\cong \frac{0.27\times10^{-33}}{1.6\times10^{-19}} \text{ eV} = \mathbf{1.68 \times 10^{-15}\ eV} \quad \textbf{Ans.}$$

Example 7. *What would be the wavelength of quantum of radiant energy emitted, if an electron transmitted into radiation and converted into one quantum.*

Solution. According to Planck, the energy E associated with one quantum $1 = h\nu$, where ν is the frequency of radiation.

When the energy of an electron is transmitted into radiation, we have the relation

$$E = mc^2 \text{ (mass energy relation)}$$

$$\therefore \quad h\nu = mc^2 \text{ or } \frac{hc}{\lambda} = mc^2$$

$$\therefore \quad \lambda = \frac{h}{mc} = \frac{6.6\times10^{-34}}{9.1\times10^{-31}\times3\times10^{8}}$$

$$= 0.0244 \times 10^{-10} \text{ m} = \mathbf{0.0244\ Å} \quad \textbf{Ans.}$$

Example 8. *Calculate the velocity and de Broglie wavelength of a proton of energy 10^5 electron volts, given that*

Mass of proton = 1.66×10^{-24} gm
Change on the electron = 4.8×10^{-10} e.s.u.
Planck's constant = 6.62×10^{-27} erg. sec.

Solution. The energy of proton

$$\frac{1}{2}mv^2 = 10^5\text{eV} = 10^5\times1.6\times10^{-12}\text{ergs.}$$

$$(\therefore\ 1 \text{ eV} = 1.6 \times 10^{-12} \text{ ergs})$$

$$\therefore \quad v^2 = \frac{2\times10^5\times1.6\times10^{-12}}{m}$$

$$= \frac{2\times10^5\times1.6\times10^{-12}}{1.66\times10^{-24}}$$

or $$v = \sqrt{\left(\frac{2\times10^5\times1.6\times10^{-12}}{1.66\times10^{-24}}\right)}$$

The de Broglie wavelength λ is given by

$$\lambda = \frac{h}{mv} = \frac{h}{\sqrt{(2mE)}} \quad \text{because } \frac{1}{2}mv^2 = E$$

or $$mv = \sqrt{(2mE)}$$

According to problem $E = 10^5 \times 1.6 \times 10^{-12}$ ergs

$$\therefore \quad \lambda = \frac{6.62\times10^{-27}}{\sqrt{(2\times1.66\times10^{-24}\times10^5\times1.6\times10^{-12})}}$$

$$= 9.3 \times 10^{-12} \text{ cm}$$
$$= \mathbf{9.3 \times 10^{-14} \text{ m.}} \quad \textbf{Ans.}$$

Example 9. *Calculate the de Broglie wavelength for an electron which has been accelerated through at p. d. of 28.8 V, given that*

$$e = 4.8 \times 10^{-10} \text{ e.s.u.}$$
$$m = 9.1 \times 10^{-28} \text{ gm}$$
$$h = 6.62 \times 10^{-27} \text{ erg. sec.}$$

Solution. We have the relation $\lambda = h.\sqrt{\left(\frac{150}{meV}\right)}$

$$= \frac{6.62\times10^{-27}\times\sqrt{(150)}}{\sqrt{(9.1\times10^{-28}\times4.8\times10^{-10}\times28.8)}}$$
$$= \frac{12.27}{\sqrt{(28.8)}}\times10^{-8} \text{ cm.}$$
$$= \mathbf{2.4 \text{ Å}} \quad \textbf{Ans.}$$

Example 10. *Energy of a particle at absolute temperature T is of the order kT. Calculate the wavelength of thermal neutrons at 27° C. Given mass of the neutron = 1.67 × 10^{-27} kg, Planck's constant h = 6.60 × 10^{-34} Joule-sec and Boltzmann's constant k = 8.6 × 10^{-5} eV deg^{-1}.*

Solution. We know that $\lambda = \frac{h}{mc} = \frac{h}{\sqrt{2mE}}$

From the problem $E = kT$

$\therefore$ $\lambda = \frac{h}{\sqrt{(2mkT)}}$

Here $h = 6.60 \times 10^{-34}$, $m = 1.67 \times 10^{-27}$ kg.

$$T = 27° \text{ C} = 300 \text{ K}$$
$$k = 8.6 \times 10^{-5} \text{ eV deg}^{-1}$$
$$= 8.6 \times 10^{-5} \times 1.8 \times 10^{-19} \text{ joule deg}^{-1}$$
$$= 1.376 \times 10^{-23} \text{ joule deg}^{-1}$$

Putting the values, we get

$$\lambda = \mathbf{1.777 \text{ Å.}} \quad \textbf{Ans.}$$

Example 11. *What is the energy of gamma ray photon having a wavelength of 1 Å ? Given that: h = 6.62 × 10^{-27} erg. sec.*

Solution. We know that $\lambda = \frac{h}{p}$ or $p = \frac{h}{\lambda} = \frac{6.62\times10^{-27}}{10^{-8}}\left(\frac{\text{erg. sec}}{\text{cm}}\right)$

Energy of gamma ray photon $E = p \times c$ where c = velocity of light

$$E = \frac{6.62\times10^{-27}}{10^{-8}}\times3\times10^{10} \text{ ergs}$$
$$= \frac{6.62\times10^{-27}\times3\times10^{10}}{10^{-8}\times1.6\times10^{-12}} \text{ eV}$$
$$= \mathbf{1.24 \times 10^{4} \text{ eV.}} \quad \textbf{Ans.}$$

Example 12. *Calculate the de Broglie wavelength of an α-particle accelerated through 200 volts, given that:*

$$\text{mass of } \alpha = 6.576 \times 10^{24} \text{ gm.}$$
$$h = 6.62 \times 10^{27} \text{ ergs. sec.}$$

Solution. We have the definition of potential difference in e.s.u. between two points defined as the work done in moving a unit positive charge from one point to the other. In this way the work done in moving an α-particle between the two points between which the potential difference is V volts is given by

$$= e\frac{V}{300} \text{ ergs.}$$

$$\therefore \quad \frac{1}{2}mv^2 = \frac{eV}{30}$$

$$\therefore \quad mv = \sqrt{\frac{me}{150}}$$

Now $$\lambda = \frac{h}{mv}$$

or $$\lambda = h\sqrt{\frac{150}{(2meV)}}$$

$$= 6.62\times10^{-27}\sqrt{\frac{150}{(2\times6.576\times10^{-24}\times4.8\times10^{-10}\times2000)}}$$

$$= \frac{9.44}{4}\times10^{-11} \text{ cm} = \mathbf{2.362 \times 10^{-11} \text{ cm}} \quad \textbf{Ans.}$$

8.7. Heisenberg's Uncertainty Principle

In the experiments for the determination of specific charge of an electron, it was shown that electron gets deflected by the application of electrical and magnetic fields, which means that the electron behaves as a particle and it has definite mass, momentum and energy which can be measured with desired accuracy. But the experiment of Davisson and Germer, and Thomson showed the wave nature of electrons and the wavelength of the electron is given by the relation

$$\lambda = \frac{h}{mv} = \frac{h}{p}$$

where h is the Planck's constant and mv is the momentum of the electron. A wave extends through space and it is difficult to locate the exact position of an electron behaving as a wave, at any given instant of time.

This dual behaviour of electron as a particle and as a wave presents difficulty in locating the exact position and momentum at the same time. This difficulty is overcome by *Heisenberg's uncertainty principle.*

The principle states that the exact position and momentum of a particle (say electron) cannot be determined simultaneously with a desired accuracy. *Taking Δx as the error in deretmining its position and Δp the error in determining its momentum* at the same instant, these quantities are related as under :

$$\Delta x\,\Delta p \cong \frac{1}{2}\times\frac{h}{2\pi} \quad ...(i)$$

i.e., the product of the *two errors* is approximately of the order of Planck's constant.

Multiplying and dividing the left hand side of equation (*i*) by v, the velocity of the particle

$$\frac{\Delta x}{v}\Delta p\, v \approx \frac{1}{2}\times\frac{h}{2\pi}$$

or $$\Delta t \;\Delta E \approx \frac{1}{2} \times \frac{h}{2\pi} \qquad ...(ii)$$

Here ΔE represents the uncertainty in the measurement of the energy of a particle and $\Delta\ t$ the uncertainty in the measurement of time.

In equation (*i*), the product $\Delta x \Delta p$ is of the order of $h/2\pi$. If Δx is small Δp will be large and *vice versa*. It means that if one quantity is measured accurately, the other quantity becomes less accurate. A similar argument hold good for ΔE Δt as well.

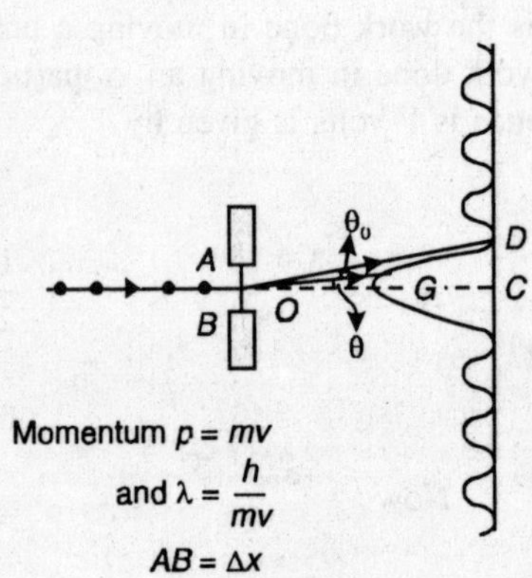

Fig. 8.9

Consider a beam of electrons travelling in the direction shown in Fig. 8.9. The slit *AB* of witdth Δx is perpendicular to the path of the electrons. Before entering the slit, the electron has a definite momentum $p = mv$. After passing through the slit the electron gets diffracted and acquires a momentum along *OG*. The angular deflection θ depends upon component of the momentum parallel to the slit, *i.e*, $\Delta\ p = p \sin\theta = p\,\theta$ for small angular deflection. The angle θ_0 corresponding to the direction of first momentum of the diffraction pattern is

$$\theta_0 = \frac{\lambda}{\Delta\ x}, \text{ for small value of } \theta_0$$

or $$\Delta\ x = \frac{\lambda}{\theta_0}$$

$\therefore$ $$\Delta\ p\,\Delta\ x \sim p\lambda \times \frac{\theta}{\theta_0}$$

But $$p\lambda = h$$

$\therefore$ $$\Delta\ p\,\Delta\ x = h\frac{\theta}{\theta_0}$$

Taking, $$\theta = \theta_0 \text{ approximately,}$$

$$\Delta\ p\,\Delta\ x \sim h$$

The probable deflection θ of the electron is less than θ_0 and according to Heisenberg, the uncertainty relation is given by

$$\Delta\ p\,\Delta\ x \geq \frac{1}{2} \times \frac{h}{2\pi}$$

In general any experiment that measures *position or momentum* accurately, introduces an uncertainty in the other. *The uncertainty is never less than the predicted value according to Heisenberg's relation.* Thus no instrument can measure the quantities more accurately than predicted by *Heisenberg's principle or indeterminancy.*

8.8. Applications of Uncertainty Principle

(*i*) Non-existence of electrons and existence of protons and neutrons in nucleus. The radius of the nucleus of any atom is of the order of 10^{-14} m. If an electron is confined inside the nucleus, then the uncertainty in the position Δx of the electron is equal to the diameter of the nucleus, *i.e.*, $\Delta x \approx 2\times10^{-14}$ m. Using the Heisenberg's uncertainty relation, the uncertainty in momentum of electron is given by the relation

$$\Delta p_x \geq \frac{h}{2\pi.\,\Delta x} \geq \frac{1.055\times10^{-34}}{2\times10^{-14}} \text{ kg ms}^{-1} \quad \therefore\ \Delta x \approx 10^{-14}$$

where $$\frac{h}{2\pi} = 1.055\times10^{-34}$$

$$\therefore \qquad p_x \geq 0.527\times10^{-20} \text{ Ns}$$

It means that the momentum component p_x and hence the magnitude of total momentum $|\vec{p}|$ of the electron in the nucleus must be at least of the order of magnitude, *i.e.*,

$$|\vec{p}| \sim p_x \sim \Delta p_x \sim 0.527\times10^{-20} \text{ Ns.}$$

Now the mass of the electron is 9.1×10^{-31} kg, the order of magnitude of momentum (0.527×10^{-20} kg ms^{-1}) is relativistic. Using the relativistic formula for the energy E of the electron, we have

$$E = p^2c^2 + m_0^2c^4$$

As the rest energy m_0c^2 of the electron is of the order of 0.511 MeV, which is much smaller than the value of first term, so it can be neglected. Thus, we have

$$E^2 = p^2c^2 \quad \text{or} \quad E = pc$$

$$\therefore \qquad E = (0.527 \times 10^{-20}) \times (3 \times 10^8) \text{ joules}$$

$$\approx \frac{(0.527\times10^{-20})\,(3\times10^8)}{1.6\times10^{-19}} \text{ eV}$$

$$= 10 \text{ MeV}$$

which means that if the electrons exist inside the nucleus, their energy must be of the order of 10 MeV. However, we know that the electrons emitted by radioactive nuclei during beta decay have energies only 3 to 4 MeV. So, *in general electrons cannot exist in the nucleus.*

For protons and neutrons, $m_0 = 1.67 \times 10^{-27}$ kg. This is a non-relativistic problem as $|\vec{v}| \sim |\vec{p}|/m_0 \sim 3\times10^6$ ms^{-1}. The kinetic energy K in this case is given by the relation

$$K = \frac{p^2}{2m_0} = \frac{(0.527\times10^{-20})^2}{2\times1.67\times10^{-2}} \text{ joules}$$

$$\approx \frac{(0.527\times10^{-20})^2}{2\times1.67\times10^{-27}\times1.6\times10^{-19}} \text{ eV} = 52 \text{ keV}$$

Since the energy is less than the energies carried by these particles emitted by nuclei, *both these particles can exist inside the nuclei.*

(*ii*) **Binding energy of an electron in atom.** In an atom, the electron is under the influence of electrostatic potential of the positively charged nucleus. It is confined to the linear dimensions equal to the diameter of electronic orbit. The uncertainty in the position Δx of the electron is of the order or $2R$ where R is the radius of the orbit and the corresponding uncertainty in momentum component Δp_x is given by

$$\Delta p_x \geq \frac{h}{4\pi.\,2R}$$

which shows that the momentum of an electron in atomic orbit is at least

$$|\vec{v}| \sim p_x \sim \Delta p_x \sim \frac{h}{4\pi . 2R}$$

$$\sim 0.527 \times 10^{-24} \text{kg ms}^{-1} \qquad \because R = 10^{-10}\ m$$

This is non-relativistic momentum for an electron. The kinetic energy K is given by

$$K = \frac{p^2}{2m_0} = \left(\frac{h}{4\pi R}\right)^2 \frac{1}{2m_0} = \frac{h^2}{32\pi^2 m_0 R^2}$$

The potential energy of an electron in the field of nucleus with atomic number Z is given by

$$V = \frac{-Ze^2}{4\pi\,\varepsilon_0 R}$$

So the total energy of the electron in its orbit will be

$$E = K + V = \frac{h^2}{32\pi^2 m_0 R^2} - \frac{Ze^2}{4\pi\,\varepsilon_0 R}$$

$$= \frac{(1.055\times10^{-34})^2}{8\times9.1\times10^{-3}\times R^2} - \frac{Z(1.6\times10^{-19})^2}{4\times3.14\times8.85\times10^{-12}R}$$

where $\qquad \frac{h}{2\pi} = 1.055\times10^{-34}$

or $\qquad E = \frac{10^{-20}}{R^2} - \frac{15\times10^{-10}Z}{R}\text{eV}$

Taking $\qquad R = 10^{-10}\ m$ we have

$$E = (1-15\ Z)\ \text{eV}$$

Now the binding energies of outermost electrons in H and He are –13.6 and –24.6 eV respectively, so the value of binding energy derived on the basis of uncertainty principle is acceptable as these are comparable in magnitudes.

(*iii*) **Finite width of the spectral lines.** From Heisenberg's Principle of energy and time relation, we have

$$\Delta E . \Delta t \cong h$$

Since the life time of electron in an excited orbit is finite ($\sim10^{-8}$ sec), so the energy levels of an atom given by $\Delta E = \frac{h}{\Delta t}$ must have a finite width which means that the excited levels must have a finite energy spread *i.e.*, the radiation given out when an electron jumps must be truly monochromatic. In other words, it means that the spectral lines can never be very sharp but must have a natural finite width.

(*iv*) **Strength of nuclear force.** If we assume that the nuclear radius is of the order

$$r_0 = 1.2 \times 10^{-13} \text{ cm.}$$

From uncertainty principle, momentum will be of the order

$$p = \frac{\hbar}{r_0},$$

∴ K.E. will be of the order

$$\text{K.E.} \simeq \frac{p^2}{2M} \simeq \frac{\hbar^2}{2Mr_0^2} = 10 \text{ MeV}$$

where M corresponds to the mass of a nucleus (proton or neutron).

Since the nucleus is bound, so the *B.E.* should be greater than *K.E.* with a negative sign. So the B.E. of a nucleus is of the order of 10 MeV.

8.9. Complementary Principle

The most important consequence of the uncertainty principle is that it is impossible to determine the wave and particle properties exactly at the same time.

According to the complementary principle the wave and particle aspects of matter and light are complementary rather than contradictory, i.e., both aspects are necessary to have a complete picture of the same system. For example, consider an experimental arrangement in which light is diffracted by a double slit and is detected on a screen that consists of many adjacent photoelectric cells. The photoelectric cells responds to photons (having particle properties). However, if we plot the number of photons each cell counts in a certain period of time against the location of cell, a wave like pattern (as in the case of interference of a pair of coherent wave trains) is obtained. It follows that only experiment which can be devised displays either the particle like characteristics or wave like characteristics of the system, moreover the wave and particle pictures give complementary description of the same system.

Example 13. *An electron has a speed of 500 ms⁻¹ correct upto 0.01% with what minimum accuracy can you locate the position of this electron.*

Solution. The momentum of electron $p = mv$

$$= 9.11 \times 10^{-31} \times 500 = 4.55 \times 10^{-28} \text{ kg ms}^{-1}$$

The uncertainty in momentum

$$\Delta p = m\Delta v = mv\frac{\Delta v}{v} = \frac{0.01}{100}$$

$$= 4.55 \times 10^{-28} \times 0.0001$$

$$= 4.55 \times 10^{-32} \text{ kg ms}^{-1}$$

The minimum uncertainty in position is given by

$$\Delta x = \frac{h}{2\pi}\frac{1}{\Delta p} = \frac{6.625\times10^{-34}}{2\pi\times4.55\times10^{-32}}$$

$$= \mathbf{2.318\ mm.} \quad \textbf{Ans.}$$

Example 14. *An electron has a speed of 600 m/s with an accuracy of 0.005%. Calculate the certainty with which we can locate the position of the electron. Given that*

$$h = 6.6 \times 10^{-34} \text{ joule-sec}$$

$$m = 9.1 \times 10^{-31} \text{ kg}$$

and $v = 600$ *m/sec.*

Solution. The momentum of electron $= mv$

$$= 9.1 \times 10^{-31} \times 600 \text{ kg m/s}$$

$$\Delta\, px = \left(\frac{0.005}{100}\right)mv = 5 \times 10^{-5} \times 9.1 \times 10^{-31} \times 600 \text{ kg m/s}$$

$$\therefore \quad \Delta x \approx \frac{h}{\Delta\, px} \approx \frac{6.6\times10^{-34}}{5\times10^{-5}\times9.1\times10^{-31}\times600} \approx \mathbf{0.02253\ m.} \quad \textbf{Ans.}$$

Example 15. *10 keV electrons are passed through a thin film of a metal for which atomic spacing is 5.5 × 10⁻¹¹ m. What is the angle of deviation of the first order diffraction maximum ?*

Solution. In this case the kinetic energy of electron is small as compared with its rest mass energy (500 keV)

$$\therefore \quad m = m_0 = 9.11 \times 10^{-31} \text{ kg.}$$

K.E. of electrons $\left(\frac{1}{2}mv^2\right) = 10 \text{ keV} = 10^4 \text{ eV}$

$$= 10^4 \times 1.6 \times 10^{-19} \text{ joule}$$

$$v = \sqrt{\frac{2\times10^4 \times 1.6\times10^{-19}}{m}}$$

de-Broglie wavelength, $\lambda = h / mv$

Putting the values, we get $\lambda = \frac{6.625\times10^{-34} \times \sqrt{m}}{m \times \sqrt{2\times10^4 \times 1.6\times10^{-19}}}$

$$= \frac{6.625\times10^{-34}}{\sqrt{3.2\times10^{-15} \times m}}$$

or
$$\lambda = \frac{6.625\times10^{-34}}{\sqrt{3.2\times10^{-15} \times 9.11\times10^{-31}}}$$

$$= 1.23 \times 10^{-11} \text{ m.}$$

Now using Bragg's *eq*. We have

$$2d \sin\theta = \lambda$$

In this case $n = 1$ for first order diffraction

$$d = 5.5 \times 10^{-11} \text{ m}$$

$$\therefore \quad 2 \times 5.5 \times 10^{-11} \sin\theta = 1 \times 1.23 \times 10^{-11}$$

or
$$\sin\theta = \frac{1.23\times10^{-11}}{2\times5.5\times10^{-11}} = \mathbf{0.1118}$$ **Ans.**

Example 16. *Determine the de Broglie wavelength of an electron, having kinetic energy of 1 eV. (Given m_0 of an electron = 9.11×10^{-31} kg and h = 6.625×10^{-34} J-s).*

Solution. Here the kinetic energy of the electron is very small as compared with its rest mass energy, *i.e.*, 0.512 MeV and hence negligible.

$$m = m_0 = 9.11 \times 10^{-31} \text{ kg}$$

K.E. of the electron $\left(\frac{1}{2}mv^2\right) = 1 \text{ eV} = 1.6\times10^{-19} \text{ J}$

$$v^2 = \frac{2 \times 1.6 \times 10^{-19}}{m}$$

de Broglie wavelength, $\lambda = h / mv$

Putting the values, we have

$$\lambda = \frac{6.625\times10^{-34} \sqrt{m}}{m\sqrt{2\times1.6\times10^{-19}}}$$

$$\lambda = \frac{6.625\times10^{-34}}{\sqrt{3.2\times10^{-19} \times m}}$$

$$= \frac{6.625\times10^{-34}}{\sqrt{3.2\times10^{-19} \times 9.11\times10^{-31}}}$$

$$= 1.23 \times 10^{-9} \text{ m}$$
$$= \mathbf{12.3 \text{ Å}} \quad \textbf{Ans.}$$

Example 17. *In jumping from an excited to a stationary state, an atom takes 10^{-9} second. What is the uncertainty in the energy of the emitted radiation ?*

Solution. $$\Delta E . \Delta t = \frac{h}{2\pi}$$

An electron having momentum of this order is non-relativistic in behaviour and its kinetic energy is given by the relation.

$$T = p^2/2m$$
$$= \frac{(2.2 \times 10^{-24} \text{ kg ms}^{-1})^2}{2 \times 9.1 \times 10^{-31} \text{ kg}}$$
$$= 2.7 \times 10^{-18} \text{ joule} = 17 \text{ eV}$$

We can consider the limit to the accuracy with which we can determine the frequency of the radiation emitted by an excited atom as under :

The excited atom gives up its excess energy by emitting one or more photons of characteristic frequency. The average time gap between the excitation of an atom and the time it radiates energy is 10^{-8} sec. So the photons energy is uncertain by an amount given by

$$\Delta E = \frac{h}{\Delta t} = \frac{1.054 \times 10^{-34} \text{ J}}{10^{-8} \text{ s}}$$
$$= 1.054 \times 10^{-26} \text{ J}$$

and the frequency of the light is uncertain by an amount

$$\Delta v = \frac{\Delta E}{h} = \frac{1.054 \times 10^{-26}}{6.62 \times 10^{-34}} = \mathbf{1.6 \times 10^7 \text{ Hz.}} \quad \textbf{Ans.}$$

and this limit cannot be reduced.

Example 18. *If the uncertainty in position of an electron is 4×10^{-10} m, calculate the uncertainty in its momentum.*

Solution. From the uncertainty principle

$$\Delta x \, \Delta p_x \approx h$$

$$\therefore \quad \Delta p_x \approx \frac{h}{\Delta x} \approx \frac{6.6 \times 10^{-34}}{4 \times 10^{-10}} \approx \mathbf{1.6 \times 10^{-24} \text{ kgm/sec.}} \quad \textbf{Ans.}$$

Example 19. *An electron is confined to a box of length 10^{-9} m, calculate the minimum uncertainty in its velocity.*

Given mass of the electron $m = 10^{-31}$ kg and Planck's constant $h = 6.6 \times 10^{-34}$ joules-sec.

Solution. We have the relation

$$\Delta x . \Delta p \cong h$$

If Δx is maximum, Δpx must be minimum

i.e., $\quad (\Delta x)_{max} \, (\Delta p_x)_{min} \cong h$

According to the problem

$$(\Delta x)_{max} = 10^{-9} \, m$$

$$\therefore \quad (\Delta p)_{min} = \frac{h}{(\Delta x)_{max}} = \frac{6.6 \times 10^{-34}}{10^{-9}}$$

$$= 6.6 \times 10^{-25} \text{ kg } m/s.$$

Again $\quad (\Delta p)_{min} = m(\Delta vx)_{min}$

$$\therefore \quad (\Delta p)_{min} = \frac{(\Delta px)_{min}}{m} = \frac{6.6\times10^{-25}}{9\times10^{-31}} = \mathbf{7.3 \times 10^5}\ \boldsymbol{m/s} \quad \textbf{Ans.}$$

Example 20. *Using the uncertainty relation* $\Delta E.\Delta t = \frac{h}{2\pi}$, *calculate the time required for the atomic system to retain rotation energy for a line of wavelength 6000 Å and width* 10^{-4} *Å.*

Solution. Here, the width of special line $d\lambda = 10^{-4}$ Å

$$= 10^{-4} \times 10^{-10} \text{ m and } \lambda = 6 \times 10^{-7} \text{ m}$$

We know that $\quad E = h\nu = \frac{hc}{\lambda}$

or $\quad \Delta E = -\frac{hc}{\lambda^2}\Delta\lambda$

Using uncertainty relation

$$\Delta E.\,\Delta t = \frac{h}{2\pi}, \text{ we have}$$

$$\Delta t \approx \frac{h}{2\pi\Delta E} \approx \frac{h}{2\pi\left(\frac{hc}{\lambda^2}\right)\Delta\lambda} \approx \frac{\lambda^2}{2\pi c\Delta\lambda}$$

$$= \frac{(6\times10^{-7})^2}{2\times3.14\times3\times10^8\times10^{-14}} \text{ second}$$

$$(\because d\lambda = \Delta\lambda = 10^{-14} \text{ m})$$

$$= \mathbf{1.9 \times 10^{-8}} \textbf{ second} \quad \textbf{Ans.}$$

Example 21. *If the uncertainty in the location of a particle is equal to its de-Broglie wave length, what is the uncertainty with velocity ?*

Solution. We have from uncertainty principle,

$$(\Delta p)\ (\Delta x) = h$$

$$\therefore \quad \Delta(mv)\ (\Delta x) = h$$

or $\quad \Delta v = \frac{h}{m\Delta x}$

but $\quad \Delta x = \lambda$

$$\therefore \quad \Delta v = \frac{h}{m\lambda} = \frac{h}{m.\ h/mv} = v$$

$$\therefore \quad \Delta v = v \qquad \textbf{Ans}$$

Example 22. *A nucleon is confined to a nucleus of diameter* 5×10^{-4} *m. Calculate the minimum uncertainty in the momentum of the nucleon. Also calculate the minimum kinetic energy of the nucleon.*

Solution. We have the relation

$$(\Delta p)_{min}\ (\Delta x)_{max} = h$$

$$\therefore \qquad (\Delta p)_{min} = \frac{h}{(\Delta x)_{max}}$$

In this case, $(\Delta x)_{max} = 5 \times 10^{-14}$ m

and $h = 6.626 \times 10^{-34}$ J-s

$$\therefore \qquad (\Delta p)_{min} = \frac{6.626 \times 10^{-34}}{5 \times 10^{-14}} = 1.33 \times 10^{-20} \text{ kg.m/sec}$$

Since p cannot be less than $(\Delta p)_{min}$, we have

$$P_{min} = (\Delta p)_{min}$$

$$\therefore \qquad E_{min} = \frac{(P_{min})^2}{2m}$$

Putting the values of $P_{min} = 1.33 \times 10^{-20}$ kg-ms^{-1}

and $m = 1.675 \times 10^{-27}$ kg

$$E_{min} = \frac{(1.33 \times 10^{-20})^2}{2 \times 1.675 \times 10^{-27}} = 0.529 \times 10^{-13} \text{ J}$$

$$= \frac{0.529 \times 10^{-13}}{1.6 \times 10^{-19}} \text{eV} = \mathbf{0.33 \text{ meV}} \textbf{ Ans.}$$

8.10. Electron Microscope

The electron microscope, like the optical microscope, is an instrument used principally in the research laboratory for magnifying small objects to such an extent that their minutest parts may be observed and studied in detail. *The importance of this device in the field of medical research cannot be overestimated.* To illustrate, many viruses known to medical science as being responsible for certain human diseases lie beyond the range of the optical microscope. With the electron microscope, magnifications from 10 to 100 times that of finest optical microscopes make many of these viruses, and some of their detailed structure, visible to the eye. While the highest magnification obtained with the best optical microscope is about **200 ×**, electron microscopes have already been made to give magnifications as high as **100,000 ×**.

Principle. It is based upon the principle that

1. particles such as electrons exhibit wave properties like rays but have much shorter wavelength; and
2. electrons can be focussed by suitable electric and magnetic fields just as light rays can be focussed with the help of lenses.

By Definitions. *The resolving power of an optical instrument like a microscope is directly proportional to the aperture (diameter) of the lens used and inversely proportional to the wavelength of the light used.* So the smaller the wavelength, the greater shall be the magnifying power. If we accelerate electrons through a potential difference of say, 60000 volts, the velocity v attained by them can be calculated from the relation $eV = \frac{1}{2}mv^2$ where m, the electron mass = 9×10^{-31} kg. Calculating λ from de Broglie wave equation $(\lambda = h/mv)$, the wavelength comes out to be about 5×10^{-12} m which is 10^5 times smaller than that on visible light.

Construction. In figure 10.10, the schematic diagram of electron microscope has been shown by the side of an optical microscope for comparison. In an optical microscope, Fig. 10.10 (*b*), the light from a source S such as sunlight or an electric lamp is made to pass through a narrow slit and then made to fall on a condenser lens. The condensed light is then made to illuminate the object to be seen magnified. An objective lens gives the first magnified image of the object. This image is again magnified by a projection lens. the final magnified image is either seen directly or is obtained on a photographic plate for permanent record.

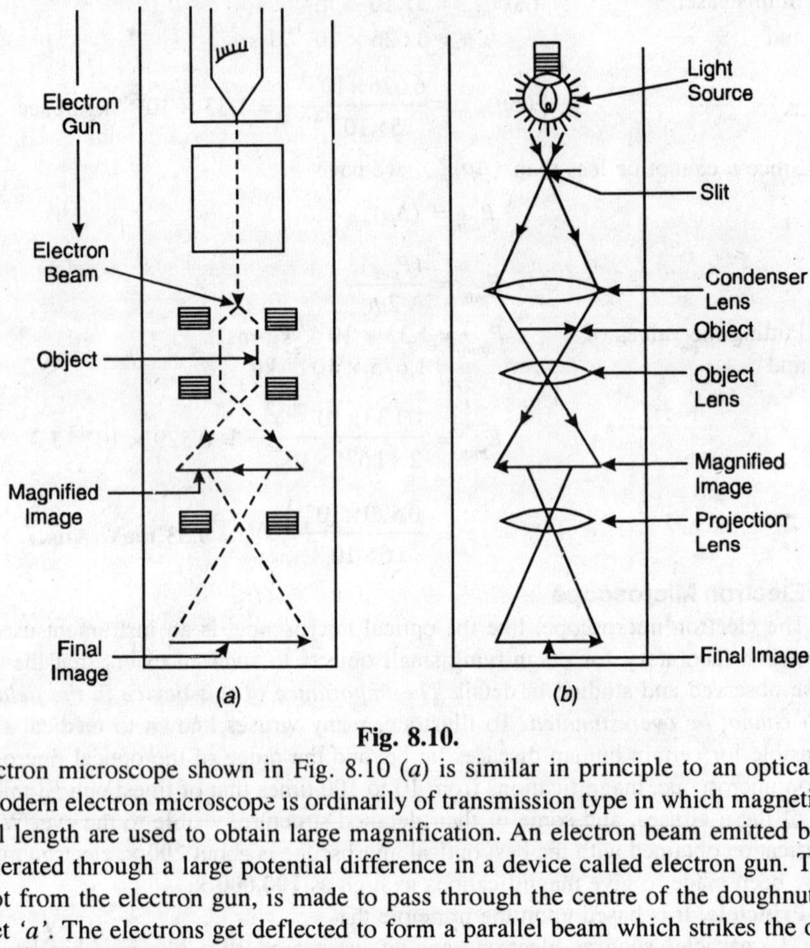

Fig. 8.10.

An electron microscope shown in Fig. 8.10 (*a*) is similar in principle to an optical microscope. The modern electron microscope is ordinarily of transmission type in which magnetic lenses of short focal length are used to obtain large magnification. An electron beam emitted by a filament is accelerated through a large potential difference in a device called electron gun. The electron beam got from the electron gun, is made to pass through the centre of the doughnut shaped electromagnet '*a*'. The electrons get deflected to form a parallel beam which strikes the object to be magnified. It should be noted that the electrons will be transmitted more through the transparent parts of the object and transmitted in less number through comparatively denser portions. The transmitted beam will thus have the likeness of the object transversed by it. The second electromagnet '*b*' causes the electron beam to diverge to produce enlarged image of the object. The electromagnet '*c*' focusses the electron beam from part of the enlarged image on the fluorescent screen producing still greater magnification. The image obtained on the fluorescent screen is made visible by scintillation for direct view. It can also be obtained on a suitable photographic plate for a permanent record. Sharp focussing is obtained by adjusting the intensity of magnetic fields produced by electromagnets.

The object can be given fine micro-movement to bring different parts of object under the incident beam for detailed examination. Since the electron beam operates in vacuum, the apparatus is mounted in a glass tube, completely evacuated. Accelerating voltage of the order of 3,00,000 volt are usually employed for the examination of thicker objects. With the electron microscope, it has been possible to magnify the objects 100,000 times with adequate resolving power.

Uses. The resolving power of an optical microscope is limited by the wavelength of visible light and the smallest size of the object that can be examined with it is of the order of 2000 Å. The electron microscope has very large magnifying power and with the help of the best type of instrument objects of the order 50 Å in size have been made visible. An electron microscope is put to valuable uses in all those fields where the optical microscope has shown the way, e.g., in biology, medicine, physics, chemistry, metallurgy and so on.

1. In medicine and biology, it is used to study *virus,* the disease causing agent beyond the reach of ordinary microscope. The bacteria are shown in greater detail. So that suitable means may be used for their distruction.
2. It has been used in the investigation of atomic structure and structure of crystals in details.
3. Small particles forming colloids have become open to study and analysis.
4. It has also been used in the study of the sturcture of the textile fibres, purification of lubricating oils, composition of paper and paints, surface of metals and plastics.

Example 23. *Calculate the de-Broglie wavelength of neutron of energy 28.8 eV. Given $h = 6.62 \times 10^{-34}$, $m = 1.67 \times 10^{-27}$ kg.*

Solution. The rest mass energy of neutron is given by

$$m_n c^2 = 1.67 \times 10^{-27} \times (3.0 \times 10^8)^2 = 1.503 \times 10^{-10} \text{ joule}$$

or
$$m_n c^2 = \frac{1.503 \times 10^{-10}}{1.6 \times 10^{-19}} \text{eV} = \mathbf{939.4\ MeV\ \ Ans.}$$

as the K.E. of the neutron *i.e.* 28.8 e.V is very small compared to its rest mass energy, So we may ignore the relativistic mass.

Hence de-Broglie's wavelength for non-relativistic particle of mass m_0 may be related to the wavelength as.

$$\lambda = \frac{h}{\sqrt{2m_0 E}}$$

In this case
$$E = 28.8 \text{ eV} = 28.8 \times 1.6 \times 10^{-19} \text{ joule}$$

$\therefore$
$$\lambda = \frac{6.62 \times 10^{-34}}{\sqrt{2 \times 1.67 \times 10^{-27} \times 28.8 \times 1.6 \times 10^{-19}}}$$

$$= 5.336 \times 10^{-12} \text{ m} = \mathbf{0.05336\ Å\ \ Ans.}$$

Example 24. *Calculate the de-Broglie wavelength associated with a proton moving with a velocity equal to 1/20 velocity of light.*

Solution. We have the relation.

$$\lambda = \frac{h}{mv}$$

In this case $h = 6.62 \times 10^{-34}$ J-sec, $mp = 1.67 \times 10^{-27}$ kg, $v = c/20 = \frac{3 \times 10^8}{20} \text{m/sec}$

$\therefore$
$$\lambda = \frac{6.62 \times 10^{-34} \times 20}{1.67 \times 10^{-27} \times 3 \times 10^8} = \mathbf{2.643 \times 10^{-14}\ m\ \ \ Ans.}$$

Example 25. *Can a photon and an electron of the same momentum have the same wavelengths ? Compare their wavelengths if the two have the same energy.*

Solution. From the de-Broglie concept of matter wave the momentum of the electron may be written as

$$p_e = \frac{h}{\lambda_e} \quad ...(1)$$

and the momentum of photon may be written as $p_{ph} = \frac{h\nu}{c} = \frac{h}{\lambda_p}$ ...(2)

So, if the electron and photon have same momentum, then we have

$$\lambda_e = \lambda_p$$

So, the photon and electron of the same momentum have the same wavelength.

The de-Broglie wavelength of electron is given by

$$\lambda_e = \frac{h}{mv}, But \frac{1}{2}mv^2 = E \quad \text{or} \quad mv = \sqrt{2mE}$$

where E is the energy of the electron

So $$\lambda_e = \frac{h}{\sqrt{2mE}} \quad ...(3)$$

The de Broglie wavelength of photon is given by

$$\lambda_{ph} = \frac{h}{p}$$

but $$E = h\nu = \frac{hc}{\lambda} = pc$$

$\therefore$ $$\lambda_{ph} = \frac{hc}{E} \quad ...(4)$$

Divide eq. (4) by eq. (3)

$$\frac{\lambda_{ph}}{\lambda_e} = \frac{hc}{E}.\frac{\sqrt{2mE}}{h} = c\sqrt{\frac{2m}{E}} = \sqrt{\frac{2mc^2}{E}} \quad \textbf{Ans.}$$

Example 26. *What will be the kinetic energy of an electron if its de-Broglie wavelength equals the wavelength of the yellow line of sodium (5896Å). The rest mass of electron is $m_o = 9.1 \times 10^{-31}$ kg and $h = 6.63 \times 10^{-34}$ joule-sec.*

Solution. The de Broglie0's wavelength of a wave associated with a material particle of mass m moving with a velocity v may be written as

$$\lambda = \frac{h}{mv} \quad \text{or} \quad v = \frac{h}{m\lambda} \quad ...(1)$$

If we ignore relativistic considerations, $m = m_0$

So $$\lambda = \frac{h}{m_0 v} \quad \text{and K.E. } K = \frac{1}{2}m_0 v^2 \quad ...(2)$$

Putting the value of v from equation (1) in equation (2), we obtain.

$$K = \frac{1}{2}m_0\left(\frac{h}{m_0\lambda}\right)^2 = \frac{h^2}{2m_0\lambda^2}$$

Now $h = 6.63 \times 10^{-34}$ joule-sec, $m_o = 9.1 \times 10^{31}$ kg and $\lambda = 5896 \times 10^{-10}$ m

$\therefore$ $$K = \frac{(6.63\times10^{-34})^2}{2\times9.1\times10^{-31}\times(5896\times10^{-10})^2} = 6.95 \times 10^{-25} \text{ joule}$$

or $$K = \frac{6.95\times10^{-25}}{1.6\times10^{-19}} = \mathbf{4.34 \times 10^{-6}\ eV\ \ Ans.}$$

Example 27. *Calculate the de Broglie wavelength of a neutron having kinetic energy of 1 eV.*

Solution. The rest mass energy of the neutron is given by

$$m_n c^2 = 1.67 \times 10^{-27} \times (3.0 \times 10^8)^2 = 1.503 \times 10^{-10} \text{ joule}$$

or $$m_n c^2 = \frac{1.503\times10^{-10}}{1.6\times10^{-19}} \text{ eV} = 939.4 \text{ MeV}$$

Since K.E. 1 eV, is very small compared to rest mass energy, so the relativistic consideration may be ignored is the present case.

The de Broglie's wavelength λ for non-relativistic particle of mass m_0 may be written as

$$\lambda = \frac{h}{\sqrt{2m_0E}}$$

In this case $E = 1$ eV $= 1 \times 1.6 \times 10^{-19}$ joule, $m_0 = 1.67 \times 10^{-27}$ kg and $h = 6.63 \times 10^{-34}$ J-sec

$$\therefore \quad \lambda = \frac{6.63\times10^{-34}}{\sqrt{2\times1.67\times10^{-27}\times1.6\times10^{-19}}}$$

$$= 2.87 \times 10^{-11} \text{ m} = \mathbf{0.287\ \text{Å}\ \ Ans.}$$

Example 28. *Show that the de-Broglie wavelength for a material particle of rest mass m_0 and charge q accelerated from rest through a potential difference of V volts relativistically is given by as*

$\lambda = h/\{2m_0\, qv[1 + qV/2m_0c^2]\}^{1/2}$

Solution. We have the De Broglie wavelength as

$$\lambda = \frac{h}{p} \qquad ...(1)$$

Now as the particle has charge q and accelerated to a potential difference V, So its kinetic energy, $E_k = qV$. As velocity is relativistic, we find momentum p by using relativistic formula for E a

$$E^2 = p^2c^2 + m_0^2c^4, \qquad E = E_K + m_0c^2 = qV + m_0c^2$$

$$\therefore \quad p^2c^2 = E^2 - m_0^2c^4 = (qV + m_0c^2)^2 - m_0^2c^4$$

$$= q^2V^2 + m_0^2c^4 + 2m_0c^2qV - m_0^2c^4$$

$$= q^2V^2 + 2m_0c^2qV$$

or $$p^2 = 2m_0qV + \frac{q^2V^2}{c^2} = 2m_0\, qV\left(1 + \frac{qV}{2m_0c^2}\right)$$

or $$p = \sqrt{2m_0\, qV\left(1 + \frac{qV}{2m_0c^2}\right)} \qquad ...(2)$$

Put the value of p from equation (2) in equation (1). we obtain

$$\boldsymbol{\lambda = \frac{h}{\sqrt{2m_0\, qV\left(1 + \frac{qV}{2m_0c^2}\right)}}} \quad \textbf{Ans.}$$

Example 29. *An electron has a speed of 1.05 × 10⁴ m/sec within the accuracy of 0.01%. Calculate the uncertainty in the position of the electron.*

Solution. From Heisenberg uncertainty principle, $\Delta x.\ \Delta p \geq \hbar$

Now momentum of the electron, $p = mv = 9.0 \times 10^{-31} \times 1.05 \times 10^4 = 9.45 \times 10^{-27}$ kg-m/sec

In this case uncertainty in the value of p is 0.01% of this value, which means.

$$\Delta p = \frac{0.01}{100} \times 9.45 \times 10^{-27} = 9.45 \times 10^{-31}$$

∴ The uncertainty in the position of this electron may be written as

$$\Delta x = \frac{h}{2\pi\ \Delta p} = \frac{6.62 \times 10^{-34}}{2 \times 3.14 \times 9.45 \times 10^{-31}} = \mathbf{1.115 \times 10^4\ m}\ \textbf{Ans.}$$

Example 30. *An electron is confined to a box of length 1.1×10^{-8} m. Calculate the minimum uncertainty in its velocity. Given $m = 9.1 \times 10^{-31}$ kg and $h = 1.05 \times 10^{-34}$ J-sec.*

Solution. From Heisenberg's uncertainty principle, we have.

$$\Delta x . \Delta p \geq h \text{ or } \qquad \Delta p \simeq h / \Delta x$$

Let Δv be the uncertainty in the velocity of a particle of mass m. so we have.

$$\Delta p = m\ \Delta v \qquad \text{or} \qquad \Delta v = \Delta p / m$$

Now $\Delta x = 1.1 \times 10^{-8}$ m, $\hbar = 1.05 \times 10^{-34}$ J-sec and $m = 9.0 \times 10^{-31}$ kg

Hence $$\Delta v = \frac{\hbar}{m\ \Delta x} = \frac{1.05 \times 10^{-34}}{9.0 \times 10^{-31} \times 1.1 \times 10^{-8}} = \mathbf{1.06 \times 10^4\ m/sec}\ \textbf{Ans.}$$

Example. 31. *Calculate the smallest possible uncertainty in the position of an electron moving with velocity $v = 3 \times 10^7$ m/sec.*

Solution. Let Δx be the smallest possible uncertainty in the position of an electron and Δp the maximum uncertainty in the momentum of the electron. From principle of uncertainty, we have

$$(\Delta p)(\Delta p) = \hbar \qquad \text{or} \qquad \Delta x = (\hbar / \Delta p) \qquad ...(1)$$

In this case. $$(\Delta p) = p = mv = \frac{m_0 v}{\sqrt{1 - v^2 / c^2}}$$

Put this value of Δp in equation (1), we have.

$$\Delta x = \frac{h}{2\pi m_0\ v / \sqrt{(1 - v^2 / c^2}} = \frac{h\sqrt{(1 - v^2 / c^2}}{2\pi m_0\ v}$$

Now $h = 6.62 \times 10^{-34}$ J-sec, $v = 3.0 \times 10^7$ m/sec, $m_0 = 9 \times 10^{-31}$ kg and $c = 3 \times 10^8$ m/sec

$$\therefore \quad \Delta x = \frac{6.62 \times 10^{-34} \times 0.995}{2 \times 3.14 \times 9.0 \times 10^{-31} \times 3.0 \times 10^7} = 3.88 \times 10^{-12}\ \text{m} = \mathbf{0.0388\ Å}\ \textbf{Ans.}$$

Example 32. *The speed of an electron is measured to be 5.00×10^3 m/sec to an accuracy of 0.003%. Find the uncertainty in determining the position of this electron.*

Solution. From Heisenberg uncertainty principle $\Delta x . \Delta p \geq \hbar$

The momentum of the electron, $p = mv = 9.0 \times 10^{-31} \times 5.00 \times 10^3 = 4.5 \times 10^{-27}$ kg-m/sec

The uncertainty in the value of p is 0.003% of this value, which means

$$\Delta p = \frac{0.003}{100} \times 4.5 \times 10^{-27} = 1.35 \times 10^{-31} \text{ kg-m/sec}$$

So, the uncertainty in the position of this electron is may be written as

$$\Delta x = \frac{h}{2\pi \Delta p} = \frac{6.63 \times 10^{-34}}{2 \times 3.14 \times 1.35 \times 10^{-31}} = \mathbf{7.82 \times 10^4 \text{ m. Ans.}}$$

Example. 33. *What is the minimum uncertainty in the frequency of a photon whose life time is about 10^{-8} sec ?*

Solution. From Heisenberg uncertainty principle, we have.

$$\Delta E . \Delta t \geq \hbar, \quad \text{But} \quad E = h\upsilon \text{ or } \Delta E = h\Delta\upsilon,$$

$$h\Delta\upsilon . \Delta t \geq \hbar \quad \text{or} \quad \Delta\upsilon \geq \frac{1}{2\pi \Delta t}$$

In this case $\Delta t = 10^{-8}$ sec

$$\therefore \quad \Delta\upsilon \simeq \frac{1}{2 \times 3.14 \times 10^{-8}} = \mathbf{15.92 \times 10^6 \text{ sec}^{-1} \text{ Ans.}}$$

Example 34. *A certain excited state of hydrogen atom is known to have a life time of 2.5×10^{-14} sec, what is the minimum error with which the energy of this state can be measured ? ($h = 6.63 \times 10^{-34}$ joule sec)*

Solution. The excited state of energy can not be measured accurately as the atom stays in the excited state for a very short times *i.e.,* 10^{-14} sec, in case it has to come to ground state by emitting a photon. From energy conservation principle the uncertainty of energy to the photon is equal to the uncertainty in energy of excited state.

Form Heisenberg's uncertainty principle $\Delta E . \Delta t \geq \hbar$

where Δt is the mean life time of the excited state which may he write as

$$\Delta E \geq \frac{\hbar}{2\pi \Delta t} \equiv \frac{6.63 \times 10^{-34}}{2 \times 3.14 \times (2.5 \times 10^{-14})} \equiv 4.22 \times 10^{-21} \text{ joule}$$

$$\text{or} \quad \equiv \frac{4.22 \times 10^{-21}}{1.6 \times 10^{-19}} \equiv \mathbf{0.026 \text{ eV} \quad \text{Ans.}}$$

Example 35. *Find the uncertainty in the momentum of a particle when its position is determined within 0.01 cm. find also the uncertainty in the velocity of an electron when it is located with in 5×10^{-8} cm.*

Solution. Let Δx be the uncertainty in the position and Δp in the momentum of a particle, then from Heisenberg's uncertainty principle, we have

$$\Delta x . \Delta p \geq \hbar \quad \text{or} \quad \Delta p \approx \frac{\hbar}{\Delta x} \approx \frac{h}{2\pi \Delta x}$$

In this case $h = 6.62 \times 10^{-34}$ J-sec and $\Delta x = 0.01$ cm $= 0.01 \times 10^{-2}$ m

$$\Delta p = \frac{6.62 \times 10^{-34}}{2 \times 3.14 \times 0.01 \times 10^{-2}} = \mathbf{1.054 \times 10^{-30} \text{ Kg-m/sec}}$$

Let Δv be the uncertainty in the velocity of a particle of mass m, so we have

$$\Delta p = m\Delta v = \frac{h}{2\pi \Delta x} \quad \text{or} \quad \Delta v = \frac{h}{2\pi m \Delta x}$$

For electron, $m = 9.0 \times 10^{-31}$ kg and $\Delta x = 5 \times 10^{-8}\ cm = 5 \times 10^{-10}\ m$

$$\Delta\nu = \frac{6.62\times10^{-34}}{2\times3.14\times9.0\times10^{-31}\times5\times10^{-10}} = \mathbf{2.34\times10^5\ m/sec}\quad \textbf{Ans.}$$

QUESTIONS

1. Explain de Broglie's concept of matter wave. Find the de Broglie wavelength of an electron of energy V electron volts.
2. Find the de Broglie wavelength of an electron of energy VeV.
3. Give de-Broglie theory of matter waves. Derive an expression for the de Broglie wavelength of matter waves. Show that electrons accelerated through a potential difference of V volt have a wavelength $12.27/\sqrt{V}$ Å associated with them.
4. A particle of charge q and mass m is accelerated from rest through a potential difference V. Find it, de-Broglie wavelength.
5. What are de-Broglie matter waves ?
6. Discuss the dual nature of matter and Waves.
7. What are de Broglie's matter waves ?
8. Describe Division-Germer experiment to demonstrate the wave nature of particle.
9. What are de Broglie waves ? How do they help in the interpretation of Bohr's quantisation rule ?
10. Deduce an expression for de Broglie wavelength of helium atom having energy at temperature $T°$ K.
11. Distinguish between the phase velocity (V_p) and the group velocity (V_g) of a wave paper. Prove that $V_pV_g = c^2$
12. Show that the phase velocity of de-Broglie wave is greater than the velocity of light, but the group velocity is equal to the velocity of particle with which the waves are associated.
13. State and explain Heisenberg's uncertainty principle ? Using this principle show that electron cannot reside in an atomic nucleus.
14. What is uncrtainity principle ? Apply it to prove the non-existence of the electron in the nucleus.
15. State and discuss Heisenberg's uncertainty principle. Give two examples to illustrate the principle.

NUMERICAL PROBLEMS

1. Calculate the de broglie wavelength of an α-particle (mass = 6.576×10^{-27} kg and charge = 3.2×10^{-19} coulombs) accelerated through 2000 volts. **[Ans.** 2.36×10^{-13} m]
2. Calculate the de Broglie wavelength of an electron whose kinetic energy is 50 eV. ($h = 6.62 \times 10^{-34}$ joule-sec, $m_0 = 9.1 \times 10^{-31}$ kg, 1 eV = 1.6×10^{-19} joule). **[Ans.** 1.73Å]
3. Compute the de broglie wavelength of a proton whose kinetic energy is equal to the rest energy of an electron. Mass of the proton is 1836 times that of the electron. **[Ans. 0.0004Å]**
4. An electron is accelerated between two points of potentials 50 volt and 70 volt from rest. Calculate the kinetic energy, velocity and de Broglie wavelength for the electron **[Ans.3.2 × 10^{-18} J, 2.6 × 10^{6} ms^{-1} 2.7 Å]**
5. What is the uncertainty in the momentum of an electron which is restricted to a region of linear dimension equal to 1 Å **[Ans.**1.06 × 10^{-24} kg $-ms^{-1}$]
6. What is the minimum uncertainty in the energy state of an atom, if an electron remains in the state for about 10^{-8} sec ? ($h = 6.6\ z \times 10^{-34}$ joule-sec, 1 eV = 1.6×10^{-19} joule) **[Ans.** 6.6 ×10^{-8} *e.v.*]
7. The accuracy with which the frequency of the atomic radiation can be determined by an amount $\Delta\upsilon \sim 1.6\times10^7$ cycles /sec. What percentage of the frequency of a 5000 Å photon is this ? **[Ans.** 2.67 × 10^{-6} %]
8. A beam of silver atoms for Stern-Gerlach experiment is produced by collimating atoms that vaporize from a silver held in a furnace at 1200 °C. If the beam travels 1 m, use the uncertainty relation to find the order of magnitude of the smallest spot that can be obtained at the detector. [h (Planck' constant)

= 6.5×10^{-34} joule-sec, k (Boltzmann constant) 1.4×10^{-23} joule-deg^{-1}, mass of hydrogen atom = 1.66×10^{-27} kg] **[Ans.** 7.07×10^{-7} m]

9. Calculate the time required for the atomic system to retain the excitation energy for a line of wave-length 6000 Å and width 10^{-4} Å **[Ans.** 1.9×10^{-8} sec]

10. An electron has a speed of 4×10^7 cm\sec accurate to 0.01%. With what accuracy can we locate the position of the electron ? ($h = 6.6 \times 10^{-34}$ joule-sec, $m = 9.1 \times 10^{-31}$ kg). **[Ans.** 2.8×10^{-6} m]

11. What voltage must be applied to an electron microscope to produce electrons of wavelentgth 1.0 Å ? **[Ans.** 150.5 V]

12. Calculate the value of de-Broglie wavelength associated with 27°C. helium atom **[Ans.** 1.26 Å]

13. The life time of an excited state of a nucleus is 10^{-12} sec. What is the uncertainty in energy of γ-ray photon emitted ? ($\hbar = 1.05 \times 10^{-34}$ joule-sec, 1 eV = 1.6×10^{-19} joule **[Ans.** 6.6×10^{-4} eV]

14. What is the energy of gamma photon having a wavelength of 1 Å. given Planck's constant, $h = 6.62 \times 10^{-34}$ joule sec. **[Ans.** 1.24×10^4 eV]

15. Show that if a component of angular momentum of an electron in a hydrogen atom in known to be $2\hbar$ within 5% error, its angular orbital position in a plane perpendicular to that component can not be specified at all. **[Ans.** 10 Radians]

9

QUANTUM THEORY

9.1 Newtonian Mechanics and Quantum Mechanics

The fundamental difference between *Newtonian mechanics and quantum mechanics* lies in what it is that they describe. *Newtonian mechanics* is concerned with the motion of particle under the influence of applied forces, and it takes for granted that such quantities as the particle's *position, mass, velocity and acceleration can be measured*. This assumption is, of course, completely valid in our everyday experience, and *newtonian mechanics* provides the "correct" explanation for the behaviour of moving bodies in the sense that the values it predict for observable magnitudes agree with the measured values of those magnitudes.

Quantum Mechanics, too, consists of relationships between observable magnitudes, but the *uncertainty principle* radically alters the definition of "observable magnitude" in the *atomic realm*. According to the uncertainty principle, the position and momentum of a particle cannot be accurately measured at the same time, while in newtonian mechanics, *both are assumed to have definite, ascertainable values at every instant*. The quantities whose relationships quantum mechanics explores are *probabilities*. Instead of asserting, for example, that the radius of the electron's orbit in a ground-state hydrogen atom is always exactly 5.3×10^{-11}, quantum mechanics states that this is the most probable radius; if we conduct a suitable experiment, most trials will yield a different value, either larger or smaller, but the value most likely to be found will be 5.3×10^{-11} m.

At first glance Quantum mechanics seems a poor substitute for newtonian mechanics, but closer inspection reveals a striking fact : *Newtonian mechanics is nothing but an approximate version of Quantum mechanics. The certainties proclaimed by newtonian mechanics are illusory,* and their agreement with experiment is a consequence of the fact that macroscopic bodies consist of so many individual atoms that departures from average behaviour are *unnoticeable*. Instead of two sets of physical principles, one for the macroscopic universe and one for the microscopic universe, there is only a single set and *quantum mechanics represents our best effort to date in formulating it.*

9.2. The Wave Function (Born interpretation)

The quantity with which quantum mechanics is concerned is the wave function Ψ of a particle. While Ψ itself has no physical interpretation the square of its absolute magnitude $|\Psi|^2$ (or $\Psi^* \Psi$ if Ψ is complex) evaluated at a particular point at a particular time is proportional to the *probability* of experimentally finding the particle there at that time. The problem of quantum mechanics is to determine Ψ for a particle when its freedom of motion is limited by the action of external forces.

Even before we consider the actual calculation of ψ, we can establish certain requirements it must always fulfill. For one thing, since $|\Psi|^2$ evaluated at a point is proportional to the probability P of finding the particle described by Ψ at that point, the integral of $|\Psi|^2$ over all space must be finite-the particle is somewhere, after all. If

$$\int_{-x}^{x} |\psi|^2 \, dV \qquad ...(1)$$

is 0, the particle does not exist, and if it is ∞, the particle is everywhere simultaneously $|\psi|^2$ cannot be negative or complex because of the way it is defined, and so the only possibility left is that its *integral be a finite quantity if* ψ *is to describe properly a real body.*

It is usually convenient to have $|\Psi|^2$ be *equal* to the probability P of finding the particle described by Ψ, rather than merely be proportional to p. If $|\Psi|^2$ is equal to P, then it must be true that

$$\int_{-x}^{x} |\psi|^2 \, dV = 1$$

$$\int_{-x}^{x} P dV = 1 \qquad ...(2)$$

is the mathematical statement that the particle exists somewhere at all times. A wave function that obeys eq. {2} is said to be *normalized.* Every acceptable wave function can be normalized by multiplying by an appropriate constant.

Besides being normalizable, Ψ must be single-valued, since P can have only one value at a particular place and time. A further condition that Ψ must obey is that it and its partial derivatives $\partial\Psi/\partial x$, $\partial\Psi/\partial y$, $\partial\Psi/\partial z$ be continuous everywhere.

Schrodinger`s *equation*, which is fundamental equation of quantum mechanics in the same sense that *the second law of motion* is the fundamental equation of *newtonian mechanics* in a wave equation in the variable Ψ.

9.3. Schrödinger's Wave Equation for a Free Particle

We now want to establish some of the mathematical properties of the wave describing a quantum-mechanical particle, such as an electron. In this case we will assume that the quantum-mechanical particle is free *i.e.* external forces are absent. Furthermore, for the sake of simplicity, we will assume the motion is one-dimensional say, along the *x*-axis.

The wave describing a particle of momentum p and energy E has a wavelength $\lambda = h/p$ and frequency $v = E/h$. For convenience, we define the wave number*.

$$k = \frac{2\pi}{\lambda} = \frac{p}{h} \qquad ...(1)$$

and the angular frequency

$$\omega = 2\pi v = \frac{E}{\hbar} \qquad ...(2)$$

Since for our free particle $E = p^2/(2m)$,we can also write Eq.(1)as

$$k = \frac{\sqrt{2mE}}{\hbar} \qquad ...(3)$$

If the particle travels in the $+x$ direction, the wave will also have to travel in that direction. The possible harmonic wave satisfying this condition are

$$\sin(\omega t - kx) \qquad \cos(\omega t - kx) \qquad ...(4)$$

$$e^{-i(\omega t-kx)} \qquad e^{i(\omega t-kx)} \qquad ...(5)$$

The waves (4) are *real* and the waves (5) are *complex*. According to the mathematical rules for evaluating complex exponentials we have the identities

* In three dimensions we would define the wave vector as $k = p/\hbar$

$$e^{ix} = \cos x + i \sin x \quad ...(6)$$

and $$\cos x = \frac{e^{ix} + e^{-ix}}{2}, \qquad \sin x = \frac{e^{ix} + e^{-ix}}{2i} \quad ...(7)$$

Thus $$e^{-i(\omega t - kx)} = \cos(\omega t - kx) - i \sin(\omega t - kx) \quad ...(8)$$

and $$e^{+i(\omega t - kx)} = \cos(\omega t - kx) - i \sin(\omega t - kx) \quad ...(9)$$

This shows that the complex waves are linear combinations of the real waves, with a complex coefficient. To settle which of the waves (4) or (5) we want, let us consider the superposition principle of quantum mechanics : whenever ψ_1 and ψ_2 are possible quantum mechanical waves, then so is $\psi_1 + \psi_2$. Suppose we try to describe a particle travelling in the positive x-direction by the wave $\sin(\omega t - kx)$, then a particle travelling in the negative x-direction will be described by $\sin(\omega t + kx)$. According to the Superposition Principle, the sum

$$\sin(\omega t - kx) + \sin(\omega t + kx) \quad ...(10)$$

should then be a possible quantum-mechanical wave function. But eq.(10) equals

$$2 \sin \omega t \cos kx \quad ...(11)$$

Obviously this is not a satisfactory wave function because at $t = 0$ it vanishes identically everywhere, *i.e., the probability for finding the particle is zero everywhere.* Since the particle cannot disappear, we must regard the wave function (10) as *unacceptable* ,and we conclude that $\sin(\omega t - kx)$ is not the wave function we want. By a similar argument, we rule out $\cos(\omega t - kx)$.

Let us now consider $e^{-i(\omega t - kx)}$. The superposition of this with $e^{-i(wt + kx)}$ gives

$$e^{-i(\omega t - kx)} + e^{-(\omega t + kx)} = 2e^{-\omega t} \cos kx$$

This never vanishes identically everywhere, and is an acceptable wave function. Thus, the wave function $e^{-i(\omega t - kx)}$ is consistent with the Superposition Principle. A wave function of the type $e^{i(\omega t - kx)}$ is also consistent with the Superposition Principle ; *but according to convention laid down by the founders of the mechanics*, this wave function is to be discarded in favour of $e^{-i(\omega t - kx)}$*.

Thus, we are led to the conclusion that the wave function is

$$\psi(x, t) = e^{-i(\omega t - kx)} \quad \text{for motion in the positive } x\text{-direction} \quad ...(13)$$

and $$\psi(x, t) = e^{-i(\omega t + kx)} \quad \text{for motion in the negative } x\text{-direction} \quad ...(14)$$

Next, we want to discover the **wave equation** obeyed by these wave functions. In general, the wave function describing any physical system obeys a wave equation, which permits us to calculate the time evaluation of the system. Mathematically, the wave equation is differential equation involving ψ and the derivatives of ψ with respect to x and t. To discover this differential equation for ψ, we begin by expressing ω and k in terms of p:

$$\psi = e^{-i(\omega t - kx)} = e^{-i\left(\frac{Et}{\hbar} - \frac{\sqrt{2mEx}}{\hbar}\right)} \quad ...(15)$$

If we take the first derivatives of ψ with respect to time and the second derivatives of ψ with respect to x, we find

$$\frac{\partial \psi}{\partial t} = -i\frac{E}{\hbar}\psi \quad ...(16)$$

*We have to chose one or the other of these. Any attempt at using both leads to trouble since the superposition $e^{-1(\omega t - kx)} + e^{i(\omega t - kx)}$ equals $2\cos(\omega t - kx)$, which is an unacceptable wave function.

and $$\frac{\partial^2\psi}{\partial x^2} = -\frac{2mE}{\hbar^2}\psi \qquad ...(17)$$

Comparing these expressions, we see that these derivatives are proportional. By adjusting the constant of proportionality, we obtain an equality :

$$-\frac{\hbar^2}{2m}\frac{\partial^2}{\partial x^2}\psi(x,t) = i\ \hbar\frac{\partial}{\partial t}\psi(x,t) \qquad ...(18)$$

This is the wave equation for a free particle: it is called **Schrödinger's equation.** Next we will discover how to include forces - or more precisely, potential-in Schrödinger's equation. This equation plays the *same role in quantum mechanics as Newton`s Second Law does in classical mechanics.* From Schrödinger's equation we can calculate the motion of a quantum-mechanical wave just as from Newton`s Law we can calculate the motion of a classical particle.

9.4. Potential in Schrödinger's Equation

(*i*) Schrödinger assumed that de Broglie wavelength holds good for any particle moving in any field of force with potential energy *V*. The sum of the potential energy and potential energy is given by

$$E = \text{kinetic energy + potential energy (K.E. + P.E.)}$$

$$E = \frac{1}{2}mv^2 + V$$

$$E = \frac{p^2}{2m} + V \qquad ...(1)$$

$$\therefore \qquad p^2 = 2m\,[E - V]$$

$$p = [2m\,(E - V)]^{1/2}$$

So de Broglie wavelength may be given by

$$\lambda = \frac{h}{p} = \frac{h}{[2m\,(E-V)]^{1/2}} \qquad ...(2)$$

For large value of V, λ is large and for a small value of V, λ is also small.

(*ii*) It is assumed that the wave function ψ is governed by the equation

$$\psi = \psi_0\, e^{-i\omega t}$$

The wave function ψ is also called **probability amplitude** for the position of the particle. The wave equation in **Cartesian coordinates** may be written as

$$\frac{\partial^2\psi}{\partial x^2} + \frac{\partial^2\psi}{\partial y^2} + \frac{\partial^2\psi}{\partial z^2} = \frac{1}{u^2}\frac{\partial^2\psi}{\partial t^2} \qquad ...(3)$$

where $$\psi = \psi_0\, e^{-i\omega t} \qquad ...(4)$$

and ψ is function of (x, y, z and t) and represents the amplitude of the wave associated with the particle and u is the velocity of motion.

Differentiating equation (4) twice w.r.t. t, we get

$$\frac{\partial^2\psi}{\partial t^2} = -\omega^2\psi_0 e^{-i\omega t} = -\omega^2\psi$$

Putting in (3), we get

$$\frac{\partial^2\psi}{\partial x^2} + \frac{\partial^2\psi}{\partial y^2} + \frac{\partial^2\psi}{\partial z^2} = -\left(\frac{\omega^2}{u^2}\right)\psi$$

Now $$\frac{\omega^2}{u^2} = \frac{(2\pi v)^2}{(v\lambda)^2} = \frac{4\pi^2}{\lambda^2}$$

So we have $$\frac{\partial^2\psi}{\partial x^2} + \frac{\partial^2\psi}{\partial y^2} + \frac{\partial^2\psi}{\partial z^2} = \left(\frac{4\pi^2}{\lambda^2}\right)\psi \qquad ...(5)$$

Now $\frac{\partial^2\psi}{\partial x^2} + \frac{\partial^2\psi}{\partial y^2} + \frac{\partial^2\psi}{\partial z^2} = \nabla^2\psi$, where ∇^2 is a **Laplacian operator.**

∴ Eq. (5) becomes $$\nabla^2\psi + \frac{4\pi^2}{\lambda^2}\psi = 0 \qquad ...(6)$$

Equation (6) is a *general equation and is independent of time.*

Now let us introduce concept of wave mechanics, from de Broglie wavelength, and also from Eq. (2)

$$\lambda = \frac{h}{mv} = \frac{h}{[2\,m(E-V)]^{1/2}} \qquad ...(7)$$

Substituting this value of λ in equation (6), we get

$$\nabla^2\psi + \frac{4\pi^2[2m\,(E-V)]}{h^2}\psi = 0$$

$$\nabla^2\psi + \left(\frac{8\pi^2 m}{h^2}\right)(E-V)\,\psi = 0 \qquad ...(8)$$

This represents Schrödinger *time-independent wave equation.*

Also $\hbar = \frac{h}{2\pi}$, so eq. (8) may be written as

$$\therefore \qquad \nabla^2\psi + \left(\frac{2m}{\hbar^2}\right)(E-V)\,\psi = 0 \qquad ...(9)$$

Equation (9) shows that the wave function ψ is a function of coordiantes, also V is a function of coordinates.

Special Case. In the case of a free particle (*i.e.*, a particle not acted by external force) $V = 0$ and so total energy is kinetic energy only.

$$\therefore \qquad \nabla^2\psi + \left(\frac{2m}{\hbar^2}\right)(E)\,\psi = 0 \qquad ...(10)$$

Equation (10) represents Schrodinger equatiön for a particle in the absence of external field, *i.e.*, **for a free particle.**

9.5. Time Dependent Schrödinger Equation

The wave function may be written as

$$\psi = \psi_0\, e^{-i\omega t} \qquad ...(1)$$

Differentiating w.r.t. time, we get

$$\frac{\partial\psi}{\partial t} = -i\omega\psi_0\, e^{-i\omega t}$$

$$\frac{\partial \psi}{\partial t} = -i(2\pi\nu)\,\psi \qquad ...(2)$$

also

$$E = h\nu$$

∴ putting $\nu = \dfrac{E}{h}$ in (2), we get

$$\frac{\partial \psi}{\partial t} = -\,i\,(2\pi)\left(\frac{E}{h}\right)\psi$$

$$E\psi = \frac{-1}{i}\left(\frac{h}{2\pi}\right)\frac{\partial \psi}{\partial t}$$

But

$$E\psi = \frac{i^2}{i}\left(\frac{h}{2\pi}\right)\frac{\partial \psi}{\partial t} \qquad [\because i^2 = -1]$$

i.e.,

$$E\psi = -i\hbar\frac{\partial \psi}{\partial t} = -i\hbar\frac{\partial}{\partial t}(\psi) \qquad ...(3)$$

Now Schrodinger equation is given by

$$\nabla^2\psi + \left(\frac{2m}{\hbar}\right)(E - V)\psi = 0$$

$$\nabla^2\psi + \left(\frac{2m}{\hbar^2}\right)(E\psi) - \left(\frac{2m}{\hbar^2}\right)V\psi = 0 \qquad ...(4)$$

Substituting the value of $E\psi = i\hbar - \dfrac{\partial \psi}{\partial t}$ from (3) in equation (4), we get

$$\nabla^2\psi + \left(\frac{2m}{\hbar^2}\right)i\hbar\,\frac{\partial \psi}{\partial t} - \frac{2m}{\hbar^2}V\psi = 0$$

$$\nabla^2\psi - \left(\frac{2m}{\hbar^2}\right)V\psi = -\left(\frac{2m}{\hbar^2}\right)i\hbar\frac{\partial \psi}{\partial t}$$

or

$$-\left(\frac{\hbar^2}{2m}\right)\nabla^2\,\Psi + V\,\Psi = i\hbar\frac{d\psi}{dt} \qquad ...(5)$$

This equation represents *Schrodinger time dependent* equation because it invloves time Equation (5) can be written as

$$\left(\frac{\hbar^2}{2m}\,\nabla^2 + V\right)\Psi = i\hbar\frac{\partial \Psi}{\partial t}$$

If we take $\left(-\dfrac{\hbar^2}{2m}\nabla^2 + V\right)$ as operator H known as Hamiltonian and from equation (3)

$$i\hbar\frac{\partial}{\partial t} = E, \text{ we have}$$

$$H\psi = E\psi \qquad ...(6)$$

Equation (6) represents the motion of a material particle in terms of Hamiltonian operator (non-relativistic particle).

9.6. Normalization of a wave function

As stated earlier, probability density is proportional to the square of magnitude of wave function. So the probability of finding a particle in a small region of volume dv around the (x, y, z) point may be given as

$$dp = N|\Psi|^2 \, dV \quad \text{... (1)}$$

The total probability of finding the particle in whole space in given by

$$P = \int_{-\infty}^{+\infty}\int_{-\infty}^{+\infty}\int_{-\infty}^{+\infty} dp$$

$$= N\int_{-\infty}^{+\infty}\int_{-\infty}^{+\infty}\int_{-\infty}^{+\infty} |\Psi|^2 \, dx.dy.dz \quad \text{... (2)}$$

(Where $dV = dx.\ dy.\ dz.$)

Now if the particle is present in the whole space, then the probability P must be unity i.e.,

$$P = N\int_{-\infty}^{+\infty}\int_{-\infty}^{+\infty}\int_{-\infty}^{+\infty} |\Psi|^2 \, dx.dy.dz = 1 \quad \text{... (3)}$$

Eq. (3) is called normalization condition and value of N can be found from the above equation. If the value of N – 1, then

$$\int_{-\infty}^{+\infty}\int_{-\infty}^{+\infty}\int_{-\infty}^{+\infty} |\Psi|^2 \, dx.dy.dz = 1 \quad \text{... (4)}$$

A wave function is said to be normalized if its normalization constant is unity. If means that probability density instead of being proportional to $|\Psi|^2$ is exactly equal to $|\Psi|^2$ and equation (4) is satisfied.

In case wave function is not normalized, then N will not be unity. However the value of N can be found using the equation (3) and redefining the wave function as

$$\Psi = \sqrt{N}\Psi$$

So that the new wave function ψ is normalized.

This process of finding the normalization constant and then redefining the wave function is called the **normalization of wave function.**

Importance : Normalization of a wave function eusures the presence of the particle in space and so all the calculation made will not become waste as the presence of particle is ensured by normalization of the wave function.

9.7. Orthogonal Wave Functions

If two different wave functions ψ_m and ψ_n both satisfy the Schrodinger wave equation and of these functions are such that the integral $\int_V \psi_m \psi_n \, dV$ or $\int_V \psi_m \psi_n{}^* \, dV$ vanishes over the total volume V, then these are called orthogonal wave functions *i.e.*,

$$\int \psi_m{}^* \psi_n dV = \int \psi_m \psi_n{}^* dV = 0 \quad \text{...(1)}$$

$$\text{for } m \neq n$$

Where the integral in taken over the whole space.

Again if the wave functions are simultaneously normalized, then we may write

$$\int \psi_m \psi_m{}^* dV = \int \psi_n \psi_n{}^* dV = 1 \quad \text{... (2)}$$

The set of wave functions which are both normalized and are also orthogonal are called **ortho-normal wave functions.**

Eqs. (1) and (2) may collectively be written as

$$\int \psi_m{}^* \psi_n dV = \begin{cases} 0 \text{ is } m \neq n \\ 1 \text{ if } m = n \end{cases}$$

As the limit to the integration is not given, it will correspond to the integration over whole space.

Importance

If we have a complete set of orthogonal wave functions, then any other wave function which may give solution to Schrodinger wave function may also be expanded as a set of orthogonal wave functions.

9.8. Time Independent Schrödinger Equation

In time dependent Schrödinger equation the potential energy of a moving particle is both function of time as well as position. In many problems, the potential energy of a particle does not depend explicitly on time. It means that the forces that act upon the particle and thereby the potential energy V changes with position of the particle only. In such cases, time parameter is eliminated from Schrödinger time dependent wave equation and the equation thereby obtained is called *time independent Schrodinger equation.*

We may write the one dimensiönal Schrödinger's time dependent wave equation as

$$\psi = Ae^{-(i/\hbar)(Et-px)}$$

$$= Ae^{-2'\frac{E}{\hbar}.t}\; e^{+(ip/\hbar)x}$$

$$= \psi_0 e^{-\frac{iE}{\hbar}.t} \qquad ...(1)$$

i.e., ψ is the product of a time dependent function $e^{-\frac{iE}{\hbar}.t}$ and a position dependent function ψ. We take up the variations of all function of particles acted upon by stationary forces have the same form as that of an unrestricted particle.

Putting ψ from (1) into the time dependent form of Schrodinger equatiön, we note that

$$E\psi_0 e^{-\frac{iE}{\hbar}.t} = -\frac{\hbar^2}{2m} e^{-\frac{iE}{\hbar}.t} \frac{\partial^2 \psi_0}{\partial x^2} + V\psi_0 e^{-\frac{iE}{\hbar}.t}$$

Dividing by $e^{-\frac{i}{\hbar}E.t}$, We get

$$E\psi_0 = -\frac{\hbar^2}{2m}\frac{\partial^2 \psi_0}{\partial x^2} + V\psi_0$$

or

$$\frac{\partial^2 \psi_0}{\partial x^2} + \frac{2m}{\hbar^2}(E-V)\psi_0 = 0 \qquad ...(2)$$

Equation (2) is the *steady state from of Schrödinger's equation.*

We may write the equation in **three dimensions** as

$$\frac{\partial^2 \psi_0}{\partial x^2} + \frac{\partial^2 \psi_0}{\partial y^2} + \frac{\partial^2 \psi_0}{\partial z^2} + \frac{2m}{\hbar^2}(E-V)\psi_0 = 0$$

9.9. Expectation Values

Dynamic quantities *viz., position, potential energy, kinetic energy, momentum* etc. which form the basis of physical measurements are called *observable* in Physics. In *quantum mechanics* each observable is represented by an *operator* which acts on a *wave function* ψ to give a new wave function.

We define the *average or expectation* value $<\alpha>$ of any observable α as

$$<\alpha> = \frac{\int_{-\infty}^{+\infty} \psi^*(r,t)\,\alpha_{op}\psi(\vec{r},t)\,dV}{\int_{-\infty}^{+\infty} \psi^*(\vec{r},t)\,\psi(r,t)\,dV}$$

From Normalization condition

$$\int_{-\infty}^{+\infty} \psi^*(\vec{r},t)\,\psi(\vec{r},t)\,dV = 1$$

$$\therefore \qquad <\alpha> = \int_{-\infty}^{+\infty} \psi^*(\vec{r},t)\,\alpha_{op}\psi(\vec{r},t)\,dV \qquad \text{... (1)}$$

We may give the expectation value of the vector $\vec{r}$

$$<r> = \int_{-\infty}^{+\infty} \psi^*(\vec{r},t)\,\vec{r}\,\psi(\vec{r},t)\,dV \qquad \text{...(2)}$$

The expectation value of position vector is also a vector but it is a function of time only since the space coordinate disappears on integration. Eq. (2) may be written in terms of component equations as

$$<x> = \int_{-\infty}^{+\infty} \psi^* x\,\psi\,dV$$

$$<y> = \int_{-\infty}^{+\infty} \psi^* y\,\psi\,dV \qquad \text{... (3)}$$

$$<z> = \int_{-\infty}^{+\infty} \psi^* z\,\psi\,dV$$

In the same way, we may write the expectation value of potential energy V as

$$<V> = \int_{-\infty}^{+\infty} \psi^*(r,t)\,V(\vec{r},t)\,\psi(r,t)\,dV \qquad \text{...(4)}$$

Some other examples value of momentum is given by the relation

$$<p> = \int_{-\infty}^{+\infty} \psi^*(x,t)\,p\,\psi(r,t)\,dx \qquad \text{...(5)}$$

To solve the integral (5), we have to express $\psi^*(x, t)\, p\, \psi(x, t)$ in terms of variables x ant t.

Classically we may express p as a function of x and t. However, in quantum mechanics, due to uncertainty principle, p *cannot be expressed in terms of* x because p and x cannot be known simultaneously.

In the same way we cannot write p as a function of t. So we have to find some other way to solve the integral.

In order to evaluate $<p>$ and $<E>$, we can differentiate the free particle wave function $\psi = Ae^{-(i/k)(Et-px)}$ *w.r.t.* x and t. Then we get

$$\frac{\partial\psi}{\partial t} = i\hbar\,P\psi \qquad \text{and} \qquad \frac{\partial\psi}{\partial t} = -\frac{i}{\hbar}E\psi.$$

which can be written as

$$P\psi = \frac{\hbar}{i}\frac{\partial}{\partial x}\psi \qquad \text{...(6)}$$

and

$$E\psi = i\hbar\frac{\partial}{\partial t}\psi \qquad \text{...(7)}$$

which shows that the dynamical quantity P in some sense corresponds to *differential operator* $\frac{\hbar}{i}\frac{\partial}{\partial x}$.

Similarly dynamical quantity E corresponds to *differential operator* $t\hbar\frac{\partial}{\partial x}$, where an operator tells us about the operation to be carried out on a dynamical quantity.

For example operator $t\hbar\frac{\partial}{\partial x}$ tells us to take the partial derivation of what comes after it *w.r.t.*'*t*' and multiply the result by $t\hbar$.

Generally the expectation value is written in **Bold face letters**, so that P and E correspond to momentum and Energy operators.

Momentum operator $$\mathbf{P} = \frac{\hbar}{i}\frac{\partial}{\partial x} \qquad ...(8)$$

Energy operator $$\mathbf{E} = i\hbar\frac{\partial}{\partial t} \qquad ...(9)$$

Derivation of Schrödinger equation. Though we have only shown that the correspondences expressed in Eqs. (8) and (9) hold for free particles, they are entirely general results whose validity is the same as that of Schrödinger's equation. To support this statement, we can replace the equation $E = T + V$ for the total energy of a particle with the operator equation

$$E = T + V \qquad ...(10)$$

Writing K.E. T in terms of momentum P, we have

$$T = \frac{p^2}{2m}$$

Which gives

Kinetic energy operator $$T = \frac{p^2}{2m} = \frac{1}{2m}\left(\frac{\hbar}{i}\frac{\partial}{\partial x}\right)^2 = \frac{-\hbar^2}{2m}\frac{\partial}{\partial x^2} \qquad ...(11)$$

So eq. (10) may be written as

$$i\hbar\frac{\partial}{\partial t} = -\frac{\hbar^2}{2m}\frac{\partial^2}{\partial x^2} + V \qquad ...(12)$$

Multiplying the identity $\psi = \psi$ by Eq. (12), we get

$$i\hbar\frac{\partial\psi}{\partial t} = -\frac{\hbar^2}{2m}\frac{\partial^2\psi}{\partial x^2} + V\psi$$

which is Schrodinger's equation. It shows that Eqs. (8) and (9) are equivalent to postulating Schrödinger's equation.

Operator and expectation value. Since p and E can be replaced by their corresponding operators in an equation, we can use these operators to get expectation values for p and E. So the expectation value for p is

$$< p > = \int_{-\infty}^{+\infty} \psi^* \, dx = \int_{-\infty}^{\infty} \psi^*\left(\frac{\hbar}{i}\frac{\partial}{\partial x}\right)\psi \, dx = \frac{\hbar}{i}\int_{-\infty}^{\infty} \psi^*\frac{\partial\psi}{\partial x}dx \qquad ...(13)$$

and the expectation value of E is given by

$$< E > = \int_{-\infty}^{\infty} \psi^* E\,\psi \, dx = \int_{-\infty}^{\infty} \psi^*\left(i\hbar\frac{\partial}{\partial t}\right)\Psi \, dx = i\hbar\int_{-\infty}^{\infty} \psi^*\frac{\partial\Psi}{\partial x}dx \qquad ...(14)$$

Both Eqs. (13) and (14) can be evaluated for any acceptable wave function Ψ (x, t). So it is necessary to expose expectation value involving operators in the form

$$(p) = \int_{-\infty}^{\infty} \Psi^* p\,\Psi\, dx$$

Every observable quantity G characteristic of a physical system may be represented by a suitable quantum-mechanical operator G. To obtain this operator, it is only necessary to express G in terms of x and p and then replace p by $(\hbar / i)\ \partial / \partial x$. If the wave function Ψ of the system is known, the expectation value of $G\ (x, p)$ is

$$<G\,(x, p)> = \int_{-\infty}^{\infty} \Psi^*\, G\, \Psi\, dx \qquad ...(15)$$

This result substantiates the statement made earlier that from Ψ can be obtained all the information about a system that is permitted by the *uncertainty principle.*

Conditions for quantization of a variable. The condition that a certain dynamical variable G be restricted to the discrete values G_n – in other words, that G be quantized–is that the wave functions Ψ_n of the system be such that

$$G\Psi_n = G_n \Psi_n$$

where G is the operator that corresponds to G and each G_n *is a real number*. When Eq. (16) holds for the wave functions of a system, it is a fundamental postulate (in fact, the fundamental postulate) of quantum mechanics that any measurement of G can only yield one of the values of G_n. If measurements of G are made on a number of identical systems all in states described by the particular eigen function Ψ_k, each measurement will yield the single value G_k.

Hamiltonian operator for total energy. The total-energy operator E of Eq. (10) is usually written in the form

$$H = -\frac{\hbar^2}{2m}\frac{\partial^2}{\partial x^2} + V \qquad ...(17)$$

and is called the *Hamiltonian operator* because it is reminiscent of the hamiltonian function in advanced classical mechanics, which is an expression for the total energy of a system in terms of coordinates and momenta only. So the steady-state Schrödinger equation can be written simply as

$$E_n \Psi_n = H\Psi_n \qquad ...(18)$$

so we can say that the various E_n are the eigen values of the Hamiltonian operator H. This kind of association between **eigen values and quantum-mechanical operators** is quite general.

9.10. Application of Schrödinger Equation : Particle in a Box (Motion in one dimension)

Let us consider a free particle in one dimensional box. The particle has a mass m and is restricted to move in a straight line along the x-axis. The range of the particle is $0 < x < a$ and the particle is reflected back whenever it reaches the end of the range. We have the general equation

$$\frac{\partial^2 \psi}{\partial x^2} + \left(\frac{2m}{\hbar^2}\right)(E - V)\Psi = 0 \qquad ...(1)$$

For a free particle $V - 0$ for $0 < x < a$.

Also the wave function ψ vanishes at $x = 0$ and $x = a$. For one dimension, the modified equation is given by

$$\frac{\partial^2 \Psi}{\partial x^2} + \left(\frac{2m}{\hbar^2}\right)E\Psi = 0$$

$$\frac{\partial^2 \Psi}{\partial x^2} + \left(\frac{8\pi^2 mE}{h^2}\right)E\Psi = 0 \qquad \left[\because \hbar = \frac{h}{2\pi}\right] \qquad ...(2)$$

Let us take $\qquad \dfrac{8\pi^2 mE}{h^2} = k^2$

$$\therefore \qquad E = \frac{k^2h^2}{8\pi^2 m}$$

So Eq. (2) may be written as

$$\frac{\partial^2\psi}{\partial x^2} + k^2\psi = 0 \qquad \text{...(3)}$$

The general solution for equation (3) is given by

$$\Psi(x) = A\sin kx + B\cos kx \qquad \text{...(4)}$$

and from the boundary condition

$$\Psi = 0 \text{ at } x = 0$$

So from equation (4), B must be zero

$$\Psi(a) = A\sin kx = 0$$

Since we cannot take A to be zero because it does not yield any solution, so we take

$$\sin ka = 0$$

or

$$ka = n\pi$$

$$k = \frac{n\pi}{a} \quad \text{where} \quad n \equiv 1, 2, 3, \text{ etc.}$$

Hence the only permissible solutions of the wave equations are

$$\Psi_n(x) = A\sin\left(\frac{n\pi}{a}\right)x \qquad \text{...(5)}$$

and the value of energy is given by $E = \dfrac{k^2h^2}{8\,\pi^2 m}$ ...(6)

Putting the value of $k = \dfrac{n\pi}{a}$ in equation (6), we get

$$E_n = \frac{n^2h^2\pi^2}{8\,\pi^2 ma^2} = \frac{n^2h^2}{8\,ma^2} \qquad \text{...(7)}$$

It shows E_n depends upon the value of n. Each value of E_n is called–**Eigen value or proper value** and each of Ψ_n is called an **Eigen function or proper function.**

From normalization condition

$$\int_0^a |\Psi_n(x)|^2\,dx = A^2\int_0^a \sin^2\left(\frac{n\,\pi\,x}{a}\right)dx = 1 \qquad \text{...(8)}$$

or

$$\frac{A^2a}{2} = 1$$

$$\therefore \qquad A = \sqrt{\frac{2}{a}}$$

$$\therefore \qquad \psi_n(x) = \sqrt{\frac{2}{a}}\,\sin\left(\frac{n\pi}{a}\right)x \qquad \text{...(9)}$$

Now for, $n = 1$

$$\psi_1(x) = \sqrt{\frac{2}{a}}\ \sin\left(\frac{n\pi}{a}\right)x$$

for $n = 2$

$$\psi_2(x) = \sqrt{\frac{2}{a}}\ \sin\left(\frac{2\pi}{a}\right)x$$

for $n = 3$

$$\psi_3(x) = \sqrt{\frac{2}{a}}\ \sin\left(\frac{3\pi}{a}\right)x$$

Graphically, the first three functions are shown in Fig. 9.1.

Equation (7) shows that energy E_n of the particle is quantized, we have

$$E_n = \frac{p_n^2}{2m}$$

or $$p_n = (2mE_n)^{1/2} = \left(\frac{2\,m\,n^2 h^2}{8\,m\,a^2}\right)^{1/2}$$

or $$p_n = \frac{nh}{2a} \qquad ...(10)$$

$\therefore$ $$p_n \propto n$$

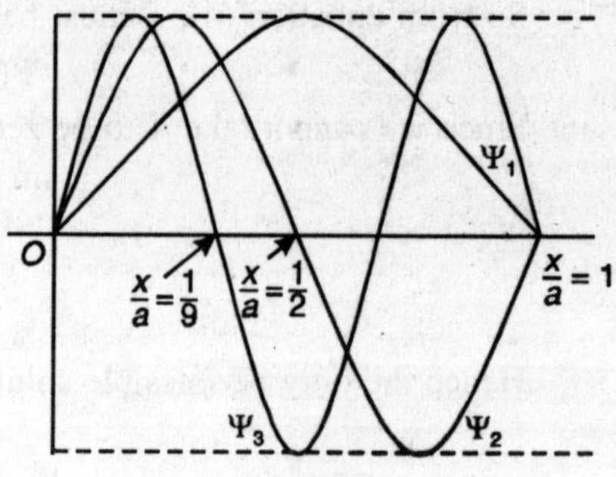

Fig. 9.1

which shows that the momentum of the particle is also quantized. Moreover, for each value of E_n, there is corresponding value of p_n.

From quantum mechanics the probability is given by

$$|\psi_n(x)|^2\,dx = \frac{2}{a}\sin^2\left(\frac{n\,\pi\,x}{a}\right)dx$$

This is wave like function and for $n = 3$ (say)

$$|\psi_n(x)|^2\,dx = \frac{2}{a}\sin^2\left(\frac{3\,\pi\,x}{a}\right)dx$$

The graph between probability and x is given in Fig. 9.2 (*a*).

In the graph the dotted line represents the constant probability according to classical mechanics. But the curve represents the probability variation with variation of x according to quantum mechanics probability is zero for values of

$$x = 0,\ x = \frac{a}{3},\ x = \frac{2a}{3} \text{ and } x = a$$

and so on.

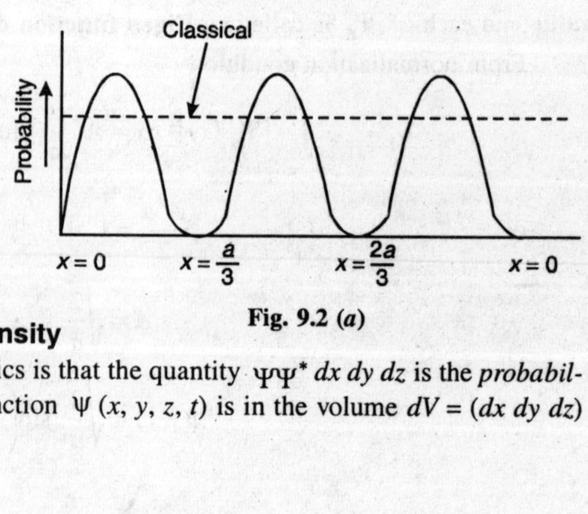

Fig. 9.2 (*a*)

9.11. Wave Function and Probability Density

The basic postulate of quantum mechanics is that the quantity $\psi\psi^*\,dx\,dy\,dz$ is the *probability* that a particle represented by the wave function $\psi\,(x, y, z, t)$ is in the volume $dV = (dx\,dy\,dz)$ at the time t.

$$P(t) = \psi\psi^* = |\psi\ (x, y, z, t)|^2$$

and the wave function Ψ corresponds to the probability amplitude for the position of the particle.

Since the particle must be somewhere in the space, it is necessary that the probability density must satisfy the normalisation condition *i.e.*,

$$\int_{-\infty}^{-\infty}\int_{-\infty}^{+\infty}\int_{-\infty}^{+\infty} \psi\,\psi^* dx\ dy\ dz = 1$$

In this case $\psi = \psi_0\ e^{-i\omega t}$ ...(1)

or $\psi = \psi_0\ e^{-e\,2\pi\,Et}/h \quad \left(\because\ E = h\nu = \dfrac{h\omega}{2\pi}\ \therefore\ \omega = \dfrac{2\,\pi\,E}{h}\right)$

or $\psi = \psi_0\ e^{iEt/\hbar}$ ...(2)

The complex conjugate is given by

or $\psi^* = \psi_0^*\ e^{iEt/\hbar}$ ...(3)

Substituting these value of Ψ and ψ^* in equation (1)

$$\int_{-\infty}^{-\infty}\int_{-\infty}^{+\infty}\int_{-\infty}^{+\infty} \psi_0\,\psi_0^*\ dx\,dy\,dz = 1 \qquad ...(4)$$

So far a wave function satisfying the time independent Schrödinger equation, the probability is also independent of time.

However, ψ has the following limitations:

(*i*) The wave function ψ partial being normalizable must be single valued as $P\ (t)$ can have only one value at a particular place at a particular time.

(*ii*) Wave function ψ and its partial derivation $\dfrac{d\psi}{dx}, \dfrac{d\psi}{dy}$ and $\dfrac{d\psi}{dz}$ must be continuous everywhere.

(*iii*) ψ must be finite for all values of x, y and z.

9.12. Probability Current Density

In order to find the quantum mechanical relation to express the *particle flow*, let us first begin with the classical relation.

Let $\rho(x, y, z, t)$ be the charge density and let the charge be concentrated in a small volume element $\Delta x\ \Delta y\ \Delta z$ [Fig. 9.2 (*b*)]. Let us first consider the case when all the velocities are along x-axis. Net charge per unit time $\Delta x\ \Delta y\ \Delta z$ at left is $\rho\ \nu$ evaluated at $x = a$ times $\Delta y\ \Delta z$, and that leaving right is

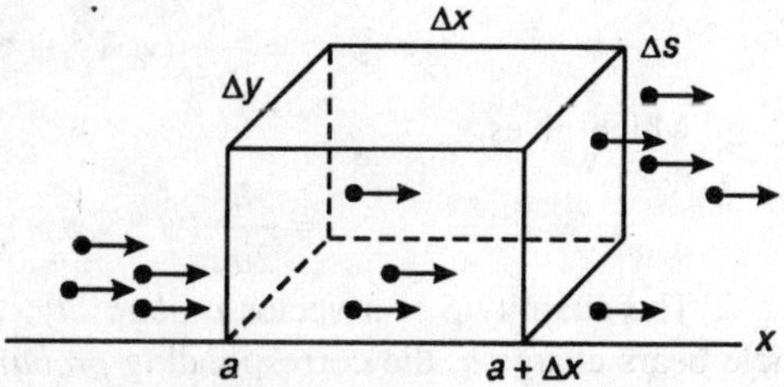

Fig. 9.2 (*b*). Charge entering and leaving a small cubical volume element *Dx Dy Dz* for a one-dimensional flow.

$$(\rho v_x)_{a+\Delta x}\,\Delta y\ \Delta z = [\rho v_x + \partial(\rho v_x)\,/\,\partial x\ \Delta x]\ \Delta y\ \Delta z \qquad ...(1)$$

The charge in $\Delta x\ \Delta y\ \Delta z$ is, or course, $\rho\ \Delta x\ \Delta y\ \Delta z$. If more charge leaves per second than arrives ρ decreases in time, and conservation of charge gives the relation

$$(\rho v_x)_{a+\Delta x}\,\Delta y\,\Delta z = (\rho v_x)_a\,\Delta y\,\Delta z\frac{\partial(\rho v_x)}{\partial x}\,\Delta x\,\Delta y\,\Delta z = -\frac{\partial \rho}{\partial t}\,\Delta x\,\Delta y\,\Delta z \qquad ...(2)$$

which gives

$$\frac{\partial\,(\rho v_x)}{\partial x} = -\frac{\partial \rho}{\partial t} \qquad ...(3)$$

in one dimension. If we extended the case to three dimensions, we get

$$\nabla.\rho v = -\frac{\partial \rho}{\partial t}$$

which is the *equation of continuity*, applicable to any flow process in which ρ *represent the density of something that is conserved.*

In order to find the quantum mechanical analogue of Eq. (2), we replace ρ with $q\ \psi^*\psi$ where q is the charge associated with the particle and $\psi^*\psi\ dx\,dy\,dz$ is the *probability* at time t to locate the particle in volume element $dx\,dy\,dz$ (where we have considered $\int\int\int \psi^*\psi\ dx\,dy\,dz$ over all space to be normalized so as to give total number of particles enclosed). We assume that there is a quantity S which represents the probability stream density of the particles in such a way that qS plays the role of ρv in the equation of continuity. We seek S in terms of ψ by writing the equation of continuity is the form

$$\nabla.qS = \partial\,(q\psi^*\psi)/\partial t$$

or

$$\nabla.S = \frac{\partial}{\partial t}(\psi^*\psi) \qquad ...(5)$$

using the eq. $\frac{\hbar^2}{2m}\nabla^2\psi + v\psi = \frac{\hbar}{i}\frac{\partial\psi}{\partial t}$

$$\nabla.S = \frac{\partial\psi^*}{\partial t}\psi - \psi^*\frac{\partial\psi}{\partial t}$$

$$= \frac{\hbar}{2mi}(\psi\,\nabla^2\,\psi^* - \psi^*\nabla^2\psi) \qquad ...(6)$$

which gives

$$S = \frac{\hbar}{2mi}(\psi^*\nabla\psi - \psi\nabla\psi^*) \qquad ...(7)$$

The quantity S is a vector *probability current density* for the flow of particles; if each particle bears charge q, the corresponding *probability current density is qS*. If A denotes any surface area enclosing a volume v, the integral of the normal component of S over the surface is equal to the time rate of decrease of probability of finding a particle within the volume; so the left-hand integral represents the rate at which probability is streaming outward across the closed surface A, while the right-hand integral represents the rate at which the probability is decreasing within the volume.

When ψ has the form, $\psi = Ae^{i(p_x{}^x - Et)/\hbar}$, then $\psi^* = A^* e^{i(E_t - P_x{}^x)/\hbar}$

so we have,

$$|S| = \frac{p}{m}|A|^2$$

where | S | corresponds to the magnitude of S. So the significance of the beam represented by eq. $\Psi = e^{i(p_x x - Et)\hbar}$ can be based on the vectors $\vec{S}$ by assuming the number of particles crossing/unit area per second is $n = \frac{p|A|^2}{m}$, where $\frac{p}{m}$ corresponds to the **speed of a particle.**

9.13. Particle in a Box (Motion in three dimensions)

Consider a single particle *e.g.*, a glass molecule of mass *m*, confined within a rectangula box with edges parallel to *X*, *Y*, and *Z* axes as shown in Fig. 9.3. Let the sides of rectangular box be *a*, *b* and *c* respectively. The particles can move freely within the region $0 < x < a$, $< 0 < y < b$ and $0 < x < c$, *i.e.*, inside the box where potential *V* is zero *i.e.*,

$$V(x, y, z) = 0, \quad 0 < x < a$$
$$V(x, y, z) = 0, \quad 0 < y < b$$
and
$$V(x, y, z) = 0, \quad 0 < z < c$$

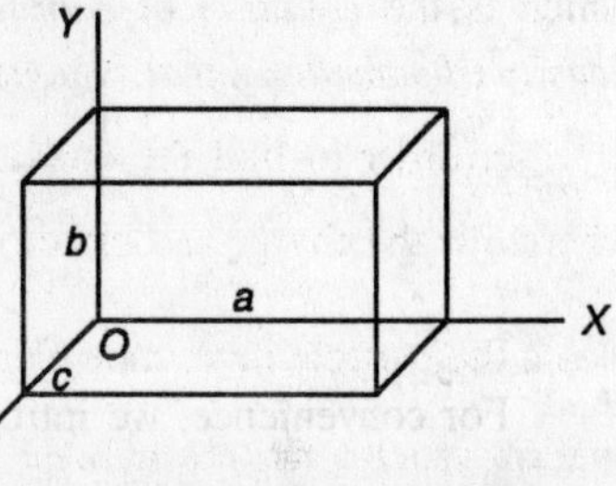

Consider that the potential rises suddenly to have a very large value at the boundaries i.e., the potential outside the box is infinite. We have Schrödinger wave equation inside the box is given by

Fig. 9.3

$$\frac{\partial^2 \psi}{\partial x^2} + \frac{\partial^2 \psi}{\partial y^2} + \frac{\partial^2 \psi}{\partial z^2} + \frac{2m}{\hbar^2} E\psi = 0 \quad \text{...(1)}$$

which is a partial differential equation in three independent variables and may be solved by the method of separation of variables. The solution of Eq. (1) we have the form

$$\psi(x, y, z) = X(x)\, Y(y)\, Z(z) = XYZ \quad \text{...(2)}$$

where $X(x)$ is a function of *x* alone. $X(y)$ is a function of *y* alone and $Z(z)$ is a function of *z* alone.

Putting the value of ψ from Eq. (2) in Eq. (1) and dividing by $X(x)\, Y(y)\, Z(z)$ we have

$$\frac{1}{X}\frac{\partial^2 Y}{\partial x^2} + \frac{1}{Y}\frac{\partial^2 Y}{\partial y^2} + \frac{1}{Z}\frac{\partial^2 Z}{\partial z^2} + \frac{2mE}{\hbar^2} = 0$$

or
$$\frac{1}{X}\frac{\partial^2 Y}{\partial x^2} = -\frac{1}{Y}\frac{\partial^2 Y}{\partial y^2} - \frac{1}{Z}\frac{\partial^2 Z}{\partial z^2} - \frac{2mE}{\hbar^2} \quad \text{...(3)}$$

The left hand side of Eq. (3) is a function of *x* alone, while the right hand side is a function of *y* and *z* and is independent of *x*. Since both sides are equal to each other. So each side should separately be equal to a constant equally, *i.e.*,

$$\frac{1}{X}\frac{\partial^2 Y}{\partial x^2} = k_x \quad \text{... (4)}$$

$$\therefore \quad -\frac{1}{Y}\frac{\partial^2 Y}{\partial y^2} - \frac{1}{Z}\frac{\partial^2 Z}{\partial z^2} - \frac{2mE}{\hbar^2} = k_x$$

or
$$\frac{1}{Y}\frac{\partial^2 Y}{\partial y^2} = -\frac{1}{Z}\frac{\partial^2 Z}{\partial z^2} - \frac{2mE}{\hbar^2} = k_x \quad \text{...(5)}$$

Again in eq. (5) the left hand side is a function of *y* alone while right hand side is a function of *z* and is independent of *y*. If above equation is to be satisfied, both sides must be equal to a constant say k_y *i.e.*,

$$\frac{1}{Y}\frac{\partial^2 Y}{\partial y^2} = k_x \qquad ...(6)$$

$$-\frac{1}{Z}\frac{\partial^2 Z}{\partial z^2} - \frac{2mE}{\partial \hbar^2} - k_x = k_y$$

or
$$\frac{1}{Z}\frac{\partial^2 Z}{\partial z^2} = -\frac{2mE}{\partial \hbar^2} - k_x - k_y \qquad ...(7)$$

Again we may have

$$\frac{1}{Z}\frac{\partial^2 Z}{\partial z^2} = k_z \qquad ...(8)$$

$$-\frac{2\,mE}{\hbar^2} - k_x - k_y = k_z$$

$$-\frac{2\,mE}{\hbar^2} = k_x + k_y + k_z \qquad ...(9)$$

For convenience, we introduce

$$k_z = -\frac{2\,mE_x}{\hbar^2},\ k_y = \frac{2\,mE_y}{\hbar^2}$$

and
$$k_z = -\frac{2\,mE_z}{\hbar^2}$$

and write differential equation in x, y and z coordinates

$$\frac{\partial^2 X}{\partial x^2} + \frac{2\,mE_x}{\hbar^2} X = 0 \qquad ...(10)$$

$$\frac{\partial^2 Y}{\partial y^2} + \frac{2\,mE_y}{\hbar^2} Y = 0 \qquad ...(11)$$

$$\frac{\partial^2 Z}{\partial z^2} + \frac{2\,mE_z}{\hbar^2} Z = 0 \qquad ...(12)$$

The general solution of eq. (10) will be a sine function of arbitrary amplitude frequency and phase, *i.e.*,

$$X(x) = A \sin (Bx + C) \qquad ...(13)$$

where A, B and C are constants whose are determined by boundary condition.

Also $|\psi|^2$ represents the probability of finding the particles at any point within the box. So $|X(x)|^2$ which is a function of x coordinates only represents the probability of finding the particles at any point along the X-axis. As the potential is very high at the walls of the box, the probability of finding the particles at the wall will be zero, *i.e.*,

$$|X(x)|^2 = 0 \text{ when } x = 0 \text{ and } x = a$$

$$X(x) = 0 \text{ when } x = 0 \text{ and } x = a$$

Using these boundary conditions in eq. (13), we get

$$0 = A \sin (0 + C)\ A \neq 0$$

$$\therefore \qquad \sin C = 0 \text{ and } 0 = A \sin (Ba + C)$$

which gives $\sin Ba = 0$ or $Ba = n_x\pi$

or $$B = \frac{n_x\pi}{a}$$

where n_x is a positive integer.

$$\therefore \quad X(x) = A\sin\frac{n_x\pi\, x}{a} \qquad ...(14)$$

Applying the normalization condition between $x = 0$ to $x = a$, we have

$$\int_0^a |X(x)|^2\, dx = 1$$

or $$\int_0^a \left| A\sin\frac{n_x\pi x}{a} dx\right|^2 = 1$$

or $$A^2\int_0^a A\sin^2\frac{n_x\pi x}{a} dx = 1$$

or $$A^2\frac{a}{2} = 1 \text{ or } A = \sqrt{\left(\frac{2}{a}\right)}$$

Therefore, $$X(x) = \sqrt{\left(\frac{2}{a}\right)}\sin\frac{n_x\pi\, x}{a} \qquad ...(15)$$

Similarly, we can solve equations (11) and (12) to get

$$Y(y) = \sqrt{\frac{2}{b}}\sin\frac{n_y\pi\, y}{a} \qquad ...(16)$$

and $$Z(z) = \sqrt{\frac{2}{c}}\sin\frac{n_z\pi\, z}{a} \qquad ...(17)$$

Now the complete wave function $\psi_{nx, y_y, n_z}(x, y, z)$ has the form

$$\psi_{n y, n_y, n_z}(x, y, z) = X(x)\, Y(y)\, Z(z)$$

$$= \sqrt{\left(\frac{2}{a}\right)}\sin\frac{n_x\pi x}{a}\sqrt{\left(\frac{2}{b}\right)}\sin\frac{n_y\pi y}{b}\sqrt{\left(\frac{2}{c}\right)}\sin\frac{n_z\pi z}{c}$$

$$= \frac{2\sqrt{2}}{\sqrt{abc}}\sin\frac{n_x\pi x}{a}\sin\frac{n_y\pi y}{b}\sin\frac{n_z\pi z}{c} \qquad ...(18)$$

From Eq. (15), we get

$$\frac{\partial^2 X}{\partial^2 X} = -\left(\frac{n_x\pi}{a}\right)^2\sqrt{\frac{2}{a}}\sin\frac{n_x\pi x}{a} = -\left(\frac{n_x\pi}{a}\right)^2 X(x) \qquad ...(19)$$

Putting the value from Eq. (19) in Eq. (10) we get

$$-\left(\frac{n_x\pi}{a}\right)^2 X(x) + \frac{2m}{\hbar^2}E_x X(x) = 0$$

or $$E_x = \frac{1}{2m}\left(\frac{n_x\pi\hbar}{n}\right)^2 = \frac{n_x^2 h^2}{8ma^2} \qquad ...(20)$$

Similarly we have

$$E_y = \frac{n_y^2 h^2}{8mb^2} \qquad \text{... (21)}$$

and

$$E_z = \frac{n_z^2 h^2}{8\, m\, c^2} \qquad \text{... (22)}$$

The allowed values of total energy are given by

$$E = E_x + E_y + E_z$$

$$= \frac{h^2}{8m^2}\left[\frac{n_x^2}{a^2} + \frac{n_y^2}{b^2} + \frac{n_z^2}{c^2}\right] \qquad \text{...(23)}$$

where n_x, n_y, n_z denote any set of three positive numbers.

When the box is a cube, *i.e.*, $a = b = c$, the energy expression may be written as

$$E = \frac{h^2}{8m\,a^2}\,[n_x^2 + n_y^2 + n_z^2] \text{ with } n_x, n_y, n_z = 1, 2, 3, \ldots$$

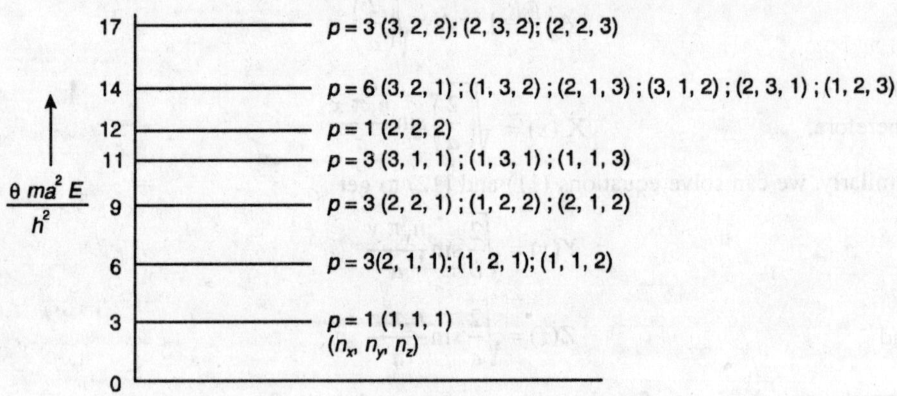

Fig 9.4.

Shows the energy levels of few states of a particle enclosed in cubical box.

Now when there is only one wave function corresponding to a particular eigen value, the level is known as *non-degenerate* but when there are a number of wave functions corresponding to a single eigen value, the level is known as *degenerate*. From Fig. 9.4 it is evident that the energy level corresponding to the energy $h2/8\ ma^2$ when $n_x = n_y = n_z = 1$ is non-degenerate. Similar level with $n_x = n_y = n_z = 2$, $n_x = n_y = n_z = 3$ and so on will be non-degenerate. In case of energy $(h^2/8\ ma^2 \times 6)$, n_x, n_y, and n_z may have values (2, 1, 1); (1, 2, 1); (1, 1, 2,) *i.e.*, energy is the same but there will be 3 wave functions hence this is a degenerate level with degeneracy 3. Similarly we can consider the degeneracy of all other levels.

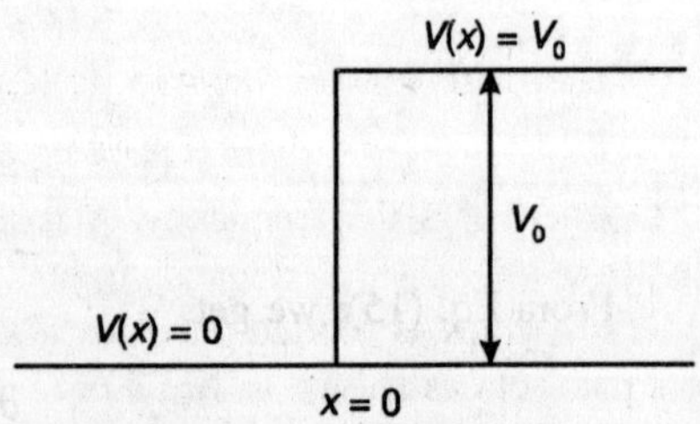

Fig. 9.5

9.14. Harmonic Oscillator

A particle undergoing simple harmonic motion is called a harmonic oscillator. In harmonic oscillator, the force applied is directly proportional to the displacement and is always directed towards the mean position.

If applied force displaces the particle through x, then Restoring force F is given by

$$F \propto -x$$

or
$$F = -kx$$

The potential energy of the oscillator is ... (1)

$$V = -\int F dx$$

or
$$V = k\int x dx = \frac{1}{2}kx^2 \quad \text{... (2)}$$

Classical view

If m is the mas of the particle, the

$$F = m\frac{d^2x}{dt^2}$$

Using equ. (1), we get

$$-kx = m\frac{d^2x}{dt^2}$$

or
$$\frac{d^2x}{dt^2} + \frac{k}{m}x = 0$$

or
$$\frac{d^2x}{dt^2} + \omega^2 = 0$$

where $\omega = \sqrt{\frac{k}{m}}$ is called angular frequency.

The classical frequency of the oscillator ν is given by

$$\nu = \frac{\omega}{2\pi} = \frac{1}{2\pi}\sqrt{\frac{k}{m}}$$

There are various ways to write the solution to Eq. (3). A common one is

$$x = A\cos(\omega t + \phi) = A\cos(2\pi\nu t + \phi)$$

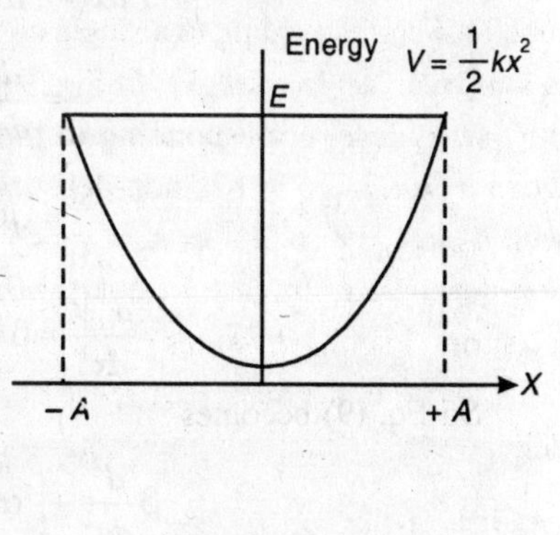

where ν is the frequency, A is the amplitude and ϕ is the phase angle which depends upon what x is at time $t = 0$.

If a graph is plotted between $V(x)$ and x, the graph will be a parabola as shown in fig. 9.6.

If E is the total energy and A is the amplitude of the motion, we find the at points $x = \pm A$, where $V = E$ are called turning points

At
$$x = \pm A$$

$$E = \frac{1}{2}kA^2$$

Since amplitude A can have any value, so in classical mechanics, the oscillator can have any value of total energy ranging from zero to any positive number.

Quantum mechanical view.

Since potential energy V does not depend on time some can use time independent Schrodings equation to find motion of simple harmonic oscillator.

Schrodinger time independent one dimensional wave equation (along x-axis) is given by

$$\frac{d^2\psi}{dx^2}+\frac{2m}{\hbar^2}(E-V)\,\psi=0 \quad \text{... (5)}$$

Substituting $V=\frac{1}{2}kx^2$, in above equation, we get

$$\frac{d^2\psi}{dx^2}+\frac{2m}{\hbar}(E-\frac{1}{2}kx^2)\,\psi=0 \quad \text{... (6)}$$

Since $\omega=\sqrt{k/m}$, so $k=m\omega^2$

So equation (6), becomes

$$\frac{d^2\psi}{dx^2}+\frac{2m}{\hbar^2}\left(E-\frac{1}{2}m^2\omega^2x^2\right)\psi=0$$

or
$$\frac{d^2\psi}{dx^2}+\left(\frac{2mE}{\hbar^2}-\frac{m^2\omega^2}{\hbar^2}x^2\right)\psi=0 \quad \text{... (7)}$$

For convenience

Let
$$\frac{2mE}{\hbar^2}=\alpha \text{ and } \frac{m\omega}{\hbar}=\beta \quad \text{... (8)}$$

So. Eq. (7) becomes

$$\frac{d^2\psi}{dx^2}+(\alpha-\beta^2x^2)\,\psi=0 \quad \text{... (9)}$$

Again for convenience, let us introduce a new variable q related to x such that

$$q=\sqrt{\beta x}$$

Now,
$$\frac{d\psi}{dx}=\frac{d\psi}{dq}.\frac{dq}{dx}=\sqrt{\beta}\frac{d\psi}{dq}$$

and
$$\frac{d^2\psi}{dx^2}=\frac{d}{dx}\left[\frac{d\psi}{dx}\right]=\frac{d}{dx}\left[\sqrt{\beta}\frac{d\psi}{dq}\right]$$

$$=\frac{d}{dq}\left[\sqrt{\beta}\frac{d\psi}{dq}\right]\frac{dq}{dx}$$

$$=\sqrt{\beta}\frac{d^2\psi}{dq^2}\sqrt{\beta}$$

or
$$\frac{d^2\psi}{dx^2}=\beta\frac{d^2}{dq^2}$$

So Eq. (9) becomes

$$\beta\frac{d^2}{dq^2}+\left(\alpha-\beta^2\frac{q^2}{\beta}\right)\psi=0$$

$$\frac{d^2\psi}{dq^2}+\left(\frac{\alpha}{\beta}-q^2\right)\psi=0$$

Asymptotic Solution.

To solve Eq. (10), let us make an attempt to obtain an expression for the asymptotic solution (the value of ψ (*q*) for very large value of *q*), when $q^2 >> \frac{\alpha}{\beta}$. In this case, equation (10) becomes

$$\frac{d^2\psi}{dq^2} = q^2\psi = 0 \qquad \text{... (11)}$$

The solution of this equation is of the from $\psi = e^{\pm q^1/2}$

But here, the solution $e^{+q/2}$ is not acceptable because, it increases rapidly with increasing q, but ince $|\psi|^2$ is the probability of finding the particle and it must decrease as x increases

∴ The particular acceptable solution is $\psi = -q^2/2$

The general solution at any distance can be written as

$$\psi = H(q)^{e-q^2/2} \qquad \text{... (12)}$$

where *H*(*q*) is a finite Hermite polynomial in q and can be written in a series of the from

$$H(q) = \alpha_0 + \alpha_1 q + \alpha_2 q^2 + ... + \alpha_n q^n$$

From Eq. (12), we have

$$\frac{d\psi}{dq} = -qe^{-q^2/2}H + e^{-q^2/2}\frac{dH}{dq}$$

and
$$\frac{d^2\psi}{dq^2} = q^2 e^{-q^2/2} H - e^{-q^2 n/2} H - qe^{-q^2/2}\frac{dH}{dq} - qe^{-q^2/2}\frac{dH}{dq} + e^{-q^2/2}\frac{d^2H}{dq^2}$$

or
$$\frac{d^2\psi}{dq^2} = \left(\frac{d^2H}{dq^2} - 2q\frac{dH}{dq} + q^2H - H\right)e^{-q^2/2}$$

Putting the value of $\frac{d\psi}{dq}$ and $\frac{d^2\psi}{dq^2}$ in Eq. (10), we get

$$\left(\frac{d^2H}{dq^2} - 2q\frac{dH}{dq} + q^2H - H\right)e^{-q^2/2} + \left(\frac{\alpha}{\beta} - q^2\right)H = 0$$

$$\frac{d^2H}{dq^2} = 2q\frac{dH}{dq} + \frac{\alpha}{\beta}H - H = 0$$

or
$$\frac{d^2H}{dq^2} - 2q\frac{dH}{dq} + \left(\frac{\alpha}{\beta} - 1\right)H = 0 \qquad \text{... (13)}$$

The solution of this equation determines H(*q*). Only those solutions are acceptable for which

$$\frac{\alpha}{\beta} = 2n + 1, \quad n = 0, 1, 2, ...$$

where *n* is called quantum number.

Putting the values of α and β from Eq. (8), we get

$$\frac{2mE/\hbar}{m\omega/\hbar} = 2n + 1$$

or
$$\frac{2E}{\hbar\omega} = (2n + 1)$$

or
$$E = \left(n + \frac{1}{2}\right)\hbar\omega$$

since energy E depends upon n, so we replace E by E_n.

$$\therefore \quad E_n = \left(n + \frac{1}{2}\right)\hbar\omega \qquad \text{... (14)}$$

The allowed integral values of n lead to certain discrete values of energy represented by eq. (14), known as eigne values of harmonic oscillator. Moreover eq. (14) indicates that the energy levels of harmonic oscillator *are equally spaced.* These energy levels are shown in Fig. 11.12 "Significance of Zero Point energy."

For ground state (or lower state) put $n = 0$, we have $E_0 = \frac{1}{2}\hbar\omega$

This is called "*Zero Point energy*". A comparison with the result $E = n\hbar\omega$ obtained by old quantum theory shows that the only difference is that all the equally energy levels are shifted upward by an amount equal to half the separation of energy levels i.e., $\frac{1}{2}\hbar\omega$ equal to zero point energy. Thus it is clear that even in the lowest state, the harmonic oscillator has energy greater than that it would have if it were at rest in its equilibrium position. The existence of zero point energy is in agreement with experiment and it important feature of quantum mechanics and is related to the uncertainty principle.

Figure 9.7 is a comparison of the energy levels of a *harmonic oscillator* with those of a *hydrogen atom* and of a *particle in a box* with infinitely hard walls. The shapes of the respective potential energy curves are also shown. The spacing of the energy levels in constant only for the harmonic *oscillator.*

Wave functions of harmonic oscillator. For each choice of the parmeter α_n there is different wave function ψ_n. Each function consists of a polynomial $H_n(y)$ (called a *Hremite Polynomial*) in either odd or even powers of y, the exponential factor $e^{-y^2/2}$, and a numerical coefficient which is needed for ψ_n to meet the normalization condition given by

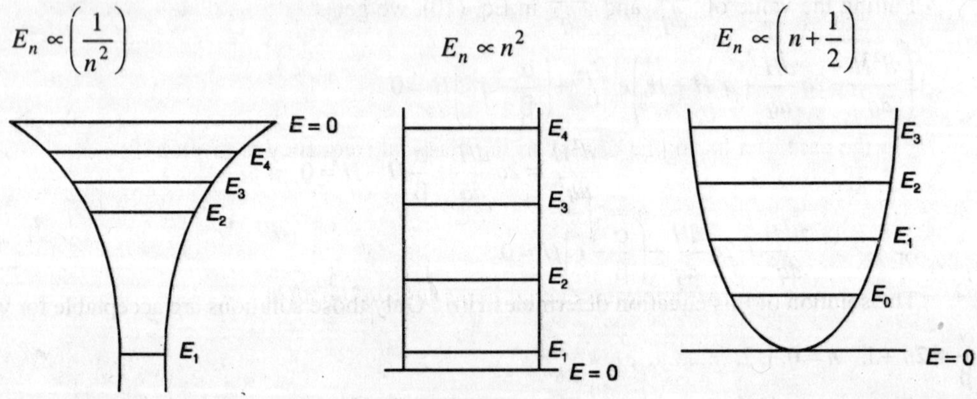

(*a*) Hydrogen atom (*b*) Particle in box (*c*) Harmonic oscililator

Fig. 9.7. Potential wells and energy levels.

$$\int_{-\infty}^{\infty} |\psi_n|^2 \, dx = 1 \qquad n = 0, 1, 2,...$$

The general formula for the nth wave function in terms of hermite polynomials is given by

$$\psi_n = \frac{1}{\sqrt{2^n . n! \sqrt{\pi}}} e^{-y^2/2} H_n(y)$$

A table of first size Hermite polynomials is given as under :

Table 9.1 Some Hermite Polynomials

n	$H_n(y)$	α_n	E_n
0	1	1	$\frac{1}{2}h\nu$
1	$2y$	3	$\frac{3}{2}h\nu$
2	$4y^2 - 2$	5	$\frac{5}{2}h\nu$
3	$8y^3 - 12y$	7	$\frac{7}{2}h\nu$
4	$16y^4 - 48y^2 + 12$	9	$\frac{9}{2}h\nu$
5	$32y^5 - 160y^3 + 120y$	11	$\frac{11}{2}h\nu$

9.15. Three Dimensional Harmonic Oscillator

In the case of a particle attracted towards a fixed point by a force which is proportional to its distance from the fixed point, the potential energy is given by

$$V(r) = \frac{1}{2}kr^2 = \frac{1}{2}k(x^2 + y^2 + z^2) \qquad ...(1)$$

which has spherical symmetry. Schrödinger's wave equation for such a system is given by

$$\frac{-\hbar^2}{2m}\nabla^2\psi + V\psi = E\psi$$

or
$$\frac{-\hbar^2}{2m}\nabla^2\psi + \frac{1}{2}kr^2\psi = E\psi \qquad ...(2)$$

In the case of a harmonic oscillator, the classical frequency is given by

$$\omega = \sqrt{\frac{k}{m}}$$

For convenience, we measure energy is the units of $\frac{1}{2}\pi\hbar\,\omega$

$$\therefore \qquad E = \frac{1}{2}\pi\hbar\,\omega$$

where λ is an eigen function of the energy operator.

Let us measure r in the units of $\left(\frac{\hbar}{n\omega}\right)^{1/2}$, so we have schrödinger equation as

$$\nabla^2\psi + (\lambda - r^2)\,\psi = 0 \qquad ...(3)$$

In the cartesion coordinates, this equation is separable with the substitution

$$\psi = X(x)\ Y(y)\ Z(z) \qquad ...(4)$$

We can write the results in separate equation in X, Y and Z as

$$\frac{d^2X}{dx^2} + (\lambda_x - x^2)\,X = 0 \qquad ...(5)$$

$$\frac{d^2Y}{dy^2}+(\lambda_y-y^2)\,Y=0 \quad ...(6)$$

$$\frac{d^2Z}{dz^2}+(\lambda_z-z^2)\,Z=0 \quad ...(7)$$

where constants λ_x, λ_y and λ_z are related by

$$\lambda_x+\lambda_y+\lambda_z=\lambda \quad ...(8)$$

Each of the equation (5), (6) and (7) is a differential equation for *one dimensional Harmonic oscillator,* the normalized wave function is given by

$$\psi_n=\frac{1}{\sqrt{2^n.n!\sqrt{\pi}}}e^{-y^2/2}H_n(y) \quad ...(9)$$

We can get the integrable solutions of the form (3) if the constants $\lambda_x,\lambda_y,$ *and* λ_z are of the form

$$\lambda_x=2n_x+1,\quad \lambda_y=2n_y+1,\quad \lambda_z=2n_z+1 \quad ...(10)$$

where n_x, n_y and n_z are positive integers.

The corresponding normalized wave function is given by,

$$\psi_{n_x,n_y,n_z}=\frac{1}{2^n n_x!n_y!n_z!\pi^{3/2}}e^{-4^{3/2}}\quad Hn_x(x)Hn_y(y)Hn_z(z) \quad ...(11)$$

where $\qquad n=n_x+n_y+n_z \qquad ...(12)$

So, form Eq. (8), the total energy is

$$E_n=(n_x+n_y+n_z+3/2)\,\hbar\,\omega$$

$$=\left(n+\frac{3}{2}\right)\hbar\,\omega$$

Which is only dependent on integer n.

For a given value of n, the wave function (11) correspond to a set linearly independent function, each of which will correspond to one of the integers n_x, n_y, n_z that satisfy the condition (12).

So the energy states of E_n are *degenerate* and the degree of degeneracy is $\frac{1}{2}(n+1)(n+2)$; which corresponds to the number of ways in which equation (12) can be satisfied by positive integers (non-negative integers) n_x, n_y and n_z.

The same results can be attained for the three dimensional harmonic oscillator if *spherical harmonic* are used. The separation of the various components is performance in the same way as above.

Example 1. *An electron is put in a cubical box of each side 1 Å. Find the value of its momentum and energy for the ground state and the first excited state.*

Solution. For the ground state $n_x=n_y=n_z=1$

$$\therefore \quad \text{Momentum } p_{1,1,1}=\frac{\pi\hbar}{l}\sqrt{3}=\frac{\pi h}{2\pi l}\sqrt{3}$$

$$\frac{\hbar}{2l}\sqrt{3}=\frac{0.63\times10^{-27}}{2\times10^{-8}}.\sqrt{3}$$

$$= (3.315 \times 10^{-19})\sqrt{3}$$
$$= 5.74 \times 10^{-19} \text{ gm cm/sec}$$

$$\text{Energy } E_{1,1,1} = \frac{\pi^2\hbar^2}{2ml^2}3 = \frac{\pi^2.\hbar^2}{4\pi^2 ml^2}3 = \frac{\hbar^2}{8ml^2}3$$

$$= \frac{(6.63\times10^{-27})^2\times3}{8\times9.1\times10^{-20}\times10^{-18}}$$

$$= 60.4 \times 3 \times 10^{-12} = 181 \times 10^{-12} \text{ erg}$$
$$= 113 \text{ eV}$$

For the first excited state one of the quantum number is = 2

$\therefore$ Momentum $p_{2,1,1} = p_{1,2,1} = p_{1,1,2} = \frac{\pi\hbar}{l}\sqrt{6}$

$$= \frac{h}{2l}\sqrt{6} = 3.315\times10^{-18}\sqrt{6}$$

$$= 8.13 \times 10^{-19} \text{ gm cm/sec.}$$

Energy $E_{2,1,1} = E_{1,2,1} = E_{1,1,2} = \frac{\pi^2\hbar^2}{2ml^2}6 = \frac{h^2}{8ml^2}\sqrt{6}$

$$= 60.4 \times 6 \times 10^{-12} = \mathbf{362 \times 10^{-12}} \textbf{ erg. Ans.}$$

Example 2. *Find the energy of an electron moving in one dimension in an infinitely high potential box of width 1 Å, given mass of the electron 9.11 × 10^{-31} kgm and h = 6.63 × 10^{-34} Js.*

Solution. We have the relation

$$E = \frac{n^2h^2}{8ma^2} \quad (n = 1,2,3,.....)$$

The least energy of the particle can be obtained by putting $n = 1$. The energy is given by

$$E = \frac{h^2}{8ma^2}$$

In this case $m = 9.11 \times 10^{-31}$ kgm, $a = 1$ Å $= 10^{-10}$ m and $h = 6.63 \times 10^{-34}$ Js.

$$E = \frac{(6.63\times10^{-34})^2}{8\times9.11\times10^{-31}\times(10^{-10})^2} \text{ joule}$$

$$= 9.1 \times 10^{-19} \text{ Joules} = \frac{9.1\times10^{-19}}{1.602\times10^{-19}} \text{ eV} = \mathbf{5.68\ eV} \textbf{ Ans.}$$

Example 3. *Think of the nucleus as a box with a size of 10^{-14} m across. Compute the lowest energy of a neutron confined to the nucleus.*

Solution. Take the nucleus to be a cubical box of size 10^{-14} m across.

$\therefore$ $x = y = z = a = 10^{-14}$ m

The neutron will have the lowest energy in the quantum state given by

$$n_x = n_y = n_z = 1$$

Now energy $E_{1,1,1} = \frac{\pi^2\hbar^2}{2m}\left[\frac{n_x^2}{l_x^2} + \frac{n_y^2}{l_y^2} + \frac{n_z^2}{l_z^2}\right]$

$$= \frac{\pi^2\hbar^2}{2m}\left[\frac{3}{l^2}\right]$$

$$= \frac{3\pi^2\hbar^2}{4\pi^2 \times 2ml^2} = \frac{3\hbar^2}{8ml^2}$$

$$= \frac{3\times[6.62\times10^{-34}]^2}{8\times1.6\times10^{-27}\times10^{-28}}$$

$= 10.29 \times 10–23$ J

= 6.43 MeV. Ans.

Example 4. *Show that* $\Psi(x) = e^{icx}$, *where c is some finite constant is acceptable eigen function. Also normalize it over the region* $-a \le x \le a$.

Solution. The given wave function is

$$\Psi(x) = e^{icx} = \cos cx + i \sin cx$$

Its derivative is given by

$$\frac{d\Psi(x)}{dx} = ice^{icx}$$

$$= -c \sin cx + ic \cos cx$$

(*i*) Since six *cx* and cos *cx* are priodic functions with maximum value 1 and *c* is finite constant, both $\Psi(x)$ and $\frac{d}{dx}\Psi(x)$ are finite for all values of *x*.

(*ii*) The function $\Psi(x)$ and $\frac{d}{dx}\Psi(x)$ are single valued because cos *cx* and sin *cx* are single-valued functions for all values of *x*.

(*iii*) The continuity of sine and cosine function implies that $\Psi(x)$ and $\frac{d}{dx}\Psi(x)$ are also continuous for all values of *x*.

Hence $\Psi(x)$ is an acceptable form of the eigen function. For normalization we write

$$\Psi(x) = Ne^{icx}$$

and determine *N* such that

$$\int_{-a}^{a} \Psi^*(x)\Psi(x)\,dx = 1$$

or $$N^2 \int_{-a}^{a} e^{-icx}.e^{icx}dx = 1$$

or $$N^2 = |x|_{-a}^{+a} = 1$$

or $$N^2 (2a) = 1$$

or $$N = \frac{1}{\sqrt{2a}}.$$

Hence the normalized wave function is

$$\Psi(x) = \frac{x}{\sqrt{2a}}e^{icx} \quad \textbf{Ans.}$$

Example 5. *An electron is bound by a potential which closely approaches an infinite square well of width* 2.5×10^{-10} *m. Calculate the lowest three permissible quantum energies the electron can have.*

Solution. In case of infinite potential well

$$E_n = \frac{n^2h^2}{8\,ma^2}$$

Here $h = 6.63 \times 10^{-34}$ Js, $m = 9.1 \times 10^{-31}$ kg

and $a = 2.5 \times 10^{-10}$ m.

Now,
$$E_1 = \frac{h^2}{8ma^2} = \frac{(6.63\times10^{-34})^2}{8\times9.1\times10^{-31}\times(2.5\times10^{-10})^2}$$

$$= 9.63 \times 10^{-19} \text{ joules} = \frac{9.63\times10^{-19}}{1.6\times10^{19}} = 6 \text{ eV.}$$

Also $E_2 = 2^2\ E_1 = 4 \times 6 = \mathbf{24\ eV}$,

and $E_3 = 9\ E_1 = 9 \times 6 = \mathbf{54\ eV}$. **Ans.**

Example 6. *A particle is moving in one dimensional potential box (of infinite height) of width 25 Å. Calculate the probability of finding the particle within an interval of 5 Å at the centres of the box when it is in its state of least energy.*

Solution. We know that the wave function of a particle enclosed with an infinite potential well is given by

$$\Psi(x) = \sqrt{\frac{2}{a}}\sin\frac{n\pi x}{a}$$

When the particle is in the least energy state $n = 1$, hence in this case

$$\Psi(x) = \sqrt{\frac{2}{a}}\sin\frac{\pi x}{a}$$

At the centre of the box $x = a/2$. The probability of finding the particle in the unit interval at the centre of box is given by

$$|\Psi(x)|^2 = \left[\sqrt{\frac{2}{a}}\sin\frac{\pi(a/2)}{a}\right]^2 = \frac{2}{a}\sin^2\frac{\pi}{2} = \frac{2}{a}$$

The probability P in the interval Δx is given by

$$P = |\Psi(x)|^2\,\Delta x = \frac{2}{a}\Delta x$$

In this case $a = 25$ Å $= 25 \times 10^{-10}$ m and $\Delta x = 5$ Å $= 5 \times 10^{-10}$ m

$$\therefore \qquad P = \frac{2\times5\times10^{-10}}{25\times10^{-10}} = \mathbf{0.4} \qquad \textbf{Ans.}$$

Example 7. *A proton or a neutron in a nucleus can roughly be regarded as a particle in a box, the surface of the nucleus plays the role of the wall of the box and proton moves more or less freely between these walls. What is the energy released when a proton makes a transition from the first excited state to the ground state of a box of nuclear size i.e., 1.0×10^{-14} m.*

Solution. We have $a = 1 \times 10^{-14}$ m

and $m = m_p = 1.67 \times 10^{-27}$ kg.

The energy of ground state is got from relation

$$E_n = \frac{n^2\pi^2\hbar^2}{2m_p a^2};$$

$$E_1 = 1.\frac{\pi^2\hbar^2}{2m_p a^2} = \frac{\pi^2\times(1.05\times10^{-34}\ \text{J}-\text{s})^2}{2\times1.67\times10^{-27}\ \text{kg}\times(1.0\times10^{-14}\ \text{m})^2}$$

$$= 3.3 \times 10^{-13}\ \text{J}$$

Energy of first excited state will be

$$E_2 = 2^2.\frac{\pi^2\hbar^2}{2m_p a^2} = 4\times3.3\times10^{-13}\,\text{J}$$

$$= 13.2 \times 10^{-13}\,\text{J}$$

$\therefore$ $$E_2 - E_1 = 13.2 \times 10^{-13}\,\text{J} - 3.3 \times 10^{-13}\,\text{J}$$

$$= 9.9 \times 10^{-13}\,\text{J} = \frac{9.9\times10^{-13}}{1.6\times10^{-23}}$$

$$= \mathbf{6.2\ MeV.\ Ans.}$$

9.14 Energy levels of a particle in one dimensional potential well of infinite height

Consider a particle of mass m moving along x-direction between rigid walls w_1 and w_2 at positions $x = 0$ and $x = a$ (Fig. 9.7). The particle moves freely between the walls. *As no force is acting on the particle so its potential is constant between the walls.* For simplicity this constant potential may be taken as zero.

When the particle strikes any one wall, it gets reflected back since we consider the wall to be rigid, force acting on the particle abruptly changes from zero to some finite value, say F. Also, since $|F| = \frac{\partial V}{\partial x}$, we may have $\Delta V \to \infty$ when $\Delta x \to 0$ such that $\frac{\partial V}{\partial x}$ has some finite value $|F|$. It means that potential energy of the particle becomes ∞ at the walls. The above statement may be written in mathematical form as

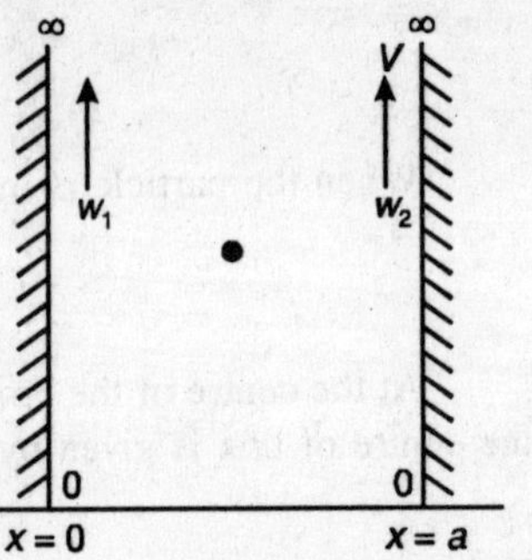

Fig. 9.8

$$V(x) = \infty \quad \text{for } x < 0 \text{ and } x > 0$$

and $$V(x) = 0 \quad \text{for } 0 < x < a$$

Now Schrödinger wave equation for a particle has the form

$$\frac{d^2\Psi}{dx^2} + \frac{8\pi^2 m}{h^2}(E - V)\Psi = 0 \qquad \text{... (1)}$$

As $V = 0$ between the walls, so the equation may take the form

$$\frac{d^2\Psi}{dx^2} + \frac{8\pi^2 m}{h}E\Psi = 0$$

Put $\frac{8\pi^2 m}{h^2}E = k^2$, so Equation (2) may be written as

$$\frac{d^2\Psi}{dx^2} + k^2\Psi = 0 \qquad \text{... (3)}$$

the general solution of Equation (3) has the form

$$\Psi(x) = A \sin kx + b \cos kx \qquad \text{... (4)}$$

where A and B are constants whose values can be found from boundary conditions.

Since the particle is enclosed between two rigid walls and ψ^2, by definition, represents *probability of finding particle at any instant between the walls.* Also, the particle can not penetrate the walls, so

$$\left.\begin{aligned} \psi &= 0 \text{ at } x = 0 \\ \text{and } \psi &= 0 \text{ at } x = a \end{aligned}\right] \qquad \text{... (5)}$$

relations given by Eq. (5) represent the boundary conditions.

Applying the said conditions, in eq. (4), we get

$$0 = A \sin 0 + B \cos 0$$

$\therefore$ $$B = 0 \qquad [\because \sin 0 = 0 \text{ and } \cos 0 = 1]$$

Putting in Eq. (4), we get

$$\psi(x) = A \sin kx$$

Again applying second boundary condition, we get $0 = A \sin ka$, which means, either $A = 0$ or $\sin ka = 0$ $[\because \psi = 0, \text{ when } x = a]$

Now $A \neq 0$ because in that case since $B = 0$; entire function $\psi(x)$ will be zero (from Eq. (4))

$\therefore$ $$\sin ka = 0 \text{ or } ka = n\pi$$

or $$k = n\pi/a \qquad \text{... (6)}$$

So the wave function becomes

$$\psi(x) = A \sin \frac{n\pi x}{a} \qquad \text{... (7)}$$

also we have taken $k^2 = \frac{8\pi^2 m}{h^2} E$ in Eq. (2).

$\therefore$ $$\frac{n^2\pi^2}{a^2} = \frac{8\pi^2 mE}{h^2}$$

Let us denote E by E_n for general case, then we have

$$\frac{n^2\pi^2}{a^2} = \frac{8\pi^2 m}{h^2} E_n \qquad \text{... (8)}$$

or $$E_n = \frac{n^2 h^2}{8ma^2} \qquad (\text{where } n = 1, 2, 3, \ldots)$$

From Eq. (8) it is clear that in an infinite potential well, the particle can have only discrete energy values *i.e. particle energy is quantized.*

We can get the constant of Eq. (7) from normalization condition *i.e.*

$$\int_0^a |\psi(x)|^2 = 1$$

$\therefore$ $$\int_0^a A^2 \sin^2 \frac{n\pi x}{a} dx = 1$$

or $$A^2 \int_0^a \sin^2 \frac{n\pi x}{a} dx = 1$$

or $$A^2 \int_0^a \frac{1}{2}\left[1 - \cos\frac{2n\pi x}{a}\right] dx = 1$$

or $$\frac{A^2}{2}\left[x - \frac{a}{2\pi n}\sin\frac{2n\pi x}{a}\right]_0^a = 1$$

or $$\frac{A^2}{2} a = 1 \quad \therefore A^2 = 2/a \text{ or } A = \sqrt{2/a}$$

So wave function may be written as

$$\psi(x) = \sqrt{2/a}\,\sin\frac{n\pi x}{a} \qquad \text{... (9)}$$

Equation (7) gives the wave function of the particle enclosed in finitely deep potential well.

Example 8. *A particle is confined to one dimensional infinite potential well of width 0.2 nm. It is given that where the energy of the particle is 230 eV, its eigen function has 5 antinodes. Find the mass of the particle and show that it can never have energy equal to 1 keV.*

Solution. Since the eigen function corresponding to energy 230 eV has 5 antinode, the particle is in a quantum state with $n = 5$. For this case

$$E_5 = 5^2 E_1 = 25 E_1 = 230 \text{ eV}$$

So that

$$E_1 = 9.2 \text{ eV} = 9.2 \times 1.6 \times 10^{-19} \text{ J}$$
$$= 14.7 \times 10^{-19} \text{ J}$$

Also

$$a = 0.2 \text{ nm} = 0.2 \times 10^{-9} \text{ m} = 2 \times 10^{-10} \text{ m}$$

Now

$$E_1 = \frac{\hbar^2\pi^2}{2\,ma^2} = \frac{\hbar^2}{8\,ma^2}$$

or

$$m = \frac{\hbar^2}{8\,E_1 a^2} = \frac{(6.625\times10^{-34})^2}{8\times4\times10^{-20}\times14.7\times10^{-19}}\text{ kg}$$

For $E_n = 1$ keV, n should be such that

$$n^2 = \frac{E_n}{E_1} = \frac{10^3}{9.2} = 108.7$$

i.e. $n =$ **10.4 Ans.**

Since n is not an integer, $E_n = 1$ keV is not a permitted value of energy.

Example 9. *Using the operator representation of the x component of momentum of a particle, prove that*

$$(xp_x - p_x x)\,\psi = i\hbar\psi$$

where ψ is an arbitrary wave function.

Solution. We have the relation in operator form as $(xp_x - p_x x)\psi$

$$= x\left[\frac{\hbar}{i}\frac{\partial\psi}{\partial x} - \frac{\hbar}{i}\frac{\partial}{\partial x}(x,\psi)\right]$$

$$= \frac{\hbar}{i}\left[x\frac{\partial\psi}{\partial x} - x\frac{\partial\psi}{\partial x} - \psi\right]$$

$$= \frac{-\hbar}{i}\psi = i\hbar\psi \quad \textbf{Ans.}$$

Example 10. *A particle is in motion along a line between $x = 0$ and $x = a$ with zero potential energy. At points for which $x < 0$ and $x > a$, the potential energy is infinite. The wave function for the particle in the nth state is given by*

$$\psi_n = A\sin\frac{n\pi x}{a}$$

Find the expression for the normalized wave function.

Solution. The probability that the particle is between x and $x + dx$ for the nth state is given

$$P_n dx = |\psi_n(x)|^2\,dx$$

$$= A^2\sin^2\frac{n\pi x}{a}\,dx$$

Since the particle has to be in the region $x = 0$ and $x = a$ for all times, so we have

$$\int P_n dx = \int_0^a |\psi_n(x)|^2 dx = 1$$

$$= \int_0^a A^2 \sin^2 \frac{n\pi x}{a} dx = 1$$

or
$$A^2 \int_0^a \frac{1}{2}\left[1 - \cos\frac{2\pi nx}{a}\right] dx = 1$$

or
$$\frac{A^2}{2}\left[x - \frac{a}{2\pi nx}\sin\frac{2\pi nx}{a}\right]_0^a = 1$$

$\therefore$
$$\frac{A^2}{2}a = 1$$

Because the second term in the brackets is zero for both the limits.

So we have

$$A = \sqrt{2/a}$$

$\therefore$ Normalized wave function is

$$\psi n(x) = \sqrt{\frac{2}{\text{a}}}\sin\frac{n\pi x}{a}$$

Example 11. *For a particle in a one divisional box, show that the value of uncertainity product is*

$$\nabla x.\nabla b = h\left(\frac{n^2\pi^2}{12} - \frac{1}{2}\right)^{1/2}$$

Solution. To calculate uncertainity of any measurable quantity, we must have knowledge of Expectation Value Wave function.

The expectation value of average value of any quantity α as *defined earlier is*

$$<\alpha> = \frac{\int_{-\infty}^{\infty} \psi^* \hat{\alpha}\, \psi\, dx}{\int_{-\infty}^{\infty} \psi^* \psi\, dx} =$$

Where $\hat{\alpha}$ is an operator for the quantity α.

We know that uncertainity in a quantity is nothing but the standard deviation, and so are uncertainity in position may be written as

$$\Delta x = \sqrt{<x^2> - <x>^2}$$

Similarly uncertainity in momentum may be represented as

$$\Delta p = \sqrt{<p^2> - <p>^2}$$

Now we can calculate the uncertainity product for particle in a one dimensional box of length – $\angle$ for which wave function is given as $\psi_n(x) = \sqrt{\frac{2}{1}}\sin\left(\frac{n\pi x}{\text{L}}\right)$ (1)

$\therefore$
$$<x> = \frac{\int_{-\infty}^{\infty} \psi^* x\, \psi\, dx}{\int_{-\infty}^{\infty} \psi^* \psi\, dx}$$

$$= \frac{2}{L}\int_0^L x\sin^2\left(\frac{n\pi x}{L}\right)dx \quad \left(\because \int_0^L \psi^*\psi\, dx = 1 \text{ as wave function is normalised}\right)$$

$$= \frac{2}{L}\int_0^L x\left[\frac{1-\cos\left(\frac{2n\pi}{L}x\right)}{2}\right]dx$$

$$= \frac{1}{L}\left[\int_0^L x\,dx - \int_0^L x\cos\left(\frac{2n\pi}{L}x\right)dx\right] = \frac{1}{L}\left[\frac{L^2}{2} = 0\right] = \frac{L}{2}$$

Again
$$<x^2> = \frac{\int_{-\infty}^{\infty}\psi^* x^2\,\psi\, dx}{\int_{-\infty}^{\infty}\psi^*\,\psi\, dx} = \frac{2}{L}\int_0^L x^2\sin\left(\frac{2n\pi}{L}x\right)dx$$

$$= \frac{1}{L}\int_0^L x^2\left(1-\cos\left(\frac{2n\pi}{L}x\right)\right)dx$$

$$= \frac{1}{L}\left[\int_0^L x^2\,dx - \int_0^L x^2\cos\left(\frac{2n\pi}{L}x\right)dx\right]$$

$$= \frac{1}{L}\left[\frac{L^3}{3} - \frac{L^3}{2n^2\pi^2}\right] = \frac{L^2}{3} - \frac{L^2}{2n^2\pi^2}$$

Also
$$< p > = \frac{\int_{-\infty}^{\infty}\psi^*\left(-i\hbar\frac{\partial}{\partial x}\right)\psi\, dx}{\int_{-\infty}^{\infty}\psi^*\psi\, dx}$$

$$= \frac{2}{L}\int_{-\infty}^{\infty}\sin\left(\frac{n\pi}{L}x\right)\left(-i\hbar\frac{\partial}{\partial x}\sin\left(\frac{n\pi}{L}x\right)\right)dx$$

$$= i\hbar\times\frac{L}{n\pi}\times\frac{2}{L}\int_0^L\sin\left(\frac{n\pi}{L}x\right)\cos\left(\frac{n\pi}{L}x\right)dx$$

$$= \frac{i\hbar}{n\pi}\times\int_0^L\sin\left(\frac{2n\pi}{L}x\right)dx = 0$$

and
$$< P^2 > = \frac{\int_{-\infty}^{\infty}\psi^*\left(-i\hbar\frac{\partial}{\partial x}\right)^2\psi\, dx}{\int_{-\infty}^{\infty}\psi^*\,\psi\, dx}$$

$$= \hbar^2 \times \frac{2}{L}\int_0^2 \sin\left(\frac{n\pi}{L}x\right)\left[\frac{d^2}{dx^2}\left(\sin\left(\frac{2\pi}{L}x\right)\right)\right]dx$$

$$= -\hbar^2 \times \frac{2}{L} \times \left(\frac{n^2\pi^2}{L^2}\right) \times \int_0^L \sin\left(\frac{n\pi}{L}x\right)\left[-\sin\left(\frac{n\pi}{L}x\right)\right]dx$$

$$= \frac{n^2\pi^2\hbar^2}{L^3}\int_0^L\left(1-\cos\left(\frac{2n\pi}{L}x\right)\right)dx$$

$$= \frac{n^2\pi^2\hbar^2}{L^3}[L-0] = \frac{n^2\pi^2\hbar^2}{L^2}$$

$\therefore$ $$(\Delta x)^2 = \langle x^2\rangle - \langle x\rangle^2$$

$$= \frac{L^2}{3} - \frac{L^2}{2n^2\pi^2} - \left(\frac{L}{2}\right)^2 = \frac{L^2}{12} - \frac{L^2}{2n^2\pi^2}$$

Similarly $$(\Delta P)^2 = \langle P^2\rangle - \langle p\rangle^2$$

$$= \frac{n^2\pi^2\hbar^2}{L^2} - 0 = \frac{n^2\pi^2\hbar^2}{L^2}$$

Now $$(\Delta x\,\Delta p)^2 = \left(\frac{L^2}{12} - \frac{L^2}{2n^2\pi^2}\right)\left(\frac{n^2\pi^2\hbar^2}{L^2}\right)^2 = \left(\frac{n^2\pi^2}{12} - \frac{1}{2}\right)\hbar^2$$

Hence $$(\Delta x\,\Delta p) = \hbar\left(\frac{n^2\pi^2}{12} - \frac{1}{2}\right)^{\frac{1}{2}}$$

As minimum value of n is 1, so, minimum value of uncertainity product is $\hbar\left(\frac{\pi^2}{12} - \frac{1}{2}\right)^{\frac{1}{2}}$

$= 0.57\,\hbar$

Which means that $0.57\hbar > \frac{\hbar^2}{2}$. Hence condition $\Delta x\,.\,\Delta p \geq \hbar/2$ is valid and so this may be considered as a proof of uncertainity for a particle is a box.

Example 12. *A particle is moving in one dimensional box and its wave function is given by*

$$\psi_n = A\sin\frac{n\pi x}{a}$$

Find the expression for the normalized wave function.

Solution : The probability that the particle in between x and $x + dx$ for the nth state is given by

$$P_n dx = |\psi_n(x)|^2\,dx$$

$$= A^2\sin^2\frac{n\pi x}{\alpha}dx$$

Since the oparticel has to be in the region $x = 0$ and $x = a$ for all times, so we have

$$\int P_n dx = \int_0^a |\psi_n(x)|^2\,dx = 1$$

$$= \int_0^a A^2 \sin^2 \frac{n\pi x}{a} dx = 1$$

or $$A^2 \int_0^a \frac{1}{2}\left[1 - \cos\frac{2\pi nx}{a}\right] dx = 1$$

or $$\frac{A^2}{2}\left[x - \frac{a}{2\pi nx}\sin\frac{2\pi nx}{a}\right]_0^a = 1$$

$\therefore$ $$\frac{A^2}{2}a = 1$$

Because the second term in the brackets is zero for both the limits.

So we have

$$A = \sqrt{2/a}$$

$\therefore$ Normalized wave function is

$$\psi n(x) = \sqrt{\frac{2}{a}} \sin \frac{n\pi x}{a} \quad \textbf{Ans.}$$

Example 13. *A proton is confined to a nucleus or radius 5.0×10–15 m. Calculate the minimum uncertainity in its momentum. Also calculate the minimum kinetic energy of the proton. Hence, show that the presence of protons in the nucleus in permitted (mass of proton = 1.675 × 10–27 kg).*

Solution. Diameter of the nucleus,

$$\Delta x = 2 \times 5 \times 10^{-15} = 10^{-14} \text{ m}$$

(*i*) From uncertainity principle, we have

$$\Delta p \,.\, \Delta x = \frac{h}{2\pi}$$

$\therefore$ $$\Delta p = \frac{h}{2\pi \Delta x} = \frac{6.626 \times 10^{-34} \text{ J.s}}{2\pi \times 10^{-14} \text{ m}} = 1.055 \times 10^{-20}$$

$\therefore$ $$\Delta p = 1.055 \times 10^{-20} \text{ kg m s}^{-1}$$

Which means the minimum value of momentum $p = 1.055 \times 10^{-20}$ mg m s^{-1}. **Ans.**

(*ii*) Corresponding minimum K.E. is given by

$$(\text{K.E.})_{min} = \frac{p^2}{2\text{m}}$$

$$= \frac{(1.055 \times 10^{-20})^2}{2 \times 1.67 \times 10^{-27}}$$

$$= 0.332 \times 10^{-13} \text{ J}$$

$$= \frac{0.332 \times 10^{-13}}{1.6 \times 10^{-19}} \text{ eV}$$

$$= \textbf{0.23 Me V.} \quad \textbf{Ans.}$$

(*iii*) Now rest mass Energy of proton is $m_0 c^2$,

i.e., $$m_0 c^2 = \frac{1.675 \times 10^{-27} \times 9 \times 10^{16}}{1.6 \times 19^{-19} \times 10^6}$$

$$= \textbf{942 Mev}$$

which is very large an compared to K.E.

So, we presence of protons with such kinetic energies is quite plausible and hence its presence in the nucleus is permitted.

Example 14 : *An electron has a de Broglie wavelength equal to that of a photon. Show that the ratio of the K.E. of the electron to the energy of the photon =* $\dfrac{(m^2c^4+h^2\nu^2)^{1/2}-mc^2}{h\nu}$.

What is the limiting value of this ratio for low energy photons? If the photon has energy of 100 KeV, calculate the K.E. of the electron.

Solution. (*i*) We have the de Broglie relation

$$\lambda = \frac{h}{mv}$$

or
$$v = \frac{h}{m\lambda} = \frac{h\nu}{m.\,c}$$

Now,
$$\text{K.E.} = \frac{1}{2}mv^2 = \frac{1}{2}m.\left[\frac{h\nu}{mc}\right]^2$$

$$= \frac{h^2\nu^2}{2mc^2} \qquad \text{... (1)}$$

which may be written in the from

$$E_e = mc^2 + \frac{h^2\nu^2}{2mc^2} - mc^2$$

$$= mc^2\left[1+\frac{h^2\nu^2}{2m^2c^4}\right] - mc^2$$

$$= mc^2\left[1+\frac{h^2\nu^2}{2m^2c^4}\right]^{1/2} - mc^2 \qquad \text{[From Binomial Theorem]}$$

$$= mc^2\left[\frac{m^2c^4+h^2\nu^2}{m^2c^4}\right]^{1/2} - mc^2$$

$$= \left[m^2c^4+h^2\nu^2\right]^{1/2} - mc^2$$

In case of photon, $E_p = h\nu$

$$\therefore \qquad \frac{E_e}{E_p} = \frac{(m^2c^4+h^2\nu^2)^{1/2}-mc^2}{h\nu} \quad \textbf{Ans.}$$

(*ii*) For limiting case, $v = c$

i.e.,
$$mc = \frac{h}{\lambda} - \frac{h\nu}{c}$$

which means,
$$p = mc = \frac{h\nu}{c}$$

So, in the case, K.E. of erlectron is

$$E_e = \frac{1}{2}mc^2 = \frac{1}{2m}(mc)^2 = \frac{p^2}{2m}$$

$$= \frac{1}{2m}\left(\frac{h\nu}{c}\right)^2$$

i.e.,
$$E_e = \frac{h^2\nu^2}{2mc^2}$$

also,
$$E_p = h\nu$$

$$\therefore \qquad \frac{E_e}{E_p} = \left[\frac{h^2v^2}{2cm^2}\right]\frac{1}{hv} = \frac{hv}{2mc^2} \quad \textbf{Ans.}$$

(*iii*) Photon energy is given by

$$hv = 100 \text{ keV}$$
$$= 100 \times 10^3 \times 1.6 \times 10^{-19}$$
$$= 1.6 \times 10^{-14} \text{ J}$$

Also K.E. of Eleelectron from eq. (1) is given by

$$E_e = \frac{h^2v^2}{2mc^2}$$

$$= \frac{(1.6\times10^{-14})^2}{2\times9.1\times10^{-31}\times9\times10^{-16}}$$

$$= 0.0156\times10^{-13} \text{ J}$$

$$= \frac{0.0156\times10^{-13}}{1.6\times10^{-19}} eV$$

$$= 9.77 \times 10^3 \text{ eV}$$

$$= \mathbf{9.77 \text{ keV}} \qquad \textbf{Ans.}$$

Example 15 : *Consider a grain dust* (m = 1μ gm) *confined move between two rigid walls separated by* 10^{-4} *m. It requires 100 sec to cross the gap. Calculate the quantum number that describes this motion.*

Solution. Velocity of motion (v) $= \dfrac{10^{-4}}{100} = 10^{-6}$ ms^{-1}

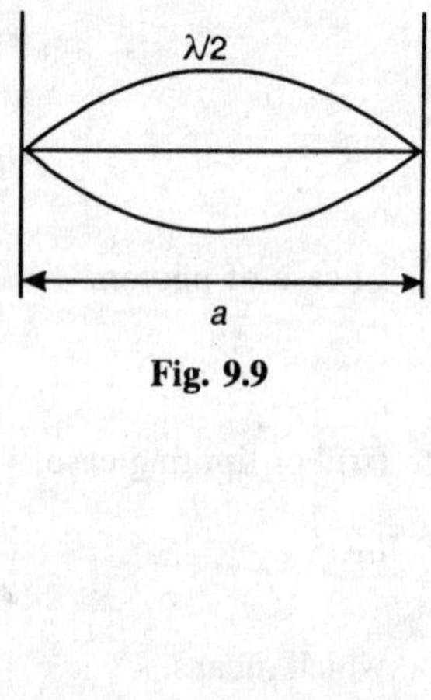

Fig. 9.9

Kinetic energy $\qquad E = \dfrac{1}{2}mv^2$

with m = 1 μ gm = 10^{-9} kg and g $v = 10^{-6}$ ms^{-1},

we have, $\qquad E = \dfrac{1}{2}\times10^{-9}\times(10^{-6})^2$

$$= 5 \times 10^{22} \text{ J}$$

Also, the energy of a particle in a rigid box in given by

$$E = \frac{n^2h^2}{8ma^2}$$

$$n^2 = \frac{8\,ma^2}{h^2}E$$

using $m = 10^{-9}$ kg, $a = 10^{-4}$ m,

$$E = 5\times10^{-22} \text{ J}, \; h = 6.62\times10^{-34} \text{ J.s}$$

we have, $\qquad n^2 = \dfrac{8\times10^{-9}\times(10^{-4})^2\times5\times10^{22}}{(6.62\times10^{-34})^2}$

$$= 9\times10^{28}$$

$$\therefore \qquad n = \mathbf{3 \times 10^{14}} \qquad \textbf{Ans.}$$

Example 16. *For a free particle show that Schrodinger wave equation leads to the de Broglie relation* $\lambda \frac{h}{p}$.

Solution. The time-independent solution of Schrodinger wave equation may have the from,

$$E_n = \frac{n^2 h^2}{8ma^2}$$

$$\therefore \quad E_1 = \frac{h^2}{8ma^2}$$

In this case, $$a = \frac{\lambda}{2}$$

$$\therefore \quad E_1 = \frac{4h^2}{8m \cdot \lambda^2}$$

$$= \frac{\lambda^2}{2m\lambda^2} \quad \text{... (1)}$$

Also, $$E_1 = \frac{p^2}{2m} \quad \text{... (2)}$$

∴ From (1) and (2),

$$\frac{h^2}{2m\lambda^2} = \frac{p^2}{2m}$$

or $$p^2 = \frac{h^2}{\lambda^2}$$

or $$p = \frac{h}{\lambda}$$

i.e., $$\boldsymbol{\lambda = \frac{h}{p}}$$ **Ans.**

Example 17. *Calculate the value of the difference between the energy of a hydrogen molecule having back and forth motion on a path of* 1 cm *long when* $n = 2$ *and* $n = 3$ *at* 298 K.

Solution. Mass of hydrogen molecule

$$= \frac{2}{6.02 \times 10^{26}} = 0.33 \times 10^{-26} \text{ kg}$$

we have the energy of a particle confined in a box given by the relation,

$$E_n = \frac{n^2 h^2}{8\ ma^2}$$

$$\therefore \quad E_2 = \frac{4h^2}{8\ ma^2}$$

and $$E_3 = \frac{9h^2}{8\ ma^2}$$

$$E_1 - E_2 = \frac{h^2}{8\ ma^2}(9-4)$$

$$= \frac{5\ h^2}{8\ ma^2}$$

in this case, $m = 0.33 \times 10^{-26}$ kg

$$a = 0.01 \text{ m}, h = 6.626 \times 10^{-34} \text{ J.s}$$

$$\therefore \quad E_3 - E_2 = \frac{5 \times (6.626) \times 10^{-34})^2}{8 \times 0.33 \times 10^{-26} \times (0.01)^2}$$

$$= 83.1 \times 10^{-38} \text{ J}$$

$$= \frac{83.1 \times 10^{-38}}{1.6 \times 10^{-19}} \text{ J}$$

$$= \mathbf{5.2 \times 10^{-18}} \textbf{ eV} \quad \textbf{Ans.}$$

Example 18. *A particle is confined to a one-dimensional infinite potential well of width 0.2 nm. It is given that where the energy of the particle is 230 eV, its eigen function has 5 antinodes. Find the mass of the particle and show that it can never have energy equal to 1 keV.*

Solution. Since the eigen function corresponding to energy 230 *eV* has 5 antinodes, the particle is in a quantum state with $n = 5$. For this case.

$$E_5 = 5^2 \; E_1 = 25 \; E_1 \; 230 \; eV$$

so that $E_1 = 9.2 \; eV = 9{,}2 \times 1.6 \times 10^{-19}$ J

Also, $a = 0.2 \; nm = 0.2 \times 10^{-19} \quad m = 2 \times 10^{-10} \; m$

Now, $$E_1 = \frac{\hbar^2 \pi^2}{2ma^2} = \frac{h^2}{8ma^2}$$

or $$m = \frac{h^2}{8a^2 E_1} = \frac{(6.626 - 10^{-34})^2}{8 \times 4 \times 10^{-20} \times 14.7 \times 10^{-19}} \text{ kg}$$

$$= 9.3 \times 10^{-31} \text{ kg}$$

For $E_n = 1 \; keV$, n should be such that

$$n^2 = \frac{E_n}{E_1} = \frac{10^3}{9.2} = 108.7$$

i.e., $n = 10.4$

Since n is not an integer, $E_n = 1 \; ke \; V$ is not a permitted value of energy. **Ans.**

Example 19. *Calculate the value of lowest energy of an electron moving in a one-dimensional force free region of length 4 Å.*

Solution. We have m = 9.1×10^{-31} kg, L = 4×10^{-10}m

Since energy of electron in *n*th state is given by

$$E_n = \frac{n^2 \pi^2 \hbar}{2ma^2} = \frac{n^2 h^2}{8ma^2}$$

For ground state $n = 1$

$$E_1 = \frac{h^2}{8ma^2} = \frac{(6.625 \times 10^{-34})}{8 \times 9.1 \times 10^{-31} \times (4 \times 10^{-10})^2}$$

or $E_1 = 0.376 \times 10^{-18}$ J

Example 20. *Calculate the value of momentum and energy of an electron in a box of wavelength 1Å, for n = 2.*

Solution. We have L = λ = 1 Å = 10–10 *m*, *n* =2

Since momentum, $p_n = \frac{n\pi\hbar}{2} = \frac{nh}{2a}$

For $n = 2$

$$p_n = \frac{2\times 6.625\times 10^{-34}}{2\times 10^{-10}} \text{ kg ms}^{-1}$$

$$p_2 = 6.625\times 10^{-24} \text{ kg ms}^{-1}$$

Ans energy, $E_n = \frac{n^2\pi^2\hbar}{2ma^2} = \frac{n^2h^2}{8ma^2}$

For $n = 2$

$$E_2 = \frac{4\times(6.625\times 10^{-34})^2}{8\times 9.1\times 10^{-31}\times 10^{-20}}$$

$$E_2 = 150.8 \ eV$$

Example 21. *Evaluate the expectation value of kinetic energy of a particle in one dimensional rigid box in the nth quantum state.*

Solution. We know that the wave function of a particle enclosed in one dimensional box is given by

$$\psi(x) = \sqrt{\frac{2}{a}}\sin\left(\frac{m\pi x}{a}\right)$$

The expectation value of kinetic energy E is given by

$$<E> = \int_{-\infty}^{\infty} \psi^* \ E\psi \ dx = \int_{\infty}^{\infty} \psi^* \frac{p^2}{2m}\psi \ dx$$

$$= \frac{1}{2m}\int_{\infty}^{\infty} \psi^* \ (-i\hbar)^2 \frac{\partial^2}{\partial x^2}\psi \ dx$$

$$= \frac{\hbar}{2m}\int_{\infty}^{\infty} \psi^* \frac{\partial^2}{\partial x^2} dx$$

$$= \frac{\hbar^2}{2m}\int_0^L \frac{2}{a}\sin\left(\frac{n\pi x}{a}\right)\frac{\partial^2}{\partial x^2}\left[\sqrt{\frac{2}{a}}\sin\left(\frac{n\pi x}{a}\right)\right]dx$$

$$= \frac{2\hbar^2}{2ma}\int_0^a \sin\left(\frac{n\pi x}{a}\right)\frac{\partial^2}{\partial x^2}\left[-\frac{n^2\pi}{a^2}\right]\sin\frac{n\pi x}{a} dx$$

$$<E> = \frac{n^2\pi^2\hbar^2}{ma^3}\int_0^a \sin^2\frac{n\pi x}{a} dx \qquad \left[\begin{array}{c}-\text{ Ve sign} \\ \text{is omitted}\end{array}\right]$$

$$= \frac{n^2\pi^2\hbar^2}{2ma^3}\int_0^a\left[1-\cos\frac{2n\pi x}{a}\right]dx$$

$$<E> = \frac{n^2\pi^2\hbar^2}{2ma^3}\cdot a = \frac{n^2\pi^2\hbar^2}{2ma^2}$$

$$\therefore \qquad <E> = \frac{n^2\hbar^2}{8ma^2} \quad \textbf{Ans.}$$

Example 22. *A particle of mass 10^{-6} kg is moving with a speed of 10^{-4} m s^{-1} in a box of length 10^3 Å. Assuming this to be one dimensional square well problem, calculate the value of n.*

Solution.

We have $m = 10^{-6}$ kg, $v = 10^{-4}$ ms^{-1}, $a = 10^3$ Å $= 10^{-7}$ m

Since energy of a particle is *n*th state is given by

$$E_n = \frac{n^2\pi^2\hbar^2}{2ma^2}$$

$$n^2 = \frac{8\,ma^2 E_n}{h^2} \qquad \left[\because \hbar = \frac{h}{2\pi}\right]$$

or

$$n^2 = \frac{4m^2L^2v^2}{h^2} = \frac{4m^2\,a^2\,v^2}{h^2} \qquad \left[\because E_n = \frac{1}{2}mv^2\right]$$

or

$$n = \frac{2mav}{h} = \frac{2\times10^6\times10^{-7}\times10^{-4}}{6.625\times10^{-34}} = \mathbf{3\times10^{16}} \quad \textbf{Ans.}$$

QUESTIONS

1. Derive Schrödinger's (*i*) time-independent and (*ii*) time-dependent equations of matter waves. Give the physical interpretation of the wave function.
2. Derive Schrödinger's time independent wave equation. What is the physical significance of stultification 'ψ' used in this equation? What condition must it fulfil?
3. Drive time dependent and time independent Schrödinger wave equation.
4. Can you measure the position and momentum of a particle simultaneously with perfect accuracy? Obtain an expression relating uncertainties Δx and Δp.
5. Derive time-dependent time-independent Schrödinger equations for a non-relativistic free particle. Give the stationary state solution of Schrödinger wave equation.
6. Derive time independent Schrödinger wave equation.
7. Find the expression for the energy state of a particle in one dimensional box.
8. Solve the Schrödinger equation for a particle enclosed in a one-dimensional rigid box of side *L*. Obtain its eigen values. Draw a graph of its first three eigen functions. Discuss the probability of finding the particle at different points in the box when it is in different states.
9. Solve the Schrodinger equation in three dimensions for a free particle and discuss the salient features of the wave function.
10. Find the eigen functions of a free particle using cartesian coordinates and show that the energy values are continuous.
11. Discuss quantum mechanically the problem of a particle in a finite square potential well. Draw diagrams showing the amplitude wave and probability density for the same. What will be the effect of increasing the width of the square well on energy levels?
12. Explain the problem of the leaking of a particle through a rectangular potential barrier of finite width.
13. Solve the Schrödinger equation for a particle in one dimensional square well potential. Derive an expression for the transmission coefficient and explain its significance.
14. Give steady state form of the Schrödinger's equation for a particle in three dimensional rigid box. Solve this equation to obtain the energy eigen values and eigen functions of the particle.
15. Obtain Schrödinger wave equation for a particle in a square well potential and discuss energy levels when the well is infinitely deep.

NUMERICAL PROBLEMS

1. Consider a particle of mass m trapped in a two dimensional box L long and w wide. Starting from Schrodinger equation, show that the permitted energies of the particle are given by

$$E = \frac{h^2}{8m}\left(\frac{a^2}{L^2} + \frac{b^2}{w^2}\right) \quad \text{where } a \text{ and } b \text{ are positive integers.}$$

2. Calculate the first two energy levels of an electron confined to a box 1 A wide. (h = 6.254 × 10–34 J sec, $m_e = 9.1 \times 10^{-31}$ kg). **[Ans.** *(i)* 37.7 eV *(ii)* 1.56.8 eV]
3. An important property of eigen function of an operator is that they are orthogonal to one another, *i.e.*,

$$\int \psi_n \psi_m \, dv = 0 \quad n \neq m$$

Verify this relation for a particle in one dimensional box using the relation $\sin\theta = (e^{2\theta} - e^{-2\theta})/2i$
4. Find the lowest energy of neutron confined to a box 10^{-14} m across (nucleus has size of this magnitude). **[Ans.** 2.1 MeV]
5. Show that the lowest ground state energy level of a particle in one dimensional potential box with rigid walls is in agreement with the uncertainty principle.
6. Find the energies of the six lowest energy levels of a particles in a cubical box of length-∠.

$$\left(3,6,9,11,12 \text{ and } 14 \times \frac{\pi^2 \hbar^2}{2mL^2}\right)$$

7. A spherical dust particle of radius r = 10 = 5 m density $\rho = 10^4$ kg/m^3 and moving with a velocity v = 10^{-2} ms^{-1} encounters a potential step of height equal to twice the kinetic energy of the particle. Estimate the penetration depth d_p of the particle inside the step. **[Ans.** 2.5 × 10^{-22} m]
8. Calculate the transmission probability for an electron of total energy 2 eV incident upon a rectiongular potential barrier of height 4 eV and width 10^{-9} m. **[Ans.** 2 × 10^{-6}]
9. Calculate the probability of transmission for a proton of energy 1 MeV through a 4 MeV high rectangular potential energy barrier of width 10^{-2} cm. **[Ans.** 0.0015]
10. Find the height of the potential barrier in eV for α -particle emitted from an element (At. No. Z = 86). Assume the effective nuclear radius = 10^{-14} m. **[Ans.** 24.16 MeV].

10

SUPER CONDUCTIVITY

10.1. Introduction

The electrical resistivity of all metals and alloys decreases when they are cooled. For perfectly pure metal, the resistivity should approach zero as the temperature is reduced to OK. This zero resistance which a hypothetical perfect specimen would acquire if it could be cooled absolute zero in not, however the phenomena of superconductivity. Any real specimen of metal cannot be perfectly pure. As a result there will be a certain residual resistively (P_o) which remains at the lowest temperatures. The more impure the metal, the larger will be its residual resistivity.

Certain metal, however, show a very remarkable behaviour, when they are cooled, their electrical resistivity decreases in the usual way, but at certain temperature called critical temperature (T_c) the resistivity abruptly drops to zero. This new state is called 'superconducting state'. This phenomenon can even occur in impure specimen.

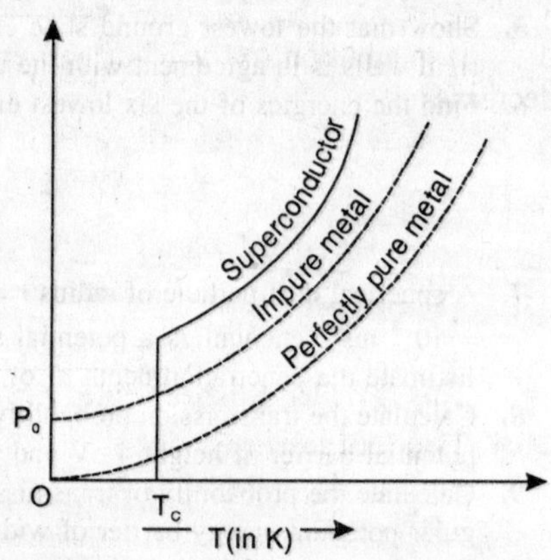

Fig (10.1) Variation of resitivity with temp.

This phenomenon was first discovered by K Onnes in 1911, when he was measuring the resistivity of mercury at low temperature. He observed that the electrical resistivity of pure mercury suddenly drops to zero at 4.15K.

10.2. Effect of Magnetic Field

In addition to the current, magnetic field can also destroy the superconductivity. The magnetic field at which superconductivity vanishes is called critical magnetic field H_c. Its value depends upon the nature of the material and its temperature. The variation of Hc with temperature is nearly parabolic and can be represented

by the equation $H_c(T) = H_c(o)\left[1 - \frac{T^2}{T_c^{\,2}}\right]$ ——— (10.1)

where H_c (T) is the critical magnetic field at temperature T and H_c (o) be its value at OK.

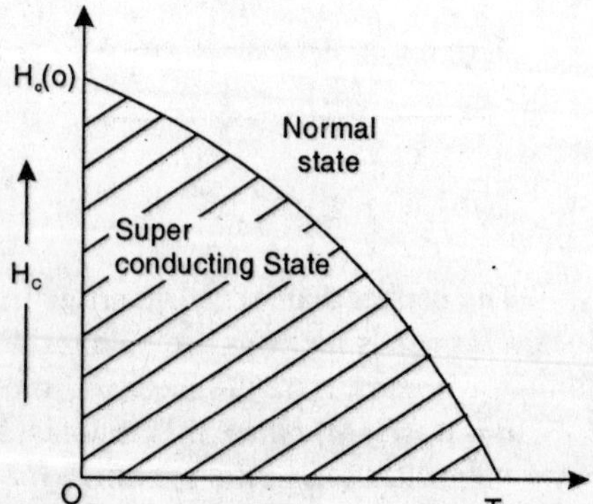

Fig (10.2) Variation of H_C with temp.

10.3. Critical Currents

The magnetic field which destroy the superconductivity is not necessarily an external applied field, it may also be the magnetic field due to the current flowing in the superconducting wire etc. The maximum value of current flowing through the superconductor at which superconductivity is destroyed is called critical current.

If super conducting wire of radius r carries current I, then using Ampere's law

$$\oint \vec{H} \cdot \vec{dl} = I$$

$$\oint H dl = I$$

$$H.2\Pi r = I$$

$$I = 2\pi r H$$

at $H = H_c \quad I = I_c$

$\Rightarrow \boxed{I_c = 2\pi r H_c}$ It I becomes Ic superconductivity will be destroyed.

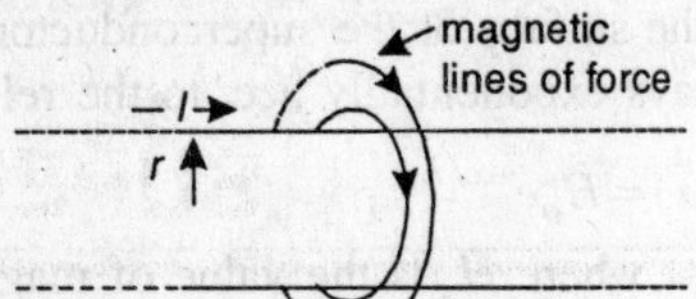

Fig (10.3) Superconducting wire carrying current

If in addition to current, transverse magnetic field H is applied, the value of critical current decreases

Now $H_c = H_I + 2H$ where H_I is the field due to current

$$H_I = H_c - 2H$$

$$H_I = \frac{I_c}{2\pi r} = H_c - 2H$$

$$\boxed{I_c = 2\pi r\,(H_c - 2H)} \quad \text{——— (10.2)}$$

This is called Silsbee's rule.

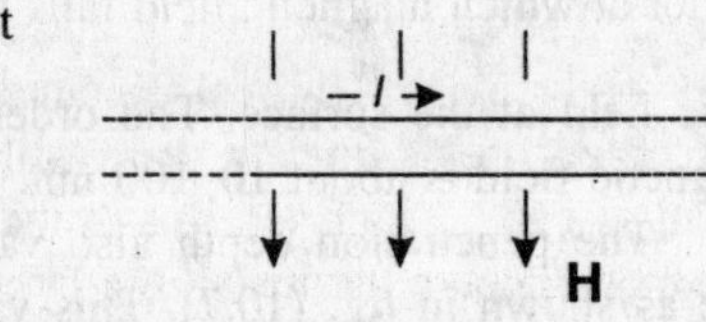

Fig (10.4) Current carrying superconductor in transverse magnetic field.

10.4. The Meissner Effect

When superconducting material cooled below its critical temperature, it not only becomes resistanceless but perfectly diamagnetic also. That is to say there is no magnetic field inside a superconductor, whereas inside a merely resistanceless metal there may or may not be a magnetic field, depending on the circumstances.

This effect, when superconductor placed inside a magnetic field in cooled below its critical temperature, all the magnetic flux is expelled out of it, is called Meissner effect.

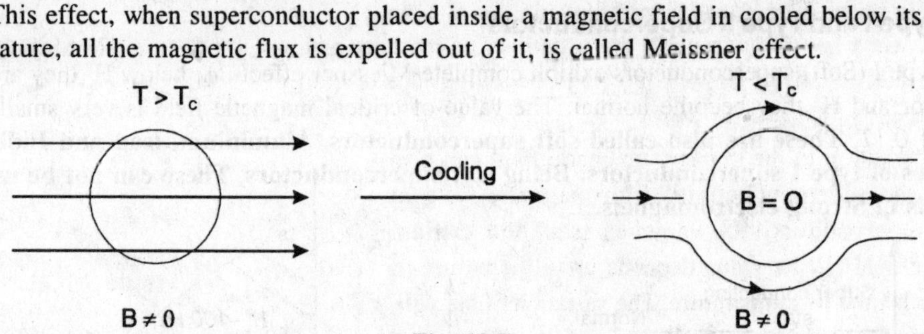

Fig (10.5)

The perfect diamagnetism arises from some special bulk magnetic property of the superconductor. If there is no magnetic field inside the superconductor it can be said that its relative permeability μr is zero. Here the mechanism of diamagnetism is not considered we know

$B = \mu_0 (H+M)$ where B is resultant magnetic induction H is the extend magnetic field, and M is the magnetization of the specimen for superconductor $B = O$

$$O = \mu_0 (H+M)$$

$\Rightarrow M = -H$ and the magnetic susceptibility

$$\chi_m = \frac{M}{H} = -1$$

It must be noted that superconductivity is not only a strong diamagnetism, but also a new type of diamagnetism.

10.5. The Penetration Depth

In 1935 F. London and H London described that during superconding state, the field does not suddenly drop to zero at the surface of the superconductor, but decays exponentially acc. to the relation

$$H(x) = H_o e^{-x/\lambda}$$

where H_o is the value of magnetic field at the surface and $H_{(x)}$ is the magnetic field inside the specimen at a distance x from the surface.

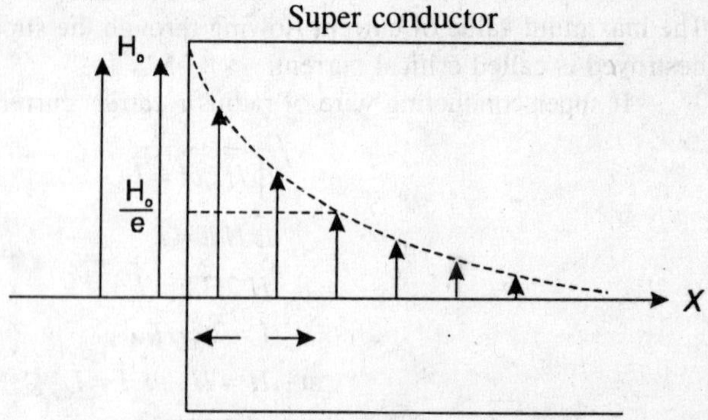

Fig. 10.6

λ is called London penetration depth. It is defined as the distance inside the superconductor at which magnetic field falls to $\frac{1}{e}$ time the magnetic field at the surface. The order of penetration of magnetic field is about 10–100 nm.

The penetration depth also varies with temperature as shown in fig. (10.7). This variation can be expressed as

$$\lambda(T) = \frac{\lambda(0)}{\sqrt{\left(1 - \frac{T^4}{T_C{}^4}\right)}} \qquad \text{—— (10.3)}$$

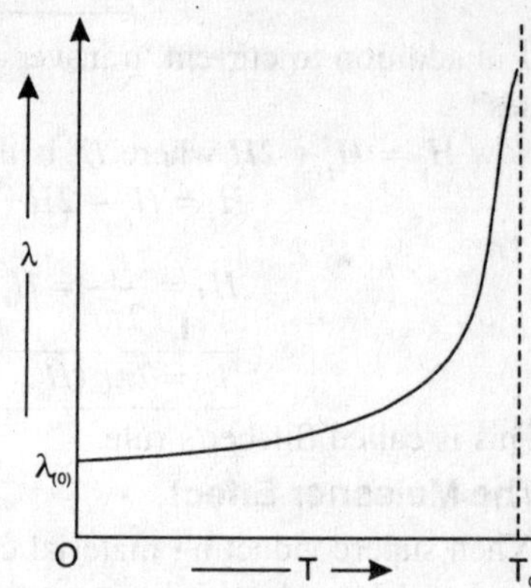

Fig. 10.7

where $\lambda(T)$ is value at temp. T, and $\lambda(0)$ is the value at absolute zero.

10.6. Type I and Type II Superconductors

Type I (Soft) superconductors exhibit complete Miessner effect *i.e.,* below H_c they are super conductor and H_c they become normal. The value of critical magnetic field is very small of the order of $0.1T$. **These are also called soft superconductors. Aluminium, lead and Indian are examples of type I superconductors. Being soft superconductors, These can not be used for the Coils of Strong electromagnets.**

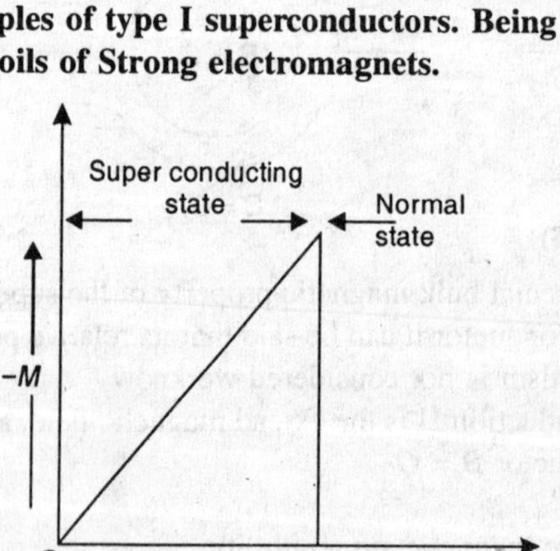

Fig. 10.8 a

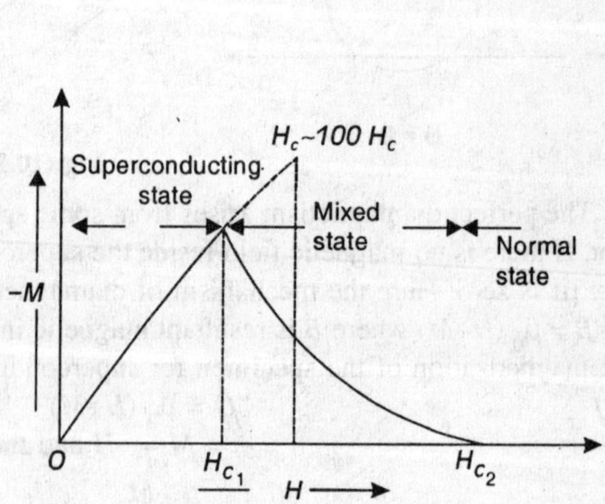

Fig. 10.8 b

Type II (Hard) superconductors exhibit complete Miessner effect below a certain critical field H_{C1}, at this value magnetic field begins to penetrate and its diamagnetism and superconductivity starts decreasing. It is in the mixed state also called vortex state. At certain value of magnetic field H_{c2}. The superconductivity disappears. Transition metals and alloys consisting of niobium, Aluminium and silicon exhibit type II superconductivity. Ceramic superconductors also belong to this category i.e. Hard Superconductors.

10.7. London Equations

This is macroscopic theory, based on the two-fluid-model. London brothers F London and H London derived two equation which explains the zero resistance state and perfect diamagnetism of superconducting state.

According to London's theory, it is assumed that there are two types of conduction electrons in a superconductor, namely the super electrons and normal electrons. At OK, a superconductor contains only super electrons, but as temperature increase super electrons decrease and normal electrons, increase.

The superconding electrons are not subjected to any lattice scattering and are therefore accelerated in an electric field.

Let u_n, u_n and n_s u_s respectively the no. density and drift velocity of normal and super electrons.

The equation of motion of superconducting electrons under the action of electric field is given by

$$m\frac{d\vec{u_s}}{dt} = -e\vec{E} \quad \text{———— (10.4)}$$

Now current and drift velocity are related as

$$I_s = n_s e A u_s$$

$$J_s = n_s e\, u_s$$

$$\vec{J_s} = -n_s e\,\vec{u_s}, \text{ the equation (10.4) becomes}$$

$$m\left(\frac{d}{dt}\left(-\frac{\vec{J_s}}{n_s e}\right)\right) = -e\vec{E}$$

$$\boxed{\frac{d\vec{J_s}}{dt} = \frac{n_s e^2\vec{E}}{m}} \quad \text{———— (1.05)}$$

This the London's first equation which describes the absence of resistance. If $\vec{E}$ is made zero then $\frac{d\vec{J_s}}{dt} = 0$

$\vec{J} =$ constant, *i.e.*, steady currents exists even if electric field is zero.

Now from Maxwell's equation

$$\vec{\nabla}\times\vec{E} = -\frac{\partial\vec{B}}{dt}$$

$\vec{B}$ in terms of magnetic vector potential is given as $\vec{B} = \vec{\nabla}\times\vec{A}$

$$\vec{\nabla}\times\vec{E} = -\frac{\partial}{\partial t}\left(\vec{\nabla}\times\vec{A}\right)$$

$$\vec{\nabla}\times\vec{E} = -\vec{\nabla}\times\frac{\partial \vec{A}}{\partial t}$$

$$\vec{E} = -\frac{\partial \vec{A}}{\partial t}$$

From equation (10.5)

$$\frac{m}{n_s e^2}\frac{d\vec{J}}{dt} = -\frac{\partial \vec{A}}{\partial t}$$

$$\frac{d}{dt}\left(\frac{m}{n_s e^2}\vec{J_2}\right) = \frac{\partial \vec{A}}{\partial t}$$

$$\frac{m}{n_2 e^2}\vec{J_s} = -\vec{A}$$

$\boxed{\vec{J_s} = -\frac{n_s e^2}{m}\vec{A}}$ —— (10.6) This is London's second equation Again from Maxwell's equation for steady state

$$\vec{\nabla}\times\vec{B} = \mu_0\vec{J} \text{ from equation 10.6}$$

$$\vec{\nabla}\times\vec{B} = \mu_0\left(-\frac{n_s e^2}{m}\vec{A}\right)$$

Take curl on both sides

$$\vec{\nabla}\times\vec{\nabla}\times\vec{B} = -\mu_0\frac{n_s e^2}{m}\vec{\nabla}\times\vec{A}$$

$$\vec{\nabla}\times(\vec{\nabla}.\vec{B}) - \nabla^2\vec{B} = -\frac{\mu_0 n_s e^2}{m}\vec{B}$$

$$\nabla^2\vec{B} = \frac{\mu_0 n_s e^2}{m}\vec{B}$$

$\nabla^2\vec{B} = \frac{1}{\lambda^2}\vec{B}$ where $\lambda = \sqrt{\frac{m}{\mu_0 n_s e^2}}$ is constant at given temperature and is called London penetration depth in one dimension the question can be written as

$$\frac{d^2 B}{dx^2} - \frac{1}{\lambda^2}B = 0$$

The solution of this equation is of the form

$$\boxed{B(x) = B_0 e^{-x/\lambda}} \quad \text{—— (10.7)}$$

Inside the superconductor, B decreases exponentially and within very small penetration it becomes zero, which explains Miessner effect.

10.8. BCS Theory of Superconductivity

Bardeen Cooper Scruffier offered an explanation for the superconductors with a theory called **BCS theory** (after the name of the inventors).

They proposed that by using the knowledge of **solids** and applying the **quantum mechanics,** it is possible to explain the superconducting state in which the motions of all the electrons in conduction band are **locked together.** In this state the electrons are coupled together in pairs at **very low temperatures.** One electron interacts with a **crystal lattice** and perturbs it and this perturbed lattice then interacts with the second electron in the pair in such a way that there is an **attraction** between the two electrons and this attraction overcomes the **coulomb repulsion** between them. The electron interaction *via* **lattice** produces an energy gap between the superconducting state in which electrons act **collectively** while in non-superconducting (normal) state these electrons act **individually.**

The energy required to excite the system to the higher non-superconducting state and **destroy superconductivity** is not available at lower temperature. Thus the material becomes **superconductor** at such a low temperature.

BCS theory has been successfully able to explain all the experimental facts of superconductivity. In the latest research efforts are being made to have **underground** power lines to **transport** electricity from generating plants to consumers. These lines will have **negligible resistance** and so **energy losses** and **heating effects** will altogether be prevented. However, for this we will require liquid helium to make the wires act as **superconductors** at low temperatures and it is quite a difficult job.

Josephson Effect. In 1962, Brian Josephson, an English Physicist predicted that if a very thin layer (about 2 nm) of an insulator (normally an oxide) is sandwiched between two **superconductors** and a voltage *V* is applied across the two superconductors, a current starts flowing. This current has two components (*i*) continuous dc component which persists even when the voltage is removed (*ii*) ac component which is present only when the voltage V is present.

An important thing to note is that the ac component is always having frequency exactly $\nu = \frac{2eV}{h}$, no matter which materials are used for the so called **Josephson function.**

So this phenomenon is useful for finding accurately the ratio e/h using the equation.

$$\frac{e}{h} = \frac{\nu}{2V}$$

Since frequency can be measured very accurately, the value of can be measured precisely.

The results obtained are so accurate that volt can now be defined in terms of frequency using equation (1). So instead of **standard cell** or other **measuring instruments,** this method *gives very good calibration of voltmeter.*

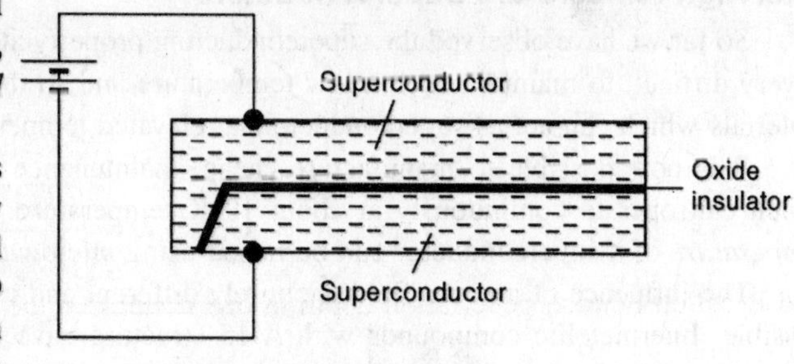

Fig. 10.9

Recently, Josephson junctions have led to the development of very sensitive detectors and fast and efficient memory computers. SQUID (superconducting quantum interference device) detectors have been successful used to measure poison levels of iron of liver, so that iron build up can be treated before much harm is done to the body.

10.9. Application of Superconductors

1. They form a new generation of energy saving power systems. Superconducting generators are very small and compact in size and weight as compared to the conventional generators. In U.S.A. many companies have built superconducting generators.
2. Electric power companies are doing research on superconductors to eliminate the power losses in transmission lines.

3. In Japan and France, superconducting magnets have been used to raise the trains above the railway tracks and attain very high speeds as frictional losses are minimized. A similar magnetic propelsion system is being used to launch the satellites into orbits directly from the earth without the use of rockets.
4. As the current in a superconducting ring can flow without change in its value, it can be used as a memory or a storage element in computers.
5. High efficiency ore-separating machine may be built using super conducting magnets. Also super conducting magnets may be used to separate tumour cells from the healthy cells and is finding great use in medical sciences.
6. Superconductors can act as a relay or switching system in a computer. The basic principle can be understood from the Fig. 10.10. The fact that when we apply a magnetic field whose value is more than the *critical field of the superconductor, the superconducting properties disappear* is used in gating circuits or in cyclotrons. Consider a wire A of superconducting material, inside a superconducting material B. Let the temperature of the two be lowers than the transition temperature of the two materials.

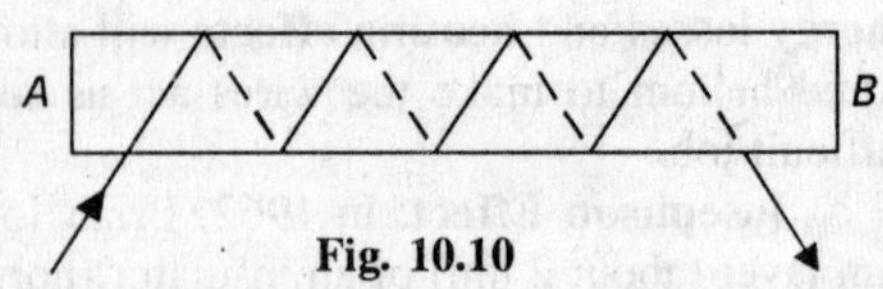

Fig. 10.10

Current in central wire *A* can be controlled by a current in coil B as the magnetic field produced by B may exceed the critical field of the core material (wire) *A* at the operating temperature. It means that *A* will become a normal conductor and system can be used as *relay or a switch* in computers.

7. Very fast and accurate computers can be constructed using superconductors and the power consumption is also very low ~ 0.5 watt.
8. Superconducting solenoids are used in *magneto hydrodynamic* power generation to maintain plasma and also NMR (Nuclear Magnetic Resonance) Technique to have the image of total body which can be scanned to find the unhealthy cells or tumours.
9. Superconductors are also used to detect the brain wave activities, for building potent beam weapons and also in harnessing the various forms of nuclear energies.

10.8. High Temperature Superconductors

So far we have observed the superconducting property at very low temperatures. However, it is very difficult to maintain such a low temperature and so these days research is on to find new materials which can act as superconductors at elevated temperatures.

It is now possible to manufacture cheap, maintenance free and and compact refrigerators which can operate continuously at about 10 K temperature. Also it is observed that *transition temperature* of a superconductor can be raised using *microwaves and lasers.*

The influence of an e.m. field is entirely different and a macroscopic explanation is not yet possible. Intermetallic compounds with A-15 structure have been found to have transition temperatures ~ 25 K. There are theoretical predictions that some *special organic compounds* may exhibit super conductivity even above room temperatures. *Sodium, Tungesten, Bronzes and Barium, Lead-bismuth oxides are important intermetallic compounds with high superconducting transition temperatures.*

In Bell laboratory, during 1986, it was discovered that a crystalline complex of lanthanum, strontium and copper grown in an atmosphere of oxygen has a high *transition temperature* ~ 37 K.

Following this discovery, a chinese group of Scientists found superconducting state in an octahedral crystalline complex of *lead, Lanthanum, copper and oxygen at 77 K.*

In 1987, Nobel prize was awarded to **Karl Alexander Muller** and Johnes Georg Bednorz of IBM Zurich Research Laboratory for their discovery of **high temperature (35 K) super conduc-**

tivity in certain Lanthanum-Copper oxide ceramic materials. After that C.W. Chu and co-workers found super conductivity at 90 **K** in **yittrium, Barrium oxide** compounds. In these days the researches are going on to study the superconductivity at even the room temperature. The day is not far off when **superconductors** may enter the various fields in a big way and completely change the total concepts of so many things that we see around.

10.11. Squids

SQUID may be abbreviated as Super Conducting Quantum Interference Device. Basically, SQUID is a **super conducting loop** which has *week link* that can measure change in magnetic flux within the loop. As can be visualized, a *weak link* corresponds to a region that has lower critical current than the rest of the superconducting ring. As soon as the current in the link is more than the critical current, link becomes normal and allows fluxons (A flux quantum) to penetrate the link and current falls to critical value and the link goes back to the super conducting state. So the weak link behaves like a gate (*g* in the figure), which can be adjusted in such a way that only one fluxon may enter through it.

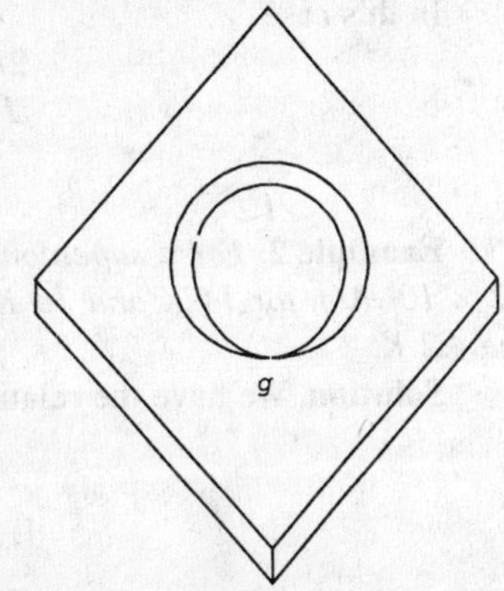

Fig. 10.11

A graphical representation of the critical current in the week link is shown in the Fig. 10.11 from which it is clear that the critical current varies periodically with the change of total flux through the area enclosed by superconducting loop and the current undergoes one cycle for each change of fluxon by factor ϕ_0 $\left(\text{where, } \phi_0 = \frac{h}{2e}\right)$.

Periodic variation of SQUID current can be studied by a neighbouring coil which may act as a sensor as the changing magnetic field due to change in squid current induces current and e.m.f. in that coil. This induced e.m.f. may be used to drive an **electronic counting circuit.**

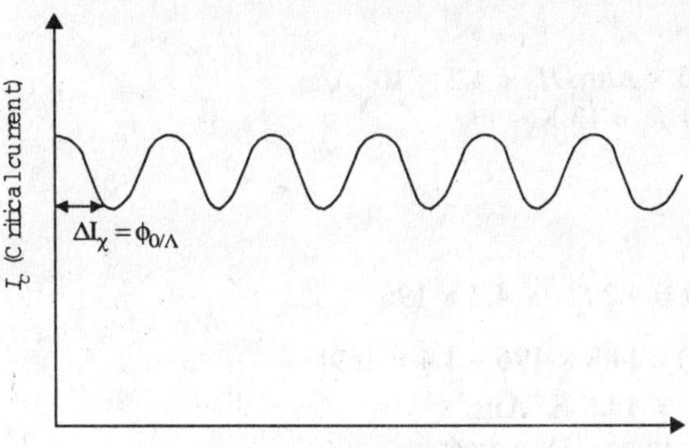

Fig. 10.12

Applications

Squids act as very sensitive magnetometers and find applications in various electric and magnetic field measurements. They are highly useful in the medicine field as **heart and brain** generate very feeble electric current which may correspond to weak magnetic fields ~ 10^{-14} to 10^{-15} Tesla. As SQUIDS can respond to very weak fields, so thay can increase the intensity of these fields and help in proper diagnosis of the heart and brain disorders.

10.12. Isotope Effect

It is observed that the critical temperature of super conductors varies with isotope mass. This observation was fissh studied by *Maxwell* and others using mercury isotopes. An idea of magnitude of the effect can be had from the case of mercury in which T_c varies from 4.1895 to 4.146 k as the isotope mass varies from 199.5 to 203.4.

The explanation given is that isotope mass can enter in the process of formation of superconducting phase of electron-Phonon interaction.In the early years of BCS theory development, a simple relation was given as

$$T_c \propto M^{-\beta}$$

where β was *through to be in most* case given by

$$\therefore \quad T_c . m^{+\beta} = \text{constant.}$$

However, later on some departure from the above relation was also observed.

Example 1. *Calculate the critical which can flow through a long thin super conducting wire of Aluminium of diameter 10^{-3} m. Critical magnetic field for Aluminium is 7.9×10^3 Amp/metre.*

Solution : There critical current flowing through a wire is given by

$$I_c = 2\pi r \, H_c$$

In this case

$$2r = 10^{-3} \, m \text{ and } H_c = 7.9 \times 10^3 \text{ A/m.}$$

$$\therefore \quad I_c = \pi \times 2r \times H_c$$

$$= \pi \times 10^{-3} \times 7.9 \times 10^3$$

$$= \mathbf{24.806 \text{ Ampere} \quad Ans.}$$

Example 2. *For a superconducting secimen, the values of critical fields are 1.4×10^5 and 4.2×10^5 A/m for 14 K and 13 K. Calculate the transtion temperature and critical fields at OK and 4.2 K.*

Solution. We have the relation

$$\text{H}_c = \text{H}_c \left[1 + \left(\frac{T_1}{T_c}\right)^2\right] \quad \text{... (1)}$$

$$\text{H}'_c = \text{Ho} \left[1 + \left(\frac{T_2}{T_c}\right)^2\right] \quad \text{... (2)}$$

$$\therefore \quad \frac{H_{c_1}}{H_2} = \frac{T_c^2 - T_2^2}{T_c^2 - T_2^2}$$

we are given $\quad \text{H}_c = 1.4 \times 105 \times \text{A/m}, \; H_c = 4.2 \times 10^5 \text{ A/m}$

$\text{T}_1 = 14 \text{ K and } T_2 = 13 \text{ K.}$

$$\therefore \quad \frac{1.4}{4.2} = \frac{T_c^2 - 14^2}{T_c^2 - 13^2}$$

or $\quad 1.4 \; T_c^2 - 1.4 \times 169 = 4.2 T_c^2 \; \times 4.2 \times 196$

$\therefore \quad T_c^2 \; (4.2 - 1.4) = [4.2 \times 196 - 1.4 \times 169]$

or $\quad \mathbf{T_c = 14.5 \text{ K} \quad Ans.}$

Putting 14.5 for Tc and 1.4 ö 105 for Hc in eq. (1), we get

$$\text{H}_0 = \frac{1.4 \times 10^5}{\left[1 - \left(\frac{14}{14.5}\right)^2\right]}$$

$$= \frac{1.4 \times 10^5}{1 - 0.9322} = 20.7 \times 10^5 \text{ A/m}$$

Critical fields at 0 K and 4.2 K are calculated as under :

AT 0 K $\quad H_c = H_0 \left[1 - \left(\frac{T}{T_c}\right)^2\right] = H_0 \, (1 - 0) = \text{H}_0$

$$\text{H}_c = \mathbf{20.7 \times 10^5 \text{ A/m} \quad Ans.}$$

AT 4.2 K $H_c = H_0\left[1-\left(\frac{T}{T_c}\right)^2\right]$

$$= 20.7 \times 10^5\left[1-\left(\frac{4.2}{14.5}\right)^2\right]$$

$$= \mathbf{18.9 \times 105\ A/m}$$ **Ans.**

Example 3. *Calculate the critical current density for 1 mm diameter wire of leat at (a) 4.2 K and (b) 7 K. A parabolic dependence of H_c on T may be assumed. Given T_c for lead = 7.18 K and H_c for lead = 6.5 × 10⁴ A/m.*

Solution. The formula is

$$H_c = H_0\left[1-\left(\frac{T}{T_c}\right)^2\right]$$

In this case $H_c = 6.5 \times 10^4$ A/m; $T = 4.2$ K and 7 K; $T_c = 7.18$ K.

$T = 4.2$ K

$$H_c = 6.5 \times 10^4\left[1-\left(\frac{4.2}{7.18}\right)^2\right]$$

$$= 4.28 \times 10^4 \text{ A/m}$$

Now $I_C = (2\pi r)\,H$

With $H_c = 4.28 \times 10^4$ A/m

and $r = 0.5 \times 10^{-3}$ m.

Critical current density is given by

$$g_C = \frac{I_c}{A} = \frac{2\pi r\, H_c}{\pi r^2} = \frac{2H_c}{r}$$

$$= \frac{2\times 4.28\times 10^4}{0.5\times 10^{-3}}$$

$$= \mathbf{1.71\times 10^8\ A/m^2}$$ **Ans.**

10.13. Thermodynamics of Superconductors

We can study the thermodynamic behaviour of superconductors and how they behave differently as compared to normal conductors by taking up some of the following properties.

(*i*) **Latent heat of transmission to superconducting state:** If the superconductivity of a superconductor destroyed with change of temperature (isotopically) using a magnetic field. Work is down on superconductor, thereby absorption of heat takes place. If the magnetic field is reduced to region superconductivity, work is done by the superconductor and thereby heat dissipation takes place.

This latent heat is given by

$$Q = \frac{\rho}{4\pi}\cdot\frac{H_C . T}{\omega}\cdot\frac{dH_C}{dt}.10^{-7}$$

Q is latent heat of transition

ρ is the density of specimen

W is the At. Wt. of the specimen

As $H_c = 0$ at T_c there will not be any latent heat of transition in the absence of a magnetic field.

(*ii*) **Specific heat.** In normal metals, electronic specific heat varies linearly with absolute temperature. In case of super conductors, the variations is exponential and given by

$$\text{Sp. } heat\ C_{el} \propto e^{-\Delta T/KB.T}$$

curve below show the variation of heat capacity of gallium in normal and superconducting states.

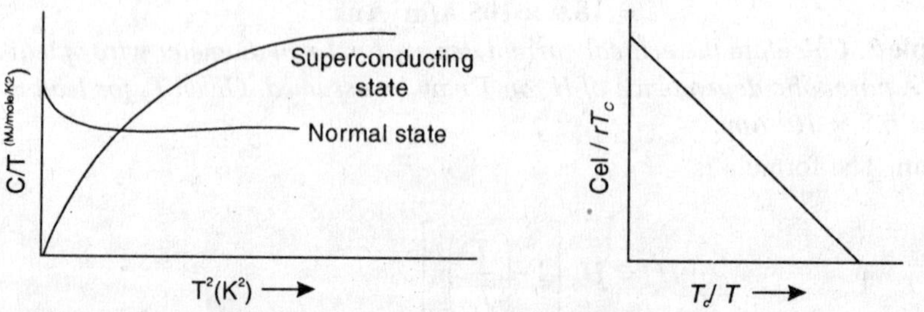

(a) Heat capacity of galluim in the normal and superconducting states (b) electronic heat capacity part in the super conducting state

Fig. 10.13

In the normal state (which may be restored by a magnetic field of 200 gauss) has electronic. lattice and nuclear quadraupole contribution. In Fig. 10.9 (b), electronic part C_{el} of specific heat is plotted in superconducting state on a log scale T_C/T. As is clear from the graph, there is exponential dependence on $\frac{1}{T}$ as shown by the straight line. It depicts the excitation of electrons through an energy gap.

(*iii*) **Entropy.** *Entropy corresponds to the disorder in a system.* In all the superconductors entropy decreases substantially below the critical temperature T_C.

A comparison between the normal state and superconducting state shows that superconducting state is more ordered than normal state, *i.e.*, the entropy decreases as a substance enters into super conducting state. We can conclude from this that most of the electrons excited in the normal state are ordered in the superconducting state.

In case of type 1, superconductors the order extends to a distance of 10^{-16} m and this range is called coherence length.

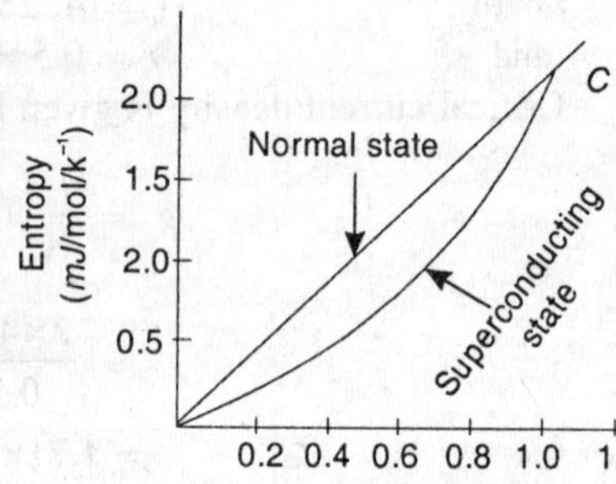

Fig. 10.14

Curves in Fig. 10.14 show the entropy of aluminium in normal and superconducting state.

(*iv*) **Energy gap in superconductors.** As stated earlier, sp. heat in a superconducting state varies exponentially with temperature which means that an energy gap may exist in the superconducting electron levels separating the lowest excited state from the ground state.

Energy gap will be a primary important property of superconducting state which also accounts for thermal properties and high frequency response of the superconductor.

In the case gallium. sp. heat is found to vary as $e^{-Eg/2k_B.T} = e^{-\Delta/kBT}$ where $\Delta = Eg/2$.

Δ is found to be equal to $1.4k_B.T_C$.

So, Energy Gap $E_g = 2\Delta = 2\times 1.4k_B.T_c$

$$\therefore \qquad \frac{E_g}{T_c} = 2.8k_B$$

$$\text{or} \qquad \frac{E_g}{k_B.T_c} = 2.8$$

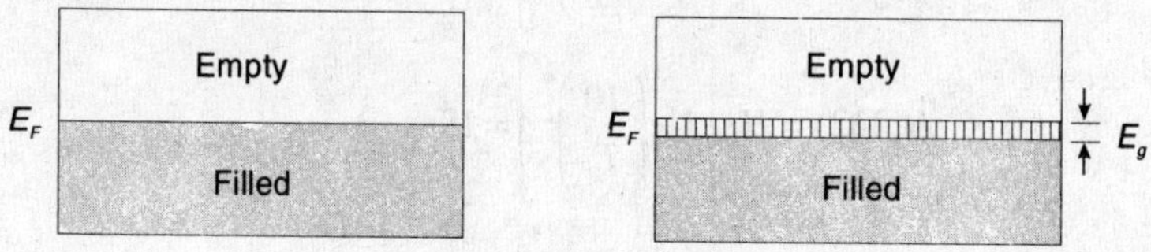

Fig. 10.15

Δ is also called Energy Gap Parameter. Further as shown in the curve below, energy gap decreases continuously to zero as the temperature is increased and reeches transition temperature.

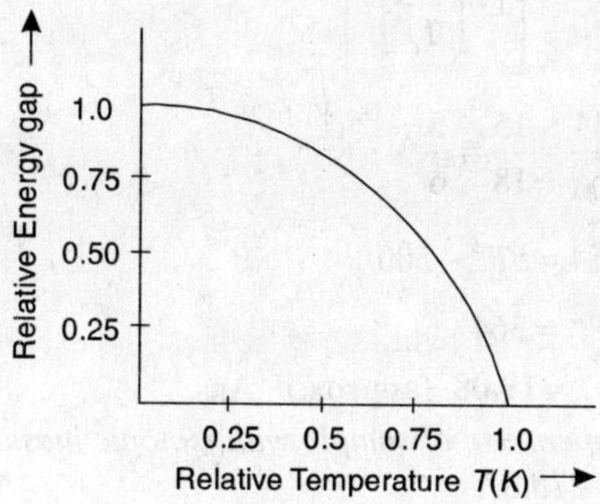

Fig. 10.16

Example 4. *Critical temperature of a superconductor when no magnetic field is present is T_c. Find the temperature at which the critical field becomes half of its value at 0.K.*

$$H_c = H_0\left[1-\left(\frac{T}{T_c}\right)^2\right]$$

In this case, $$H_c = \frac{H_0}{2}$$

$$\therefore \qquad \frac{H_0}{2} = H_0\left[1-\left(\frac{T}{T_c}\right)^2\right]$$

$$\text{or} \qquad \left(\frac{T}{T_c}\right)^2 = \frac{1}{2}$$

$$\therefore \qquad T = \frac{T_c}{\sqrt{2}} = 0.24\,T_c \quad \textbf{Ans.}$$

Example 5 : *Experiments on a sample show that at 12 k a critical field of 15T exist and at 10 K the critical field is 18T. Calculate the transition temperature and critical field at o K.*

Solution. We have the relation,

$$H_c = H_0\left[1-\left(\frac{T}{T_c}\right)^2\right]$$

At 12 K, $H_c = 15T$

At 10 K, $H_c = 18T$

$\therefore$ Above relation may be written as

$$H_c(12) = H_0\left[1-\left(\frac{12}{T_c}\right)^2\right] = 15$$

and $$H_c(12) = H_0\left[1-\left(\frac{10}{T_c}\right)^2\right] = 18$$

$$\therefore \quad \frac{H_c(12)}{H_c(10)} = \frac{\left[1-\left(\frac{12}{T_c}\right)^2\right]}{\left[1-\left(\frac{10}{T_c}\right)^2\right]} = \frac{15}{18}$$

$$\therefore \quad \frac{T_c^2-144}{T_c^2-100} = \frac{15}{18} = \frac{5}{6}$$

or $$6T_c^2 - 864 = 5T_c^2 - 500$$

$$T_c^2 = 364$$

or $T_c = \mathbf{19.08}$ **(approx.)** Ans.

Example 6. *Critical temperature of sample with isotopic mass of 204.87 is 19.2 K. Find T_C when isotopic mass change to 218.87.*

Solution. Isotopic relation is given by

$$T_c . m^{1/2} = \text{constant} \qquad \text{... (1)}$$

Now, $T_c = 19.2$ for $M = 204.87$

and $T_c = ?$ for $M = 218.87$

So, we have

$$(19.2)\,(204.87)^{1/2} = \text{constant} \qquad \text{... (2)}$$

$$T_c = (218.87)^{1/2} = \text{constant} \qquad \text{... (3)}$$

From (2) and (3),

$$T_c = \frac{19.2 \times (204.87)^{1/2}}{(218.87)^{1/2}}$$

$$= \mathbf{18.58\ K}\ \ \textbf{Ans.}$$

Example 7. *Penetration depth of S_n at absolute zero is 3.4×10^{-6} cm. Find the value of penetration depth at 3.5 K. Transition temperature of S_n is 3.72 K.*

(Given : $\mu_0 = 4\pi \times 10^{-7}$ H/m, $m = 9.1 \times 10^{-31}$ kg and $e = 1.6 \times 10^{-19}$ C)

Solution : In this case,

$$\lambda\,(0) = 3.4 \times 10^{-6} \text{ cm} = 3.4 \times 10^{-8} \text{ m}$$

$$T_c = 3.72 \text{ K}$$

We have relation

$$\frac{\lambda(T)^2}{\lambda(0)} = \left[\left(1-\frac{T}{T_c}\right)^4\right]^{-1/2}$$

$$\therefore \quad \lambda(T) = \lambda(0).\left[1-\left(\frac{T}{T_c}\right)^4\right]^{-1/2}$$

or

$$\lambda(T) = \frac{\lambda(0)}{\left[1-\left(\frac{T}{T_c}\right)^4\right]^{1/2}}$$

$$\therefore \quad \lambda(3.5K) = \frac{3.4\times10^{-8}}{\left[1-\left(\frac{3.5}{3.72}\right)^4\right]^{1/2}}$$

$$= \mathbf{7.30 \times 10^{-8}\ m} \quad \textbf{Ans.}$$

Example 8. *The critical temperature T_C for Hg with isotopic mass 199.5 is 4.185 k. Find the critical temperature for an isotopic mass of 202.3.*

Solution. We have in relation

$$T_c\sqrt{m} = \text{const.}$$

$$\therefore \quad \frac{T_{C_1}}{T_{C_2}} = \sqrt{\frac{m_2}{m_1}} = \sqrt{\frac{202.3}{199.5}}$$

$$T_{C2} = \sqrt{\frac{199.5}{202.3}} \times 4.185$$

$$= 4.156\ \text{K}.$$

Example 9. *London pevetration depth for a sample at 6K and 7K are 41.2 nm and 180.3 nm respectively. Find its transition temperature and pevetration depth at 0K.*

Solution. We have the relation

$$\lambda(T) = \frac{\lambda(0)}{\left[1-(T/T_C)^4\right]^{1/2}} \quad \text{... (1)}$$

In this case $\lambda(6) = 41.2$ nm

$\lambda(7) = 180.3$ nm

So eq. (1) may be written as

$$\frac{\lambda(5)}{\lambda(7)} = \frac{\left[1-\left(\frac{6}{T_C}\right)^4\right]^{1/2}}{\left[1-\left(\frac{7}{T_C}\right)^4\right]^{1/2}}$$

$$\therefore \quad \left(\frac{42.1}{180.3}\right)^2 = \frac{T_C{}^4-(6)^4}{T_C{}^4-(7)^4}$$

which gives $T_C = 5.92$ K,

Putting in eq. (1), we get

$$\lambda(5) = \frac{\lambda(0)}{\left[1-\left(\frac{5}{T_C}\right)^4\right]^{1/2}}$$

$$\therefore \quad \lambda(0) = 41.1\, nm \times \left[1-\left(\frac{5}{T_C}\right)^4\right]^{1/2}$$

$$= \textbf{29.5 nm.} \quad \textbf{Ans.}$$

QUESTIONS

1. What is superconductivity? Give the BCS theory of super conductivity.
2. Explain Meissner effect of super conductivity.
3. Distinguish between type I and type II superconductors.
4. What is superconductivity? How it can be destroyed.
5. What is Meissner effect? State and explain London's equations of Superconductivity.
6. Treating the transition term normal to superconducting state as a reversible thermodynamic effect, find the effect of specific heat on superconductors.
7. Outline the various theories and experimental facts of superconductivity.
8. Explain isotropic effect on superconductors.
9. What is the effect of magnetic field on a super conductor.
10. Give the thermodynamics of super conductors.
11. What are soft and hard superconductors. How the energy gap concept is explained in super con ductivity.
12. Give some important applications of super conductors.
13. What is SQUID? Give its importance is super conductivity.
14. What are London equation? How the penetration depth is explained from these equation?
15. Explain why good conductors can not be super conductors.

NUMERICAL PROBLEMS

1. Critical temperature for mercury isotrope 199.5 is 4.185 K. Calculate the critical temperature for isotropic mass of 203.4. (Ans. 4.14K)
2. For a certain specimen, the critical fields are 2.8×10^5 A/m and 5.4×10^5 A/m for 14 K and 13 K respectively. Calculate the transition temperature and critical fields at (i) O K and (ii) 5.0 K. (**Ans.** (i) 2.25×10^6 A/m (ii) 2.0×106
3. A type I super conductor with T_c = 7 K has slope $\frac{dB_C}{dt}$ = – 25 mt/ K at T_c. Find its critical field at 6 K. (**Ans.** 23.21 mT)
4. Estimate the London penetration depth from the following data.
   T_c = 3.7 K ; density = 7.3×10^3 kg /m^3
   At. wt. = 118.7 ; Effective mass = 1.9 m_0
5. London penetration depth for Pb. at 3k and 7.1 K are respectively 29.6 nm and 173 nm. Calculate its critical temperature and the penetration depth at OK. (**Ans.** 7.193 K; 39 nm)
6. At what temperature is H_C (T) = 0.1 H_C (0) for Pb having T_C = 7.2 K. (**Ans.** 6.83 K)
7. Find the energy gap Eg for *Nb* (T_C = 9.25 K) at T = 0 and find the minimum Photon wavelength needed to break the cooper pair and show the region of that wavelength in e.m. spectrum) (**Ans.** $4.3 \times 1\text{Ɔ}^{-m}$; it is in the for I.R. region of e.m. spectrum)
8. Find the London penetration depth λ_0 at OK for Pb whose density is 11.3×10^3 kg / m^3 and atomic weight 207. It is T_C is 7.22, find the percentage increase in penetration depth at 3.61 K from the value at OK. (**Ans.** 32.8%)

9. Find the London penetration depth from the following data

Density = 7.3 × 10³ kg/m3; *At. wt.* = 118

effective mass = 1.9 m_0 ; T_C = 3.7 K

(Ans. λ = 0.0016 nm)

10. A voltage of 5.0 μ*V* is applied across Jofeptson Junction. Find the frequency of radiation from the junction. **(Ans. 2.4 × 10⁹ *H*$_z$)**

11. If Cs and C_N are the heat capacities for super conducting and normal state and critical field

$$H_C = H_0\left(1 - \frac{T^2}{T_C^{\,2}}\right)$$; then prove in relation $C_S - C_N = \dfrac{4\,H_0^{\,2}}{\mu_0 T_C}$

12. Calculate in critical current which can flow through a long then super conducting wire of diameter 1 m.m. The critical magnetic field for the material of wire in 7.9×10³ A/M. **(Ans. 24–81A)**

Physics–101
(B.Tech. 1st/2nd Semester, 2124)

Time : 3 Hours **Maximum Marks : 60**

Note : Section A is compulsory. Candidates are required to attempt Five questions from Section B and C selecting at least two questions from each Section.

Section –A **Marks : 2 Each**

1. (a) Write the main laser sources or optical fibre systems.
   (b) Explain magnetic scalar potential.
   (c) Give the importance of mass-energy relationship.
   (d) What is the importance of coherence in laser?
   (e) Explain the interactions of x-rays with matter.
   (f) What are the various conditions for the acceptability of wave function?
   (g) Explain how the continuous and characteristics X-rays are produced.
   (h) What is meant by polarization in a dielectric material?
   (i) Define the term expectation value for the energy 'E' of a particle in a box.

Section –B **Marks : 8 Each**

2. (a) Deduce Gauss's law in a differential from and further deduce the Coulomb's law from Gauss's law.
   (b) Three charges are situated at the corners of a square of side 'a'. How much work does it take to bring in another charge +q, from far away and place it in the fourth corner? How much work does it take to assemble the whole configuration of four charges?
3. (a) By using Maxwell's equations develop wave equation for transverse electric and magnetic field in free space.
   (b) A battery of emf ε and internal resistance 'r' is hooked up to a variable load resistance 'R'. In order to deliver a maximum possible power to the load, what is the value of 'R'?
4. (a) What is active medium, population inversion and optical pumping? Give their importance in the study of lasers.
   (b) Describe holography. Why is laser needed for holography?
5. (a) What are the various modes of an optical fibre? Give their importance.
   (b) How optical fibres and lasers are useful in medical science?

Section –C **Marks : 8 Each**

6. (a) Derive an expression for the dependence of relative mass on its velocity. Give arguments to show that a particle with finite mass can never attain velocity equal to that of light.
   (b) A woman leaves the earth in a spacecraft that makes a round trip to the nearest star, 4 light years distant, at a speed of 0.9 c. How many days younger is she upon her return then her twin sister who remained on earth?
   (i) linear absorption coefficient of aluminum,
   (ii) the thickness of the absorber needed to cut down the intensity of x-rays to 1/20 of its initial value.
8. (a) What is group velocity and phase velocity? Show that the group velocity of particle is equal to the velocity of the particle.
   (b) Show that the fractional energy difference $\Delta E/E$ between two adjacent levels of a particle in a box in given by, $(2n + 1)n^2$.
9. Discuss the following according to BSC theory of superconductivity:
   (a) Electron-lattice-electron interaction
   (b) Cooper pairs
   (c) Existence of energy gap
   (d) Coherence length.

Physics–101
(B.Tech. 1st/2nd Semester, 2054)

Time : 3 Hours **Maximum Marks : 60**

Note : Section A is compulsory. Attempt any Five questions from Sections B and C, selecting at least Two questions from each Section.

Section – A **Marks : 2 Each**

1. (a) Why are electric field lines normal to an equipotential surface at all points?
   (b) Explain how a dielectric inserted between the plates of a capacitor increases its capacitance.
   (c) Write down the Maxwell's equations.
   (d) What is the difference between an ordinary image and a hologram?
   (e) Distinguish between Step index and Graded index optical fibres.
   (f) Define an inertial frame of reference. Does an inertial frame exist?
   (g) State and explain Moseley's law.
   (h) Is it possible to observe Compton Scattering of visible light? Explain.
   (i) Can every physical system be described by the time-independent Schrodianger equation? Comment.
   (j) Metals which are very good conductors at normal temperatures doe not show superconducting behaviour. Why?

Section –B **Marks : 8 Each**

2. (a) State and explain Gauss's law of electrostatics. Use it to find the capacitance of a parallel plate capacitor. 6
   (b) The electric potential in a certain region is given by $V= 10x^2 + 20y + 5z^3$. Calculate the electric field at the point P (2, 3, 1). Is the field uniform? 2
3. (a) Write down the Biot Savart's law of magneto-statics, and with is help find the magnetic induction at a point due to a long straight conductor carrying current. 6
   (b) Discuss Faraday's law of electromagnetic induction and express it in integral from. 2
4. (a) Explain with suitable diagrams, the difference between spontaneous and stimulated emission. How will you achieve higher probability of stimulated emission? 3
   (b) Describe the construction and working of a He-Ne Laser. 5
5. (a) Discuss the propagation of light through a step-index multimode fibre. Explain the meaning of acceptance angle and numerical aperture. Also derive expressions for these. 6
   (b) The core of a glass fibre has a refractive index of 1.5 while its cladding is doped to given a fractional changer in refractive index of 0.005. Find:
   (i) the refractive index of cladding
   (ii) critical internal reflecting angle
   (iii) acceptance angle, and
   (iv) numerical aperture. 2

Section –C **Marks : 8 Each**

6. (a) Derive an expression for variation of mass of a body with its speed. 5
   (b) Define 'proper length' and 'proper time'. A space crew has a life support system that will last them for 1000 hours. Find the minimum speed for safe travel between two space stations at a 'proper distance' of 8×10^{11} km from each other. 3
7. (a) Give an elementary theory of Raman effect. How is it useful in the study of molecular structure? 5
   (b) The wavelength of L_α X-ray line of platinum (atomic number 78) is 1.321 Å. An unknown substance emits L_α X-ray of wavelength 4.174 Å. Calculate the atomic number of the unknown substance. Given screening constant for L_α lines = 7.4. 3
8. (a) Show that the energy lost by a photon of frequency ν is compton interaction with a stationary electron (rest mass m_0) is given by

$$\frac{h\nu\alpha\,(1-\cos\theta)}{1+\alpha\,(1-\cos\theta)}$$

where $\alpha = h\nu/m_0C^2$ and θ is the angle of scattering of photon. Also obtain an expression for the maximum kinetic energy of the recoil electron. 6

(b) Find the ground state energy of an electron confined to a one-dimensional rigid box of width a Å. 2

9. (a) Distinguish between Type I and Type II superconductors. Briefly discuss the BCS theory of superconductivity. 5

(b) State and explain Meissner effect. How do London equations accountant for this effect? 3